# Maya® 8

## The Complete Reference

# About the Authors

**Tom Meade** is a San Francisco–based 3D artist, instructor, and writer. Meade earned his B.A. in film production from Boston University and worked on several interactive CD-ROM games as a 2D production artist before making the leap to 3D. For three years, he worked as an interactive character technical director at Pulse Entertainment. Since then, Meade has served as a contributing artist for *WIRED* magazine, taught Maya in San Francisco State University's Multimedia Studies Program, and developed a certification course for NVIDIA's Film Division, which he personally deployed in India. Meade is currently a full-time faculty member at the Academy of Art University in San Francisco.

**Shinsaku Arima** is a 3D artist actively working in the visual effects industry. After earning his B.F.A. from the University of Southern California in sculpture and photography, Arima worked for several years as a professional sculptor, which eventually led him into 3D modeling. During this time, he began teaching Maya and 3D animation at such well-respected institutions as the Academy of Art University, Mesmer Animation Labs, and DeAnza College. In 2003, he joined ESC Entertainment as a trainer and technical director for the production of *The Matrix: Revolutions*. Since then, he has worked for several of the world's top studios. His film credits include *The Day After Tomorrow*, *Sky Captain and the World of Tomorrow*, *Madagascar*, and *Over the Hedge*. He is currently working as an effects technical director at Dreamworks on *Bee Movie*. Arima is co-author (with Anthony Rossano) of *XSI Illuminated: Foundation 2* (Mesmer Press, 2000).

# About the Technical Editor

**Shannon Thomas** began 3D modeling in 2003. Prior to that, he received his B.S. in pre-veterinary medicine and worked in the veterinary field for three years in Tampa, Florida. Feeling unfulfilled and yearning for a creative challenge, he retired from the medical field to pursue a career in the film industry. Shannon then moved to California in 2003 where he attended the Academy of Art University in San Francisco. Shortly prior to graduation, Shannon accepted a modeling position at Rhythm & Hues Studios where he worked on *Night at the Museum*, *Evan Almighty*, and the Academy Award–winning *Happy Feet*. Since then, Shannon has been modeling at Weta Digital on such titles as *Avatar*, *Fantastic Four 2*, and *The Water Horse*.

# Maya® 8

# The Complete Reference

Tom Meade
Shinsaku Arima

New York  Chicago  San Francisco
Lisbon  London  Madrid  Mexico City
Milan  New Delhi  San Juan
Seoul  Singapore  Sydney  Toronto

The **McGraw·Hill** Companies

**Cataloging-in-Publication Data is on file with the Library of Congress**

McGraw-Hill books are available at special quantity discounts to use as premiums and sales promotions, or for use in corporate training programs. For more information, please write to the Director of Special Sales, Professional Publishing, McGraw-Hill, Two Penn Plaza, New York, NY 10121-2298. Or contact your local bookstore.

**Maya® 8: The Complete Reference**

1234567890 DOC DOC 01987

IISBN-13: Book p/n 978-0-07-149829-6 and DVD p/n 978-0-07-149827-2
of set 978-0-07-148596-8

ISBN-10: Book p/n 0-07-149829-X and DVD p/n 0-07-149827-3
of set 0-07-148596-1

| | | |
|---|---|---|
| **Sponsoring Editor**<br>Judy Bass | **Copy Editor**<br>Bill McManus | **Production Supervisor**<br>Jean Bodeaux |
| **Project Editor**<br>Janet Walden | **Proofreader**<br>Francesca Ferrie | **Composition**<br>Lucie Ericksen |
| **Technical Editor**<br>Shannon Thomas | **Indexer**<br>Claire Splan | **Art Director, Cover**<br>Jeff Weeks |

**Illustration Credits:**
*Front:* Chris Bostjanick *(top)*; Landis Fields *(bottom)*
*Back (l–r):* Glen Pitman and Chris Logan; Tom Meade; Digitrove, Inc.; Jimmy Wu; Timothy Odell
*Interior:* John Levans *(Part I)*; Roger Ridley *(Part II)*; Tom Meade *(Part III)*; Roger Ridley *(Part IV)*;
Landis Fields *(Part V)*; Roger Ridley *(Part VI)*; Tom Meade *(Part VII)*

# Contents at a Glance

# Contents

# II  Modeling

## 3  Polygonal Modeling . . . . . . . . . . . . . . . . 57

# VII  Postproduction

# Foreword

## by Alex Lindsay

Maya is not only one of the most powerful 3D software packages available, it's one of the most complicated. The possibilities are as limitless as the paths to get to them. This maze of tools can be very difficult to traverse for those looking to begin "high-end 3D" work. I think your guides, Tom and Shin, will make this journey much easier.

I met Tom when he was my student. I was still with Industrial Light & Magic, and Tom was just beginning his 3D work. Along with a handful of other students, we began to work on projects that required us to approach the work not from a student or teacher perspective, but as a production artist. While this may seem like a subtle difference, production is a completely different world. The saying, "The map is not the territory" perfectly describes the 3D market. So many things seem to make sense in theory and work in isolated cases, but they quickly fall apart under the weight of the "Real World."

When thinking about graphics, production, or life in general, this is one of the most important things to remember. We often think of learning as a set of skills, a collection of "tricks up our sleeve." Many classes and books provide many of these tidbits. But this rarely leads us to a profound understanding of our art that truly provides a foundation for inspiring work. Small problems become days of frustration or failure. We follow what we are "supposed" to do for weeks when the most effective, and often unconventional, solution is only days away.

The key is to understand the underlying principles of the process. With these principles in mind, one has the agility to handle the issue at hand with a solid footing and clear vision. Through true understanding of the art in which we are immersed, we can have a bird's-eye view of the process to help us make intelligent and grounded decisions.

So how do we get there?

- *Observe.* Would you like to know the fastest way to add realism to your work? Take a digital camera and shoot 10 photos a day of everyday items. Shoot doorknobs, street signs, old cars. Watch not only how the light interacts with the object but how the camera interacts with the scene. I still do this constantly.

- *Fail.* Push yourself outside your boundaries. If you aren't failing, you aren't pushing.

- *Study.* When something doesn't work, find out why it doesn't work. Don't just figure out how to get around the problem—find out why the problem exists.

- *Experiment.* Playing with ideas is very important. You need to keep what you do fun. Experimentation gives you the space to play and is especially useful if you just failed at something.

- *Work with others.* Working by yourself is a nearly guaranteed way to fall behind. You can't learn everything and your eye is not perfect. You need a second opinion and a different perspective constantly to progress.

- *Work on real projects.* It is important, probably more important than anything else, to find real clients with real projects—even if it's work for free at first. People rarely do their best work for themselves. A client's needs help you mold your work and push you to perform. We often think we want to be a great artist like the masters in the art museums, rarely thinking about how much of their work was for contract—not art for art's sake. If you don't have any clients, find them. Give your work to non-profits or trade your work for services, but whatever you do, *do not* hide away to "work on your demo tape." I've seen many a computer artist get lost on this path.

With that, I will leave you with Tom and Shin. Their combined experience in production will provide a real grounding for the content presented here. I have worked with Tom for years, and Shin's background speaks for itself. This book is a place to start—and a good one. The rest of the path will be up to you.

Alex Lindsay
dvGarage
San Francisco, California

# Acknowledgments

First, we'd like to thank Judy Bass for getting this project rolling. Thanks to Janet Walden, Jean Bodeaux, Bill McManus, Francesca Ferrie, and Lucie Ericksen for taking on the monstrous tasks of coordinating, editing, and producing this project. Their close attention to the smallest details amazes us and will be such a wonderful benefit to all of our readers. Thanks to Shannon Thomas for bringing some real perspective to the material in these pages during the technical editing process. Again, the readers will be grateful for your work. Finally, an extra thank you to Margie McAneny who played a pivotal role in starting this project five years ago.

*Tom*—I'd like to thank every Maya user that I've had the opportunity to talk with, teach to, and learn from. This is truly an amazing community. Thank you students and faculty at the Academy of Art University. An extra special thanks to Landis and Anna Fields for helping supply the cover artwork for the book. And thank you Mom, Dad, Kelly, Chris, and all of my friends for your love and support during this project. It would not have been possible without you. Lastly, a big kiss for my wife Pamela who put up with my absence during the production of this book.

*Shin*—I'd like to thank my mother and father for endless support; Rodney Iwashina for being the most amazing and efficient supervisor during *Matrix* production; Remo Balcells for being a very patient and wise VFX supervisor during *Day After Tomorrow* production; Rudy Grossman for being an extremely friendly and assuring supervisor for *Sky Captain*; all my students for teaching me how to teach and supplying me with beautiful 3D images; Wenchin Hsu for taking care of me with whole-hearted love; Steven, Cara, David, Jesse, Rachel, and Juan for being a great company; and Gabriela and Rachel for giving me great joy in my life.

# Introduction

If you've gone to the movies lately or played a video game on an Xbox system or Wii, you are familiar with the types of animation that Maya can produce. Maya is currently the industry-standard application for producing animation and effects for film, television, video games, and the Internet. This book was written to demonstrate how the program works and how it is used to produce content for these industries.

Not so many years ago, access to Maya by someone outside of the animation industry was limited. The software cost tens of thousands of dollars and the hardware required to run it was just as expensive. Attending a school was just about the only option for anyone interested in learning about this exciting new technology.

As computer hardware became faster and cheaper, software companies began to realize that they could sell these high-end applications to smaller studios, independent artists, and students if they made the price more appealing to those markets. Today you can purchase Maya Complete for under $2000, or you can download a free version, called the Personal Learning Edition, from the Autodesk website. The final obstacle, once you have the software installed, is learning how to use it. That's where this book comes into play.

Learning an application as large and powerful as Maya can be overwhelming, to say the least. While Maya 8's online documentation provides an excellent resource for learning about each and every little piece inside the program, it was not really designed to show you how all of the pieces fit together so that you can go about designing and completing a project. Knowing how to use a hammer does not necessarily mean that you can build a house. This book teaches you many of the tools in Maya *in the context of using them in a real-world production environment.*

This book was designed to cover the basics of the program and to demonstrate how Maya can be used to create projects. What you will gain from reading this book is a knowledge of specific workflows that are common to most 3D productions. Whether it be modeling a head for animation or rendering a vehicle for integration into a live-action plate, this book will direct you through processes that have been tried and tested in production environments. Successfully building a project that is free of errors and that can be easily edited at any time during a production is beyond the scope of what any manual can teach you, but it is something that reading and working through the tutorials in this book can help you achieve.

## Who Should Read This Book

This book is intended for CG artists, engineers, filmmakers, or hobbyists who are serious about using Maya to create high-quality images and animations. Some of the information here is very basic, while much of it is more advanced. Beginners will be happy to know that we've detailed where to find each and every tool or window as you follow along the tutorials. Advanced users will find plenty of information within these pages that you won't find in other books. We believe we have created a book that will be useful to anyone with a genuine interest in Maya.

While the book covers the fundamental aspects of Maya, such as the user interface and navigation, some knowledge or experience with 3D animation techniques will be necessary before you attempt to complete the tutorials. If you are migrating to Maya from another 3D package, it is suggested that you complete the introduction and Project One tutorials in the *Learning Maya Foundations* book that ships with the software.

## How the Book Is Organized

We have organized this book into seven parts that begin with an overview of Maya and then follow the order of a typical production workflow. However, this does not mean that you have to read this book in order. Users with previous experience in Maya will have no problems jumping right to the character-rigging chapters in Part III or building explosions in Part VI.

For people with less experience, it is recommended that you try and follow the book in the order in which it has been laid out. We took extra care to explain how to execute basic tasks in the beginning of the book. These tasks, such as setting tool options and executing commands, are used throughout the entirety of the book but are not explained in such detail in later chapters.

Part I is designed especially for people new to the 3D production process or to Maya. It outlines the different processes used in production and explains how they fit together. The overview of Maya's user interface is followed by a tutorial that directs you through a simple project, showing how to navigate through the interface and introducing some basic 3D concepts.

Part II covers modeling with all three of the geometry types available in Maya: polygons, NURBS, and subdivision surfaces. The chapters covering these topics include techniques for both hard surface and organic modeling.

Part III covers character setup and deformation techniques. The chapters contained within this part are geared toward the aspiring character technical director/animator. It is in these chapters that you will learn how to harness the real power of Maya as the concepts used here can be applied to any aspect of a 3D production.

Part IV is dedicated to animation, with a strong emphasis on character animation. After covering the basics of keyframe animation, the principals of animation are applied to animating a biped character. Some additional features, such as animation referencing and motion capture data management, are also introduced here.

Part V begins with an in-depth study of how to use Maya's Hypershade interface to build materials and control the lighting in a Maya scene. Advanced texture-mapping techniques are demonstrated, including UV mapping techniques for organic surfaces. Workflows for using Maya's Paint Effects toolset are also covered in this section. Finally, we practice different rendering techniques in both Maya's software renderer and the Mental Ray renderer for Maya.

Part VI explores Maya's Dynamics toolset. This includes particles, fields, expressions, and rigid and soft body dynamics for creating realistic effects and animation by running simulations based on natural phenomena. Maya's hardware renderer is also covered here.

Part VII focuses on completing a single project of a spaceship landing. Camera mapping, multipass rendering, command-line rendering, and compositing techniques are all explained in detail.

Also, the 16-page insert section contains color renderings from some of the tutorials in this book, as well as the work from several 3D artists.

## Conventions Used in This Book

A few conventions are used in this book. When we direct you to choose a command from the menu, we use the | symbol to signify a submenu. For example, if we say "create a polygonal sphere by choosing Create | Polygon Primitives | Sphere," that means you need to click the Create menu in the menu bar, mouse over Polygon Primitives from the resulting list, and then choose Sphere from the submenu.

Most of the tools and commands in Maya have options that can be edited and set before executing the tool or command. These options are usually accessed by clicking the little box next to a tool or command in the menu items. This book uses the □ symbol to denote the Options window command. For example, to open the Options window for the Sphere command, we say "choose Create | Polygon Primitives | Sphere ❑."

Another convention used throughout this book are icons appearing in the margin of some pages. Maya offers several ways to execute tools and commands. We will always direct you to execute these tools and commands via the main menus, but you may also choose to click a button on a shelf or toolbar. For this reason, we provide the button icon in the margin when introducing a tool or command for the first time. (The margin icons that look like a disc are explained next.)

Finally, keyboard combinations are included for both the Windows and Mac platforms. Windows combinations appear first, and Mac combinations follow in parentheses.

## About the DVD

This book includes a DVD that contains Maya scene files and other resources that are used throughout the book. Many tutorials will instruct you to open a certain DVD file to use at the start of a project. This is emphasized with a disc icon in the text margin.

You will also find QuickTime movies demonstrating some of the tutorials in the book. In order to view these movies you must have QuickTime 7 installed on your computer. QuickTime 7 can be downloaded from Apple's website (www.apple.com/quicktime).

In order to keep files up to date, get corrections on any errors, or discuss topics from this book, we have set up a website to support this book. Please visit:

**www.datasynthi.com/maya-complete-reference/**

> NOTE   *Most of the files included with the book's DVD were designed to open in Maya 8 or later.*

# Introduction
# to Maya

# Core Concepts

**Maya is the state-of-the-art, industry-** standard application that is widely used for 3D modeling, animation, and effects. Users of the program produce content for film and television production, video game development, architectural design, and web and print production.

Maya is a culmination of technologies created by some of the most prominent computer graphics developers of the past two decades. It is the largest commercial computer application ever written, with levels of complexity and functionality exceeding other high-end 3D animation packages.

Since its release, film effects companies such as Weta Digital, Industrial Light & Magic (ILM), Pixar, Sony Imageworks, and Digital Domain have adopted Maya as their standard application for producing 3D animated effects. Recognizing its technological superiority to its competitors, Sony and Microsoft have helped to define Maya as the industry standard for video game production as well, with companies such as Electronic Arts and LucasArts developing content for the PlayStation and Xbox platforms. Web, print, and industrial designers are also poised to adopt Maya as an industry standard for 3D graphics because of the unlimited functionality and price point not offered by any other 3D application on the market today.

In 2003, the Academy of Motion Picture Arts and Sciences recognized these achievements and the software's impact on the movie industry by awarding Alias with an Oscar for Technical Achievement. In 2006, Alias was purchased and is now owned by Autodesk.

Maya comprises a complete, integrated set of practical, easy-to-use tools for creating complicated special effects. These tools enable 3D modeling; animating; texturing, lighting, and rendering capabilities; and dynamics. Maya's scripting language, MEL (Maya Embedded Language) allows users the flexibility to create and modify existing toolsets to create their own custom functions and streamline their production processes. The unique level of integration of Maya's tools prevents compatibility problems often caused in competing applications that rely on plug-in technologies or that require additional software packages to create all elements of an animated 3D scene.

So now that you have Maya installed on your computer, how can you begin realizing your creative potential on the screen in front of you? Before we get into the specifics, it is important that you understand the concepts and processes behind a 3D production. In this chapter, we define these processes and explain how Maya handles them.

# Production Workflow

In almost everything we do, we can choose from a variety of available options to accomplish the task. Maya is no different. In fact, Maya often offers so many options that it can take days, months, or even years to discover the most efficient processes to use for completing a project. This book is designed to share and suggest certain processes, often called *workflows*, that are used in many professional 3D productions.

To illustrate the importance of an efficient workflow, consider the manufacture of an automobile. During the design and engineering process, the most critical decisions are made: What will this car need to do? Who is it for? How will it be used? After answering these questions, the designers and engineers get to work, the design and mechanics of the car begin to take shape, and eventually the car is ready for production.

When in production, it is common that different tasks will be tackled by different teams. It is essential that, even while they work separately, the teams remain in constant communication. Any change at one end, no matter how insignificant it may seem, may have a dramatic effect on another team's production. Furthermore, if the pieces are assembled in the wrong order, the next piece may not fit and all of the work will have to be disassembled and reworked. It is very easy for the production to fail if each team does not understand the production as a whole piece or cannot communicate outside of its own circle.

A 3D production is no different. If you build and animate a character while the initial designs are still being worked out, chances are that all of the model, rigging, and animation will have to be redone. If a texture map is painted before the model is completed, for example, the features will probably not match up. Even within each aspect of a production, there is an order that each individual must follow to accomplish a task efficiently.

Although the 3D production technology, Maya included, attempts to provide solutions to these order-of-operations issues, it is best to avoid them altogether by being educated on how everything fits together *before* you begin production. It is crucial that every single element in a 3D production—be it modeling, rigging, animation, or lighting—not attempt to outdo or become more prevalent than the others. Rather, all elements of production should make the best effort to support each other. The most successful productions are those for which the technology aids in clearly communicating a story.

Figure 1-1 shows a block diagram of a general production pipeline used in many movie studios today. While individual studios handle everything a little bit differently, all of them follow this same general path.

In general, a good workflow should begin with a story. It becomes more and more evident everyday that just because our culture is blessed with all of this amazing technology, realizing its potential is impossible without great ideas for how to use it. Once a story is defined, the design of every element must be developed. It is very common that the design and story feed back on one another, exposing new possibilities and pushing the visual design further. This whole process falls under the preproduction phase of a production pipeline.

The next major phase of the production pipeline is the modeling phase. While a modeling department might constantly be delivering and updating models throughout a production, it is imperative that the hero objects, the main characters of a scene, be at least roughed out and delivered to the rigging department so that they can be set up for the animators.

**FIGURE 1-1** *A basic pipeline for a 3D film production*

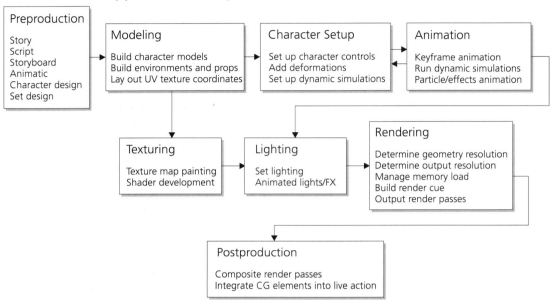

A lot happens during the animation phase. In addition to the keyframed animation techniques you'll learn about in Chapters 14 and 15, effects such as fire, smoke, and water can be added into the scene through the use of particles. Finally, the objects are textured, the scene is lit, and the animation is rendered to create the final sequence of images.

Depending on the type of production, the last phase of the workflow involves some kind of postproduction process. In film and television, this might include using compositing software to combine live-action footage shots with the elements rendered from Maya. In a video game production, postproduction might involve the programming of the game within a game engine.

As we begin to examine these different processes of a production pipeline, we want to stress that these are very general assessments of production workflows. Every company has its own structure that defines how the pipeline tasks are arranged and the specifications that each team must adhere to so that every single piece can fit into the complete, final output. As an individual, you will adopt your own personal style of working with Maya, just as you would for any other tool. So, as you carve your way through this book, just remember that the features and processes demonstrated are not the only way to accomplish any task. Workflows are being evaluated and improved every day. As a result, new tools become available in the software. Never forget your place as a user! Think outside of the box. If a specific process does not fit your production, write your own tool. *That* is the real power of this software.

Let's look at some specific workflows within the production process to learn more about them.

# Preproduction

Preproduction is not necessarily a workflow particular to Maya, but it is one of the most important phases of any 3D production workflow. Even in a world of advanced technology, a 3D project should begin on pencil and paper. Storyboarding, conceptual sketches, and character design are truly essential to any successful 3D project.

> *NOTE  Remember that the object of your animation is to tell a story visually. Use traditional filmmaking techniques to enhance your idea. Color, lighting, camera angles, and composition are all important elements that help to convey a mood. Plan to incorporate these elements into your shots so that every frame tells a story of its own.*

For *Star Wars: Episode I,* George Lucas and his team of computer graphics artists at JAK Films pioneered the idea of using computers to previsualize scenes in the movie once the storyboards were complete. The resulting *animatics* enabled Lucas to use basic geometry, animation, and lighting to set up and experiment with the shots that the storyboard artists had conceived. It also provided the perfect channel for communication with the visual effects artists at Industrial Light & Magic. These artists not only received the drawn storyboards, but also had an idea of exactly how the objects and the camera moved though the scenes. Since then, previsualization has become popular in the film industry as a method for designing complicated shots involving effects. The movie *Sin City* was approved for production by a major studio based on an impressive 3D animatic. Even as a solo artist, "previz" can help you quickly figure out if your shot, or sequence of shots, is working.

> *NOTE  Many books are available to aid you in designing your animations through a series of shots. One in particular,* **Film Directing Shot by Shot,** *by Steven D. Katz (Michael Wiese Productions, 1991), is full of information about previsualization and visual storytelling techniques. You can also visit www.thestoryboardartist.com for more information on storyboarding.*

# Modeling

Before you can begin animating, you need to have created objects to animate. Modeling is the process of building the characters, props, and environments in the scene. These objects are constructed from 3D geometrical surfaces so that they can be rotated around and viewed from all angles. One of the biggest advantages of using 3D over more traditional 2D techniques for animation is that the 3D objects, a character for example, needs to be built only once, while a character created in 2D needs to be re-created for every frame of the animation. Figure 1-2 shows an example of a 3D model in its *wireframe* view.

**FIGURE 1-2**    *A 3D model displayed as a wireframe (image courtesy of Landis Fields).*

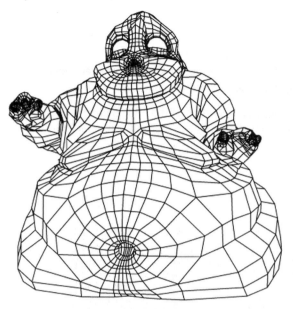

In terms of workflow, the most important thing for you to know before you begin is how much of the model will be seen in the animation and how close the camera will get to the model. It is important that you figure out in the preproduction process how a model will be used in the project. For example, spending a week to model a detailed cell phone that will only be seen sticking out of someone's pocket from a distance is a waste of time. If you plan ahead in preproduction, you can avoid spending time on irrelevant details and focus on the important ones.

Once you have determined the types of scenes and camera shots that will be needed for a model, you then need to consider what will happen with the model downstream. In the case of a character, its geometry must be modeled so that it will *deform*, or bend, properly when it is animated. In most cases, the geometry must be complete before the texture coordinates, known as *UV coordinates*, are laid out for texture mapping. The model's complexity will also have a direct effect on the amount of time it takes to render.

> **NOTE**    *Maya 8's new Transfer Attributes command can be used to exchange UV information between similar pieces of geometry with different topologies. This offers substantial flexibility when making small changes to a model after the UV coordinates have been laid out.*

Maya offers three different modeling toolsets: NURBS (Non-Uniform Rational B-Splines), polygonal, and subdivision surface modeling. (NURBS is a type of geometry in Maya where the surfaces are defined mathematically.) Chapters 3 through 7 cover these toolsets and demonstrate workflows for using them together.

## Character Setup

Character setup, also called *rigging*, is the process of preparing a character so that it can be animated. Typically, you begin by creating a skeleton that matches the scale and features of a model. For example, the character's hip joint needs to be placed at the hip, the knee joint at the knee, the ankle joint at the ankle, and so on. Figure 1-3 shows an example of a character rigged with controls.

Control objects are created and connected to the skeleton. These controls enable the animator to make a 3D character perform similar to the way a puppeteer makes a puppet perform. If a skeleton is properly rigged, it can be handed off to an animator who lacks any computer knowledge and that animator will be able to intuitively pose and animate the character.

Apart from the animation controls of a character rig, a large portion of the rigging process involves setting up how the model will bend or *deform* with the rig. For characters that are intended for realistic purposes, many combinations of deformation utilities must be used to maintain

**FIGURE 1-3** *This worm character has been rigged with controls that will enable an animator to achieve a good performance without having to dig around Maya's graphical user interface (GUI). (Image courtesy of Landis Fields.)*

the volume of the mesh as it bends and reacts to the animated behaviors. Occasionally, physically based muscle systems are created and dynamically simulated to add an extra element of realism to the final product.

You will learn techniques for character rigging in Chapters 8 through 13.

# Animation

The animation process is what will finally let your character zip through time and space. Effective animation is achieved through orderly *keyframing*—the process of recording an object's position, rotation, scale, shape, and such at a specific time.

The most efficient animation workflow is using what is known as a *block and refine* technique. The first pass at the animation is focused on timing: what poses the character is in at specific frames. At this point, usually no keyframe *interpolation* is involved—that is, no movement exists to transition between the keyframes, and the character merely pops into position, holds, and then pops into the next pose.

During the next few passes, you block out some of the secondary poses that fall between those you set in the previous pass. Once all of the significant poses have been set, interpolation is enabled and the process of refining the motion between these poses begins. Because this process can take a long time, much patience is required as you do pass after pass, adding more and more detail to the motion.

Chapter 14 will introduce you to some basic animation techniques on simple objects. You will then apply these techniques to a character in Chapter 15.

# Shading and Texturing

Shading and texturing is the process that adds realistic or stylized surface elements to your models; otherwise, all 3D elements would render out to be the same textureless, flat color. For every surface or group of surfaces, a material is created that determines the surface's characteristics—what color it is, or how transparent, shiny, bumpy, or reflective it is. 3D artists usually say that the material determines how the object *shades*. The worm character in Figure 1-4 has texture maps that add details, such as spots and small wrinkles.

Bitmapped images, such as those that you might create in an external image editing or illustration program such as Adobe Photoshop or Pixologic's ZBrush application, can be used to control the various shading characteristics. In most cases, much of the finer detail of an object's surface can be added via these texture maps. The wrinkles in skin or the panels on an airplane's wing are usually added with texture maps.

**FIGURE 1-4**   *The worm character with texture maps applied to the geometry (image courtesy of Landis Fields).*

Chapter 17 will teach you the basics of shading and texturing. In Chapter 18, these techniques will be put into practice in two tutorials.

# Lighting and Rendering

The final piece of the 3D production process is the lighting and rendering of the scene. Lights are added and used in Maya just as they might be in the real world. In filmmaking, lights are used not just to illuminate the scene, but also to create or enhance a mood. This is done through the placement, intensity, and color of the lights used as well as any go-betweens that may indicate elements offstage.

To see the results of your lighting, you must first render the scene. *Rendering* is the process of creating an image from all of this 3D data. For a single frame, the rendering engine draws each pixel by finding an object in front of the camera and drawing it based on the direction of the surface, the surface characteristics, and the lighting information. The goal of the rendering process is to achieve in a reasonable amount of time an image that is free of unwanted artifacts.

Once the specific calculations have been determined and the file is set up, the rest of the process is accomplished by the computer and can take anywhere from a few seconds to hours—or even days—to render, depending on the complexity of the scene.

Figure 1-5 shows a rendered scene that has been lit with various types of lights. While some lights are set up simply to illuminate the scene overall, you will also notice that some of the light beams are visible because they have been rendered with glow and fog effects. You will learn all about how to set up and use all kinds of lights in Chapter 20.

*FIGURE 1-5* *This environment was lit using a variety of different lights and light effects (image courtesy of Edwin Poon).*

## Postproduction

After every frame of the animation has been rendered, the footage will usually be brought into another software package so that the 3D elements can be combined with other elements shot on video or film. It is also common for the 3D elements in the scene to be rendered separately.

For example, each character might be rendered as a separate element and then integrated at the *compositing* stage.

In some cases, the various surface characteristics are rendered in a separate pass for each object. This gives the compositor absolute control over the entire image. With this method, the reflectivity, shininess, and color can be changed easily without the artist having to go back and correct it in 3D and re-render the image. We will explore these techniques in Chapters 22 and 23.

Now that you have an idea of the different workflows involved in the 3D production process, you should be able to map out a good plan for how to attack your animation in the most efficient and least frustrating way possible. Sometimes there may be some back and forth between the different processes to get it right. For example, the rig built for a character may need to be reworked if the first attempt was not sufficient for the shot. This is inevitable. However, you don't want to take a model all the way through the pipeline only to realize that the design is wrong. There are always ways to patch things, but in most cases you'll be starting from scratch. So, always understand how your pipeline works.

# Maya's Architecture

If you are a race car driver, you may not be expected to know how to build a car from scratch, but you should at least have a good understanding of how your car works. Knowing how the gearbox works will help you shift gears more efficiently while you are racing. A Maya user should approach Maya in the same way. The Maya user interface is designed so that it is possible for someone to use it to complete a project without knowing much about what's going on "under the hood." However, to harness its power and work more efficiently, you should understand the basics of how Maya works.

Maya's documentation describes a Maya scene as "a collection of nodes with attributes that are connected." But what does this mean, exactly?

## Nodes, Attributes, and Dependencies

A *node* is the basic building block for anything in Maya. There are many different types, including shape nodes, transform nodes, rendering nodes, and nodes that contain algorithms for certain operations. Every node contains *attributes*, or *channels*, that are specific to that particular node. Attributes are the properties of the nodes. Color is an example of an attribute on a material node. An attribute's input and output can be connected to the input and output of an attribute on another node. When an attribute's input is controlled by another attribute, we say that it is *dependant* on the incoming connection. To understand this better, let's look at what nodes make up a NURBS sphere and what attributes are connected.

Any object in Maya, be it a piece of geometry, a texture map, a light, or even an operation, can be defined in one or a group of nodes. Figure 1-6 shows the nodes that make up a primitive NURBS sphere in Maya, as displayed in Maya's Hypergraph. The Hypergraph window shows a graphical representation of the relationship between objects in a scene. In Figure 1-6, the Hypergraph is displaying the input and output connections' attributes, or dependencies, between the nodes that are used to build this sphere. Although this might not make much sense to you now, you'll learn more about using Hypergraph in almost all of the chapters in this book. For now, you should know that the lines that appear between nodes show that the attributes of the nodes are connected, and the arrows show the direction of these connections. In other words, in this example, Maya handles the information contained in the first node, makeNurbSphere1, and outputs that information through an attribute that is connected to an input attribute on the next node, nurbsSphereShape1. It then moves to the initialShadingGroup node and processes the attributes contained within it to display the final shape on your computer screen.

**FIGURE 1-6**   _Maya's Hypergraph displaying the input and output connections of a NURBS sphere_

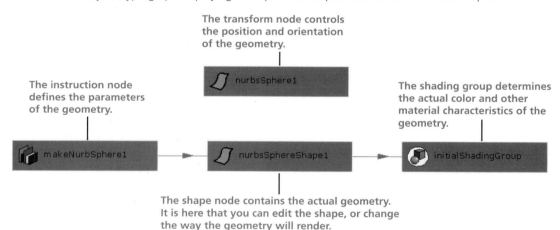

All the information about the sphere's size, _sweep_ angle (where it begins and ends), and resolution is stored in unique _attributes_ as some value inside the makeNurbSphere1 node. The values for these attributes can contain different data types: an integer (a whole number), a float (a decimal), a Boolean (on or off), or a string (text). You as the user can edit the values for the attributes on this node by selecting the node in the Hypergraph and modifying its attributes in the Attribute Editor or Channel Box.

Maya takes all of the information from this makeNurbSphere1 node and sends it out from the Output Surface attribute. This Output Surface attribute feeds into the nurbsSphereShape1 node's Create attribute. The nurbsSphereShape1 node contains attributes that deal mostly with how the rendering engine will interpret this shape at render time. For example, the shape node contains attributes that determine whether the object will cast a shadow, receive a shadow, cast a reflection, or will even be rendered at all.

Now that Maya knows what the shape is, it needs to know how the object will shade so that it can render it properly. The initialShadingGroup node takes the output from the shape node and the output from the material node to decide the surface properties for this sphere. It then outputs that information, along with any other information based on its own attributes, and feeds the sphere into the render partition—where the information from the shading group and the lights are calculated together for the renderer to do its job. (The renderPartition node is not displayed in Figure 1-6.)

Since the nurbsSphereShape1 node is dependant on the makeNurbSphere1 node, we say that the shape has *history*. You can delete the history of any object by selecting it and choosing Edit | Delete By Type | History. This will break the connections to all incoming nodes and delete them. In this case, once the history is deleted on the shape node, we can no longer go back and edit its sweep angle or resolution. If we wanted to do this, we would need to perform an additional operation that would introduce another node and connect it to the shape node.

# Node Hierarchies

If you refer back to Figure 1-6, you'll see a node named nurbsSphere1 that does not display any lines connecting it to the other nodes in the sphere's network. This is a *transform* node. A transform node contains attributes whose values determine the translation (position), rotation, scale, and visibility of the objects connected to it. The reason that the connections are displayed differently for this node is that none of the attributes from any of the other nodes shown in the Hypergraph are connected to the transform node. Instead, they share a different type of relationship, called a *hierarchical* relationship.

When you work with 3D modeling and animation, you often use a hierarchy to create relationships among objects. For example, to have multiple objects follow one object's movement, you need to create a hierarchical relationship. Say you're working with a car object. You want to be able to select the car's body and have the wheels, doors, trunk, and hood all move with it. The car body object, in this case, would be the *parent object*, and the other objects would be the *children* of the car body. The way children follow a parent is called *hierarchical transformation*.

This hierarchical concept is useful while grouping many objects into one model or for animating several objects efficiently. Figure 1-7 shows a simple hierarchy for some of the planets in our

FIGURE 1-7    *The Hypergraph displaying a scene hierarchy*

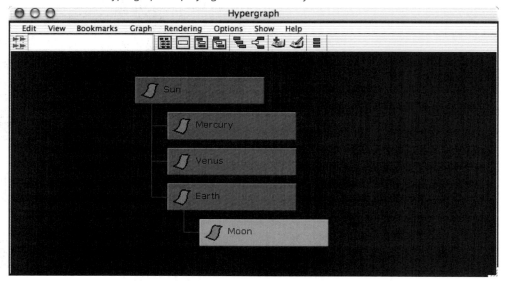

solar system. Notice how the Hypergraph is displaying the nodes in this example. No colored lines with arrows connect them. Here, the Hypergraph is displaying the scene hierarchy instead of the dependencies.

The structure of a hierarchy is known as a *tree*, because of its resemblance to that natural phenomenon—except it's upside down. The top, or base level, of the hierarchy is called the *root*, while the objects below the root hierarchy are called the *branches*. Sometimes you'll see the terms *parent* and *child* used to describe the hierarchy. Basically, the parent is the root, and the children are the branches.

# Summary

It is important that you remember the core concepts presented in this chapter as you read the rest of the book and work with Maya. They will help you understand how the application works and why it behaves the way that it does. Knowing this will be the key to troubleshooting any problems that will inevitably arise during a production. We'll take a look at the Maya interface in Chapter 2, where you will get an opportunity to apply some of the concepts discussed here.

# The Maya
# User Interface

**Maya's interface hasn't changed much**
since its conception, probably because the
development team at Alias created a functional
and easy-to-grasp user interface from the start.
In this chapter, we explore the Maya interface
by first learning the placement of the most use
ful buttons and windows and then by using
them in a simple step-by-step tutorial.

# A Tour of the Maya Interface

Figure 2-1 shows Maya's default arrangement of the user interface elements. This arrangement is called a *layout*. Here is a quick tour of each element of this layout from top to bottom, left to right.

> **NOTE** *The exact location and style of the menu bar, title bar, and the Minimize, Maximize, and Close buttons will differ between platforms. (Figure 2-1 is from a Mac.) Everything else within the Maya user interface will be the same whether you're using Windows or Mac OSX.*

**FIGURE 2-1** *The default Maya interface*

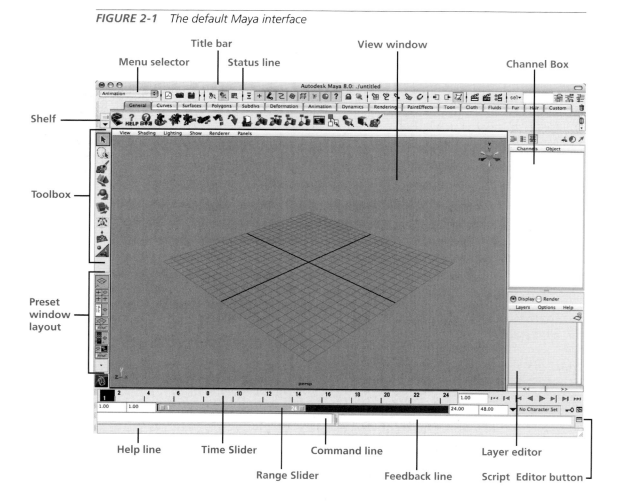

# The Title Bar

The title bar displays the Maya version number, the name of the scene you are working on, and the name of any object you have selected. It also includes standard Minimize and Maximize buttons and a Close button.

# The Menu Bar

The menu bar lets you quickly access numerous functions in Maya via pull-down menus. The contents of each pull-down menu give you access to related tools, commands, and settings, and, when available, the menu lists the keyboard shortcuts for executing tools or commands.

## Menu Sets

Because Maya has so many menus, they won't all fit in a single-row menu bar. The Maya interface solves this space problem by partitioning menus in *menu sets*. A menu set is divided up based on tools and commands related to a particular workflow.

There are five menu sets available in Maya Complete: Animation, Polygons, Surfaces, Dynamics, and Rendering. Maya Unlimited has one additional menu set for Cloth. You can access these menu sets from the menu selector's drop-down list on the status line, as shown in the illustration. After you have selected a menu set, you will notice that some of the options in the menu bar will change according to the set you chose.

> **TIP**   *By using the Hotbox, discussed later in this chapter, you can view all the menus at once.*

You can also access these menu sets by pressing keyboard commands, or *hot keys*, as shown in Table 2-1.

You can also choose a menu set by pressing the H key while you hold down the left mouse button in the view window. A *marking menu* will appear, containing each menu set, as shown in

**TABLE 2-1**   *Hot Keys for Accessing Menu Sets*

| HOT KEY | MENU SET |
|---------|----------|
| F2 | Animation |
| F3 | Polygons |
| F4 | Surfaces |
| F5 | Dynamics |
| F6 | Rendering |

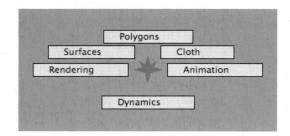

this illustration. You can choose the set you want by releasing the left mouse button when the cursor is over the appropriate set.

*TIP*   **Marking menus are used throughout the Maya interface. If you hold down a specific hot key and click in a window, or if you right-click anywhere in a window (without using a hot key), a marking menu will appear.**

Regardless of the menu sets you're using, the File, Edit, Modify, Create, Display, Window, and Help menu options will always appear in the menu bar. These items are not specific to any particular workflow. Instead, they give you access to many of the common editing commands found in most software packages (such as Cut, Copy, Paste, Save, and Close) and allow you to create new objects (the Create menu), modify these objects (the Modify menu), access different windows (the Window menu), and choose how objects are displayed in the view window (the Display menu).

### Tools and Commands

Tools and commands (also called actions) are two different things in Maya. As you browse though Maya's menus, you'll notice that some of the items listed include the word *tool* and others do not. The differences between a *tool* and a *command* are subtle. For example, the Create menu includes a tool called the CV Curve tool. When the tool is selected from the menu, Maya enters a mode in which that tool is active. In the case of the CV Curve tool, a control vertex is created each time you click your mouse button in the view window. To finish using the tool, you must press the ENTER (RETURN) key.

A command, on the other hand, can require some sort of input before it can be selected from a menu. The Edit | Duplicate command is a good example. Before you can select the Duplicate command from the Edit menu, you first must select an object or objects that you wish to duplicate. After the selections have been made, you can choose the command from the Edit menu and it is executed. The result, of course, is a duplicate of the selected object(s), and that's it—the operation executed by the Duplicate command did its job and you can move on to your next task. Some commands don't require any input—they simply create an object when they're executed. For example, the Create | Locator command simply creates a locator.

### Tool Options and Command Settings

Another element that can be found next to some of the tools and commands in the menus is a rectangular box icon (□). Selecting this icon will open either the Options window for a command or the Tool Settings window for tools. The Options window will always open as a floating window. Here you can modify the settings for the specified command and then execute it. Figure 2-2 shows the Duplicate Special Options window, where you would be able to specify how many duplicates should be created and how each duplicate should be moved, rotated, or scaled from the original position. To open this window, choose Edit | Duplicate Special ❑.

When the ❑ icon is selected for a tool, the Tool Settings window will open on the right side of the Workspace. This interface was designed to allow you to modify the tool's settings as you work with it. For example, Figure 2-3 shows the Tool Settings window for the 3D Paint tool (discussed in Chapter 19). This window opens when you choose Texturing | 3D Paint Tool ❑ (you need to pull up the Rendering menu set to find this menu). When this tool is active and you are painting an object in the view window, you can easily adjust settings such as color and brush size.

## The Status Line

The status line includes valuable tools you can use while you are working, such as selection masks, snapping modes, and a rendering button. We will explore some of these tools and buttons in the tutorial later in this chapter. We'll use others throughout the book. Figure 2-4 shows arrangements of useful buttons on the status line.

*FIGURE 2-2*   *The Duplicate Special Options window*

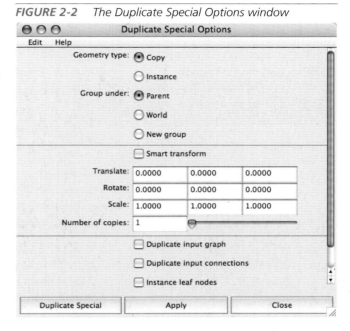

*FIGURE 2-3*   *With the 3D Paint Tool Settings visible on the right side of the UI, you can make adjustments to the setting*

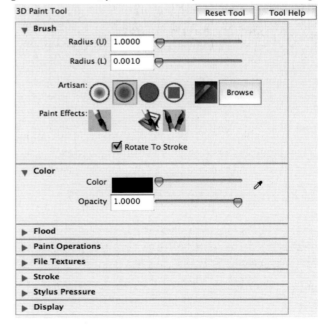

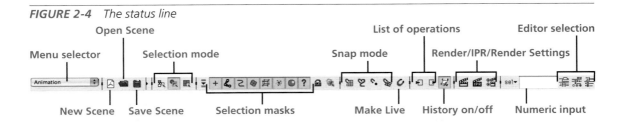

*FIGURE 2-4*   The status line

Probably the most widely used buttons on the status line are the selection modes and selection masks. The selection modes section gives you three different options for selecting objects in the view windows. From left to right, these are Select by Hierarchy, Select by Object Type, and Select by Component Type. Once one of these options is toggled on by clicking it on the status line, the selection masks section will update to display the relevant selection types for that mode.

Notice in Figure 2-4 that the middle selection mode button, Select by Object Type, is depressed. Therefore, the selection masks section displays a collection of buttons that have icons representing different object types. In this case, the selection types available for Select by Object type are (from left to right) Handles, Joints, Curves, Surfaces, Deformers, Particles, Rendering Nodes, and miscellaneous objects.

These selection modes and masks can make selecting objects in the view window much easier when the scene becomes crowded. For example, your scene may have hundreds of surfaces very close to one another with a curve tucked among them. Selecting that curve without also selecting a surface could be tricky. In such a case, you could click the Surfaces button in the selection masks section to disable surface selection. Now when you click the curve in the view window, you will be unable to select the surfaces, making it much easier to select just the curve.

## The Shelf

The shelf, located under the status line, contains buttons for accessing your most frequently used commands and tools. Figure 2-5 shows the buttons and tabs located on the shelf. These buttons have been organized into tabs respective to specific workflows. Choosing a tab displays shelf

*FIGURE 2-5*   Buttons and tabs on the shelf

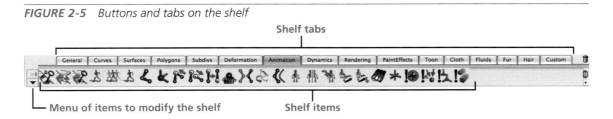

items in that tab. By clicking a button on the shelf, you can execute commands or launch tools without having to choose them from a menu. As you become more experienced in Maya, you will no doubt want to customize your shelf with tools or commands that execute custom settings.

If you find that a command or tool is available from the menu bar but is not on a shelf, you can add it to the shelf. Hold down SHIFT-CTRL and then select the command or tool that you want to add from the menu in the menu bar. As soon as you release the mouse button, the item you selected will be added to the shelf. You can quickly delete any shelf button by pressing the middle mouse button (MMB) and dragging the shelf button to the trashcan icon in the upper-right corner of the shelf.

> **NOTE** *This book uses the following shortcuts for button clicks and drags: LMB (left mouse button), RMB (right mouse button), and MMB (middle mouse button).*

You can create a new shelf, delete the currently selected shelf, load a shelf from your hard disk, or open the Shelf Editor by clicking the little black arrow to the left of the shelf. This opens a list of items that modify the shelf. Using these options, you can organize shelf buttons in individual tabs, such as a Modeling tab, Animation tab, Lighting tab, and so on, to accommodate different workflows.

We will make custom shelf buttons in many chapters in this book, and we encourage you to use the shelf whenever you find that you are often going to the menu bar for a specific tool. For a detailed example, see Chapter 4.

> **TIP** *Shelf buttons not only give you quick access to tools and commands, but also can be used to store different settings for a single command. Therefore, it is possible to have, for example, multiple shelf buttons for the Duplicate Special command, each with its own unique settings.*

# The Toolbox

The toolbox, shown in Figure 2-6, contains shortcuts to the most commonly used tools for all workflows.

The first three tools in the toolbox are the basic *selection* tools. You can select an object in the view window by clicking the Selection tool and then clicking the object in the view window. To select multiple objects, you can either hold down the SHIFT key while clicking objects in the view window or choose the Lasso tool in the toolbox and drag a selection around all of the objects that you want included in your selection. When selecting components (subobjects of a shape), you can use the Paint Selection tool. This not only is useful for selecting multiple components at

**FIGURE 2-6**  *The toolbox contains basic selection and transformation tools as well as preset panel layouts.*

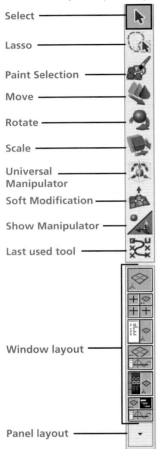

Select

Lasso

Paint Selection

Move

Rotate

Scale

Universal Manipulator

Soft Modification

Show Manipulator

Last used tool

Window layout

Panel layout

a time, but also keeps you from selecting components on the opposite side of the object you are selecting from.

The next four tools in the toolbox are called *transform* tools—the Move, Rotate, Scale, and Universal Manipulator tools. Select an object in the view window and then choose the tool from the toolbox. Figure 2-7 shows a sphere that has been selected, and the Move tool is active. Notice that a *transform manipulator* appears at the sphere's pivot point. Three arrows extend from the manipulator tool: one red, one green, and one blue. By clicking and dragging the red, green, or blue arrow, you can constrain the sphere's movements to the X, Y, or Z axis, respectively. The Rotate and Scale tools have their own unique manipulators whose colors correspond to the same three axes. The Universal Manipulator displays a manipulator around the selected objects that gives you access to all three transform functions. You will get plenty of practice using these tools in every chapter of this book.

*TIP*  **You can use – and + keys to make the manipulators appear smaller or larger, respectively.**

The next tool is the Soft Modification tool. This tool lets you quickly manipulate the shape of an object, similar to sculpting in clay. By clicking any object, you are able to move the manipulator and modify a region of the object. The range of the effect and the falloff can be adjusted in the tool's settings. We will cover this tool in great detail in Chapter 8 when we sculpt geometry or blend shapes.

The Show Manipulator tool can be used interactively to edit the attributes of certain kinds of nodes. Its exact functionality depends on what is selected. A common use for the Show Manipulator tool is to place a spotlight and set its direction. If a light is created and the Show Manipulator tool is chosen, two transform manipulators will show up in the view windows: one to control the position of the light and the other to control the light's target. We will use the Show Manipulator tool to edit attributes of various types of nodes throughout this book.

Continuing down in the toolbox is a varying option that shows the last tool used. This can be useful when you are using a tool repeatedly. It saves you from having to select the tool from the menu or from the shelf.

FIGURE 2-7    A sphere is selected with the Move tool and displays a transform manipulator.

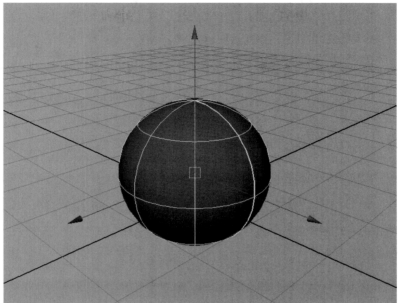

TIP    The Y key acts as a shortcut to select the last tool used.

The last part of the toolbox contains shortcut buttons to access various layouts for the Workspace. These buttons activate common preset layouts. Since we have not yet discussed all of the windows that are available in these layouts, we won't go into specifics here. However, you will be directed to use some of these buttons to change the layouts throughout many tutorials in this book.

## The Workspace

The Workspace is composed of one or more *view panels* used to access different parts of the user interface. By default, only one panel is displayed when you start up Maya. This is the Perspective view, which is shown back in Figure 2-1. Figure 2-8 shows the Maya UI with three panels loaded: the Perspective view, the Hypergraph, and the Graph Editor. The Channel Box appears on the right side of the UI. This type of window layout may suit the needs of an animator pretty well. In this example, the RT_shoulder object is selected. Its transform node is displayed in its hierarchy in the Hypergraph while the animation data for the object is displayed in the Graph Editor. With all of these panels available, an animator can select objects in the Perspective view window or the Hypergraph and use the Perspective view, Graph Editor, and

**FIGURE 2-8**   *The Maya UI with a three-panel layout containing the Perspective view (top left) the Hypergraph (top right), and the Graph Editor (bottom). Also shown is the Channel Box, along the right side of the UI.*

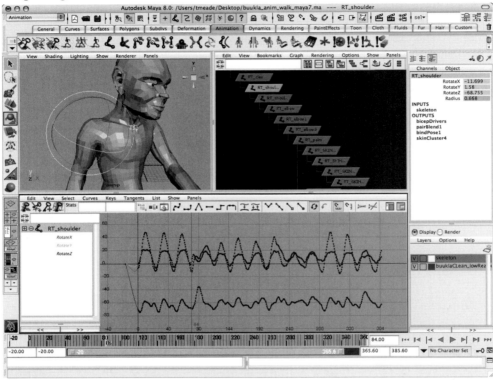

Channel Box to analyze or make edits to the animation data. The exact use of these panels is discussed later in this chapter in the tutorial section.

## The Channel Box

Recall from Chapter 1 that attributes of nodes are also referred to as *channels*. The Channel Box, shown in Figure 2-9, lets you view, edit, and keyframe all of the keyable attributes on any node of the object that is selected. A *keyable* attribute is an attribute that a keyframe can set on to enable that attribute to be animated. You will soon learn that most nodes contain a lot more attributes than those displayed in the Channel Box. However, by default, only certain at-tributes are set up to be animatable.

_**NOTE**_ _**Nonkeyable attributes can be edited in the Attribute Editor, which is discussed later in the chapter. In Chapter 12, you will learn how to edit a node's attributes in the Channel Control window to make it keyable or nonkeyable.**_

You can edit the values for the attributes in the Channel Box in two ways. You can click in the field containing the current value, type in a new value, and press ENTER (RETURN). Or you can click the attribute's name in the Channel Box to highlight it, and then middle-mouse-button-drag (_MMB-drag_ from here on) anywhere in the view window. The value will change as if it were being controlled by an invisible or virtual slider.

_**TIP**_ _**By holding down the** CTRL **key while dragging the virtual slider, you can edit the value for the attribute in smaller increments.**_

## The Layer Editor

Just below the Channel Box is the Layer Editor, shown in Figure 2-10. You can use the Layer Editor to organize your scene by grouping objects into layers. Layering provides you with a quick and easy way to hide or show groups of objects by making a layer invisible or visible, or by making it renderable or unrenderable. What happens depends on whether the objects are grouped to a _display_ layer or to a _render_ layer.

_**FIGURE 2-9**_ _The Channel Box displays the attributes for the object called nurbsSphere1._

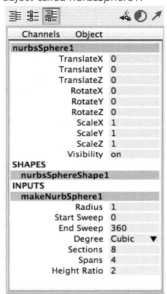

_**FIGURE 2-10**_ _The Layer Editor_

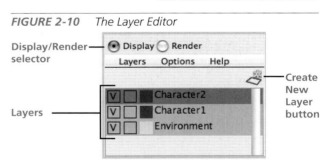

Click the Create New Layer button in the Layer Editor to create a new layer. Each layer is named and listed in the Layer Editor. Click the boxes to the left of the layer name to turn on or off the visibility of the layer (V) and to cycle through Selectable (the blank box), which makes the layer selectable, Template (T), in which the object can be seen as a wireframe and is not selectable, and Reference (R), in which the object can be seen as shaded and can be rendered but is not selectable.

You can double-click a layer in this list to bring up that layer's Properties window. This window lets you rename the layer and assign a color to it. When a color is assigned to a layer, the objects in that layer will display their wireframe in that color in the view window. This does not, however, have any effect on the rendered color of any object.

We will use display layers to group geometric types when we model objects in Chapters 3 through 7 and set up our character rigs in Chapters 9 through 12. Then, in Chapters 27 and 28 we will use the render layers to set up separate render passes.

## The Time Slider and the Range Slider

Figure 2-11 shows both the Time Slider and the Range Slider, along with other controls located along the bottom of the Maya Workspace window. You can scrub through an animation by clicking and dragging the Time Slider along the timeline. To play the animation forward or in reverse, you can use the DVD-like playback controls (the usual arrow buttons).

**FIGURE 2-11**   *The Time and Range Slider controls*

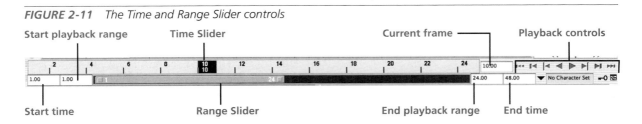

You can set the total length of the animation by keying in the start and end times in their respective fields. The Range Slider sets the range of frames that will be displayed in the Time Slider. Use the Range Slider to limit how much of the total animation time you want to play back. This is especially helpful when you want to work on a small section of a longer animation.

## The Command Line and Script Editor Button

You use the command line near the bottom of the Maya window, shown next, to type in Maya Embedded Language (MEL) commands.

If you are more of a programmer type than an artist type, the command line is a real plus. You type the commands on the left side, and any error message or feedback is displayed at the right. The tiny box icon to the right of the feedback line is the Script Editor button, which lets you create and edit long MEL scripts. The Script Editor also shows all the errors and warnings for a particular piece of script.

NOTE   *Basic MEL scripting is discussed in Chapter 13.*

## The Help Line

The light-gray bar across the bottom of the Maya window is the help line, shown next. If you move your mouse cursor around Maya's user interface, the help line displays what each part of the interface does. When you use actions and tools, the help line tells you which tool you are using and what you should do next.

> Move Tool: Use manipulator to move object(s). Use edit mode to change pivot (HOME).  Ctrl+LMB to move perpendicular.

## The Hotbox

If you hold down the SPACEBAR, the Hotbox will open, as shown in Figure 2-12. The Hotbox is a group of menus that provides an easy-to-access interface that you can place anywhere on your screen. The Hotbox is highly customizable to meet your particular needs. You can click Hotbox Controls in the Hotbox and then choose Show All to show all the menus available in Maya at once.

Because you can potentially access all of Maya's tools and commands from the Hotbox, you could conceivably hide all of the other interface elements by choosing Display | UI Elements and disabling all of the checked items in that list. This means that you could display

**FIGURE 2-12**    *The Hotbox is useful for quickly accessing menu commands while not taking your attention to another*

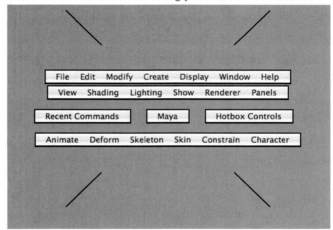

just the Workspace and nothing else. We won't be doing that in this book, but if you are short on screen real estate, it might be helpful to hide your menus and anything else you don't need and access them through the Hotbox instead.

# Tutorial: Working with the Maya Interface

Now that you have some idea of where helpful tools, buttons, boxes, and bars are located in the interface, it's time to start a project and put them to use. Here's a small exercise you can use to practice working with the user interface.

In this tutorial, we create a scene that contains three NURBS spheres that represent the Sun, Earth, and the Moon. We will create these objects and edit attributes on their transform nodes to place them in the scene. Next, we will group these objects together in a hierarchy and animate them. After that, we will create some materials for these objects and edit their colors. Finally, we'll set up lighting and render the animation.

As you work through this tutorial, remember that its real purpose is to let you practice using the interface. Don't get hung up on the specifics of keyframing the animation or editing the materials. It is more important that you pay attention to general concepts, such as where the various elements are located and how they can be used. This tutorial includes the following aspects of working with Maya:

- How to navigate a through 3D scene
- How to create projects and manage files
- How to use the Workspace window
- How to work with interface elements
- How to use the transform manipulators to edit objects' transform attributes

## Navigating a 3D Scene

This section is designed to help a Maya or 3D beginner become familiar with performing some very basic functions in Maya. We'll first create a simple 3D object, and then learn to interactively manipulate the viewing camera so that we can change the angle from which we view the scene.

### Create a Primitive Sphere

*Primitives* in Maya are objects that represent rudimentary geometric shapes. Some primitive shapes include a sphere, cube, cylinder, cone, plane, and torus. Let's create a sphere and look at some of the ways we can edit its attributes.

Choose Create | NURBS Primitives | Sphere ❑ to open the NURBS Sphere Options window to create a sphere. The NURBS Sphere Options window, shown in Figure 2-13, allows you to choose what the initial state of your sphere shape will be when you click the Create button. The data elements for all of these settings are the inputs for the makeNurbSphere node that is controlling the shape. Let's take a look at the NURBS sphere options:

- **Pivot**   By default, Object is chosen, which means that the pivot point will be located at the sphere object's center. You could also choose the User Defined option and then enter the X, Y, and Z coordinates in the fields for the Pivot Point attribute on the next line.
- **Axis**   Controls in what direction the poles of the sphere will face. All of the isoparms on the surface of the sphere meet at the poles. The default sets the Axis to Y so that

FIGURE 2-13    *The NURBS Sphere Options window*

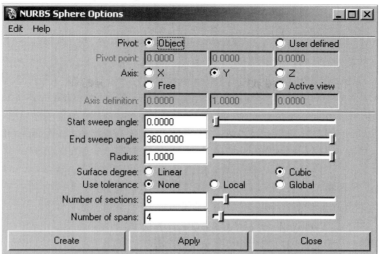

the poles of the sphere will point in the Y direction. Choosing the X or Z axis makes the poles of the sphere point in the X or Z direction, respectively. When the Free option is chosen, the user can define the direction of the axis in the Axis Definition attribute below. Finally, choosing the Active View option will set the axis to align with the current view window. This means that you could use the orientation of the Perspective view to align the axis of the sphere you are about to create.

- **Start Sweep Angle and End Sweep Angle**   Controls what angle the sweep of the sphere begins and ends. For example, by setting the Start Sweep Angle to 0 and the End Sweep Angle to 180, the resulting sphere will sweep 180 degrees total to make a half sphere. If the Start Sweep Angle were set to 90 and the End Sweep Angle were set to 270, the result would still be a half sphere, but it would appear to have been rotated 90 degrees from the sphere created with the 0, 180 settings.

- **Radius**   Indicates the size of the radius of the sphere, in Maya units. (Units can be set to represent centimeters, meters, inches, feet, and so on by choosing Window | Settings/Preferences.)

- **Surface Degree**   Determines how the control point data will be interpolated to produce a surface. Set to Linear or Cubic. This means that the degree of the surface will be 1 or 3 (Surface Degree is discussed in detail in Chapter 5). The Linear setting will result in a rigid-looking object because the edit points of the sphere will have been connected with flat surfaces.

- **Use Tolerance**   Used to improve the precision of the sphere that is represented. This has more to do with how the sphere will be tessellated at render time. By default, the sphere is defined by a number of spans in the U and V directions. However, this can be set based on a curvature tolerance, an amount by which the generated surface can deviate from the mathematical definition of the sphere. A lower tolerance value will be tessellated heavier to match the mathematical description that defines this shape.
- **Number of Sections and Number of Spans**   Settings for the number of spans in the U and V directions. Here, Maya is calling the spans in the U direction *sections*.

You may find it easier to edit the NURBS sphere's attributes after you create the primitive, because you can get visual feedback as you edit. For this reason, we'll create the sphere with the default options and then edit the attributes in the Channel Box and Attribute Editor. To make sure that you are using the default settings for this tool, you can reset it by choosing Edit | Reset Options from the NURBS Sphere Options window. Then click Create to create the sphere with these settings and click Close to close the NURBS Sphere Options window.

If it isn't already being displayed, turn on the Channel Box either by choosing Display | UI Elements | Channel Box or clicking the Channel Box button on the status line.

Let's take a look at what we have so far. With the sphere selected, the Channel Box will display a list of all of the object's keyable attributes (that is, attributes that can be animated), as shown here. Under the Inputs list is a node called makeNurbSphere1. Click that node to display this input node's attributes.

INPUTS
makeNurbSphere1
Radius 1
Start Sweep 0
End Sweep 360
Degree Cubic ▼
Sections 8
Spans 4
Height Ratio 2

Do these attributes look familiar to you? They should. These are some of the same attributes that we saw in the sphere's Options window. The "make" node of a primitive contains the algorithms used to generate a surface based on the given attributes. Clicking in the value field and typing in a new number can change any of these attributes. In the End Sweep value field, type **180**. You'll notice now that a half sphere is shown.

You can also change a value by selecting the attribute name in the Channel Box and then dragging in the view window with the middle mouse button (MMB). This acts as a kind of invisible slider. As you MMB-drag to the right, the value increases and the sphere automatically updates in the view window. As you drag to the left, the value decreases.

You might also choose to edit these values with the Show Manipulator tool. With the End Sweep attribute selected in the Channel Box, activate the Show Manipulator tool from the toolbar or press the T key.

A disc with a little yellow box will appear around the sphere. Dragging the yellow box will change the selected End Sweep value and update in the view window in real time. You can see an example of this in Figure 2-14.

**FIGURE 2-14**  *Editing the end sweep attribute with the Show Manipulator tool*

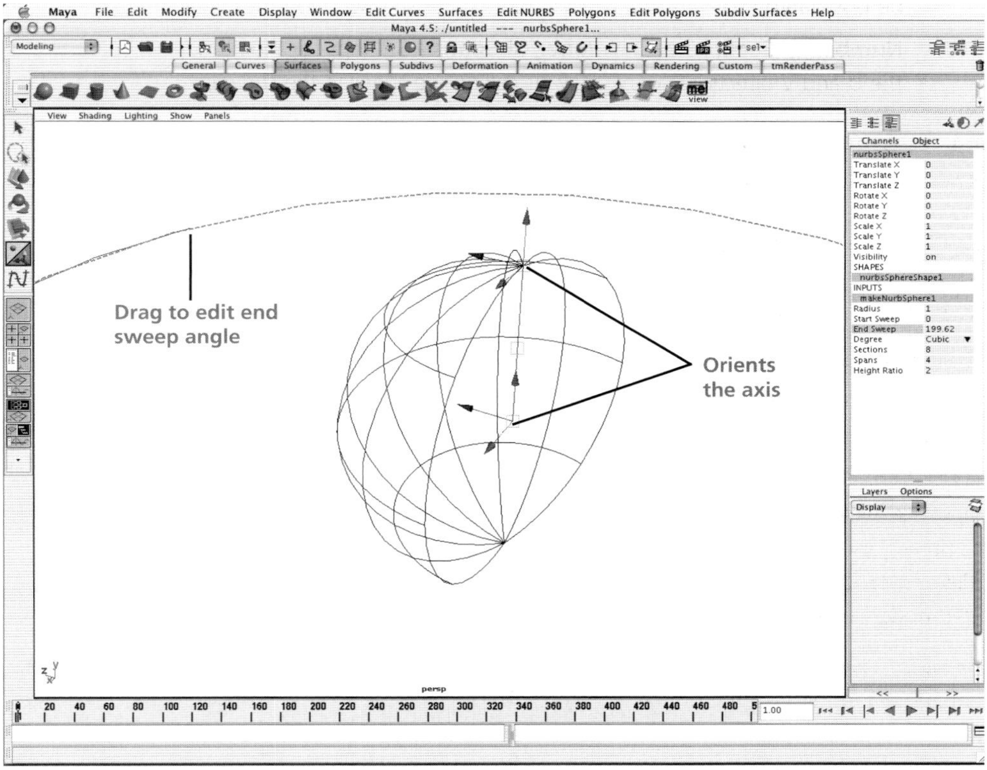

## Set Up a Project

A *project* is an assemblage of folders and subfolders that are set up on your hard drive to organize your Maya projects. It is likely that your Maya project will contain other files in addition to your Maya scene file, such as texture maps or cached animation data. It is therefore helpful for each project if you first set up a directory that contains subdirectories for all of the different elements the project includes. Fortunately, setting up this structure from Maya is easy—in fact, Maya does it for you.

# Maya's Cameras

As you look at your scene through one of the view windows, it is important to realize that you are actually looking through one of Maya's *cameras*. While cameras will be discussed in detail in Chapter 20, it is imperative that you learn right from the start how to manipulate your view as you look through them. The three most common tools used to manipulate a camera are Tumble, Track, and Dolly. The illustration here depicts these three actions as a camera that is viewing a cone object in the scene.

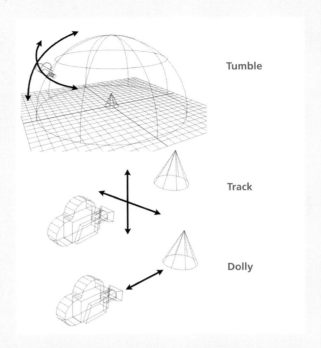

Tumble

Track

Dolly

The Tumble tool allows you to orbit, or rotate, around the camera's center of interest. This tool will work only in Perspective view, because, for example, if you were to tumble in the Front view, you would no longer be looking at the object's front—right? This tool can be accessed from the view window by choosing View | Camera Tools | Tumble Tool.

The Track tool will move, or *pan*, the camera from side to side or up and down. It can be accessed by choosing View | Camera Tools | Track Tool.

The Dolly tool moves toward or away from the center of interest. This tool can be accessed by choosing View | Camera Tools | Dolly Tool.

While these tools are all available from the view window's menu bar, you will probably find it much more efficient to use keyboard/mouse button combinations to access them. The following table describes the shortcuts for accessing these tools. (Note that the Macintosh actions appear in parentheses.)

| ACTION | TOOL | FUNCTION |
|---|---|---|
| ALT-LMB-click (OPTION-LMB-click) and drag—Perspective view only | Tumble | View your scene from all around by orbiting |
| ALT-MMB-click (OPTION-MMB-click) and drag | Track | Move a view horizontally and vertically |
| ALT-LMB-click-MMB-click (OPTION-LMB-click-MMB-click) and drag; or ALT-RMB-click (OPTION-RMB-click) and drag | Dolly | Move the view forward and back |

Let's start a new project:

1. In Maya, choose File | Project | New. Figure 2-15 shows the New Project window.

2. Type in a name for your project. For this example, enter **mcr8_ch02**. (Note that the project name is not the name of the actual scene file, but rather the name of the directory that will contain all of the subdirectories.)

3. Set the Location where you want to save your Maya projects. You can type in the exact path or click the Browse button and navigate to the folder in which you want to save your project. The default path will be in a Maya directory in your user directory. This is generally a good place to store all of your Maya projects.

*FIGURE 2-15    The New Project window is where you manage the storage space on your computer for a Maya project.*

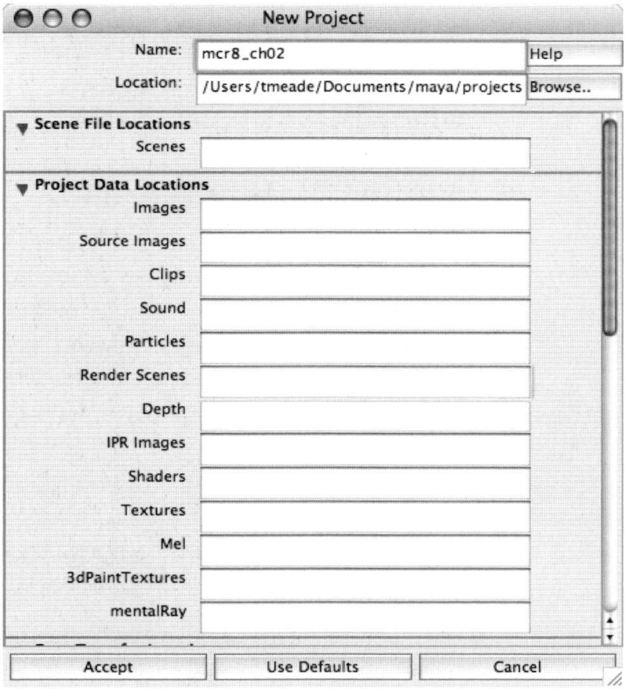

4. Click the Use Defaults button at the bottom of the window. For this project, we'll be using the predefined folder names assigned automatically by Maya. However, if you want to customize the names of these directories, you can type in a name for each folder in the text fields under Scene File Locations and Project Data Locations.

**5.** Click the Accept button to create the project.

After you create the project, the project folder will appear in the location you chose in Step 3. Figure 2-16 shows our mcr8_ch02 folder in the default location. To open this window, choose File | Project | Set. Highlight the project folder you want to access and click Choose. Now, as you save your scene, Maya will automatically place the scene file in the scene's directory inside the mcr8_ch02 directory.

> **TIP**   *It's a good idea to make a habit of creating a project folder whenever you start a new project. This will help you keep your project components intact.*

**FIGURE 2-16**   *The new project appears in the default location of your hard disk.*

## Create and Place Geometry

Now we'll create some objects and place them in the scene using some of the transform tools. You'll learn how to manipulate or view objects in the view window so that you can zoom in on objects and orbit around the scene. Since these are some of the most common actions you will use in every project, Maya offers some keyboard shortcuts to access them quickly. These will be discussed in various sections throughout the rest of this chapter.

**FIGURE 2-17**   *The Channel Box displaying the nodes and attributes that make up the Sun object*

| Channels | Object | |
|---|---|---|
| **Sun** | | |
| | TranslateX | -4.956 |
| | TranslateY | 0 |
| | TranslateZ | -2.758 |
| | RotateX | 0 |
| | RotateY | 0 |
| | RotateZ | 0 |
| | ScaleX | 1 |
| | ScaleY | 1 |
| | ScaleZ | 1 |
| | Visibility | on |
| **SHAPES** | | |
| SunShape | | |
| **INPUTS** | | |
| makeNurbSphere1 | | |

**1.** To create a NURBS sphere object from the menu bar, choose Create | NURBS Primitives | Sphere. Click and drag somewhere on the grid to place and size a sphere. It does not matter where you place it as we will move it to a precise location later on.

**2.** Currently, this object is named nurbsSphere1. Let's rename it to make sure that our project is well organized. Look on the right side of the Maya window and find the Channel Box. The first line in the Channel Box, nurbsSphere1, should be highlighted in gray. This is the name of the object's transform node. Click in that field and type in **Sun**. Then press ENTER (RETURN). The node will now be named Sun, as shown in the Figure 2-17, and Maya will update the shape node name to SunShape.

3. We will place the Sun object at the origin of the scene, just as the real Sun sits at the center of our solar system. We can do this numerically by typing a value of **0** in the Translate X and Translate Z fields in the Channel Box.

4. When we created the sphere by clicking and dragging, the position was determined by the object's transform node. However, the scale, in this case, was set by editing an input node of this object's shape node. As you will recall from Chapter 1, these are two different nodes that are hierarchically connected. To edit the scale, click the makeNurbSphere1 node listed under the Inputs section in the Channel Box. This node's attributes will appear at the bottom of the Channel Box. Set the Radius attribute to a value of **3**.

### Using the Hypergraph Window

The Hypergraph window, shown in Figure 2-18, can display all of the nodes in a Maya scene and depict how they are related. This provides us with a more graphical alternative to editing nodes and their attribute than the Channel Box does.

As mentioned in Chapter 1, nodes can share two different relationships: a dependent relationship, when the attributes are connected, and a hierarchical relationship, when the transform nodes are parented to one another. The Hypergraph can display the scene to show either of

**FIGURE 2-18**   *The Hypergraph window displaying the SunShape node's input and output connections*

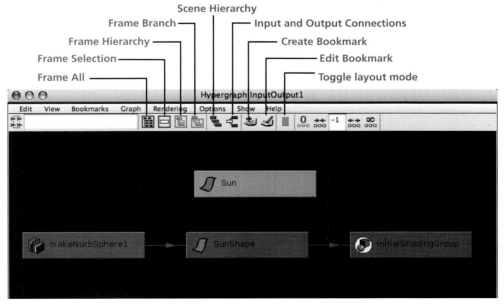

these two relationships. When you're attempting to edit the attributes of certain nodes in a scene, the Hypergraph provides an easy way to select specific nodes directly. Let's open the Hypergraph and view the dependencies on the Sun (the SunShape node):

**1.** To open the Hypergraph, choose Window | Hypergraph Input and Output Connections. The Hypergraph will appear in its own separate floating window.

**2.** Try clicking the different nodes shown in the Hypergraph and notice that the Channel Box updates to show the attributes of the selected node.

**3.** While we could continue editing some of these attributes to make only a half sphere or change its resolution, the existing, default attributes will work fine for our Sun. Since we no longer need the information here to determine the current shape of our Sun, we can delete the Sun object's history. This will get rid of all of the SunShape node's dependencies. With the Sun node selected, choose Edit | Delete By Type | History. Notice that the makeNurbSphere1 node is no longer listed in the Inputs section in the Channel Box nor is it available in the Hypergraph window. At this point, the SunShape node has no dependencies. You can close the Hypergraph window.

Now let's create Earth and the Moon. We could do this by creating another two spheres from the Create menu, but since we really don't need to modify the makeNurbSphere1 node, we can just duplicate the Sun, move it, scale it, and rename it.

**1.** To duplicate the Sun, first select it in the view window and choose Edit | Duplicate. (You can also press the shortcut, CTRL-D [COMMAND-D], on your keyboard.) At first, you may not notice anything different in your view window; don't worry—this is because the duplicate is sitting right on top of the original. Once you move the duplicate, you will see that two identical spheres now appear in the scene.

**2.** Choose the Move tool in the toolbox. You will notice that the move manipulator appears in the view window and extends from the sphere's pivot point. Click and drag the red manipulator to move the sphere out along the X axis. You will see that you indeed have two separate sphere objects.

**3.** With the sphere copy still selected, go to the Channel Box and rename this object from Sun1 to **Earth**. Now duplicate the Earth and move the duplicate out along the X axis to separate it from the original. Name this third object **Moon**.

To place and scale these objects more precisely, we will change our Workspace to display some different views. We will change the layout of the Workspace to the Four View layout.

## Using the Four View Layout

Keep your cursor on top of the Perspective view and quickly tap the SPACEBAR. The Workspace will now show the Four View layout, which includes the Top view, Perspective view, Front view, and Side view, as shown in Figure 2-19.

Let's take a moment to understand these view windows and navigate inside them. The window at the top right is the *Perspective* view. This view uses *perspective correction*, which works just like our eyes or a camera—that is, objects farther away from you look smaller than closer objects. Perspective view is good for visualizing how a final rendering will look, since that's how a rendering camera will see the scene. Moving counterclockwise from Perspective view are the Top, Front, and Side orthographic views.

**FIGURE 2-19** *The Maya Workspace showing the Four View layout*

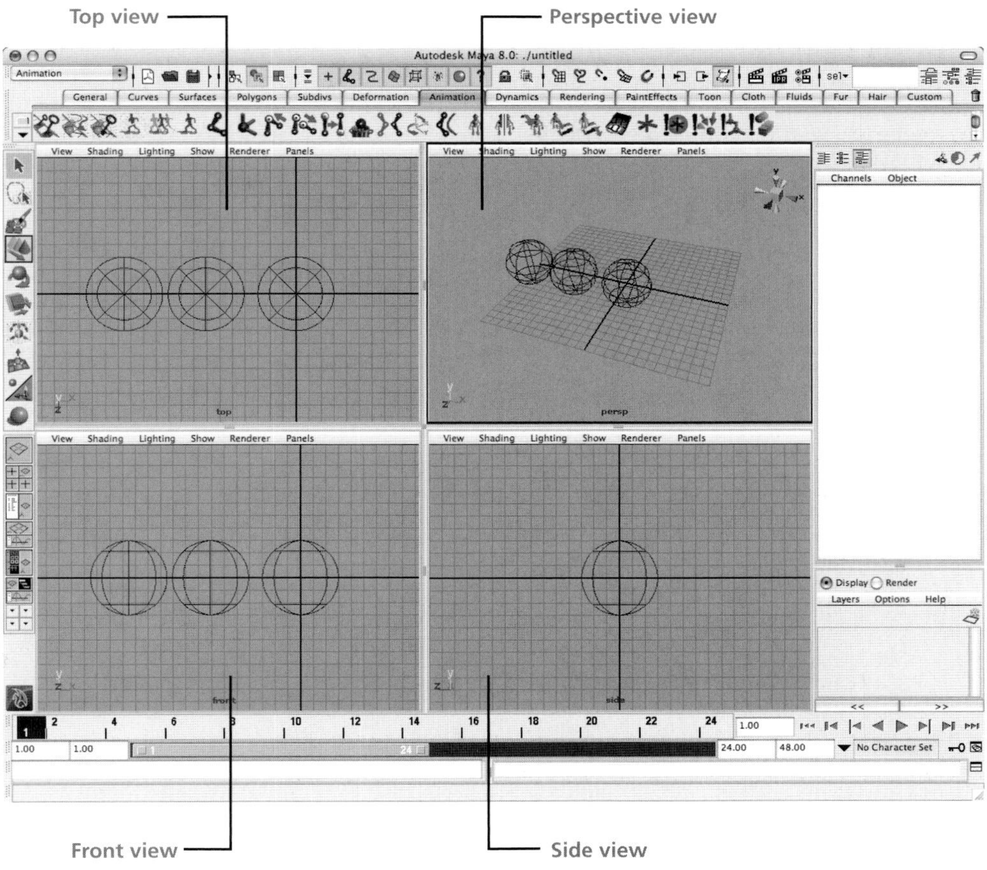

*Orthographic* views (the Front, Top, and Side) have no perspective correction. If two objects of matching size appear, and one is placed "farther away" from the object in front, they both will appear to be the same size in an orthographic view. Orthographic views are good for analytically viewing your models and scenes. For example, you can compare object sizes, place objects, and check alignments in these views without accounting for the distortion caused by perspective correction.

## Focus and Shade in the View Window

You'll often work with dozens, or even hundreds, of objects in a scene. Every time you want to adjust the view to move closer to, or *focus* on, one or more objects, you may find it cumbersome to use the Tumble, Track, or Dolly tool. Instead, you can use hot keys from the keyboard that let you focus on a selected object or group of objects. Focusing on objects will frame or fit them to the bounds of your view window. It also sets the camera's center of interest at the center of the selected object or objects. This feature is especially helpful when you select the Tumble tool in the Perspective view, because this tool orbits the camera around this center point.

The hot keys available for these focusing actions are listed and described here:

| HOT KEY | FUNCTION |
| --- | --- |
| F | Focus on a selected object in a selected window. |
| SHIFT-F | Focus on a selected object in all windows. |
| A | Focus on all objects in a selected window. |
| SHIFT-A | Focus on all objects in all windows. |

Let's focus on all of the objects in all of the windows of the view:

1. In the Perspective view, click to select a sphere (or, rather, one of the celestial elements, if you will). Hold down the SHIFT key and click the other spheres to add them to your selection.

2. Press the F key to focus on the spheres. The three spheres should zoom in closer to the camera.

3. Try tumbling the camera in Perspective view. Hold down the ALT (OPTION) key and LMB-drag. You'll notice that the camera is orbiting around the *center of the three objects*, not the scene's origin, as it was earlier.

### Display Options

So far, you have been viewing these objects in *wireframe* mode. You can change the way a piece of geometry will display in a view window by setting the display options for that window. These options can be accessed via the View menu in the view window. You can also press a hot key to choose a view mode, as shown in Table 2-2. These display modes will activate your computer's hardware shading engine built into your video card and let you view the objects in your scene as shaded; shaded and textured; and shaded, textured, and lit.

*TABLE 2-2*   *Hot Keys for Display Modes*

| HOT KEY | MODE |
| --- | --- |
| 1 | Rough, NURBS, and subdivision only |
| 2 | Medium NURBS, and subdivision only |
| 3 | Fine NURBS, and subdivision only |
| 4 | Wireframe |
| 5 | Smooth shade, default light |
| 6 | Smooth shade with hardware texturing, default light |
| 7 | Smooth shade with lighting, hardware rendering lights if you have created lights; otherwise objects shown in black |

These display options can also let you view certain types of geometry at different levels of detail. Both NURBS and subdivision surfaces, two of the three geometry types in Maya, are mathematical representations of surfaces. Although these geometric types are discussed in detail in Chapters 5, 6, and 7, for now, you should realize that these types of surfaces can be displayed at varying levels of detail. For example, if you have a lot of objects in your scene and are displaying them at their finest level, your computer's performance could suffer as you manipulate your camera or objects through the scene and your computer attempts to render the complex objects. Displaying surfaces in less detail will help to improve this performance. This is where changing view modes comes in particularly handy.

Spheres in rough, medium, fine, wireframe, smooth shade, and texturing modes are shown in Figure 2-20.

> **NOTE**   *Be aware that when you use hot keys, any changes you make affect only the objects in the active window.*

Let's take a look at the spheres with a fine level of detail and smooth shaded in the Perspective view. Press 3 (which is the default detail level) to view the spheres at fine detail, and then press 5 to see them smooth shaded. This gives you a better idea of how these objects might look when rendered. But before we start rendering, let's work on placement and scale.

# Transform Objects

Here we'll concentrate on the exact placement and scale of each object. We won't worry about the precise distances and proportions to make our planets match the scale of the solar system

**FIGURE 2-20** *Display options in Maya*

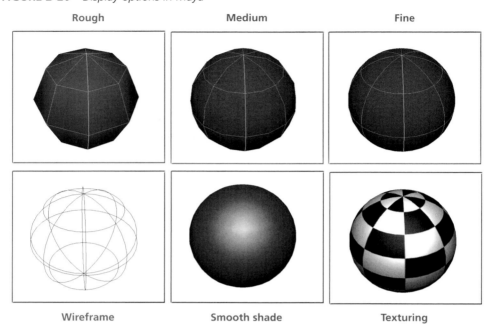

elements in our own universe. However, we do want to depict that the Sun is the largest object in our model solar system while the Moon is the smallest. We also want to arrange the objects so that the distance between the Sun and Earth is greater than the distance between Earth and the Moon.

While we already practiced moving and scaling in the first step of this tutorial, we used the transform tools in the toolbox. To get you moving along faster, you should learn and use the hot keys to select the items in the toolbox. From now on, try to get in the habit of using the hot keys to access these tools, as shown in Table 2-3.

**TABLE 2-3** *Toolbox Hot Keys*

| HOT KEY | TOOL |
| --- | --- |
| Q | Select tool |
| W | Move tool |
| E | Rotate tool |
| R | Scale tool |
| T | Manipulator tool |
| Y | Last used tool |
| V | Snap to Grid |
| X | Snap to Point |

_TIP_   **You may find it helpful to rest your left hand on the keyboard with your pinky finger on top of the Q key and the rest of the fingers on your left hand on the W, E, and R keys. This will give your fingers quick access to all of these tools, regardless of where your cursor is located on the screen.**

1. To place our three planetary objects in the scene, we will use the Top view. Select the Earth object and press the W key to activate the Move tool.

2. Let's place this object so that it sits at 10 units on the X axis. To make sure you place the object exactly at 10 units on X, turn on grid snapping to enable you to snap any object to specific grid units. On the status line, click the Snap to Grid button.

3. Now that grid snapping is enabled, return to the Top view and click inside the yellow box at the origin of the Move tool's manipulator. This will let you move the object in any direction relative to the axis of the current view. Since we are snapping to the grid, it should be easy not to slip the object out onto the Z axis.

4. Move the Earth object out in the X direction to 10 units. You can confirm that it is at 10 units in X by checking the values in the Channel Box. The field labeled Translate X should show a value of 10.

5. Now scale the Earth object down so that it is smaller than the Sun. When we originally set the scale of the Sun, it was by way of editing the Radius attribute on the makeNurbSphere1 node. Now that we have deleted the history, that node no longer exists. Therefore, we will use the Scale attributes on the Earth object's transform node to set the scale. In the Channel Box, set the Scale X, Y, and Z to 0.3. You could also enter the values separately in the fields one by one, but a quicker way is to click in one of the fields and LMB-drag down over all of the input fields in the Channel Box for Scale X, Y, and Z, as shown here.

| Channels | Object | |
|---|---|---|
| **Earth** | | |
| | TranslateX | 10 |
| | TranslateY | 0 |
| | TranslateZ | 0 |
| | RotateX | 0 |
| | RotateY | 0 |
| | RotateZ | 0 |
| | ScaleX | 0.3 |
| | ScaleY | 0.3 |
| | ScaleZ | 0.3 |
| | Visibility | on |
| **SHAPES** | | |
| **EarthShape** | | |

6. After the input fields are selected, a cursor will appear in the last field chosen. Type in a value of **0.3** and press ENTER (RETURN). Notice that all three fields will update to contain a value of 0.3.

7. With the Sun and Earth in place, focus your attention on the Moon. Use the techniques just discussed to edit the Translate and Scale attributes of the Moon in the Channel Box. Place the Moon at 12 units in X and enter a scale of **0.08** for its scale values in X, Y, and Z.

8. Now that the initial positions, orientations, and scales have been determined, we will zero out the objects' transforms by freezing their transformations. This will set the translations and rotations to 0 while offsetting the Scale attribute to 1. In case any of the objects were to get nudged or misplaced, one could easily place them into their

initial state by resetting the transform values. Select all of the objects in the scene and choose Modify | Freeze Transformations.

## Create a Hierarchy

Now that the objects are placed and scaled, we need to group them together in a *hierarchy*. The Sun will be the *root* object, Earth will be the *child* of the Sun, and the Moon will be a *child* of Earth. To monitor this hierarchical relationship, we will view the hierarchy in the Hypergraph, but instead of opening the Hypergraph as a floating window from the Window menu, we will change the Side view in the Four View layout to display the Hypergraph.

Each of the four views displayed in the current layout is inside a *panel*. In our project, each of four panels displays one of the default views (see Figure 2-19). Any panel can be set to display another window, instead of having to open the window from the Window menu and view it as a floating window.

To change a panel's display, use the Panels menu in the selected panel's menu bar. In this case, we will set the panel that is currently displaying the Side view to display the Hypergraph instead:

1. Select the Side view panel, and from that panel's menu bar, choose Panels | Hypergraph Panel | Scene Hierarchy. The Hypergraph will load into this panel and should include the three transform nodes named Sun, Earth, and Moon.

2. To parent the Earth to the Sun, MMB-drag the Earth node onto the Sun node. Notice that the Hypergraph now displays a line connecting these two nodes, which means that these two nodes are connected hierarchically.

3. Now use a different technique to link the Moon to the Earth. In the Top view panel, select the Moon object, and hold down the SHIFT key to select the Earth object. (Pressing the SHIFT key as you click, or SHIFT-clicking, lets you add objects to your selection.)

4. With the Moon and Earth objects selected *in that order*, press the P key. If you look in the Hypergraph, you will see that the Moon is now parented to the Earth (that is, the Moon is a child of the Earth). The nodes in the Hypergraph should look like those in the illustration.

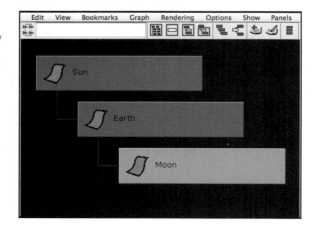

**5.** To see how these objects behave in their hierarchy, select the Sun in one of the view windows and choose the Rotate tool by pressing the E key. Click and drag the Rotate manipulator to rotate the Sun. Notice how the child objects behave. Press CTRL-Z (COMMAND-Z) to undo the rotation and return the objects to their original orientation.

## Create a Group Node

While these objects are now arranged in the correct hierarchy, we need to make some additions so that they will animate properly. Recall that back in Chapter 1 we discussed how every object's transform node transforms around its pivot. We know that our Earth spins around its own axis once every 24 hours. Using the current settings, we could animate this behavior now, because if you select the Earth and rotate it, it will rotate around its pivot, which is located at the Earth's center. However, we also need to account for Earth's orbit around the Sun, so in this case, we need the Earth to pivot around the Sun. If the Earth is already using its own pivot point to rotate around its own axis, we cannot use the same pivot point for the orbit. Therefore, we need to create another pivot point between the Earth and the Sun and place it at the center of the Sun. The easiest way to do this is to group the Earth object to itself. This will parent the Earth to a new transform node with its own pivot.

A *group node* is simply a transform node with no shape object associated with it. Often, objects are grouped together for organizational purposes—to keep similar things grouped together. In our planetary project, we will use grouping to give us an extra transform node with an extra pivot point so we can accommodate our Earth's orbit around the Sun:

**1.** In Top view, select the Earth object and choose Edit | Group. This will create a group node called *group1* as the parent of the Earth. By default, a group's pivot point will always be at the origin of the scene—the 0,0,0 point. For the purpose of our Earth sphere, this is exactly where we want this pivot point to be located, since 0,0,0 also happens to be the location of the center of our Sun sphere.

**2.** With the group1 node still selected (use the Hypergraph to select it if you have accidentally unselected it), choose the Rotate tool and rotate the node around with its Rotate manipulator. The Earth and its child (the Moon) now rotate around the Sun.

**3.** In the Channel Box, change the new group name from group1 to **earthOrbit**.

**4.** You know that the Moon does not rotate around its own axis—it rotates around Earth. Therefore, you need to edit the pivot point for the Moon's transform node so that it is at the center of the Earth. Press the W key to choose the Move tool. You will see that the transform manipulator is at the origin of the Moon.

**5.** Press the INSERT (HOME) key to edit the pivot. The transform manipulator in the view window changes, as shown here.

**6.** On the status line, turn on grid snapping. (You can also turn on grid snapping by holding down the X key while you move an object in the view window.)

**7.** Drag the pivot icon and snap it to the grid line at the center of the Earth.

**8.** Press the INSERT (HOME) key to return focus to the Move tool.

This will complete the setup of our hierarchy. The hierarchy displayed in the Hypergraph should now look like the one shown in Figure 2-21. Next, we will animate these objects.

**FIGURE 2-21** *The Hypergraph window displaying the final hierarchy of the objects*

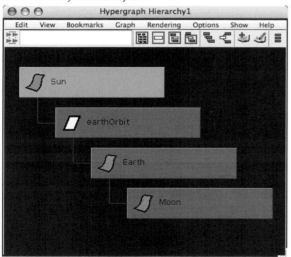

## Animate the Objects

Now we will animate these objects to resemble the behavior of the solar system. While they won't be *completely* accurate, we will attempt to animate the actions that occur during a one-month time span. What this means is that the Earth will rotate around the Sun 30 degrees (360/12 = 30); the Earth will rotate around its own axis 30 times, once each day; and the Moon will rotate around the Earth 30 times (though this really happens every 27 days, but we'll round off to 30 just to make it easy).

Now we need to decide how long this animation will last. Since *30* seems to be the magic number we are using to determine orbits and rotations, let's say that the animation will be 30 seconds long. So how many frames is this?

By default, Maya's frame rate (frames per second, or fps) is set to 24, which is the standard for film production. Since 30 seconds × 24 fps = 720 frames, we will set the range of the animation to 720 frames long. Let's get started:

**1.** Along the bottom of the Maya window you'll find the Range Slider, just below the Time Slider. Set the total animation time and Range Slider time to **720**.

**2.** Now you will set up the Workspace to use another layout that is especially useful for animating: the Outliner/Perspective layout. Choose the Outliner/Perspective layout button in the toolbox. The Workspace layout will change.

### The Outliner

The Outliner, shown in Figure 2-22, is another way of displaying all of the objects in a scene. The objects listed here show the hierarchies in a way that's similar to that of a common Explorer (Windows) or Finder (Mac) view used to browse file directory structures. To access the Outliner, choose Window | Outliner. To unfold a hierarchy, click the plus sign (+) icon to the left of the object's name. If you want to view all of the branches and subbranches of a hierarchy, hold down the SHIFT key and click the plus sign.

**1.** SHIFT-click the + icon to the left of the Sun node (and notice that the plus sign changes to a minus sign). This will expand the hierarchy to reveal all of the child objects, as shown in Figure 2-22.

**2.** Click and drag in the Outliner to select all the objects in the Sun's hierarchy. Be sure that the Time Slider is set to 1, and then choose Animate | Set Key. This will set a keyframe for all of the selected objects in the scene for their current positions at frame 1.

**3.** Move the current Time Slider to frame 720. You can do this by clicking and dragging the Time Slider to the end of the frames being displayed, or you can type **720** into the Current Frame field (see Figure 2-11).

**FIGURE 2-22** *The Outliner lists the current object in the current scene.*

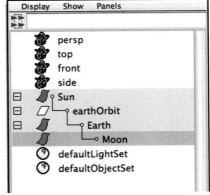

**4.** In the Outliner, select the object named earthOrbit that you created earlier. Then, in the Channel Box, set the Rotate Y attribute to **30**.

**5.** Again, choose Animate | Set Key to set a keyframe for this position.

**6.** Select the Earth object in the Outliner and set its Rotate Y attribute to **10800** (360 days × 30).

**7.** Instead of choosing Animate | Set Key, select the Rotate Y attribute in the Channel Box so that it is highlighted (make sure the actual attribute name is selected, not the value in the text field); then right-click the attribute. Choose Key Selected from the marking menu. This will key only this attribute, not all of the keyable attributes available on this transform node. This results in much more efficient keyframing.

**8.** Click the Play button (see Figure 2-11). The animation will play back. The Earth will rotate around the Sun 30 degrees, or 1/12 of the way around, while it rotates around its own axis 30 times. The Moon will follow the rotation of its parent object (the Earth) and will appear to orbit the Earth.

## Shading Objects

*Materials* give an object its shading properties. In other words, an object's material can control the color and shininess, or how reflective this object is. (We will discuss materials in depth in Chapters 17 and 18.) In this section, we will create some new materials and apply them to our geometry. Then we will edit the materials' attributes to change their colors. This section will introduce you to two new pieces of the interface: the Hypershade and the Attribute Editor.

### The Hypershade Window

The Hypershade, shown in Figure 2-23, is where materials are created and edited in a Maya scene. You can open the Hypershade by choosing Window | Rendering Editors | Hypershade. By default, the Hypershade window will contain three main sections: the Create Render Node menu, the upper tabs section, and the lower tabs section.

In the Create Render Node menu, you can browse and create various types of materials, textures, lights, cameras, and utilities. By clicking and holding on the arrow next to Create Materials, you can change this menu to show these other types of rendering nodes. The upper tabs section contains separate tabs for browsing any of the nodes in your current scene. By default, the Materials tab contains materials. Depending on what type of object is created in the scene, it will use the materials appropriate for it. By default, any geometry created in a scene will use the Lambert1 material until the user creates and assigns a new material.

The lower tabs section contains a Work Area tab (the Hypershade window) and a Shader Library tab. The work area is where you can view all of the connections on a selected material node. It is similar to the Hypergraph window, except that it is optimized for material editing in that the icons representing the node use pictorial representations, called *swatches*, in addition to their names. The Shader Library lets you browse though a collection of premade materials and texture maps that ship with Maya.

*FIGURE 2-23* *The Hypershade window with its default layout*

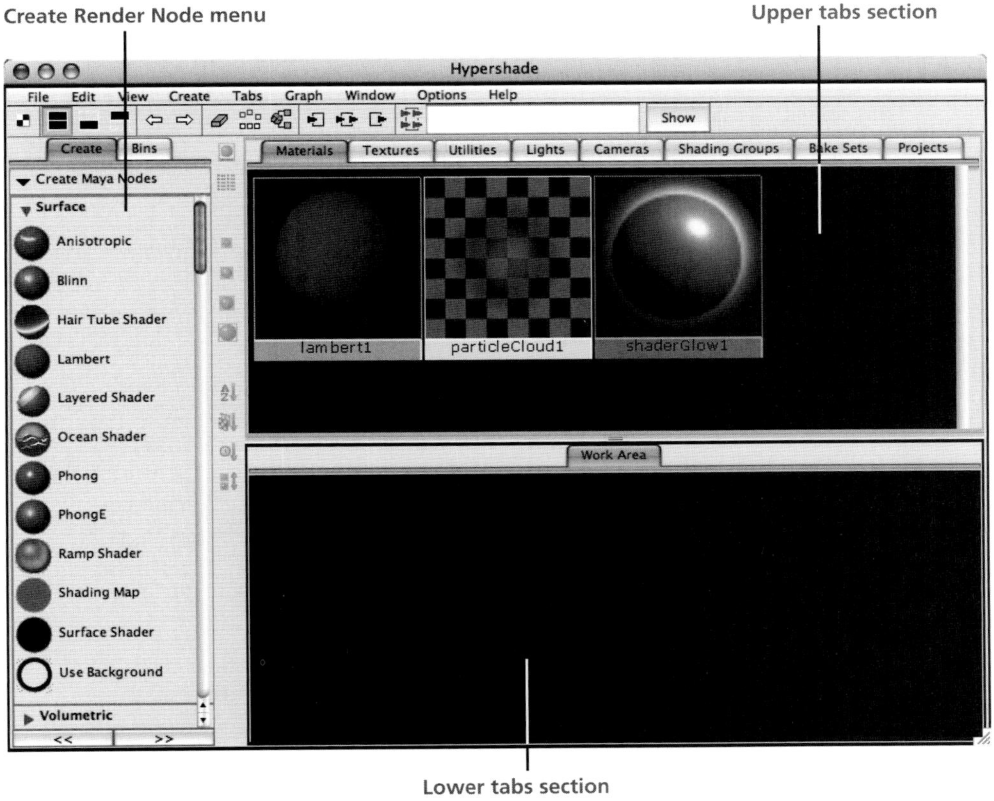

Create Render Node menu

Upper tabs section

Lower tabs section

Let's create three new materials—one for each of our spherical elements. If you have opened the Hypershade, close it now. Instead of using the Hypershade in a floating window, we will modify the Workspace to display a two-panel view—one panel showing the Perspective view and another for the Hypershade window.

**1.** In the toolbox, click the Hypershade/Perspective button to load the Workspace layout.

**2.** From the Create Render Node menu in the Hypershade window, click the icon labeled Lambert. This will create a new Lambert material. A Lambert swatch will appear in the Materials tab as well as in the Workspace and will be named Lambert2.

**3.** Select this material in the Materials tab and rename it in the Channel Box to **mSun**. The *m* denotes that this node is a *material* node. Also, because we already have a node in our scene named Sun, we need to give this node a different name, since all nodes in a Maya scene must have a unique name. (This naming convention will be used throughout the rest of the book.)

**4.** Double-click the mSun material node in the Materials tab of the Hypershade window. This will automatically change the layout of the Maya window to display the Attribute Editor instead of the Channel Box.

### The Attribute Editor

The Attribute Editor, shown in Figure 2-24, displays all of the connected nodes and their material attributes for a selected object. Each tab represents a respective node and its attributes. Figure 2-24 shows only one node, mSun, because it's the only material node we've created so far.

**FIGURE 2-24**    *Attribute Editor showing the mSun material's attributes*

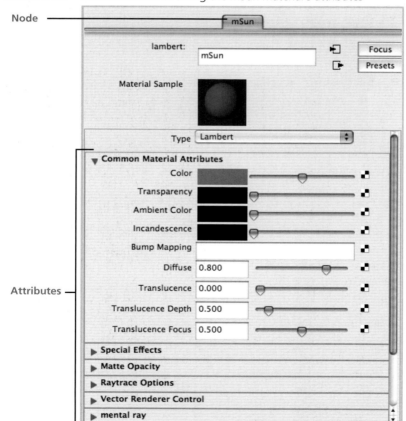

<u>NOTE</u>   *Although the Channel Box also displays attributes, it displays only the attributes of nodes that are keyable—that is, the keyframes are able to be set on these attributes to animate them. While you can make any attribute keyable, this issue will be addressed specifically in Chapter 12.*

One great aspect of the Attribute Editor is that it contains sliders, instead of text fields, that you can use to edit attributes—so you can *see* the effects of your changes. However, because it displays all of the available attributes on any one node, this may be more than you need to work with for your current process. For material editing, however, using the Attribute Editor is recommended.

Here's how you edit attributes:

1. At the top of the Common Material Attributes list, find the attribute named Color. In Figure 2-24, this is set to a gray color in the middle of the slider. Click the gray color swatch to bring up the Color Chooser, shown in Figure 2-25.

2. Click and drag in the color wheel (the hexagonal shape in the middle of the Color Chooser) to select a yellowish-orange color for the Sun. When you are happy with the color, click Accept to close the Color Chooser.

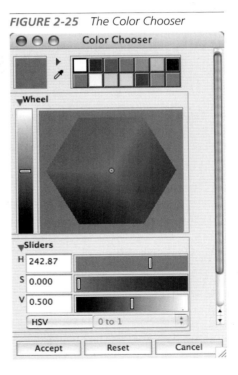

**FIGURE 2-25**   *The Color Chooser*

3. To apply this material to the Sun object, MMB-drag the material from the upper tabs section of the Hypershade window onto the Sun object in the Perspective view window.

4. Now add colors for the two other objects in the scene in the same way. Make the Earth's material blue and leave the Moon's color at its default gray.

5. You need to make one more edit to the mSun node to make the Sun appear to be illuminating. You can do this by editing the mSun's Incandescence attribute in the Attribute Editor.

Change the Incandescence attribute color to a bright yellow, using the slider as you did earlier. (This attribute will be explained in detail in Chapter 17.)

## Add Light

So far, the existing light in the scene is illuminating from the default light, which will always sit right above the camera of the view being rendered. The default light is used for nothing more than performing quick renders as you work on your scene. It is not intended to be used as your final light source.

For this scene, we will add what is called a *point light* and place it at the origin of the scene. A point light illuminates in all directions, as opposed to something like a spotlight that illuminates in one direction. Because this light source will be representing the Sun in this animation, the point light is the best choice.

**1.** Choose Create | Lights | Point Light. A light icon will appear at the scene's origin, but since the view is displaying in shaded mode, you will not be able to see it because it is inside the Sun.

**2.** To see the real-time effects of the light in Perspective view, press the 7 key; this will turn on show lighting.

**3.** Save the scene by choosing File | Save. Name the scene **solarSystem**. This scene file will be saved in the Scenes directory inside our mcr8_ch02 directory.

## Render the Animation

In this final step of the tutorial, we will configure the renderer and render the animation. The main window for configuring the renderer is the Render Settings window, shown in Figure 2-26. This window will be examined in detail in Chapter 21. For the purpose of this tutorial, we will make only a few modifications here before invoking the Batch Render command.

**1.** Choose Window | Rendering Editors | Render Settings to open the Render Settings window. You can also access this window from the status line by clicking the Render Settings button.

**2.** In the Render Settings window, click the arrow to open the pull-down menu in the Frame/Animation Ext field and select name_#.ext.

**3.** Set the End Frame of the animation to **720**.

**FIGURE 2-26**   *The Render Settings window*

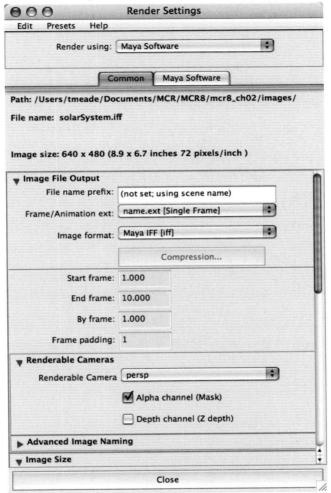

4. Click the Maya Software tab and scroll down to find the Anti-Aliasing Quality folder (the sections noted by the little black triangles are called *folders*). Click the arrow to the left of the folder to open it.

5. Set the Quality pull-down menu on the Maya Software tab to Production Quality. Click the Close button to close the Render Settings window.

**6.** Press F5 to change to the Rendering menu set. Choose Render | Batch Render. The Batch Render application will launch, and your animation will begin to render as a sequence of individual frames.

You can monitor the progress of your render by looking in the feedback line at the bottom of the Maya window. This will tell you the percentage of completion for each frame rendered as well as the path to the directory where the files are being written.

### Use Fcheck

When the rendering is complete, you can view your image sequence in the Fcheck application found in the same directory as the Maya application. You can open Fcheck through the Maya UI by choosing File | View Sequence and then navigating to the file that contains the first frame of the animation. This file will be called solarSystem_01.iff and will be located in the Images directory inside the mcr8_ch02 directory. Click Open and the image sequence will load and play back.

# Summary

By now, you should have a good idea of how to navigate the basic Maya interface and create a simple animation. While we haven't shown you everything, you should have an idea of how to select nodes, edit their attributes, change the layout of the Workspace, use some hot keys, and access windows from the Window menu. Throughout the remainder of the book, we will explore all of these elements of the user interface in greater detail.

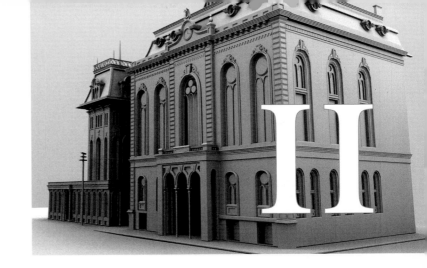

# II

# Modeling

# 3

# Polygonal Modeling

**We are going to begin our modeling**

adventures in this book by modeling in polygons.

In the past, polygonal modeling techniques were

used to create rigid, or hard-edged, models

for architecture and similar projects. Polygons

have always been the standard geometry type

used in the gaming industry for modeling

both characters and environments. Due to

technological advancements in polygonal and

texturing toolsets, as well as a computer's

ability to handle large data sets, polygonal modeling has made its way into, and is now prevalent in, the film industry, previously dominated by NURBS modeling. This chapter presents the basics of polygonal modeling and starts you on your way to building creatures, castles, city streets, and more. For those readers who are new to Maya, this chapter also walks you through some of the most basic steps to operating the software and makes you familiar with mouse clicking, marking menus, and windows that you will use throughout the rest of this book.

# Basics of Polygonal Modeling

Polygons are the building blocks of 3D modeling. Just as atoms are the basis of all matter in the universe, polygons are the smallest renderable units that make up a 3D model. Just as atoms can be broken down into smaller particles, a single polygon can be broken down into smaller *components*. At its most basic level, a polygon is a triangular-shaped *face* defined by three *vertices* that are connected by three *edges*. Any 3D model that you render inside of Maya is made up of polygons, even when it is constructed using NURBS or subdivision surfaces (more about these geometry types in Chapters 5 and 7).

> **NOTE** *As we begin this chapter, make sure that you set your menu set to Polygons so that all of the polygonal menus discussed here will appear in the menu bar.*

## Polygon Anatomy

There are three different types of components in a polygonal object: a vertex, an edge, and a face. Figure 3-1 shows the components of a polygon, which are listed here:

**FIGURE 3-1**    *Components of a polygon*

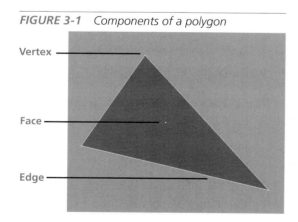

- **Vertex**    A point in space. The vertex is the most basic component of a polygonal model. When Maya stores polygonal data it assigns each vertex in a model a unique identification number and gives that point's location in 3D space. This information allows the software to reconstruct the model when a file is opened by connecting each vertex to another one with an edge. By editing the position of a vertex, you are changing the shape of the face that it creates.

- **Edge**   The polygonal components that connect two vertices. The area between at least three connected vertices creates a face.

- **Face**   A minimum of three vertices enclosed by three edges. Although you can create a face with any number of vertices, the faces will be broken up into triangles at render time. A triangular face is called a *tri*, a four-sided face is called a *quad*, and a face with more than four sides is called an *n-gon*. The collection of connected faces in a model is called a *polygonal surface* or a *polygonal mesh*.

- **UV**   Every vertex in a 3D model can be assigned a coordinate in 2D space. Once the UVs are assigned, they can be edited in the UV Texture Editor to control the placement of a texture map on the 3D model. We will discuss this type of component and the UV Texture Editor in Chapter 18.

- **Normal**   Every polygonal face has a front and a back side. A surface normal indicates the direction of the front of the face. You can display the normal of a polygonal object by selecting the mesh and choosing Display | Polygon | Face Normals. The direction of the normal can be changed by selecting a mesh and choosing Normals | Reverse. A polygonal sphere displaying its normals is shown here:

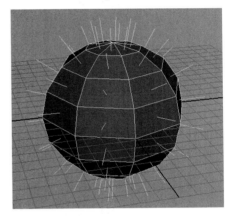

# Selecting and Editing Polygonal Components

As with many things in Maya, there are always different ways to perform one task. By default, there are two different ways to access the components of a piece of geometry. One way is through the Selection Mask located on the status line and the other is through a marking menu. This process of switching into component mode and making selections is crucial to navigating your way through the Maya UI and is used in just about any workflow along the production pipeline, not just modeling.

### Using the Selection Masks to Select Components

To practice selecting components, start with a new Maya scene (File | New) and create a polygonal cube (Create | Polygon Primitives | Cube). When you select one of the transform tools (Move, Rotate, or Scale) and manipulate the cube with the manipulator icon in the view windows, the edits that you are currently making are to the cube's transform node. Maya refers to these types of edits as edits made to the object, or in "object mode."

Remember from Chapter 1 that all geometric objects in a scene are hierarchically made up of a transform node with a shape node parented to it. If we wish to make edits to an object's shape node, we need to switch from object mode into "component mode." One way to do this is to use the selection masks located on the status line. Figure 3-2 shows the selection masks section of the status line in object mode.

**FIGURE 3-2**   *The selection masks section is set to select object. The first three buttons indicate the selection mode while the following eight buttons enable which type of objects can be selected.*

Anything we select in object mode will allow us to make edits to the object transform nodes. We can exclude any types of objects from a selection by clicking on one of the object type buttons. For example, if we wanted to select only curves, we could disable all of the buttons except for the curve type. This is helpful for excluding unwanted types of objects in your selection. An object that is selected while in object mode is indicated by a green highlight in the view windows.

To change into component mode, click the button to the right of the object mode button. The buttons in this section of the status line will change to look as they do here:

By default, the selection masks have vertices enabled. If you look in your view window, you will see that your cube is now highlighted in blue. The vertices are visible and displaying in a red color. You can now select the vertices and use any of the transform tools or an appropriate polygonal editing command to edit them.

> _TIP_   **You can toggle between object and component mode by pressing the F8 key on your keyboard.**

### Using the Marking Menu to Select Components
Undoubtedly one of the most powerful aspects of Maya is its well designed user interface. Navigating through such a large program would take time if you were left to using only the drop-down menus from the menu bar. To give the user a better experience while working, the Maya engineers implemented *marking menus*. A marking menu is one that appears at your mouse cursor by pressing a certain key or mouse button in the workspace. The advantage of this type of interface is that you don't have to go looking elsewhere in the interface to switch to other tools or selection modes. This way your eyes and attention will remain focused on your task.

The fastest method for selecting an object's components is to use a marking menu. In the view window, position your mouse cursor over the cube and RMB-click and hold. A marking menu containing all of the polygonal components will appear right at the spot where you are clicking. Figure 3-3 shows the marking menu displaying the polygonal components.

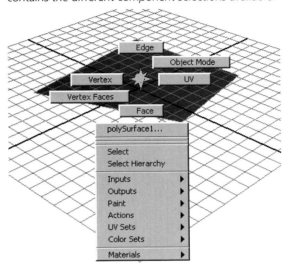

**FIGURE 3-3** *By RMB-clicking and holding over a polygonal object, a marking menu appears that contains the different component selections available.*

You can choose one of the component modes by continuing to hold down the mouse and dragging over the desired component type, Vertex, Face, Edge, or UV. When you have completed your edits to components, you can select object mode from the Component Editor. The display of your components will disappear and your object will return to its green highlighted color indicating that you are in object mode.

While using the marking menu to select the components may seem basic, it is essential that you become comfortable with the process before moving on. Really, no matter what kind of task you are doing, be it animating, rigging, or working with particles, the process of activating the marking menu with an RMB-click will provide you with tools, commands, or modes that will make you more productive.

> **NOTE** *From here on, when we direct you, the reader, to select a component or return to object mode, we expect you to use the marking menu activated by clicking and holding the right mouse button (RMB).*

## Advantages of Polygon Modeling

Polygon modeling has several advantages over NURBS modeling.

**Modeling Detailed or Branched Models Is Easy**    Because polygon models are based on the connection of independent faces, it is possible to have polygons with any number of sides and meeting any number of adjacent faces at the corners (vertices). While abusing this advantage can lead to a very sloppy model, being able to break a gridlike structure (more on surface structure in Chapter 4) to create *branching structures*, which occur in areas such as fingers emerging from a hand, can be accomplished quickly and easily while maintaining a single surface. In comparison, it is impossible to create branching structures using a single

*FIGURE 3-4* *The branching structures of the fingers extruding from the hands create a junction of five faces instead of four. NURBS topology cannot accommodate this structure in a single mesh.*

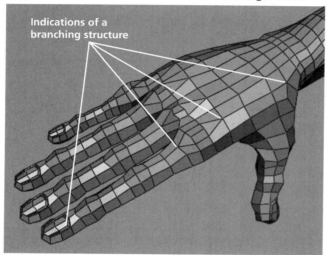

Indications of a
branching structure

NURBS surface. Instead, you are forced to use a multipatch (multiple individual NURBS surfaces) modeling technique (demonstrated in Chapter 6), and then you must manage the continuity between each patch to make the structures appear to be a single surface. This is a time-consuming effort that is of no issue when modeling with polygons. Figure 3-4 demonstrates the topology of a model in which the arms branch out from the body and the fingers emerge from the hand. Notice the structural inevitabilities of branching at the vertices that are shared between five faces.

**UV Texture Coordinates Are Editable** In a polygonal model, the artist has full control over where each vertex will be located in the UV (texture space). Maya contains an entire interface for this type of editing in the UV Texture Editor, which is discussed in Chapter 18. By having this kind of control, the artist can optimize texture maps and fix stretching or distortions. It is not possible to edit the UV of a NURBS surface without changing the layout of the surface itself in 3D space.

**Hard Edges and Corners Are Easily Accommodated** Due to the nature of polygons being made up of finite straight segments, it is very easy to create sharp edges and other rigid details. This makes polygonal modeling extremely useful for modeling architectural elements such as buildings and environments. Figure 3-5 shows a building with very hard edges, a simple task to create when using polygons.

**Simple Toolsets Are Available** While you might use several dozen commands and tools when modeling with NURBS, you can create detailed polygonal models using very few tools. This makes it much easier for someone new to 3D modeling to become productive in a short amount of time. It's also helpful to even the most experienced modeler since fewer processes need to happen to complete a task.

**Models Are Transferable Between Different Software**    Polygonal models are recognized by almost any 3D animation program or rendering engine, so, unlike NURBS models, you can transfer models between programs easily.

# Disadvantages of Polygon Modeling

The following are some disadvantages of polygonal modeling.

**Data Size of a Smooth Polygon Object Is Comparably Large**    To represent a smooth, curvy object, without noticing any faceting in the silhouette of the model, it could take many faces. This will result in slower interaction compared to that of NURBS models. The tail of a dinosaur may need several hundred or thousands of faces to define its shape and render smoothly. A similar structure in NURBS may only require a small fraction of that data.

**FIGURE 3-5**    *An object such as this building is particularly suited for polygonal modeling due to its sharp edges.*

Because of the number of vertices needed to create a smooth detailed object, the data size can be significantly larger than that of a similar NURBS model. Your computer may run into problems loading or rendering a scene if sufficient memory is not available. Also, sending large files over a computer network can clog a pipeline.

**Adding Detail to Smooth Objects Can Be Difficult**    A common polygonal modeling workflow has the modeler work on a low-resolution version of the model and then smooth, or subdivide, that model to view the final result. When the model is smoothed, the position of the vertices is interpolated to result in a much smoother form. The problem is that whenever detail is added to the model, the smoothed version will appear flat in that area, so the modeler must shuffle the position of the vertices around to obtain the smooth shape they once had. This make it difficult to model high-precision, subtly curving surfaces such as those found on a car. Figure 3-6 shows two objects. The object on the left shows a polygonal sphere that is generated by smoothing a cube. The object on the right shows what happens to the resulting sphere when an edge is added around the initial cube. To re-obtain the spherical shape, the modeler must manually reorganize the vertices.

*FIGURE 3-6* *Smoothed geometry flattens when additional geometry is added to the original mesh (shown in wireframe).*

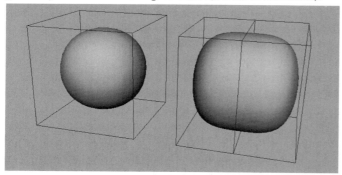

# Tips for Polygon Modeling

To avoid any problems with your model, keep in mind the following tips as you work.

**Keep Polygon Faces As Quads (Four Sided)** First of all, it's much easier to view your models with one less edge through each face. If your aim is to create an organic, smooth surface later by using the Smooth command or by converting the surface into a subdivision surface, it's a good idea to use quad faces as much as possible from the beginning. Quad faces smooth and convert to subdivision surfaces without problems. Figure 3-7 shows both quad and tri faces being converted to subdivision surfaces. You should be aware that a polygonal model with nonmanifold geometry won't convert to a subdivision surface model. This may also be of concern if you plan to render subdivision surfaces in Mental Ray (one of the rendering engines that ships with Maya).

*FIGURE 3-7* *Quad faces and tri faces smooth differently.*

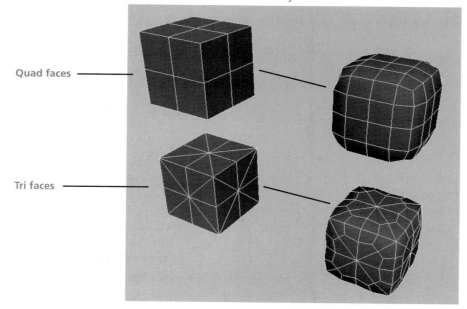

Quad faces

Tri faces

Also, some of the most useful polygonal modeling tools in Maya, such as the Split Edge Ring tool, require a series of adjacent quad faces to create a continuous split.

**Avoid Nonmanifold Geometry**  Manifold geometry relates to the arrangement of the polygonal faces in a mesh. If a surface could be unwrapped and laid out flat, then it

is described as manifold. There are several arrangements of polygons in a mesh that will result in nonmanifold cases. One type of nonmanifold arrangement would be when more than two faces share the same edge. Three faces, shown here, that share the same edge would form a T shape. This arrangement should be avoided. Another way to create nonmanifold geometry would be to have the normals of two adjacent faces pointing in opposite directions.

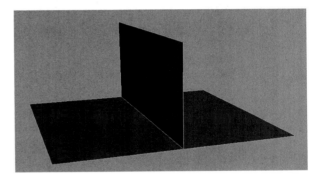

Nonmanifold geometry will cause certain modeling operations to fail. A conversion to subdivision surfaces (another type of geometry that we'll look at in Chapter 4) will fail if the geometry is nonmanifold. In some cases, Mental Ray will not be able to render your model if it contains nonmanifold geometry. If you find that your model contains nonmanifold geometry, you can use the Mesh | Cleanup options to try and fix it. But, most likely, you will need to inspect your model closely and fix it manually by reversing normals or deleting faces. This is why it is best to always be aware of the arrangement of your polygons as you move along the modeling process.

# Helpful Interfaces for Polygon Modeling

You will find that two interfaces are valuable assets as you work with polygons in Maya: the Heads Up Display (HUD) and the Custom Polygon Display.

## Heads Up Display

The HUD for poly count, shown next, displays the number of polygonal vertices, edges, faces, and UVs for the entire scene (second column), a selected object (third column), and selected

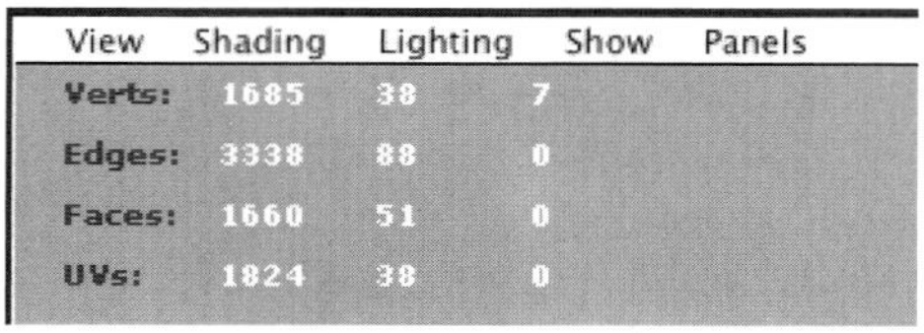

| View | Shading | Lighting | Show | Panels |
|------|---------|----------|------|--------|
| **Verts:** | 1685 | 38 | 7 | |
| **Edges:** | 3338 | 88 | 0 | |
| **Faces:** | 1660 | 51 | 0 | |
| **UVs:** | 1824 | 38 | 0 | |

components (fourth column). The HUD appears as an overlay in your view windows. The HUD is helpful in managing your polygonal models as you work. To turn it on, choose Display | Heads Up Display | Poly Count.

It is always helpful to know how many polygons are amounting in your scene. Many productions specify a limit on the number of faces that can be used for each model. Also, being able to account for the number of components selected is a real help when components overlap. Without being able to look up this information in the HUD, you may think you only have one vertex selected when you very well may have two or three.

## Custom Polygon Display

Choose Display | Polygons | Custom Polygon Display from the main menu bar to open the Custom Polygon Display Options window, shown in Figure 3-8. This window lets you set how a selected polygonal object will be displayed in the view windows. While you can turn on the

**FIGURE 3-8** *The Custom Polygon Display Options window*

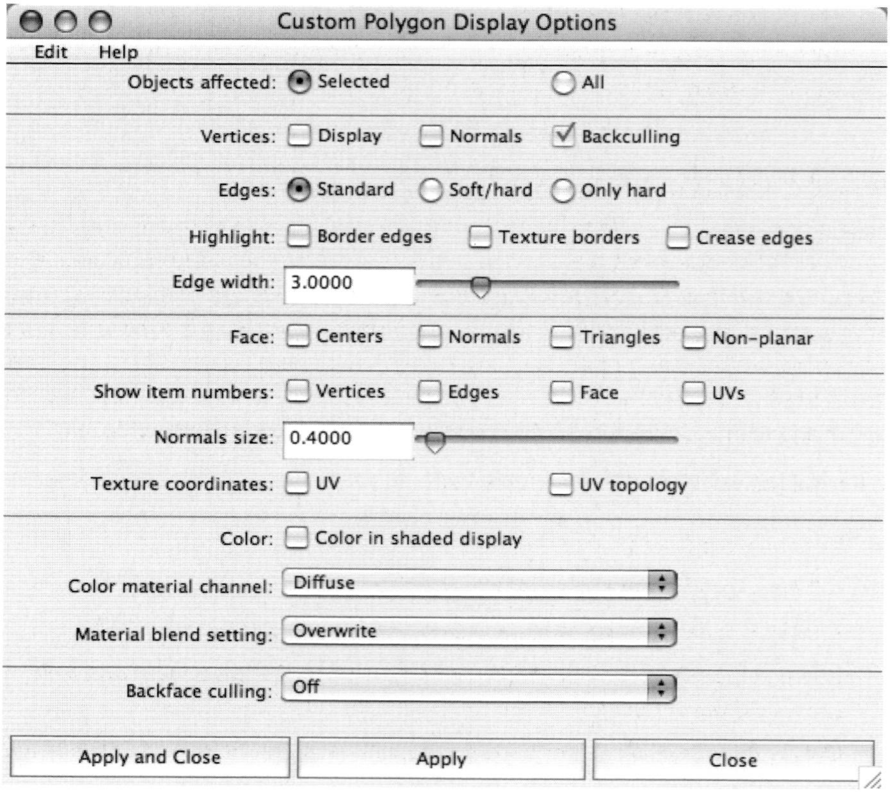

display of the normals and vertices just as you can by choosing Display | Polygon Components, this window gives you access to a whole lot more.

Some of the most useful options here control how various edges are displayed. When you are dealing with multiple polygonal surfaces, it can sometimes be difficult to determine where one surface ends and the other begins. Turning on the Highlight Border Edges option will thicken the edges along the borders of polygonal surfaces. How thick they appear is controlled via the Edge Width setting.

Backface culling (the Backface Culling option) is another useful setting that can be used as you work. By default, this is turned off. When you choose this option, the back of the polygonal faces will not render in the view window. This can speed up performance in the view window, because your computer's video hardware will not have to display the backs of any faces that it might otherwise render. For modeling, turning on backface culling can be helpful for selecting components. For example, if you drag-select several faces on a sphere, you might end up selecting many faces on the opposite side that you don't want. With backface culling selected, the faces on the opposite side will not be selected because they are not being displayed.

> _TIP_   *As of Maya 8, most of the items found in the Custom Polygon Display window can be quickly accessed in the Display | Polygons menu without having to open up a separate window.*

# Tutorial: Modeling a Church

For our first model, we are going to attempt something fairly simple, a building, specifically, a church. As we have mentioned previously, you can get pretty far in polygonal modeling by knowing only a few tools. This tutorial is designed to introduce you to the most essential tools and processes that are involved with polygonal modeling. As you will soon experience first hand, the polygonal modeling process is fairly repetitive in terms of the actions that you cycle through. It involves making a selection of components, performing an operation with a tool or command, and then transforming the geometry by using the Move, Rotate, or Scale tool.

In this tutorial you will learn about the following:

- Setting up a project directory and file system
- Editing an object's attributes in the Channel Box
- Transforming objects and components with the Move, Rotate, and Scale tools
- Using basic polygonal editing tools and commands including Extrude Face, Bevel, Extract, Boolean operations, Object Snap, and Smooth
- Making quick selections of multiple components using the Select Edge Loop command

- Duplicating objects
- Cleaning up and organizing scenes

## Set Up the Project Directory and File System

The first step in any new project is to *set up* the project from the File menu. This creates a directory structure that helps you to organize the project's assets. A directory structure helps you ensure that all of the assets related to the project are kept in one place, which makes moving projects between computers easy.

1. Open Maya and choose File | Project | New.

2. In the New Project window, click in the Name field and type in a name for the project. For this project, type in **mcr8_ch03**.

3. Click the Use Defaults button at the bottom of this window to use the default names for all of the subdirectories.

4. Click the Accept button to accept these settings and close the window.

5. Choose File | Save. Name the file **church_mdl_01**. This indicates that the file contains the church and that it is the first iteration of the model file.

Now, if you go into your My Documents/Maya/Projects/ (users\Documents\Maya\projects) directory, you should see a new folder called mcr8_ch03 that you created in Step 2. Inside that folder are several directories. One of them, the Scenes directory, contains the church_mdl_01 file that you created in Step 5. We will discuss the other folders as we move through tutorials throughout this book.

## Block Out the Shape of the Model

When drawing, most artists begin a scene by masking or blocking out their objects. This enables them to establish a scale and perspective that is shared throughout everything in the scene. Once the blocking phase is complete, the artist will take a series of passes to refine these rough shapes into lifelike details. The 3D modeling process is no different. We will begin by creating a cube and scaling it to find a nice proportion and extrude some of the main structures.

> *NOTE  While we are using this tutorial only to practice learning tools, it is absolutely imperative that you research and gather as much reference material as you can before beginning any real modeling project. Model what you see, not what you think you see.*

**1.** Create a polygonal cube (Create | Polygon Primitives | Cube). Press the 5 key to turn on smooth shading in the Perspective View window.

**2.** In the Channel Box, change the name of the object from pCube1 to **churchBase**.

**3.** We are going to move and scale this cube to create the base structure for the church building. While the transform tools would work well for this task, it will be much faster and more precise to enter the information numerically in the Channel Box. With the churchBase object still selected, set the Translate Y value to 2.5. This will move the cube up in the world 2.5 units. Change the Scale X and Scale Z attributes to 10 and the Scale Y attribute to 5. The position and scale of the churchBase object are now set.

*NOTE   The numerical values that we indicate in this tutorial are only suggestions. In fact, it is far more important for you to focus on the techniques here and rely on your own artistic intuition than to worry about plugging in specific values.*

**4.** Freeze transformations of the object by choosing Modify | Freeze Transformations. This sets all of the translations to 0 and the scale attributes to 1 but leaves the object in its current position. This is useful in case you accidentally move your object. To set it back, you can just type in 0 for all of the translations and rotations and 1 for the scale and you'll be back in business!

**5.** To create a tower emerging from the top of the base, we need to split the cube up into different sections. There are many ways to do this but the fastest is to edit the polyCube1 node that is connected to the churchBaseShape node, since we have not yet deleted the history. In the Channel Box, click the node called polyCube1 listed under the Inputs section. Selecting this node displays its attributes as shown here:

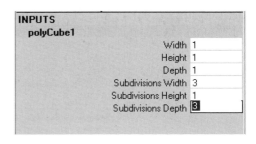

**6.** Set the Subdivisions Width and Subdivisions Depth to 3. This will divide the cube into thirds along the top and provides us with enough detail for the next operation.

Next, you will extrude one of the faces to create a tower in the left side of the building. When the tower is complete, we will duplicate it and move it to the right side. The Extrude command will create new polygonal components based on an existing selection. It is one of the most used commands in polygonal modeling. Let's try it.

1. In the Perspective window, RMB-click and hold on the churchBase object. Choose Faces from the resulting marking menu. The geometry will display a blue dot at the center of each face.

2. Select the face in the top-left corner of the churchBase object (that is, the church's left, your right!) by clicking that face's blue dot.

3. Choose Edit Mesh | Extrude. A manipulator will appear that will allow you to interactively edit the attributes of the polyExtrudeFace node that is now connected to the ChurchBaseShape node in the dependency graph. You can use this manipulator to move, rotate, and scale the extrusion face by LMB-clicking and dragging on the respective part of the manipulator.

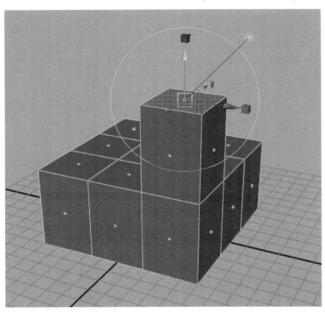

4. In this case, LMB-click the blue arrow of the manipulator and drag it upward. Make sure you click the arrow and not the cube at the tip of the arrow. As you drag the arrow, look in the Channel Box and notice that you are actually changing the Local Translate Z value of the polyExtrudeFace node. Continue dragging the blue arrow until the Local Translate Z value is 5. Your model should look like the image adjacent.

*TIP*   *A common error made by beginners using the Extrude command is that they don't pull out the selected face right after executing the command, possibly not realizing what they have done. Then they'll use the Extrude command on that face again, which results in a series of faces with a "zero length," which means that those extruded faces won't have any physical space. This can cause havoc when you start to split or smooth the polygons or render the object. If you see the blue dots on the middle of any edges when you are displaying faces, you know you have a problem. Try to fix this situation by selecting the model in object mode, choosing Mesh | Cleanup, and checking the Edges with zero length and Faces with zero geometry area. Then click the Cleanup button.*

**5.** While we are still in the blocking stage, we are going to add just a little detail on this tower while we still have the face selected. We are going to create a small stepped ledge by making a series of four extrusions. With the face at the top of our tower still selected, choose Edit Mesh | Extrude to extrude the face again. This time, we are going to scale the face. To set the manipulator to scale uniformly in X, Y, and Z, you need to first click one of the colored boxes at the tips of the arrows on the manipulator. After you click one of these boxes, a solid yellow box appears at the center of the manipulator. This lets you uniformly scale the face. LMB-click and drag the yellow box. Watch the Local Scale attributes on the new polyExtrudeFace node and scale them until the attributes read 1.05.

**6.** Now we'll extrude again. Instead of using the menu to access the Extrude Face command, press the G key on your keyboard. The G key will execute the last command that you used. Now grab the blue arrow on the manipulator and move it up until the Local Translate Z node is at a value of 0.07. One of the steps has been created.

**7.** Repeat Steps 5 and 6 to make a second step for the ledge. The result should look like this:

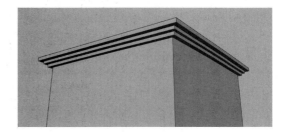

**8.** The last piece we'll block is the spire. We'll create this from a separate piece of geometry, a cylinder. Create a polygonal cylinder (Create | Polygon Primitives | Cylinder). This creates a cylinder at the world origin. Since your church is a bit larger than the cylinder, you may not be able to see it, so choose the Move tool from the toolbox or press the W key and move the cylinder up above the churchBase object.

**9.** Name the object **spire_main**.

**10.** In the Channel Box, change the Scale X and Scale Z attributes to 1.15.

**11.** Click the polyCylinder1 node in the Inputs section of the Channel Box to view its attributes. Change the Subdivisions Axis to 8 and change the Subdivisions Caps to 0.

**12.** Now we will place the spire_main object precisely on the top of the tower by using one of Maya's Snap Align Objects commands. Choose Modify | Snap Together Tool □.

**13.** The tool's settings appear on the right side of the screen. Enable the check box that is labeled Snap to Polygon Face:

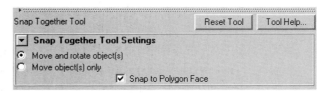

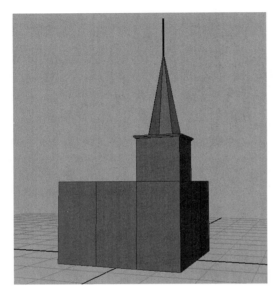

**14.** In the Perspective view, click the bottom face of the spire_main object. A blue arrow should appear. Now click the face on the top of the tower on the churchBase object. A new arrow appears that begins at the first face and ends on the second selected face. Press ENTER (RETURN) and the cylinder will snap and align to the top of the tower.

**15.** Select the top face of the spire_main object and use the Scale tool to scale this face down to a value of 0.05 in X and Z. You can monitor the scale values by looking in the help line.

**16.** With the face still selected, extrude again. This time drag the extrude manipulator along the blue arrow until the Local Translate Z attribute is 1. Your model should look like this image.

## Model Secondary Objects

Now that the rough shape of our building has been blocked, we can begin creating some smaller secondary objects that will help add some complexity to the form. We'll do this by creating a smaller spire from another cylinder. Surface details will be added by extruding faces and flattening edges with the Bevel command. Once this is in place, we can start duplicating the object and quickly populate the scene.

### Create and Place a Second Spire

With the main structures in place, we can begin creating smaller, more decorative pieces. We'll start by making a set of smaller spires that extend from the spire_main object.

**1.** Create a polygonal cylinder (Create | Polygon Primitive | Cylinder). Name it **spire_small**.

**2.** Edit the input node (polyCylinder2) so that the cylinder's shape has the Subdivisions Axis set to 8 and Subdivision Caps set to 0.

**3.** Choose the Move tool and activate the Point Snap button on the status line to turn on point snapping. Alternatively you can hold down the V key as you move the object so that it snaps to vertices. Using one of these methods, click and drag at the center of the manipulator and snap the spire_small object to the vertices at the top of one of the corners.

**4.** Move the spire_small object up along the Y and then scale it so that the height matches that of the top ledge and is wide enough that the corner of the ledge is covered by this new object, as shown here:

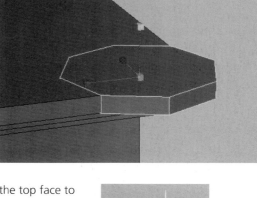

**5.** Use the same techniques that we used in Steps 5 and 6 of the previous section to create some extruded step-shaped ledges out of the bottom of the cylinder.

**6.** Extrude the top face, translate it upward, and scale it down to create a cone shape similar to the spire_main object. Also, extrude the top face to create a pole just as we did in Step 15 of the last section. Your new spire should resemble this:

### Bevel Edges

One of the most important factors of a 3D model is the treatment of the model's edges. If an edge, no matter how sharp it is supposed to be in real life, is going to be rendered properly, the edge must be beveled or rounded, depending on the amount of detail required. A round or beveled edge will help transition the shading between two adjacent faces at sharp angles. Once modeled, an edge will interact with the light sources in a scene by producing specular highlights.

Figure 3-9 shows two cubes. The cube on the left has hard edges whereas the cube on the right has beveled edges.

**FIGURE 3-9**   *A cube without and with beveled edges*

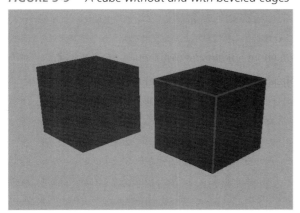

Notice that without the bevels, the shading on the cube appears quite flat. It has almost completely lost the perception of the object having any depth. In order for our model to render correctly, we need to bevel all the edges. We will begin here by beveling the edges on the spire_small object. Before we do that, however, we will separate the thin pole from the spire since the pole is thin enough that it won't require any beveling.

1. Select all nine faces in the pole—that is, eight faces that make up the sides and one on top.

2. We are going to extract these faces so that the pole becomes a separate object from the spire. It is possible to extract each face so that it ends up as its own object, thus resulting in nine objects (ten including the spire). To keep this from happening, we will make sure that the Keep Faces Together option (Edit Mesh | Keep Faces Together) is enabled. This is the default in Maya 8, but not in previous versions of Maya.

3. With the nine faces still selected, choose Mesh | Extract. The pole now is separate from the spire.

The process of beveling edges consists of you first making a selection of edges and then executing the Bevel command. We want to create continuous bevels along each of the *edge loops*. An edge loop is a series of continuous edges along a polygonal surface. Maya offers an array of tools and commands for adding and selecting edge loops. Having these kinds of tools available is really what makes modeling in polygons so fast. Let's select an edge loop and bevel it:

**FIGURE 3-10**    *The edge loop along the vertical edge of the spire is beveled.*

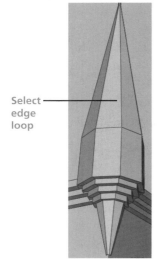
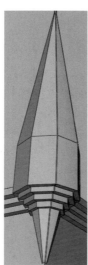

Select edge loop

1. Select one of the edges on the spire_small object that is traveling in the vertical (Y) direction.

2. Choose Select | Convert Selection | To Edge Loop. This selects the entire series of continuing edges. Sure beats picking each edge one by one!

3. Choose Edit Mesh | Bevel ❐ to open the Options window. Change the Offset Type from Fractional to Absolute. This bases the size of the bevel on its actual width instead of as a ratio between the opposite edges. Set the Offset Distance to 0.005 to indicate the size of the bevel. Push the Apply button. Figure 3-10 shows what the edge looks like before and after creating the bevel.

4. Repeat Steps 2 and 3 for the other seven vertical edges in the spire.

5. Bevel the edges along the extruded steps we made on the bottom part of the spire. Select a horizontal edge loop and, once again, execute the Bevel command.

**6.** The bevels in this horizontal direction can be a bit larger. We will edit the size of the offset in the Channel Box. Select the polyBevel node at the top of the list in the Inputs section of the Channel Box. You can experiment with values for the Offset attribute until you find something that looks appropriate. A value of 0.01 looks good. You can now return to the Bevel Options window and change the value to 0.01 and begin applying bevels with those settings on the remaining edge loops.

You can also create bevels by manually extruding faces. We will do this on the flat part of the spire that sits between the stepped section and the conal section by making a series of three extrusions. Finally we will delete the inner face to create a hole where we could put a light later on.

**1.** Select the eight faces shown in Figure 3-11a, and extrude them (Edit Polygons | Extrude Face). Use the manipulator to scale the extruded face down (see Figure 3-11b).

**2.** Press the G key to repeat the Extrude command and then scale down the new face again. Once scaled, click the blue arrow of the manipulator and push the face in toward the center of the spire (see Figure 3-11c).

**3.** Extrude one more time (press the G key) and push the face in along the Local Translate Z attribute. Press the DELETE key. The end result should look like Figure 3-11d. The modeling of this spire is now complete.

*FIGURE 3-11    This figure demonstrates the process of creating manual bevels using a series of extrudes to create openings around the base of the spire.*

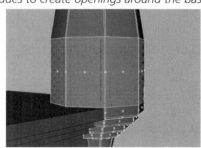
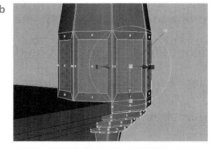
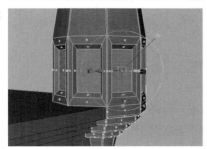
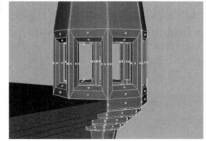

## Duplicate with Transforms

One of the fastest ways to decorate your model is by duplicating some geometry and moving it around. If you study the world around you, you'll notice that there are repeating shapes and forms everywhere you look. The Duplicate Special Options window (Edit | Duplicate Special ❑) in Maya offers options to duplicate an object a specified amount of times while offsetting each duplicate by any translate, rotate, and scale amount. We'll use this technique to complete the top of our tower. Let's start by building a small fence at the top:

**1.** Create a cube (Create | Polygon Primitives | Cube) and use the Move and Scale tools to place it along the top edge of the tower. Center it on the tower and make it long enough so that it just intersects the spire_small object. We found that a scale of 2.7, .08, .045 worked well.

**2.** Bevel the object to give it some more detail and then duplicate it and move it up in the Y direction so that it is aligned to the top part of the opening of the spires (see Figure 3-12a for an idea).

**3.** Duplicate this object again. Rotate it in Z 90 degrees and scale it down in X and Z so that it fits inside of the horizontal cubes. Move it along its Y axis to place it near the spire_small object. Figure 3-12a shows the placement for all three objects. Once the objects have been placed, select them all and freeze transformations (Modify | Freeze Transformations). This will zero out all of the objects' translations and rotations and set the scales to 1. This makes doing measurements easier.

**4.** Choose Edit | Duplicate ❑ to view the Duplicate command's options. Set the Translate X value to .15. Set the number of copies to 16. This will make 16 copies of the object. For each copy it makes, it will move the duplicate 0.15 units. Click the Apply button. Compare your result to Figure 3-12b.

**5.** We can go about adding any more geometry and details to this section of the tower. We've added another beveled cube and placed it between the spire_main and spire_small objects. Name this object **spire_support**.

---

_FIGURE 3-12_    _The polygonal object is placed (a) and then duplicated 16 times with a translation offset (b)_

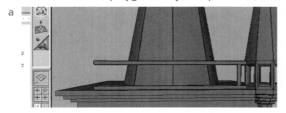

a

b

**6.** To duplicate and transform several objects at once, we can group the objects together. Grouping objects parents the selection to a group node, which is nothing more than a transform node. Select the spire_small, spire_pole, and spire_support objects and choose Edit | Group or press CTRL-G (COMMAND-G) to group them.

**7.** Once the group is created, its pivot point will be at the world origin. To move the pivot point to the location of the spire, choose Modify | Center Pivot. The pivot point is moved to the center of the bounding box of the group. A bounding box is a rectangular area that encompasses all of the objects in a hierarchy.

**8.** Duplicate the group and move it up and scale it down to create a smaller spire located near the top of the spire main. Along with the fence, we should have plenty of detail in this area to do another duplication with transform.

**9.** Open the Outliner (Window | Outliner) and select all of the geometry except for the church_base and spire_main objects. Group these objects and rename the group node to **tower_pieces_grp**.

**10.** We are going to duplicate this group three times while rotating each copy 90 degrees to fill in the detail for the other three sides of the tower. To properly do this, we need to move the tower_pieces_grp node's pivot point to the center of the spire_main object. Turn on the display of the spire_main object's selection handle by selecting the spire_main object and choosing Display | Transform Display | Selection Handle. A small cross will appear at the center of the spire_main object.

**11.** In the Outliner, select the tower_pieces_grp object. In the Perspective view, press the INSERT (HOME) key. This will let you edit the position of the pivot point. Hold down the V key to toggle on point snapping and drag the yellow dot at the center of the manipulator to the spire_main object's selection handle. The pivot point's manipulator should snap right onto the little cross when you drag near it. Once the pivot point is placed, press the INSERT (HOME) key to exit the edit pivot mode.

**12.** With the tower_pieces_grp node still selected, open the Duplicate Special Options window (Edit | Duplicate Special □). Set it to Rotate 90 degrees in Y and set the Number of Copies attribute to **3**. Click Apply. Figure 3-13 shows the upper section of the tower after the duplication.

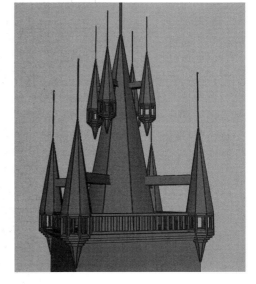

*FIGURE 3-13    The spire section of the tower is completed.*

## Boolean Operations

In this section we will add some details to the church_base object. Mainly, we need to add some windows and a main entrance. We will accomplish this by cutting holes into the church_base object using a Boolean difference operation.

Booleans let you add two polygonal objects (a Boolean *union*), subtract one polygonal bject from another (a Boolean *difference*), and extract the areas of two intersecting objects (a Boolean *intersection*). Figure 3-14 shows the results of all three Boolean operations in Maya.

---

**FIGURE 3-14**    *Results of the three Boolean operations*

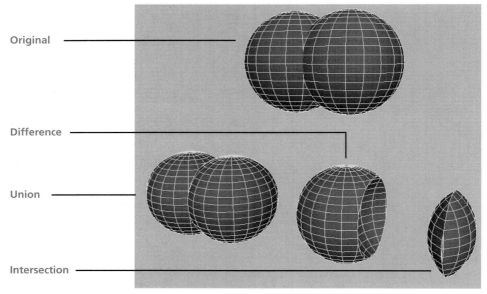

Original

Difference

Union

Intersection

---

We will begin by creating a window shape that will be used to subtract from the church_base object:

**1.** Create a cylinder (Create | Polygon Primitives | Cylinder). Move it away from the church_base object so it is not obstructed and rotate it in X 90 degrees.

**2.** In the Channel Box, edit the polyCylinder input node's Subdivisions Axis to 30. Rename the object to **window_subtract_source**.

**3.** Change to the front view and press the F key to fit the cylinder to your view panel. Choose the faces on the bottom half of the cylinder as shown here:

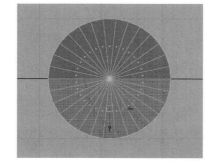

**4.** Press the DELETE key to delete the selected faces. The result is a half cylinder that has a big hole in the bottom. To fix this, select the object in object mode and choose Edit Polygons | Fill Hole. This will create a face at the base of the cylinder.

**5.** Select this new face and extrude it down along its Local Translate Z about 3 units.

**6.** We can clean up the front and back faces of this object by selecting the vertices at the poles (the point where all of the edges run together near the center of the front and back faces of the cylinder) and choosing Edit Mesh | Delete Edge/ Vertex. Figure 3-15 shows the window_subtract_source object before and after removing the vertices at the pole.

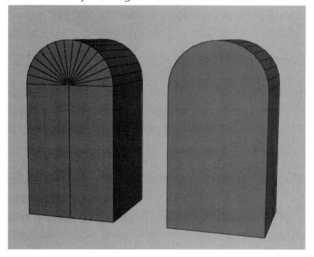

**FIGURE 3-15**  *The edges on the flat sides of the object are removed by deleting the vertices that connect them.*

**7.** Make a duplicate of the window_ subtract_source object and move the duplicate to the front face of the tower object. You can do this manually or by using the Snap Objects Together tool (Modify | Snap Align Objects | Snap Objects Together Tool).

**8.** Since Booleans require intersecting objects, push the object in so that it intersects with the tower of the church_base object.

**9.** Select the church_base object and then the window_ subtract_source1 object (the duplicate) and choose Mesh | Booleans | Difference. The result should look like this:

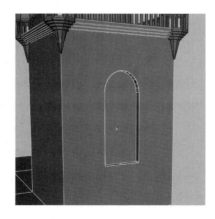

**10.** In order for this new edge to render properly, it needs a bevel. However, the Bevel command is a bit touchy sometimes, especially after performing a Boolean. The reason has to do with the fact that the face of the tower is now an *n*-gon, as it contains many edges. Therefore we will create a bevel manually by deleting the inner faces and extruding the edge around the hole.

*NOTE*    **Working with non-quads on a planar surface is one instance where n-gons will work and render without a problem.**

11. Select all of the faces around the inside of the window area and delete them. You can leave the center face that will represent the actual window, or glass piece.

12. Select an edge around the hole. Using the Convert Selection to Edge Loop command that we used earlier will not work because we are dealing with non-quad faces. Instead, we'll use Select | Select Contiguous Edges. Make sure that all of the edges around the boundary of the hole are selected. If not, you may need to manually add them to your selection by holding down the SHIFT key and clicking the unselected edges.

13. Choose Edit Mesh | Extrude. When the manipulator appears, click the green arrow and pull the edge out just a bit and then push the extruded edge in toward the center of the tower to create a small bevel.

14. Extrude the edge again (G key) and immediately choose the Move tool. Move the edge back in Z until it intersects the window. The result should look like this image.

*FIGURE 3-16    The church after Boolean differencing the windows and door*

You can use the exact same process to add windows to the other sides of the tower and even to make the front entrance. At this point, the church should resemble something like the one shown in Figure 3-16.

## Final Detailing

We'll leave the rest of the construction up to you. Using any of the techniques shown so far, you should be able to quickly decorate the model. Figure 3-17 shows the church after adding in some more details. Here are some suggestions on ways to finish:

• Create some more ledges that will separate different sections of the building. This can be done by placing some cubes, extruding them to create steps, and beveling the edges.

- Use the options in the Duplicate Special Options window to create repetitive patterns. Try and use objects of different sizes so that you have elements that fill different scales, from very small to really large.

- Look at buildings for reference and study things you could add, such as staircases, railings, window panes, and decorative designs.

## Mirror Duplicating

So far we have only been concentrating on one side of the building. As a final step in our construction, we will cut the model in half and then duplicate and mirror the completed half over.

To cut the building in half, we need to create an edge that splits down the middle. One option would be to use the Convert to Edge Ring and Split command However, since the doorway cuts through the center of the building and the main face is an *n*-gon, the edge loop tools will not work. Instead we'll use the Cut Faces tool.

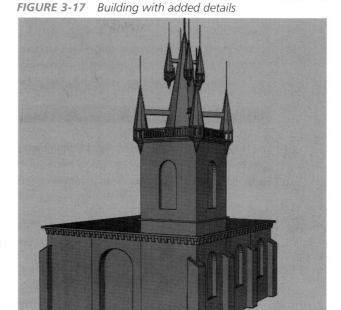

**FIGURE 3-17** *Building with added details*

1. In the Top view, select the church_base object.

2. Choose Edit Mesh | Cut Face Tool.

3. Hold down the X key and the SHIFT key. The X key will let you snap to the grid so that you can cut the piece down the middle. The SHIFT key will let you cut a face while constraining to 45-degree increments. So, with these keys held down, click near the center axis and drag downward. When you see a gray line drawn along the center line where you wish to cut, release the mouse button.

4. Select all of the faces on the left side of the church and delete them so that only the completed portion of the church remains.

**5.** Select everything that you want to mirror to the other side. This includes the church_base object, all of the spires, and any other details that you may have created. Choose Edit | Group or CTRL-G (COMMAND-G) to group all of these objects under one transform node. The transform node's pivot will be at the world origin.

**6.** Open the Duplicate Special Options window. Reset the settings and then change the Scale X attribute to –1. Click Apply and all of the pieces in the group will be mirrored in the X axis.

**7.** Select the original church_base object and then pick the mirror duplicated version. Choose Mesh | Combine. Your completed model should look like this:

## Scene Cleanup and Organization

There is nothing worse than working with a scene that is poorly organized. It's your job as a modeler to make sure that the Maya file you hand off can be read by anyone. Even if you're a one-person production, you don't want to read an error that says there is a problem on polyCube11, only to find that you have 800 objects named polyCube11. Organization is a key factor. Here are some tips for cleaning up your scene:

- Select all of the geometry in the view window and delete the history (Edit | Delete by Type | History). This breaks all of the input nodes to your geometry. Not only does it make the file size a lot smaller, it also prevents things from disappearing and errors when opening the file.

- Freeze transformations. In case anything gets moved accidentally, anyone working on the file should be able to select a piece of geometry and set the transform attributes to 0, 0, 0 and the object will move back to where it is supposed to be.

- Select the geometry and choose Polygon | Cleanup. The Options for Polygon Cleanup are shown here. Choose Nonmanifold Geometry and click Cleanup.

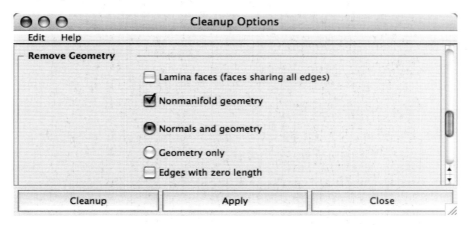

- Group and name objects. The scene will read best if each object has a unique name and is organized in a group hierarchy. Come up with a naming convention that is appropriate for your model. Start going though the model and selecting like objects. For example, select all of the pieces relating to the left front spire on the left tower and group them. Name the group LT_twr_sprLrw_01_grp. This tells us that the objects in this group are on the left tower and that they are part of the lower spires. Each of the four groups of spires can be grouped in the same way. Then all four of these nodes can be grouped under something called LT_twr_sprLwr, indicating that this group contains the entire group of lower spires in the left tower.

# Summary

The tutorial in this chapter was good practice to learn polygonal modeling methodology and for you to get used to many useful polygon tools. Yes, you've successfully learned some great techniques for modeling buildings. Many of these tools will integrate well into an organic modeling workflow as well. In addition to just modeling, you've also gone through many processes that are common to any workflow in Maya. The process of selecting components and executing commands on them and then editing their attributes should now start becoming a familiar process. Also, you've learned how important naming objects and keeping your scene clean are.

# Organic Modeling

**In Chapter 3, we began our modeling**
adventures in polygonal modeling through the
process of modeling a church. We learned several
of Maya's tools and commands for selecting and
editing polygonal surfaces. The church model was
a good example of a hard surface model. In this
chapter we are going to tackle an organic model.
While we will use many of the same tools as we
did in Chapter 3, we'll pay special attention to
the topology of the mesh as we learn to model
a human head that deforms properly when

animated. In this chapter, we will also begin to customize Maya's user interface, giving us faster access to our most frequently used tools.

# Customizing the Workspace

It only takes a few minutes of poking around in Maya to get a feeling for how large it is. While there are hundreds of commands available in the menus, each with dozens of attributes, this does not mean that getting around in the program has to be cumbersome. The ability to customize the UI is just one of the amazing things about Maya. While there are many levels of customization, even a beginner should take the time to set up the Workspace.

It is very unlikely, maybe even impossible, that you would need to use every single command in Maya within a single phase of a production pipeline. Organizing the Workspace so that the commands you need for a specific task are readily available is crucial for being productive. You want to spend your time in front of Maya creating things, not looking for the proper tools and commands. Taking a little time to customize the UI will save you weeks, maybe even months, of clicking around during your life as a 3D artist.

In this section, we look at a few ways you can begin customizing Maya's UI for polygonal modeling, mainly, by creating a custom shelf and shelf buttons that store our most often used commands with their specific attribute settings and some custom keyboard shortcuts, called hot keys. As we work our way through this book, we will continue to build our custom UI for each task. But remember...the word is customize! This means that you can set up however you'd like, regardless of what we tell you. If you find that you are using a tool or command or constantly opening a certain window, by all means, you can use the processes described here to add to your custom UI.

## Creating a Custom Shelf for Polygonal Modeling

To access tools with specific settings quickly and easily, we will create shelf buttons for some of the polygon modeling tools. As we make each button, we'll also briefly describe what each command or tool does. First, let's make a new shelf:

**1.** In the Maya Workspace find the small black triangular-shaped button on the left side of the shelf. LMB-click and hold to view the resulting marking menu.

**2.** Choose New Shelf. A Create New Shelf dialog box appears, prompting you to name the shelf. Name it **PolyModeling** and click OK. A new shelf tab appears, which contains an empty shelf all ready for us to fill.

**3.** For each of the following commands that you want to create a new shelf button for, hold the CTRL and SHIFT keys and select each command from the File menu. When these keys are held down, a shelf button is created in the current shelf.

> _NOTE_  **To delete a shelf button, MMB-drag it to the trashcan icon located near the right corner of the status line.**

### Extrude

You should already be familiar with the Extrude Face command from Chapter 3. Extrude Face (Edit Mesh | Extrude) creates a series of new faces around a user-defined selection of faces and allows the user to interactively offset the original. The resulting topology of the model is a branching structure with four "five-corner junctions" around the extruded face.

### Combine

The Combine command (Mesh | Combine) combines multiple shapes into one. Be aware that combining does not merge any components together. Combine is often used with the Merge command.

### Merge

The Merge command (Edit Mesh | Merge) welds together any selected components within a distance determined by the Tolerance attribute. This command is especially useful for cleaning up geometry once two polygonal surfaces are combined with the Combine command.

### Delete Edge/Vertex

To delete polygonal components efficiently, it is recommended that you use the Delete Edge/Vertex command (Edit Mesh | Delete Edge/Vertex). For example, if a series of edges is selected and then deleted using the BACKSPACE (DELETE) key, the edges are removed but the vertices that once connected them remain. Using the Delete Edge/Vertex command removes the "stray" vertices and ensures a cleaner mesh.

### Insert Edge Loop Tool

An _edge loop_ is a continuous series of edges that splits four-sided faces along an edge ring. The Insert Edge Loop tool (Edit Mesh | Insert Edge Loop Tool) allows you to quickly add geometry by choosing the tool and selecting an edge in the edge ring that you would like to split. An _edge ring_ is a series of edges on opposite sides of a series of four-sided polygonal faces. Figure 4-1 shows a polygonal mesh with one edge ring and two edge loops selected. The rings and loops will continue across a surface until they reach the border edge of a mesh, loop around in a circle and connect to themselves, or encounter a non-quad face.

FIGURE 4-1    *A polygonal surface showing a selected edge ring and two edge loops, one that extends from border to border and one that loops around the cylindrical extrusion and connects with itself*

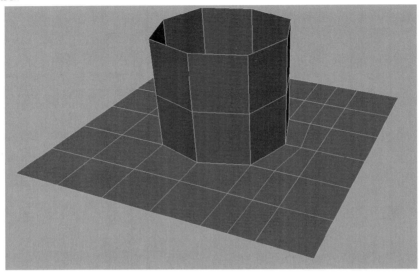

### Highlight Border Edges

Within a single polygonal shape, there can exist multiple polygonal shells. A polygonal shell is a series of faces who share common components (edges and vertices) along their borders. All of the objects we have been working with have contained a single shell. However, when you combine two shapes together with the Combine tool, you end up with many shells within a single shape. Since the edge components may overlap, it can be difficult to see where these border edges are. It is really helpful to have a visual indication of where the border edges of a surface are. Choosing Display | Polygons | Border Edges highlights the edges around the shells of a selected shape. We will use this command to ensure that all the shells have been merged into one after combining the meshes.

### Delete History

Deleting history deletes all of the input nodes connected to a selection. Deleting history keeps file sizes manageable and allows Maya to evaluate a scene more quickly. In production, you should never hand off a model with history to anyone.

When you are finished adding all of these commands to the shelf, your new custom shelf should look like the one shown here. This is just to start you out for this tutorial, but if you find that you

are using any other command quite often, go ahead and make a shelf button.

**Save the Shelf**

With the shelf now complete, you should save it. This creates a new file in your user's Maya prefs directory and will load every time you start Maya. Select the List Of Items To Modify The Shelf and choose Save All Shelves.

# Assigning Hot Keys

Another way to customize Maya's UI is to assign keyboard shortcuts, known in Maya as hot keys. Using hot keys can make executing commands even faster than executing them from the shelf because you don't have to look away from what you are focusing on. As long as your hands can locate the key, you will have instant access to the command.

To access the Hotkey Editor, choose Window | Settings/Preferences | Hotkey Editor. The Hotkey Editor is shown in Figure 4-2. The Categories column contains a list of all of the menus available in the File menu. The Commands column lists all of the commands in a selected category.

To create a hot key for the Extrude command, follow these steps:

**1.** Locate the Extrude command from the menu bar. Since it exists in the Edit Mesh menu, choose Edit Mesh from the Categories column in the Hotkey Editor.

**2.** In the Commands column, find the Extrude command and select it. Notice that every command also has a separate entry for its options. This can be helpful when you use a command and frequently need to edit its options before executing.

**3.** In the Assign New Hotkey section of the Hotkey Editor, type an **e** into the Key field. Click the Query button to see if the key is already assigned to another command. In this case, the query will return to us that the E key is assigned to the Rotate tool. While we could override it, the E key is one of the few keys that Maya users will depend on, so we don't recommend changing it. However, you have the option of using a modifier key.

**4.** Enable the Control check box and again query the hot key. It is not assigned to anything. Click Assign. Control-E now appears in the list of Current Hotkeys. It also appears in the menus next to the command, as shown here: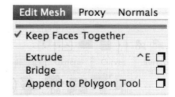

# Using Polygonal Marking Menu

The Polygon Marking menu, shown in Figure 4-3, provides you with quick access to a number of useful tools and commands for selecting and editing polygonal components. The menu is accessed by first selecting a component on a polygonal surface and then holding down the

**FIGURE 4-2**   *The Hotkey Editor*

CTRL key while also holding down the right mouse button. This marking menu contains some submenus that open another marking menu when the mouse cursor is dragged over. For instance, selecting an edge, accessing the Polygonal Marking menu, and then mousing over Edge Ring Utilities will show a marking menu with four options. Choosing To Edge Ring and Split will insert an edge loop along the selected edge ring. There are many other useful commands in here that advanced users will want to explore once they are comfortable using marking menus. In Chapter 6 we will create a custom marking menu for some NURBS commands.

FIGURE 4-3    *The Polygonal Marking menu*

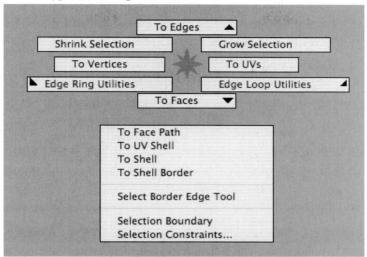

## Modeling Topology

The topology of the model is the greatest concern when building an organic model that will be used for animation. Topology refers to the *structure* of a surface, or how the vertices are connected to one another. In order for a model to deform properly, its topology must follow the musculature of the anatomy that is being modeled. The geometry in the model should be evenly spaced and only inserted where it is absolutely necessary.

> *TIP*   **While modeling, make sure that you have wireframe turned on in the view window. From the view panel's menu bar, choose Shading | Wireframe on Shaded.**

## Model Structures

As we pointed out in Chapter 3, one of the greatest strengths of polygonal models is that they can accommodate just about any arrangement of faces. Polygonal meshes can contain faces with any number of sides that then may share vertices between any number of faces. However, just because it is possible to have these topologies does not mean that it will make for a good model. In fact, we should be very careful about where and when to utilize any of these more complex topologies in our model.

*FIGURE 4-4    A polygonal sphere is
used as the starting point for a head
model by arranging the poles at the
mouth so that the surface structure
radiates from the mouth.*

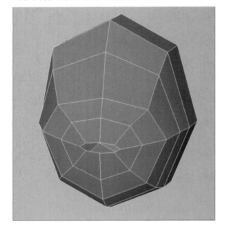

*FIGURE 4-5    Edge loops are inserted
and vertices are scaled closer together
to form the corners of the eyes but the
structure is radiating from the mouth,
not the eye.*

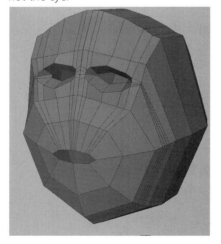

As mentioned many times already, we want to maintain four-sided faces throughout the model. This allows us to use Maya's modeling tools more effectively and prevents possible rendering issues with deforming *n*-gons. If triangles are to be used, and are approved in a pipeline's specification, it is advised to create them by collapsing quad faces in the cleanup phase of the modeling process.

It is also desirable for the model to contain as few different structures as possible. For example, when we analyze the range of motion for the mouth area, it is best to arrange the topology so that it radiates from the lips out toward the face. This way, the edge loops follow the musculature of the face and will animate accordingly. Figure 4-4 shows the beginnings of a face model that started with a sphere whose pole was positioned around the mouth so that it achieves our objective.

Currently, this model maintains this single structure as the edge loops continue around to the back of the head. Each face has four sides and shares a corner vertex with another three faces. However, this structure cannot properly accommodate the structure needed for the eye. Like the mouth, the topology of the eye area needs to radiate from the edge of the eyelid outward so that the eye can blink, squint, and expand into excitement without collapsing any other part of the model. Adding multiple edge loops around the head and sculpting the geometry will not produce an effective topology, as shown in Figure 4-5.

As you can see, both the eyes and mouth will require a radial structure to animate effectively. To fix this, we must change the structure of the surface around the eye area. A quick way to change a grid structure into a radial structure is to use the Extrude command. Any time a face or group of faces is extruded, it changes the structure of the surface by changing the direction of the loops. You can spot a changing structure by the vertices where five faces meet. Figure 4-6 shows the face with the eye structures created with the Extrude command.

Compare this to the head in Figure 4-5. Notice how much more efficiently the current geometry can be used throughout the model. Notice the points where five faces meet. This is sometimes called a "five-point junction" or "star junction." (Along the center, we currently have a six-point junction, caused by two adjacent faces that have been extruded separately.) Now, when we need to add detail to the eye socket, we can insert an edge loop radially around the eye and not affect the rest of the head.

## Planning Topologies

Now that you are aware of how structures are important, we need to plan out the topology of the entire head. Figure 4-7 shows a head that is broken up into separate four-sided structures or patches. This is the plan we will use to build our head in the following tutorial. This is a very clean structure that can be used for almost any human head of any age.

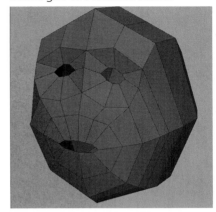

*FIGURE 4-6    The eye socket is modeled in by changing the structure from a grid to a radial structure.*

# Tutorial: Modeling a Human Head

This tutorial will take us through the process of modeling a human head. We begin by blocking, or massing out, the shape of the head with some simple geometry. Next, we create simple shapes for the separate structures of the face. These pieces will then be combined and refined through a process of adding geometry where it is needed and sculpting the shape by moving the points around. Although this tutorial focuses on the head, this same process and technique can be used to model any organic form.

## Set the Project Directory

Just as we did with the church project in Chapter 3, we begin this tutorial by setting the project. This will be even more useful for this project because we will be referencing images that will be used as templates. Since you will want to keep the file paths in order so that they are relative to your scene files, you must make sure that your project is properly set.

> ***NOTE***   *Whenever you are opening a pre-existing scene, you will want to set the current project by choosing File | Project | Set, and then browse to the project directory that contains your scene file. This will ensure that Maya looks in that directory to resolve all external file references. It is essential that you do this when you view the projects that come with the included DVD.*

**FIGURE 4-7** *The patch structures for the human head*

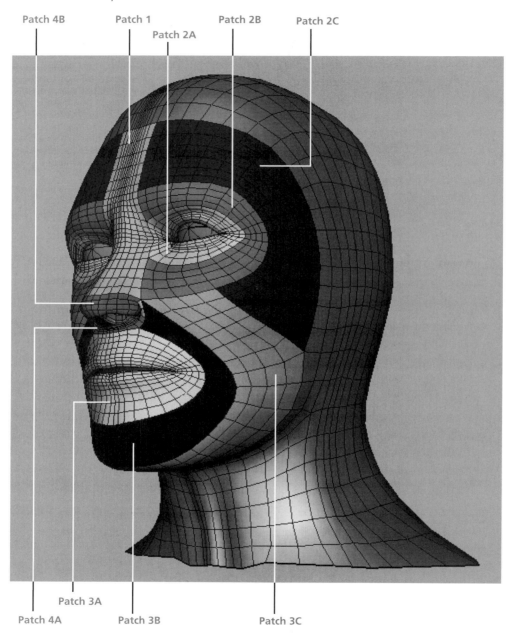

Patch 4B    Patch 1    Patch 2B    Patch 2C

Patch 2A

Patch 3A

Patch 4A    Patch 3B    Patch 3C

1. Open Maya and choose File | Project | New.

2. In the New Project window, click in the Name field and type a name for the project. For this project, type **mcr8_ch04**.

3. Click the Use Defaults button at the bottom of this window to use the default names for all of the subdirectories.

4. Click the Accept button to accept these settings and close the window.

5. Choose File | Save. Name the file **head_mdl_01**. This indicates that the file contains the head model and that it is the first iteration of the model file.

Now, if you go into your users\documents\maya\projects directory, you should see the new folder called mcr8_ch04 that you created in Step 2. Inside that folder are several directories. One of them, the Scenes directory, contains the head_mdl_01 file you created in Step 5. In the next section, we will copy some files from the DVD into one of these directories.

## Import Image Planes

We will use two sketches of a human head as visual references for this project. These images will be projected onto *image planes* and used as *underlays*. By using these as templates, we can closely match and translate the 2D drawings into 3D models.

Definition    ***image plane***   *A two-dimensional backdrop that is attached to a camera and is always perpendicular to that camera.*

1. Copy the reference images over from the accompanying DVD and place them in the sourceimages directory of this current project (mcr8_ch04). The two files are called mcr8_head_front.iff and mcr8_head_side.iff.

2. Choose Window | Saved Layouts | Four View, or click the Four View button in the toolbar. This will bring up the four panel views, with one Perspective view and three orthographic views. *Be careful not to move, pan, or scale in any of these views!* The image planes will be positioned and scaled to frame themselves within the visible region of the view window. Zooming out in one window will cause the image plane to be larger in that view than the image planes that remain at their default and will result in your underlay images being out of proportion.

*NOTE   **If you have already moved around in one of the views, you'll need to start a new project.***

**3.** In the Front view panel, choose View | Image Plane. This brings up the file browser, where you can browse for your image file. It defaults to the sourceimages directory in the current project, so if you place your image files there, they should be immediately available.

**4.** Name the imageplane to **headFront**.

**5.** Find the file named mcr8_head_front.iff and click Open. The image will appear in the Front view panel.

**6.** Choose View | Select Camera. This selects the Front View Camera node and displays its keyable attributes in the Channel Box.

**7.** Under the list of Input nodes in the Channel Box, select ImagePlane1 to display the image plane's attributes.

**8.** Set the Center Z attribute to **–15**. This pushes the image plane away from our Top view camera so that it will never obstruct the geometry and will remain aligned with the Front view.

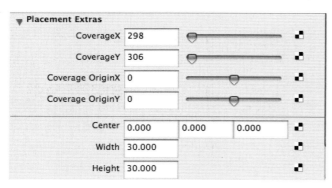

**9.** In the Front view panel, choose View | Image Plane | Image Plane Attributes to bring up the Attribute Editor for the image plane in the Top view. This illustration shows the image plane attributes in the Attribute Editor.

**10.** On the imagePlane1 tab under Image Plane Attributes, set the Display attribute to Looking Through Camera so that the image plane is displayed only in the Front view. This plane will not appear in the Perspective view.

**11.** Repeat Steps 4–8 for the Side view panel. Use mcr8_head_side.iff as the image plane for the Side view. When you set the attribute for placing the image plane (as you did in Step 6), be sure that you set the Center X rather than the Center Z value to **–15** so that it corresponds to the appropriate view.

**12.** In the Layer Editor, click the button to create a new display layer. The Layer Editor is located just below the Channel Box on the right side of the screen. Once the new layer is created, double-click it to open the Layer Options window. Name the layer **underlays** and click Save.

> _TIP_   *A good way to select an image plane is to use the Hypershade. Select the camera's tab in the upper tabs section of the Hypershade window and the image plane node will show up there. Select the node to edit its attributes in the Attribute Editor or Channel Box. You can delete the node by selecting it and pressing the* DELETE *key.*

**13.** In the Outliner (Window | Outliner) choose the Front and Side cameras. Then, in the Layer Editor, RMB-click the Underlays layer and choose Add Selected Objects from the resulting marking menu.

**14.** Test that you have successfully added the image planes to the layer by toggling the visibility of the Underlay layer. This is done by clicking the V button next to the Underlay label in the Layer Editor. The image planes should disappear in the Front and Side view windows when you disable the visibility control of the layer.

**15.** Once you have confirmed that the cameras are in the Underlays layer, set the layer to reference by LMB-clicking twice in the second column of the Layer Editor. This prevents you from accidentally selecting one of the image planes as you are modeling. The Layer Editor should look like this:

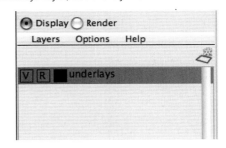

Okay, we're all set up. In the next section, we start the actual modeling process.

## Block Out the Model

In Chapter 3, we blocked out the shape of our church to get a sense of the overlay size and proportion of the building. We want to do the same thing with the head in this chapter. Blocking also helps us find any problems in the image plane reference images. For instance, if we place an edge that flows through the center of the eye in the Front view but it does not follow through in the Side view, then we know there is a problem. At that point, you can decide to create new images or work with what you have and instead use your sculpting skills to find the proper proportions. We are only going to model half of the head and mirror it over to the other side when we begin sculpting.

Create a polygonal cube with two subdivision axes in X, Y, and Z. You can do this by creating a polygonal cube and editing its attributes or by editing the poly cube options in the options window before you create the cube (Create | Polygon Primitives | Cube ❑).

**1.** In the Side view, scale the cube up so that it roughly matches the proportions in Y and Z.

**2.** Change to the Front view and scale the cube along the X axis so that it roughly matches the maximum width of the head in the underlay.

**3.** In the Perspective view, select the two bottom-rear faces and extrude them down a bit. Use the Side view to line the extruded faces up to the base of the neck. Delete these two faces by pressing the BACKSPACE (DELETE) key.

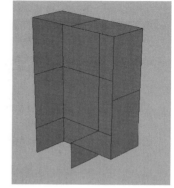

**4.** In the Front view, drag a box around the faces on the left side of the cube to select them. Press the BACKSPACE (DELETE) key to delete the left side of the cube. Your shape should look something like this:

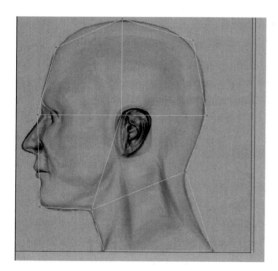

**5.** We are going to round out the form as much as we can with the geometry we have. You never add any geometry until you have done everything you can with what you have. Otherwise, you'll end up doing a lot of extra work. Start from the Side view and move the vertices to match the main points of change in the underlay. These include the base of the chin, the corner of the eye, the top of the forehead, the top of the head, and the points of intersection of the neck and head. As you select the points in the Side view, realize that the points are on top of each other. To select all of the points, marquee select them by dragging a small box around the points instead of just clicking one of them. The Side view should look like the adjacent image.

**6.** In the Front view, use the Split Edge Ring tool to split the faces down the front of the head along the width of the nose.

**7.** Round out the form of the head from the Front view by moving the vertices on the side of the head.

**8.** Continue rounding out the shape in the Perspective view by tweaking point by point, little by little. Keep orbiting the camera around to examine and evaluate the model from every possible angle. While it is still a simple shape, you need to try and imagine it capturing the volume of a complete head. The result should look like this:

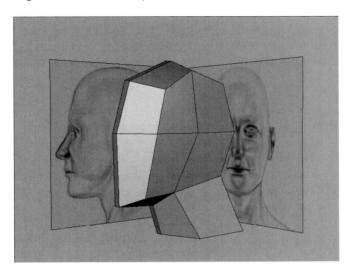

**9.** Create a new layer in the Layer Editor. Name the layer **BlockedHead**. Rename the geometry to **pBlockHead**. Add the pBlockHead to the Blocked head layer and turn off the visibility. We will come back to this piece again later.

## Set Up the Structure

Now that we have blocked out the shape of the head with some rough geometry, we have a sense of the volume of our head. We have also verified that our image planes are pretty well lined up. Now we can start placing in some geometry that defines the different structures of the head. Earlier in this chapter, we planned how the topology would flow around the head so that it corresponds to the underlying anatomy, so that the model deforms naturally during animation. In this section, we will individually create each patch that will define the overall structure of the head. While the term *patch* usually refers to a NURBS surface, we use it here to illustrate a series of four-sided polygonal faces that meet each other at a vertex that is shared with three other faces. If the geometry does not meet those requirements, it must be broken out into a separate patch, resulting in a much more complicated overall surface structure. Once the structure is set up and the patches are all combined, detailing the model is simplified since you only need to add edge loops and sculpt without any worry of creating an unwanted change in surface structure.

**NOTE**    *As you build each patch, refer to Figure 4-7 to get an idea of where these patches are located.*

### Nose Patch (Patch 1)

We'll begin laying in the patches by starting with the nose:

1. Create a polygonal plane (Create | Polygon Primitives | Plane). Set the plane's Subdivisions Width and Height attributes to 2.

2. Rotate the plane 90 degrees in X. In the Side view, move the plane forward in the Z direction so that it sits along the front of the face.

3. Select the vertices at the top of the plane and align them to the area at the top of the brow. Select the middle vertices and align them to the bridge of the nose. Select the bottom row of vertices and align them along the profile of the nose, at the top of the nostril, as shown here:

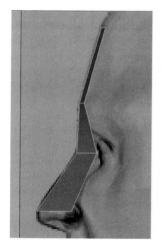

4. In the Front view, select the two faces on the left of the plane and delete them. Use the Split Edge Ring tool to split the nose patch once in the vertical direction and once in the horizontal direction, above the edge in the middle. Now align the vertices along the right side of the patch to the shape of the nose in the front underlay. Then, switch to the Side view and pull them back so that they align to where the nose intersects the face. The edge added in the horizontal direction can be used to better shape out the forehead profile. The completed nose patch should look like the image adjacent.

### Eye Patches

The first patch to define the eye area makes a complete circle around the eye. The inner bounds of the patch define the edge of the eyelid. The outer edge of the patch should be aligned with the orbital bone around the eye socket. One great thing about modeling a head is that the best reference isn't too far away. Just take your finger and feel the area around your eye! Another good idea is to get a small mirror to use to look at your own face. You can also review the topology plan we discussed and pictured in Figure 4-7.

**Patch 2A**    A quick way to generate an enclosed circular patch structure is to start with a cylinder, delete the faces on both ends, and flatten it. We'll do that here:

**1.** Create a polygonal cylinder (Create | Polygon Primitives | Cylinder) and set the Subdivisions Axis attribute to **8** and the Subdivisions Caps to **0**.

**2.** Select the face at the top of the cylinder and scale it down to about half its original size and delete it. Then select the face on the bottom and delete that one as well.

**3.** In the Channel Box, set the Scale Y attribute of the cylinder to **0**. This flattens it. Set the Rotate X attribute to **90** so that the patch is facing forward. Choose Modify | Freeze Transformations to zero out the transformations. This shape is now ready to be used for the initial eye patch. Figure 4-8(a–d) shows the process of creating this patch from a cylinder.

*FIGURE 4-8*    *An enclosed, eight-sided circular patch is created by flattening a cylinder and deleting the end caps. This shape can be used as the basis for many models with circular, looping topologies.*

a
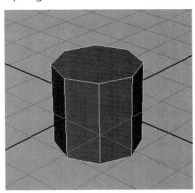

b
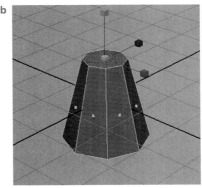

c
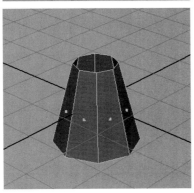

d
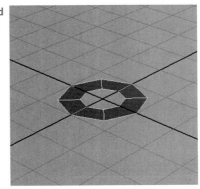

_TIP_   **Since the mouth patch will be started from a circular loop such as the one we created here, you may wish to duplicate our eye patch and use the duplicate later for the mouth.**

_FIGURE 4-9_    _The eye patch is placed and sculpted so that the inside bound lies around the eyelid and the outer bounds lie around the eye socket._

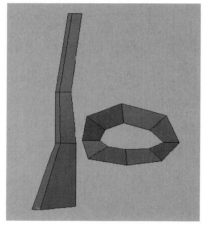

**4.** In the Front view, move the patch up and center it around the eyeball. Scale it up a bit and then start aligning the vertices to the inner and outer bounds of the eye. Again, use Figure 4-7 for reference.

**5.** Obviously, our face is not flat. Both of the inner and outer bounds of our eye patch need to curve around the eyeball. This means that we can align the patch to the front-most part of the eyeball and then start moving the vertices on the sides of the patch backward in Z to start rounding out the shape. Once you have the vertices placed correctly from the Front view, switch to the Side view and move them back. Be sure to constrain all of your translations to the Z axis by clicking and dragging the blue arrow of the manipulator. Figure 4-9 shows the eye patch after it has been placed and sculpted according to the shape of the eye.

**Patch 2B**    The second patch in the eye area will define the brow area along the top of the eye and the upper cheekbone around the bottom. The lower corner of this patch should be placed at the top of the nostril. We create this patch by extruding six of the edges from the existing eye patch:

**1.** Select the six edges along the outer boundary of Patch 2A that define the top, bottom, and side of the eye, leaving out the two edges near the nose.

**2.** Choose Edit Mesh | Extrude, use the button in your custom shelf, or use a hot key to execute the Extrude command.

**3.** When the manipulator appears, click and drag the green arrow outward to extrude the edges out along their Local Translate Y attribute. Since you will be placing each vertex individually, you need to extrude them out only just enough so that the edges are separated from the original selection.

**4.** Check and make sure that these extruded edges are connected. If you see that they are not, undo back to before the extrude and make sure that the Keep Faces Together option is enabled (Edit Mesh | Keep Faces Together).

**5.** In the Front view, select each vertex and move it into place using your underlay and Figure 4-7 as a guide. Try to keep a nice even ring that flows around the top of the brow to the top of the nostril.

**6.** Change to the Side view and start moving the vertices forward and back along the Z axis so that the vertices around the outer boundary of the patch properly flow around the shape of the head. This illustration shows the patches placed so far:

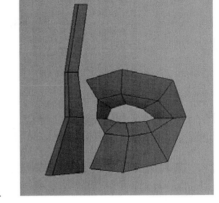

**Patch 2C**   The final eye patch defines the area from the brow to the top of the forehead and around the temples to the upper-rear part of the cheek. Refer to Figure 4-7 for guidance on where this patch is located on the head.

**1.** Select five edges starting at the top of Patch 2B and around the side of the eye.

**2.** Extrude these edges out along their Local Translate Y attribute just enough so that they are separate from the original selection.

**3.** In the Front view, move each vertex into place so that it matches the corresponding feature from the reference on the underlay. The vertices should radiate outward from the eye up along the brow but begin to thin toward the side of the face on the bottom boundary of the patch.

**4.** Use the Side view to move each vertex back along the Z axis to match the underlay.

**5.** Change to the Perspective view and begin tweaking each point one-by-one and orbiting the camera after each edit so that you can evaluate your model from all angles. While the orthographic views (Front and Side views) let you match up your geometry to the features in the face, it is imperative that you use the Perspective view to finalize each stage of your model, as this is how it will be seen during animation. The completed layout of the eye and nose patch is shown here:

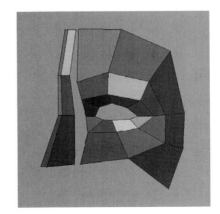

**Mouth Patch**

We'll create the mouth structure with three patches, beginning with a flattened cylinder, just as we did for the eye patch.

**Patch 3A**    We create the first mouth patch by placing a hollow, flattened cylinder at the mouth. The inner boundary should line up with the inner edge of the lip. The outer boundary will extend outward to the bottom of the nose and around to the middle of the chin.

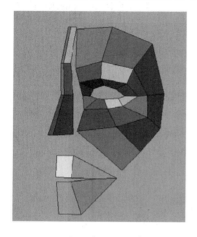

1. Use the same procedure we used to build the enclosed circular patch that we used to build Patch 2A to start the mouth patch.

2. Since the mouth crosses the axis of symmetry, we only need to use half of what we have. Select the faces on the right side of the head (that is, the character's right, your left) and delete them.

3. Use the technique of aligning the vertices along the X and Y axes in the Front view and then switching to the Side view and aligning them along the Z axis so that they match the underlay and follow the structure shown in Figure 4-7. The patch should look like the image here.

**Patch 3B**    This patch is generated by extruding three edges of the mouth patch:

1. Select three edges along the boundary of the mouth patch starting at the bottom, leaving out the edge right below the nose.

2. Extrude these edges and place the vertices flowing radially from the mouth and out under the chin.

3. To place the vertices along the top of this patch, enable point snapping on the status line (or hold down the v key) and move the top corner of this patch to the bottom corner of the eye patch, as the top of the nostril.

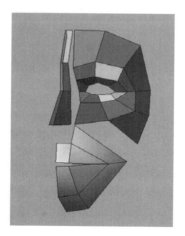

4. Switch to the Perspective view and tweak the positions of the vertices so that they work as a volume in 3D space. The result should look like this:

**Patch 3C**    Along the bottom of the face, the third mouth patch defines the area between the chin and the neck. The upper part of this patch will form the cheeks and side of the face.

1. Select the same three edges that were just extruded and extrude them again.

**2.** Place the lower vertices at around where the chin intersects the neck and work around to the corner of the mouth.

**3.** To place the vertices at the top part of this patch, you can pick the vertex at the top corner and snap it to the corner where Patch 2B meets 2C. The next vertex over on the mouth can be placed by snapping it to the outer corner of Patch 2C. After some tweaking in the Perspective view, your patch layout should look like this:

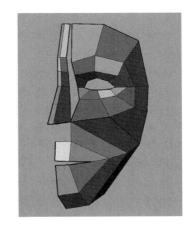

## Combine the Shapes

We now have three meshes that each contain the primary topological structures needed to model a face that will animate properly. While we did use the point-snapping function to align some of the vertices to each other, these are still separate polygonal meshes. In this section, we combine these meshes and merge the vertices together. Before we do that, though, we just want to make sure that all of the shapes have corresponding vertices to snap to. If you inspect your nose patch, you'll see that you need to add a few more edge loops so that you can match the topology to the eye patches.

**1.** Evaluate the nose patch and add edge loops so that it matches the surrounding patches and the model maintains quads.

**2.** Select all three surfaces, the nose, the eye, and the mouth.

**3.** Choose Mesh | Combine, or use the button in the shelf to combine these into a single mesh.

While we have a single polygonal shape, we still have three separate polygonal shells. A polygonal shell is a mesh that does not share any common vertices with other meshes in the shape. In order to have this shape contain a single shell, we merge the vertices:

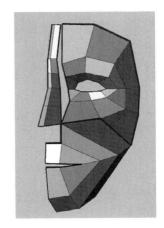

**1.** Use our custom shelf button to highlight the border edges. The border edges with unmerged vertices now display thicker. This makes it easy to find where the vertices need to be merged, as shown here:

**2.** Select all of the vertices in the mesh and click our shelf button to merge vertices with a Tolerance value of 0.001. This merges the vertices that were snapped together.

The remaining vertices are not merged because they are outside of the distance specified. We next go through the model and begin selecting pairs of vertices and merging them together using the Merge command with a high Tolerance value (10):

1. Select the vertex at the inner corner of the eye and then SHIFT-select the corresponding vertex on the edge of the nose. Click the Merge button in your shelf that is set to merge at a distance of 10. The points move together and become one. Figure 4-10 shows what this area looks like before and after the vertices are merged.

2. Continue this process for the remaining vertices. When you no longer see any highlighted edges, then you know that you are finished.

**FIGURE 4-10**   *The highlighted edges indicate that the components are not merged (a); the same area after the vertices are merged (b).*

a
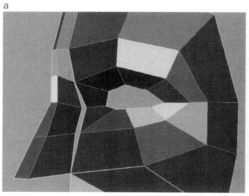

b
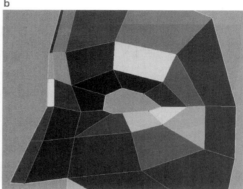

### Build the Nostril

We'll create the nostril area of the nose by extruding out the edges around the area where there is currently a hole. We'll then add some additional geometry to create the center of the nostril and then continue extruding edges to create the lip and the inside of the nostril.

**Patch 4A**   We begin by splitting edge rings on the top and bottom of the nostril area. This will give us enough geometry to create the center lip of the nostril later on.

1. Choose the Split Edge Ring tool from your shelf and split the edges shown here:

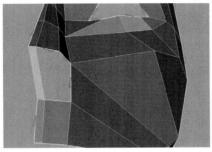

**2.** Select the six edges around the hole and extrude the mouth just a bit. Move the vertices around so that they define the outer part of the nostril of the nose.

To create the center lip of the nose, choose Mesh | Append to Polygon. Click the small edge at the top of the nostril areas and then click the small edge on the opposite, lower part of the nostril. To complete the append, you must press ENTER. The Append to Polygon command creates a new face that is automatically merged into the shape, shown in the adjacent image.

**Patch 4B**   Now we can create the nostril. This is accomplished by extruding some edges up and back into the head.

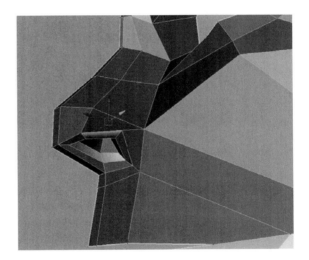

**1.** Select the five edges along the border of the hole and extrude them. Immediately after executing the Extrude command, use the Scale tool to scale the extruded edges inward to create the lip of the nostril.

**2.** Extrude one more time, this time moving the extruded edges upward to create the inside of the nose. The result should look like this:

### Outer Face Patch (Patch 5)

We will extrude the last patch from all of the outer edges around the existing model. This will define the forehead, down through the side of the head into the area where the chin joins the top of the neck.

**1.** Select all of the edges around the head and extrude them out just enough so that they are separate from the originals.

**2.** Move the vertices around so that they match the underlying image plane. The final layout of the patches is shown here:

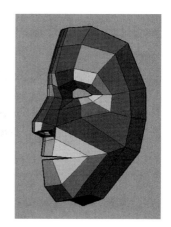

This completes the creation of the structural layout of the head. From here on, we will just be adding detail to refine the areas. Because this structure works well for just about any humanoid character, you may want to save it and use it as a starting point, or template, for all future models. Since there is very little geometry, you would be able to move the vertices around to quickly match the features of a different face.

*TIP*    *Remember to save your work often.*

## Mirror the Model

As we begin to detail the face, we'll get a better sense of the model's shape and proportion if we view it as an entire head. Maya has a Mirror Geometry command that creates a mirrored half of the geometry and combines it with the original. However, at this stage of the modeling process, we want to work on only one half of the model. Instead of using the Mirror Geometry command, we will create an instanced copy of the model with the Duplicate command. When geometry is instanced, a copy is created with the shape nodes of the geometry connected. This way, when a shape is edited on one piece of geometry, those edits propagate onto the instance as well.

**1.** Select the model and choose Edit | Duplicate Special ❑ to open the Duplicate Special Options window, shown in Figure 4-11.

**2.** Set the Geometry Type attribute to Instance. This connects the shape node of the duplicate object to the original.

**3.** To mirror the instance, set the Scale X attribute to **–1**. The settings should match those shown in Figure 4-11. Click the Apply button. The instanced copy is created.

**4.** Before we leave the Duplicate Special Options window, choose Edit | Reset Settings and close the window. This sets the options back to their defaults. This is important because duplicating objects is something that is quite common, and most of the time you want to freely create normal copies when you use the Duplicate command.

*TIP*  **Before resetting the Duplicate options, you may wish to create a shelf button that stores these settings. This makes mirror instances quicker to create in the future.**

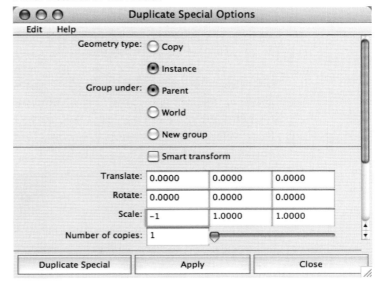

FIGURE 4-11    *The Duplicate Special options are used to create a mirror instance of the model.*

## Detail the Model

At this point, the technical part of the modeling process is complete. The structural topology of the model has been determined and shaped to match our underlays. Now it's time for the more artistic part of the process. In this section we add geometry so that the details can be modeled and refined. We'll add the geometry by inserting edge loops. Because the structure is already set up, you never have to worry about your loops going astray. Every time you add a loop, you go through a process of sculpting the model by moving the vertices around and then evaluating the model from all angles.

The most important thing to remember is that you must constantly be rounding out the shape of your model every time you add an edge loop by moving the vertices around. This gives the model a more organic form. Even moving the vertices just a tiny bit keeps the geometry from looking too square or boxy. If you forget to move the points around and continue adding more edge loops, it is much more difficult to go back and round out the form—so remember, sculpt as you go!

### Detail the Eye
We are going to start inserting edge loops in the eye area to detail the eyelid. Once we have a few loops, we can place in a sphere and start to refine the shape of the eyelid over the eyeball.

**1.** Use the Insert Edge Loop tool to insert an edge loop around the eyeball.

**2.** Push the vertices around the top of the eye back in toward the head to define the crease between the brow and the top of the eyelid.

**3.** Continue around the eye, sculpting each vertex in the edge loop that you just added.

**4.** We'll add another edge loop that defines the middle part of the upper and lower eyelids. Instead of using the Insert Edge Loop tool, as we did in Step 1, let's try the Insert Edge Loop command in the Polygon Marking menu. Select one of the edges in the edge ring that we wish to split. Now hold down the CTRL key and RMB-click and hold. The Polygon Marking menu appears. Select Edge Ring Utilities | To Edge Ring and Split.

*NOTE   You can use either method to insert edge loops. From here on, we will only instruct you to add edge loops, leaving it up to your preference as to how to go about inserting them.*

**5.** Again, round out the form of the mesh by moving all of the points around.

**6.** Insert one more edge loop and place the new vertices close to the edge to define the lip of the eyelid. The eye should look something like this:

**7.** To refine this area further, we will create a NURBS sphere and place it in for the eyeball. This allows us to more accurately sculpt the eyelid in a spherical shape. Create a NURBS sphere (Create | NURBS Primitives | Sphere) and scale it up and place it according to the underlays.

**8.** Create a new layer and add the NURBS sphere to it. Set the layer to Reference so that the sphere cannot accidentally be selected.

**9.** Sculpt the exiting geometry around the sphere so that the eyelid lies right on the surface.

**10.** To tighten the corners of the eye, we'll insert two edge loops on either side of the existing corners. Move the vertices at the corners close together. On the outer corner of the eye, shape it so that the upper vertices slightly fold over the lower ones.

**11.** Continue to round out the form by sculpting the loops that run over the nose and through the side of the head. Figure 4-12 shows the eye region up to this point.

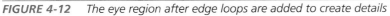

*FIGURE 4-12* *The eye region after edge loops are added to create details*

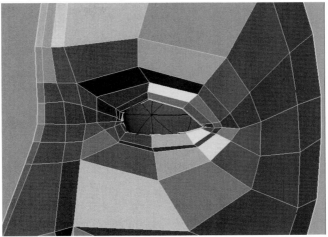

## Detail the Mouth

The process for detailing the mouth is very much the same as what we did for the eye. We start by inserting edge loops that define the lips, and then we start inserting edge loops in the other direction to define the corners of the mouth and the rounding of the lips. As these loops are added, they will run through the rest of the head, giving us the geometry necessary to sculpt the rest of the face.

**1.** Insert an edge loop that flows around the mouth. Move the vertices around so that this loop defines the outer edge of the lip.

**2.** Insert another edge loop and pull it out to begin shaping the volume of the lips.

*FIGURE 4-13* *Adding and sculpting the edge loops to define the lips*

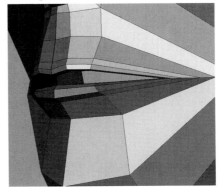

**3.** Insert one more edge loop and move the vertices very close to the inner edge to define the lip of the lip, very much like we did with the inner lip of the eye.

**4.** Insert another edge loop that flows around the mouth to define the indent around the chin below the mouth and the area between the upper lip and nose. Figure 4-13 shows the lip area after adding these four edge loops.

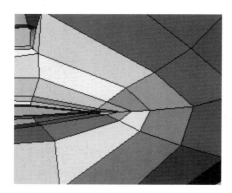

**5.** We can begin inserting edge loops running in the other direction that will define the corners of the mouth, shown here. Aside from helping us shape the mouth area, the extensions of these loops should also be sculpted to define the shape of the rest of the head.

The rest of the detailing process is very much the same. Insert an edge loop and sculpt the shape. As your model becomes more and more detailed, the goal is to try to keep the layout of edge loops relatively even. If you notice some sparse areas in the mesh, try to determine if these areas can be filled by sculpting the exiting vertices around. If not, then inset another edge loop, and sculpt. Figure 4-14 shows the face after inserting edge loops and sculpting for a while.

*FIGURE 4-14   Edge loops are added and the face is sculpted and refined.*

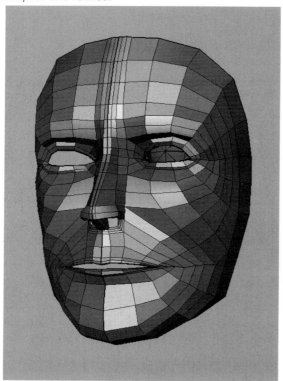

Take some time to evaluate the model by looking at it from as many angles as possible. If something looks out of place, move vertices around little by little. When it looks right, you can delete the instanced copy of the head and move onto the next section.

**Create the Inner Cavities**

If we were to animate this face right now, our audience would quickly find fault with the model when the character opens its mouth. The lips would reveal themselves to be paper thin, and if we were to look into the mouth, we'd be viewing the back side of the mesh. For this reason, we will create inner cavities for the mouth and the eyes. This can be done by extruding the existing edges:

**1.** Select the edges around the hole of the mouth. A quick way to select edges on the border of a mesh is to use the Select Border Edges tool. Click once where you'd like the selection to begin and then again where you'd like it to end. All of the edges between will become selected.

2. Choose Extrude Edge. Immediately choose the Scale tool and scale the extruded edges in the Y axis until the edges sit about half way behind the lip. Adjust the vertices around the corner of the mouth so that the inner extrusion is flowing evenly around the inside of the lip.

3. Extrude the outer edges again and scale them up to define the base and roof of the mouth.

4. Extrude them two more times and move them back and down into the throat. The result is shown here:

5. Use this same process for the eye. Since we probably won't ever see the inside of the eye socket, you only need to extrude once or twice, just enough so that the lip of the eyelids roll around, maintaining their thickness.

### Model the Back of the Head

It's time to model the rest of the head. We could do this by continuing to extrude the edges around the border of the face, but this would mean shifting a lot of vertices around. Instead, we revert to the original blocking model and work that into the detailed face:

1. Make the layer containing the blocking model visible.

2. Select the faces on the front of the model and delete them.

3. Select the model and choose Mesh | Smooth. Our blocking model rounds itself out.

4. Sculpt the smoothed version of the blocking model a bit. Analyze the distribution of the edges around the border and try to match them to flow into the face, as shown here:

5. We could start adding edge loops around the back of the head until we had enough to match the face but, again, this would take a long time because it would require a lot of sculpting to round the form. Therefore, we will smooth the back of the head again. Choose Mesh | Smooth.

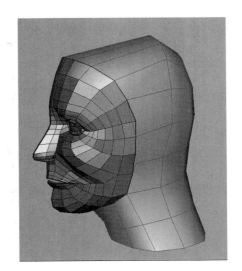

6. Now when you analyze the topology, you should start to see a closer matching number of edges between the two shapes. While it will be close, in most cases, it will not be perfect. You need to either insert an edge loop on the back of the head or delete one. When you have a one-to-one relationship of edges between the two pieces, you can combine them.

7. Choose the face and the back of the head and choose Mesh | Combine. The two pieces of geometry are now combined into one shape. However, they still exist as two separate shells. We can view the border edges of the separate shells by clicking our shelf button labeled CPD. This highlights the border edge.

8. To merge the geometry into a single shell, we select corresponding vertices on both meshes and use the Merge command. Since the default settings for the Merge command are set to always merge a selection of two vertices, there is no need to adjust the Threshold attribute. Start at one end of the head and work your way around, selecting pairs of vertices and merging them together. When you make your way around the combined edges, you should no longer see any highlighting because the border edges are now merged and living well inside of this mesh.

9. Do some cleanup. Name the model **head**. Name the eye **eye_left**. Select everything and choose Delete History. Your head should look similar to the one shown in Figure 4-15.

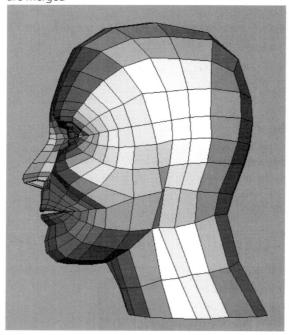

**FIGURE 4-15** *The head after the front and back sides are combined and the vertices between them are merged*

# Subdivision Surfaces

We are going to take a break from modeling our head and talk about subdivision surfaces. Recently, subdivision surface modeling techniques have become the standard modeling methods for modeling organic forms in both the game and film industries. In terms of workflow, subdivision surface modeling closely resembles polygonal modeling techniques. With subdivision surfaces, a modeler can work on a coarse, low-resolution version of a model, called a *proxy*, while viewing a higher-resolution subdivided and smoothed approximation, very similar to the way one models in NURBS (as you will soon see) but with all of the advantages of polygons.

Maya 8 offers two implementations of subdivision surfaces: the *subdiv proxy* and *hierarchical subdivision surfaces*. In addition to working with subdivision surfaces in your Maya scene, the Mental Ray renderer can create subdivision surfaces from low-resolution proxy meshes at render time, freeing up RAM and disk storage and allowing the rendering engine to determine if and how many times a model should be subdivided based on the distance from the camera. We'll dive more into this aspect of subdivision surfaces in Chapter 7. For now, let's look at the options in Maya.

# Subdivision Proxy

The subdivision proxy method is common in many 3D software packages today. In Maya, any polygonal object can be converted into a subdivision proxy by choosing Proxy | Subdiv Proxy. A smooth proxy is a subdivided or smoothed copy of the model that is still connected through the dependency graph and driven by the original, low-resolution proxy model. Each polygonal face of the original object will be divided and smoothed according to the Subdivision Levels attribute of the polySmoothProxy node that is created. Once the subdivision proxy is created, you can use any of the polygonal modeling tools to edit the proxy version and see the resulting effect instantaneously on the smoothed version.

One crucial piece of advice is to make sure that you always keep a copy of the proxy version. Many other parts of a production pipeline will prefer to work with proxy resolutions. The process of *skinning*—binding the mesh to a skeleton—is much faster to set up with fewer vertices to deal with. Also, even detail that is sculpted into high resolutions of the geometry can be extracted and stored as texture information in what is called a *normal map*. We will look at this in Chapter 18. Without this proxy information around, there really isn't any easy way to retrieve it.

One more thing to mention is that you can toggle between displaying the proxy and subdivision objects by using the Proxy | Toggle Proxy Display command. This can be especially useful in cases where the two meshes interpenetrate and therefore obstruct components that you wish to edit. To display both of the meshes again, use the Proxy | Both Proxy and Subdiv Display command.

# Hierarchical Subdivision Surfaces

In Maya, *hierarchical subdivision surface modeling* is called *subdivision surfaces* and is technically considered an entirely separate geometry type on its own. You can create various subdivision surface primitives by choosing commands from the Create | Subdiv Primitives menu. Subdivision surfaces allow you to select certain areas and refine them through hierarchical levels. This lets you add finer details to the mesh without having to add more geometry to the original "cage" polygon, ensuring a high level of detail in the resulting surface while maintaining a reasonably simple original (level 0) shape.

## Subdividing at Render Time

These days, subdivision surfaces are mainly used at the very end of the 3D production pipeline. While a modeler may use a subdivision proxy or a hierarchical subdivision surface to evaluate what the model looks like, it is most optimal to leave the subdividing up to the renderer. Rendering applications such as Mental Ray (included with Maya), Pixar's Renderman, and nVidia's Gelato are all capable of interpolating a low-resolution polygonal mesh into a smooth subdivision surface. Even some current game engines are capable of subdividing on the fly to produce a higher level of detail (LOD) at closer camera angles.

The advantages to this type of workflow are abundant in a production pipeline. First of all, the amount of storage and RAM used is much less because the amount of data is at a minimum. There isn't really any point in saving a model that has been smoothed two levels. Doing so produces 16 times more data than does storing the base mesh. Even more advantageous is that the renderer can determine the amount of subdivision necessary to produce a surface that does not display any artifacts. Combined with other rendering technologies such as normal and displacement maps, rendering heavy geometry is becoming a thing of the past in 3D production.

# Finishing the Head Model

In this section, we return to modeling the head. We start by converting the model into a subdivision proxy so that we can evaluate the smoothed model. Then, we introduce the Sculpt Geometry tool, which provides a more elegant interface for pushing and pulling vertices. Once we're happy with the model, we'll delete the subdivision proxy and return the proxy to a shaded view that is ready to be integrated into the animation pipeline.

## Convert Model to a Subdivision Proxy

We are going to convert our existing head model into a subdivision proxy with some specific settings so that both the proxy and subdivision mesh are mirrored. This eliminates any problems with a visible seam along the axis of symmetry.

**1.** Select the geometry and choose Proxy | Subdiv Proxy □ to open the Options window for this command. Figure 4-16 shows the Subdiv Proxy Options window with the settings that we'll be using here.

**2.** In the Subdiv Proxy Options window, set the mirror behavior to Full and set the Mirror Direction to –X. This mirrors both the proxy and the subdivision mesh. In addition to

FIGURE 4-16    The Subdiv Proxy Options window with the settings used

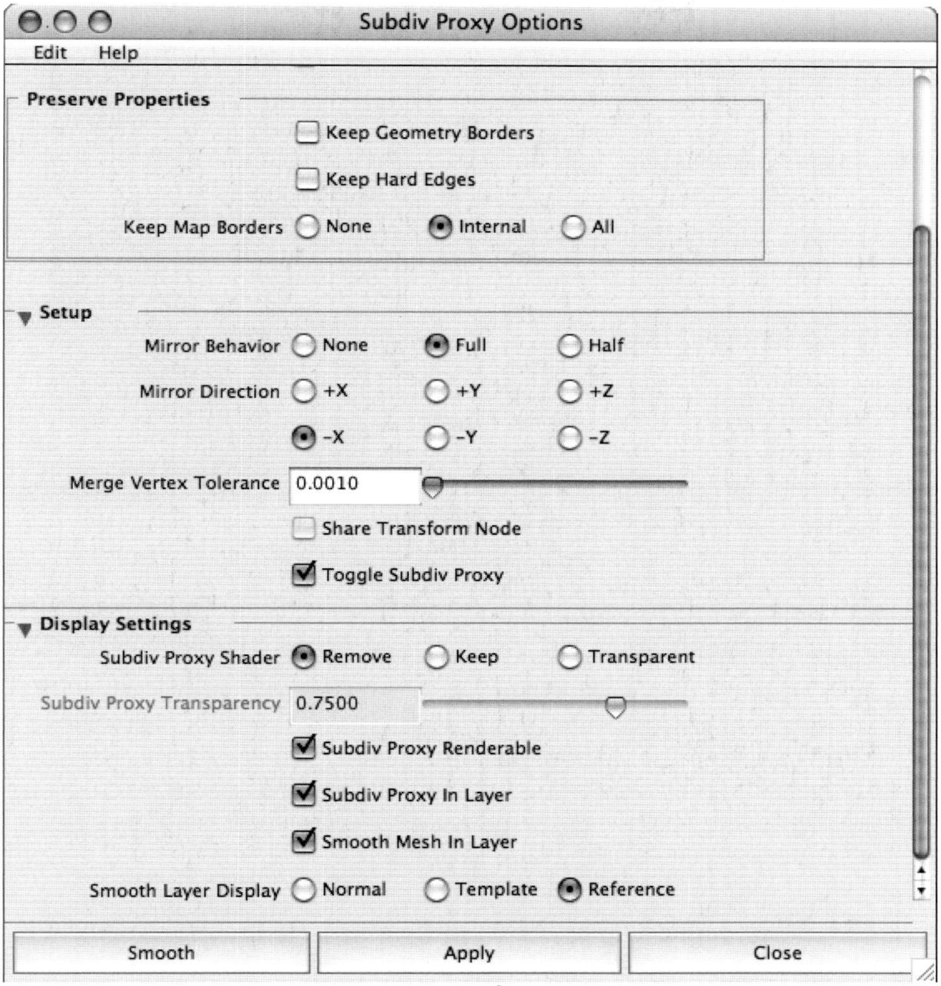

the mirroring, the new copy also behaves as an instance so that whatever edits are made to one side are also made to the other.

**3.** In the Display Settings, set the Subdiv Proxy Shader to Remove. This removes the gray shading on the proxy object, leaving only the wireframe visible.

**4.** Enable the Subdiv Proxy In Layer and Smooth Mesh In Layer check boxes. These settings place each mesh in a separate layer and set the smooth layer to Reference. Objects in a referenced layer are visible but cannot be selected in the view window.

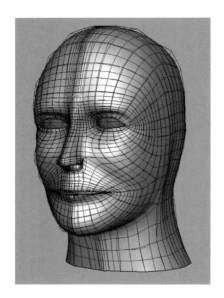

**5.** When these options have all been set, click the Smooth button. The Options window closes and the subdivision proxy is created. The result will be a smoothed model with four times the number of faces as the original. Meanwhile, the original model is still there, but the shading has been removed, leaving just the wireframe cage on top of the smoothed model, as shown here:

## Sculpt Geometry Tool

The Sculpt Geometry tool allows you to modify a model using a brush-based interface. As you will see, most of the toolsets in Maya offer this interface for performing various tasks. When it comes to modeling, the Sculpt Geometry tool provides a great way to smooth, push, or pull the positions of multiple vertices on a surface at a time without constantly having to make selections and use the transform tools.

**1.** Select your model surface in the view window and choose Mesh | Sculpt Geometry Tool ❑. The tool settings panel, shown in Figure 4-17, appears on the right side of the interface and the Sculpt Geometry tool becomes active. If you place the cursor over the head in a view window, you will notice that the cursor shows as a circular shape on the surface. This circular cursor is called a *stamp*, and painting on the surface using a stamp is called a *stroke*. You can use the stamp to push, pull, smooth, and erase operations on vertices within the bounds of the stamp.

**2.** Give it a try. When you LMB-drag across a surface, the stamp behaves much like a paintbrush, only instead of paint, it moves, or displaces, your vertices. If you are using the tool's default settings, you probably noticed that the brush stroke made a huge dent in your head. Let's adjust some of the setting to make it more ideal for sculpting our model.

**3.** Figure 4-18 shows the stamp icon over the surface of the model. This icon indicates how the tool is being used based on the setting. Let's set the Max Displacement attribute. This sets the amount of displacement each stroke applies to the surface. The value is measured in centimeters. Change this according to the size of your model. For this model, a setting of 0.01 should work well. The idea is to make subtle tweaks as you stroke the vertices. You can interactively edit the Max Displacement attribute by placing the brush over the surface and pressing the M key. A virtual slider appears.

FIGURE 4-17    *The Sculpt Geometry tool settings panel*

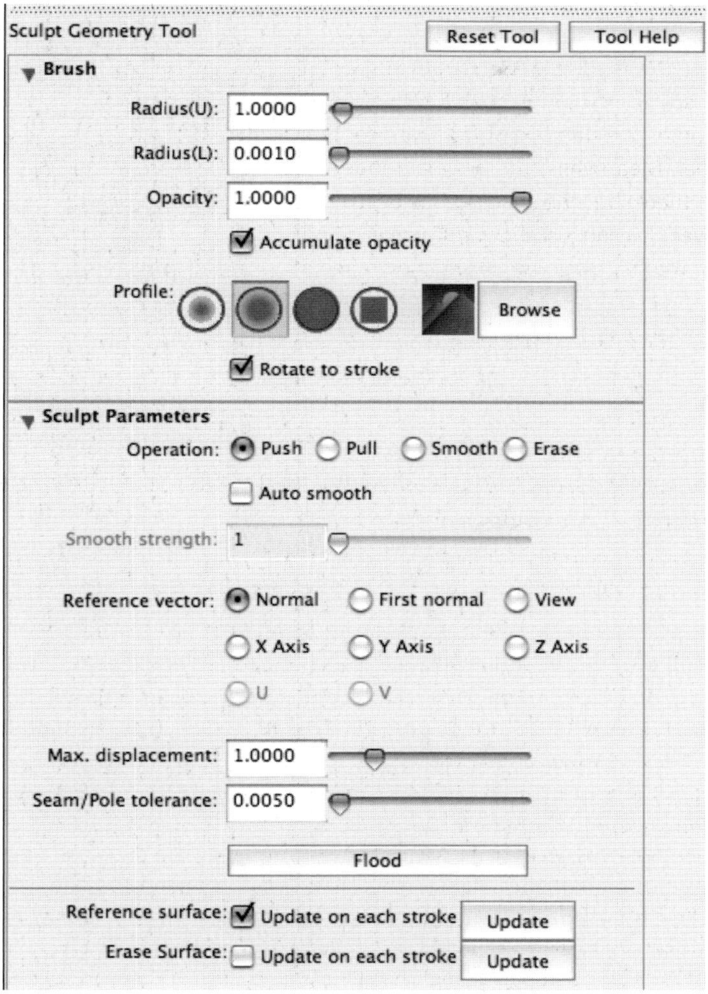

LMB-clicking and dragging edits the value. The displacement depth is indicated on the stamp by the arrow pointing out of the center (see Figure 4-18).

The next setting we'll adjust is the size of the brush. There are two attributes that control this:

- **Radius(U)**   Controls the size of the stamp. You can interactively change the radius size in the view windows by holding down the B key and LMB-clicking and dragging left and right. You will see the stamp size change.

FIGURE 4-18    *The Sculpt Geometry tool's stamp icon in the model indicates the brush size, reference vector, and operation.*

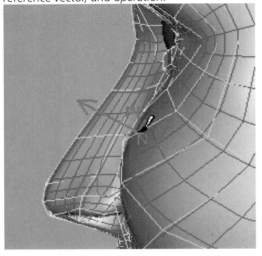

• **Radius(L)**    Controls the brush size if you are using a stylus tablet with your computer. It sets the minimum brush size when pressure on the stylus is low. The Radius(U) setting can't be smaller than the Radius(L) setting, so make sure Radius(L) is set to 0 if you are not using a stylus.

As you begin tweaking your model, you will want to push and pull vertices in different directions. The Operation attribute sets how the stamp will behave on the vertices. It will push, pull, smooth, or erase previous operations. The Pull and Push operations move vertices in the direction you set for the Reference Vector attribute. The Smooth operation averages out placement of the vertices, so the surface gets smoother. Erase erases any operations you have done to the surface. You can quickly change between these operations by holding down the U key as you LMB-click and select an operation from the marking menu.

Adjusting these settings as you sculpt should be enough to get you on your way to using this tool and finishing your model. Just realize that this tool's actions are applied to vertices, not to edges or faces. In order for it to be really effective, the head model must have enough existing geometry. If you find that an area is too sparse to edit, go ahead and add another edge loop.

## Finalizing the Geometry

We're almost finished! Before we integrate this model into the next phase of the production pipeline, we want to prepare it so that there isn't any instanced behavior. Also, it's quite possible that the model will be subdivided later on, so it isn't necessary for us to include it in this model. Luckily, Maya 8 includes a command that really speeds up this workflow.

**1.** In the view window, select the model. Choose Proxy | Remove Subdiv Proxy Mirror ❑. Set the Mirror Direction attribute to –X and click the Remove Mirror button. The subdivided surface is deleted and Maya combines the two proxy meshes into one shape while merging the vertices together.

**2.** The model is just about ready, but right now it only displays in wireframe. This is because the material was removed when we created the subdiv proxy. To quickly replace it, RMB-click over the model in the view window and choose Assign Existing Material | Lambert from the lower section of the marking menu, as shown here:

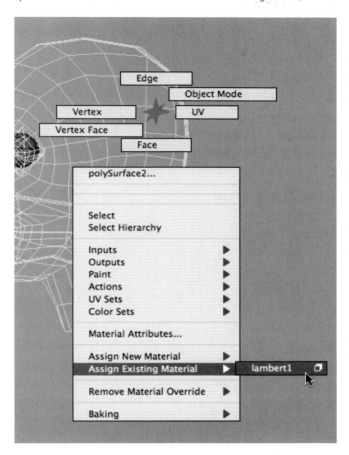

**3.** In the Channel Box, rename the model to **poly_head**.

**4.** In the Layer Editor, choose Layers | Delete Unused Layers to remove any layers that no longer contain geometry.

**5.** Save your work. Congratulations!

# Summary

In this chapter, you have learned to lay out the structure of a model so that it deforms according to the anatomy of a real human face. While we have only covered the face in this chapter, the same exact techniques can be applied to any organic structure. It all starts by laying out the framework so that the edge loops flow in the direction of the muscles of the underlying anatomy. Once you have placed the framework, you add and sculpt edge loops until the complete model has been realized.

5

# Basic NURBS
# Modeling

**The NURBS toolset is Maya's oldest**

and is an industry staple in building models

for high-end film productions and industrial

design. While NURBS are very technical in

nature, taking the time to understand how they

work is worthwhile because you can benefit

by using them in certain modeling scenarios.

*Non-Uniform Rational B-Splines* is a mathematical term that sounds complicated. To understand NURBS, you should know that a NURBS object is a mathematical description of a curve or surface. Any point on a NURBS curve or surface has an exact coordinate value, in contrast to a polygonal surface, where only the locations of the vertices can be returned.

When comparing NURBS to polygons, a good analogy to consider is the difference between vector-based graphics used in Adobe Illustrator and Flash and pixel-based images used in Adobe Photoshop and many digital video and photography applications. Because vector lines are drawn on your screen based on a mathematical description, they always remain smooth, no matter how close you zoom in on them or how big the window is in which they are played back. Pixel-based images, on the other hand, are made up of finite rectangles, called *pixels*. While these may not be visible when you're viewing an image at its normal resolution, the image starts to break up as you zoom in and see that the image is made up of thousands, or even millions, of rectangular pixels.

Polygonal geometry, which we've been working with over the last two chapters, is made up of finite segments (faces). Since a NURBS surface is mathematically described, it can be rendered to appear infinitely smooth, no matter how close up the camera gets. Figure 5-1 shows a NURBS sphere (left) and a polygonal sphere (right). Look at the edges of the spheres and notice that the polygonal sphere is made up of segments and is not smooth like the NURBS sphere.

**FIGURE 5-1**   *A close-up of a NURBS sphere (left) versus a polygonal sphere (right)*

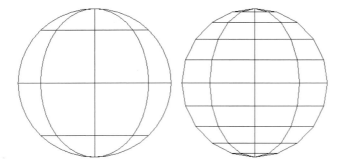

In this chapter, we'll cover the components of NURBS geometry, discuss some advantages and disadvantages of NURBS compared to polygonal modeling, and demonstrate how to work with NURBS curves and surfaces.

# Anatomy of NURBS Curves and Surfaces

Both NURBS curves and surfaces have similar components that control how they are edited. Understanding how these components relate to each other and how they are used to calculate curves and surfaces will provide you with a good foundation for transforming your simple surfaces into more complex shapes while maintaining a clean geometry that is easy to edit.

We'll first examine curves, since they have similar, but fewer, components than surfaces. You'll then learn how these components relate to surfaces and how some other components are unique to surfaces.

## Components of a NURBS Curve

Figure 5-2 shows a NURBS curve displaying its components, described next:

- **Edit points (EPs)**   When a curve is drawn in Maya, the software joins together polynomial curve segments, called *spans*. The points where these segments join are called *edit points* (sometimes called *knots*). An edit point lies on a curve and is represented by a small × that you can select and move to edit the shape of the curve. Edit points can be added to a curve by using the Insert Knot command (Edit Curves | Insert Knot).

- **Control vertices (CVs)**   CVs control how the curve is "pulled" or "weighted" between edit points. CVs basically define the shape of a curve or surface. The number of CVs between edit points depends on the *degree* of the curve. (See "Surface or Curve Degree" later in this chapter.) CVs often lie above or below a curve, in space, and are the components most commonly edited to control the shape of the curve. CVs cannot be inserted directly into the curve but rather are added when an edit point is inserted.

- **Curve points**   One of the many great aspects of NURBS technology is that, because curves are mathematically calculated, any *curve point* along a curve can be chosen

*FIGURE 5-2*   *Anatomy of a NURBS curve*

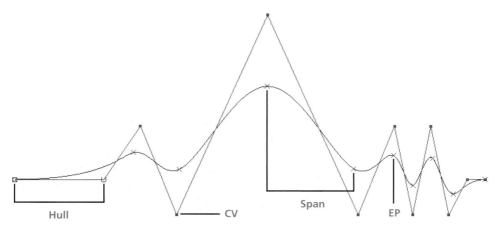

and adjusted. Once a curve point is selected, knots can be inserted or the curve can be detached (Edit Curves | Detach Curve). The shape of the curve can be edited from any curve point by using the Curve Editing Tool (Edit Curve | Curve Editing Tool).

- **Hulls**  Hulls are straight lines that connect the CVs together. Choosing a hull selects all of the CVs in a curve.

You can display and edit curve components by using the right-click marking menu, shown in Figure 5-3, which displays the available components for a NURBS curve. If you want to display and choose several components at once, you can use the selection mask buttons on the status line. To do this, click the Select by Component Type button and then activate the buttons for the components you want to use.

You can also right-click (RMB-click) a selection mask button on the status line and choose exactly what components you want it to show; these settings vary depending on what type of object is being chosen. (For example, clicking the Points Selection Mask button displays NURBS CVs, polygonal vertices, subdivision vertices, lattice points, and particles.) You might need to adjust these components, depending on what is included in your scene. A great thing about using these selection masks is that you can enable multiple selection masks at once for multiple object selections. For example, you could enable the selection masks for CVs and hulls so that you are able to pick one or both component types at the same time.

If you want to view a component but not select it, you can display it by choosing an option from the Display menu. For example, to see the hulls of your object, select the object and then choose Display | NURBS | Hulls. The hulls will be displayed for that object.

**FIGURE 5-3**  *A marking menu showing the components of a NURBS curve*

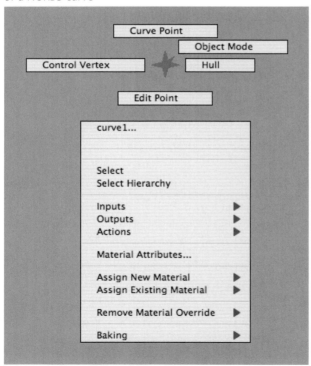

## Components of a NURBS Surface

NURBS surfaces contain some of the same components as NURBS curves. Both CVs and hulls perform the same functions that they do for curves and can be chosen and edited in the same way. However, surfaces also have some unique components, as shown in Figure 5-4. These unique components are discussed next.

*FIGURE 5-4  Components of a NURBS surface*

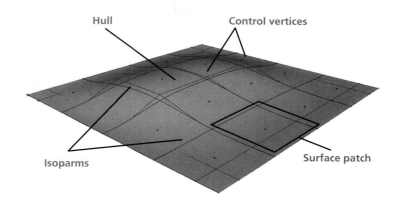

- **Isoparms**   Isoparms lie on the NURBS surface and extend across the entire surface at a constant parametric value. (See the upcoming section "Parameterization of Curves and Surfaces.") Just as edit points represent the ends of curve spans, isoparms represent the boundaries between *surface patches*—the regions on a surface enclosed by isoparms. Isoparms can be selected and added to a surface, but unlike their curve counterparts, they cannot be edited directly.

- **Surface point**   A surface point is any point or parameter on a surface. This is similar to a curve point (and is not shown in Figure 5-4).

- **Surface patch**   A surface patch is the region between the enclosed isoparms.

## Surface or Curve Degree

The *degree* of a surface or a curve refers to the exponent used in the underlying polynomial equation that defines the geometry. What this means to the Maya user is that degree equates to the number of control vertices that are needed to define one span of a NURBS curve or surface. The number of control vertices needed to define one span will always equal the degree of the curve + 1. Therefore, a one-degree (linear) curve has two control vertices—one at the beginning and one at the end. A one-degree curve is always a straight line. Maya allows NURBS objects to have one, two, three, five, or seven degrees of curvature. In Maya, a degree of three is the default and is most often used.

## Curve Direction

The direction of a curve is initially determined by the order in which you draw it, which is called the curve's *U direction*—the first point at which you draw is considered the beginning of the curve and the last point is the end point. If you don't know which direction an existing curve is going, you can select it and display the components by pressing F8 (or you can enable the Select by Component Type button on the status line) and making sure the Points Selection Mask is activated on the status line. Or you can right-click the object and choose Control Vertex. The beginning of the curve will be marked with a small box, and the second CV will be marked with a *U* to communicate that the curve is moving in the U direction. This is illustrated in Figure 5-5.

> **NOTE**  *You can change the direction of a curve by choosing Edit Curves | Reverse Curve Direction.*

Knowing the direction of a curve is especially important during certain operations, especially those that involve connecting and merging curves. You'll see an example showing why knowing the direction of curves is important later in this chapter in the section "Cutting and Filleting Curves."

## Parameterization of Curves and Surfaces

*Parameterization* describes how the numeric values are assigned to each point on a NURBS surface or curve. Each of these points has a numerical value known as a *parameter* that describes the point's location. On a curve, this parameter gives a point's location in the curve's U direction.

**FIGURE 5-5**    *A curve displaying its CVs and curve direction*

A surface has two parameters, one in U and the other in V. Any point on a surface can be queried or described using this coordinate system.

You can find the parameter value of any curve by right-clicking the curve and choosing Curve Point from the marking menu. Then click any point on the curve, and its parameter value will appear on the title bar at the top of your main Maya window, as shown here:

Autodesk Maya 8.0: /Users/tmeade/Desktop/chapter5.mb  ---  curve1.u[0.34713340320272]

You can also use the Parameter tool to find the parameter of any point on a curve (choose Create | Measure Tools | Parameter Tool) by clicking any point along a curve. The parameter value for this point appears next to the point and remains onscreen, as shown in Figure 5-6. This is perhaps a bit more convenient and useful than looking in the title bar.

### Methods of Parameterization

The parameterization, or how a parameter value is calculated across the curve, is determined by one of two methods: the uniform method or the chord-length method, both shown in Figure 5-6. When drawing a curve, the method of parameterization can be set in Knot Spacing attributes of the CV (or EP) Curve tool's settings window (choose Create | CV Curve Tool ❑).

With uniform parameterization, integral parameter values are assigned at each edit point. To demonstrate, let's consider a curve with four spans, in which the edit points are not evenly spaced. In the uniform parameterized curve shown in Figure 5-6, the first four edit points fall in the first half of the curve and the last edit point is at the end of the curve. The first edit point has a parameter value of 0, the next edit point has a value of 1, the next is 2, then 3, and the last point has a value of 4. We can use the Arc Length tool (choose Create | Measure Tools | Arc Length Tool) and the Parameter tool (Create | Measure Tools | Parameter Tool) to find a parameter value for any point on the curve by clicking it. Say, for example, you wanted to find the parameter value for the middle of the curve. Clicking at the center of the curve will return a value of 2.589839, also shown in Figure 5-6.

**FIGURE 5-6**   *Two- and four-span curves: the top curve uses uniform parameterization and the bottom curve uses chord-length parameterization.*

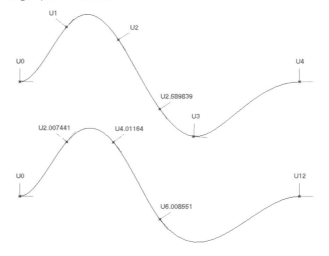

Using uniform parameterization, you can match parameterization between two curves just by figuring the total number of spans. However, determining parameters based on the length of the curves is not so straightforward. By clicking at the approximate midpoint of the four-span curve shown in Figure 5-6, the midpoint parameter shows 2.589839. This is not so intuitive if you want to insert knots at specific lengths, however—like halfway, a quarter of the way, or three-quarters of the way down the curve. When these types of calculations are necessary, it's probably more useful to use chord-length parameterization.

Using chord-length parameterization, the parameter values are assigned to the edit points based on the distance of a straight line, expressed in Maya units, extended between the edit points. While this method gives the curves irregular parameter values at their edit points, the values do reflect information that is relative to a point's position along the curve. So, for example, since this curve is 12 units long, it has a U parameter of 12. The middle of this curve could then be easily determined by dividing the parameterization in half. A U value of 6, then, would be the middle of the curve.

> **NOTE** *Since the Maya application is geared mostly toward the entertainment market, chord-length parameterization isn't necessary and in fact will cause problems when you need to match parameterization between two surfaces.*

## Parameter Range

A uniformly parameterized curve of surface always has a value of 0 at the first edit point. Exactly how the rest of the parameters are divided up over the length of the surface depends on the Parameter Range. Draw a four-span curve. In the Attribute Editor, look at the Min Max Value attributes. You should see that the Min value is set to 0 and the Max value is set to 4, as shown here. When the values are assigned this way, you can deduce that the parameter range is set from 0 to the number of spans.

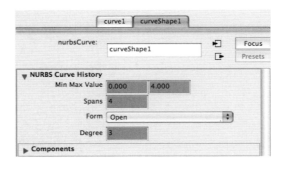

You can choose to set the parameter range to 0 to 1 as well. This has no effect on any modeling operations. This is available for many reasons. One reason is that texture coordinates are laid out on a grid that has a range from 0 to 1. Some workflows or rendering engines may only be able to properly apply textures to a surface that has a parameter range from 0 to 1. Another use for setting the parameter range from 0 to 1 is for use with particle effects. Many times a NURBS surface is used to guide particles. When the effects artist wants to run particles along a surface, from one end to the other, they would probably rather not deal with keeping track of how many spans the surfaces has. Instead, the surface's parameter range can be reset to 0 to 1 using the Rebuild Surface command (Edit NURBS | Rebuild Surface).

*NOTE**   To rebuild curves, use Edit Curve | Rebuild Curve.*

### Maintaining Parameterization While Modeling

Much of the NURBS modeling process involves the constant rebuilding of curves and surfaces so that they maintain normal parameterization and have a matching number of spans for creating clean, orderly surfaces. For example, a loft operation creates a surface between two or more curves. (Lofting is discussed in detail in the section "Surfaces" later in this chapter.) To generate the cleanest surface, the parameters of the source curves must match.

Figure 5-7 shows two surfaces that were generated by lofting two curves together. The curves used to generate the surface on the left were created with the same number of spans (four) and, as a result, they had matching parameterization. The structure of the lofted surface is clean and predictable. You can tell this by looking at the even flow of isoparms that were created by connecting the edit points on the original curves. As a modeler, you can predict exactly where these isoparms are going to intersect each curve.

The surface on the right was generated with two curves that had different parameterizations. One curve had four spans and the other had seven. When the curves were lofted together, Maya placed isoparms on the surface so that each isoparm corresponds to an edit point on each curve. The resulting surface, then, contains ten spans in the U direction to compensate for the different parameterization of the source curves. It's not as orderly as the surface on the left.

Just because two pieces of geometry have the same number of spans does not necessarily guarantee that they have matching parameterization. This mainly becomes an issue when an operation is applied to a surface. Whenever a surface is attached or detached or an isoparm is inserted, the parameterization of the geometry is affected. When this occurs, the surfaces must be rebuilt immediately in order to reparameterize that value across the geometry.

**FIGURE 5-7**   *The surface on the left was generated from two curves that had the same number of spans. The surface on the right was generated by two curves that had a different number of spans.*

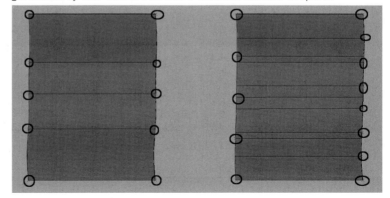

In order to understand this further, follow these steps:

**1.** Create a NURBS plane (Create | NURBS | Plane) and choose Edit NURBS | Rebuild Surface ❐ to open the Rebuild Surface Options window, shown next. Set the Parameter Range to 0 to #spans. Set the Number of Spans for the U and V attributes to 4 and click Apply. The NURBS plane now has four spans in the U and V direction and the parameter range goes from 0 to 4. You can verify this information in the Attribute Editor.

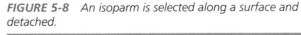

**2.** RMB-click over the surface to access the component marking menu and choose isoparm. Click and drag an existing isoparm to select an isoparm that lies somewhere between the third and fourth span in the V direction, as shown in Figure 5-8. Remember, you can look in your menu bar to see what parameter you have selected.

*FIGURE 5-8*  *An isoparm is selected along a surface and detached.*

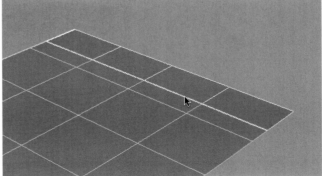

**3.** Choose Edit NURBS | Detach Surface. This splits the surface into two separate pieces at the selected isoparm.

**4.** Select the larger surface and look at its shape node in the Attribute Editor. While the number of spans is still 4, the max value in V will match the exact parameter where the surface was detached from. In this case the value is 3.345 as shown here:

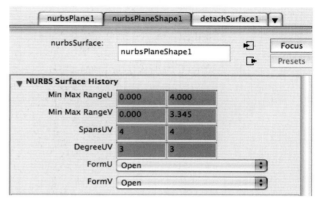

**5.** Any operation performed in the V direction will result in a poorly constructed surface. To reset the values from 0 to 4, we again use the Rebuild Surface command. In the Rebuild Surface Options window, set the Keep attribute to NumSpans and click Rebuild. The Min Max Value in the V direction will be reset from 0 to 4. The isoparms on the surface will be reorganized so that they are uniformly spread out over the surface. This surface now has uniform parameterization.

It is absolutely critical that all NURBS surfaces be rebuilt this way after attaching or detaching a surface or inserting an isoparm. Failure to do this will result in unpredictable distribution of isoparms and will possibly cause surfaces to explode down the line. In Chapter 6, we will build special shelf buttons that will rebuild surfaces automatically after executing one of these operations.

## Surface Direction

Just as polygonal surfaces have a front and back side, so do NURBS surfaces. You can display the surface normal direction of a NURBS surface by choosing Display | NURBS | Normals. You'll see a perpendicular line extending from the center of each surface patch.

Choose Display | NURBS | Surface Origins to display the surface normal as a blue line that extends from the surface *origin*. This display is a more manageable option to work with, since fewer normal lines appear that might obstruct your view of other objects in the scene. The surface normals can be reversed by choosing Edit NURBS | Reverse Surface Direction, as shown in Figure 5-9.

---

*FIGURE 5-9*   *Surface displaying normal directions of two surfaces. The U direction of the surface on the right has been reversed.*

# Advantages and Disadvantages of NURBS Modeling

NURBS modeling works well for building any smooth, curving surface—such as the body of a car or a human body. Because of its ability to represent complex shapes, NURBS is the preferred toolset used among most industrial designers. In fact, chances are that the chassis of your computer and monitor were designed using NURBS modeling.

However, the biggest drawback to using NURBS has to do with the fact that to build a complex object, you must create and link together multiple individual surfaces. Working with multiple surfaces can mean more time spent having to texture each surface individually and manage the seams between surfaces so that they match at render time.

## Advantages of NURBS

Following is a discussion of some of the benefits of NURBS modeling.

**Few Control Points**   Using NURBS, you can create and edit smooth curving objects with few control points. This results in less data needed to calculate these surface, which translates into smaller file sizes and more efficient use of your computer's RAM. In a network environment, smaller files transfer faster and do not clog the pipeline. This was a big factor back in the early years of 3D animation pipelines, when cost of RAM was at a premium. Nowadays, this aspect of NURBS is taken for granted, but is still important to consider.

**Resolution Independence**   The mathematical nature of NURBS surfaces enables smooth surfaces to remain smooth no matter how close the camera zooms in. However, when a NURBS surface is rendered, it is *tessellated*, which means that it is converted into polygonal faces, based on the user-defined *tessellation attributes* of the NURBS object. This means that while the object is being cut up into triangles, the user still has the power to assign how smooth or coarse the object will render. Having this kind of control can really help optimize a scene's rendering time. If the object is far away from the camera, it can be rendered with a low tessellation setting. The same piece of geometry can also be used in a close-up shot by increasing the tessellation setting. With a polygonal object, however, two separate models would have to be built to optimize the overall polygon count of the scene: one low-resolution model for when the object is in the background, and a high-resolution model that will work for close-up shots.

Definition    ***tessellation***   *The process of creating a faceted surface. When a NURBS surface is tessellated, it is converted into polygonal faces that define the original shape.*

**Texture Coordinates**    By its rectangular nature, any point on a NURBS surface can be defined by a coordinate in the U and V directions. Texture maps, such as bitmapped files that you might create in an image editing application like Photoshop, are also rectangular in nature, and their pixels are defined by coordinates in the X and Y directions. Because of this congruent relationship, Maya is able to fit a texture map to a surface by matching these corresponding coordinate systems, allowing the texture to fit exactly on a surface and match its curvature, no matter how many twists or turns the surface makes. With a polygon model, an entire process of laying out the UVs is necessary before any texturing can be done on an object.

The advantages are most evident in an example such as the one shown in Figure 5-10, where the surface is twisting so much that it would be difficult to map it using projections. Because the texture has to be projected from a place in 3D space, much like a projector projects a slide or piece of film, this causes stretching and overlapping of the texture on curved or overlapping surfaces.

**FIGURE 5-10**   *A texture applied to a NURBS surface in normal (UV) mode (left) versus a texture attempting to be mapped using planar projection (right)*

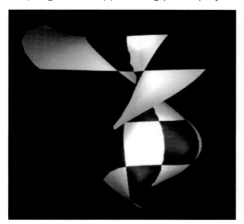 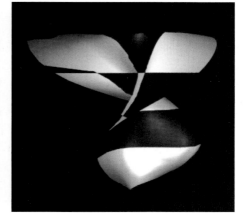

Definition   **projection mapping**   _A method used to apply a texture map to a surface or several surfaces at once. It uses primitive shapes to project the texture onto the geometry. For example, a planar projection (shown in Figure 5-10) uses a flat plane to project the texture onto the surface. However, you might also choose to use a cylindrical projection type if the object is cylindrical in shape, or a spherical projection if it is spherical in shape. Projections are discussed in detail in Chapter 17._

**Conversion-Ability**   Maya allows you to convert any geometry type to another by using the Convert command (discussed in Chapter 7). Converting a NURBS model to polygons can produce the cleanest geometry, and it requires the least amount of cleanup due to the inherent orderliness and square-patch nature of NURBS. A conversion from NURBS to a polygon is illustrated in Figure 5-11. Notice that each square surface patch on the NURBS object is converted to a square polygonal face on the polygonal object. It is even possible to specify how many polygonal faces should be created for every one surface patch on the NURBS surface so that the resulting polygonal object will appear smooth. Many modelers prefer to build the initial surfaces with NURBS and then convert them to polygons or subdivision surfaces to add more rigid detail.

**FIGURE 5-11**   _A NURBS Surface (left) is neatly converted to a polygonal approximation (right)_

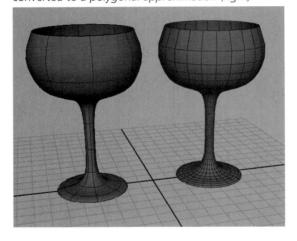

## Disadvantages of NURBS

While you can easily model just about anything using NURBS, most of the disadvantages with NURBS stem from the fact that more complex NURBS models are often made from multiple individual surfaces. As a result, you have to deal with the problems described next.

**Seams**   Probably the biggest problem with models built from NURBS has to do with the seams that can show between surfaces. When using trimmed surfaces, which are discussed in Chapter 6, the solution is to try to match the tessellation between adjacent surfaces manually. However, this usually results in heavy tessellation, which means a lot of polygons and hence longer render times.

Patch modeling, shown in Figure 5-12, makes the tessellation process simpler because of the matching parameterization, but it can still cause errors in the rendered image.

**Branching**    In modeling, *branching* refers to the parts of models that extend, or branch off, from the main surface or a group of surfaces. In these cases, the gridlike structure of the topology is broken and the five-point junctions that we saw in Chapter 4 are introduced. Since a NURBS surface can only contain a quad layout, the NURBS surface must be broken up into individual pieces and arranged to handle the change in the structure. While this can amount to a great deal of work, a similar structure in polygons can be created within one mesh by simply executing the Extrude command on a selection of faces. Figure 5-12 shows the topology of a NURBS model where the arm branches out from the body.

**FIGURE 5-12**    *Several individual NURBS surfaces (patches) are arranged to accommodate the topology of the arm branching from the body. Notice the areas where there are junctions with five patches meeting. (Image courtesy of Shannon Thomas.)*

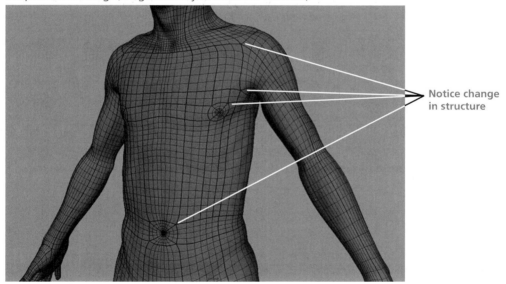

Notice change in structure

**Texture Mapping Across Multiple Surfaces**    When you're creating very detailed models using NURBS, you will often end up with multiple surfaces. While this is not necessarily a bad thing, it can make managing the scene a lot more difficult, especially when it comes to having to apply texture maps across these mismatched surfaces. And, while we mentioned having texture coordinates inherent to the model as an advantage of NURBS, it also has limitations, because the UVs cannot be edited separately from the mesh.

# Continuity

Achieving a high level of continuity between two or more curves or surfaces is one of the main challenges in NURBS modeling. The level of continuity between surfaces determines whether an obvious change occurs where one surface ends and the adjacent one begins, or whether the transition is smooth and seamless. Many times, it takes multiple, individual surfaces to build a model of moderate complexity. When these individual patches are supposed to describe a surface that is one smooth piece, care must be taken to make sure the CVs between the patches shape a continuous relationship.

Several NURBS modeling tools and commands in Maya let you achieve continuity between curves and surfaces: Align Curves, Align Surfaces, Attach Curves, Attach Surfaces, the Stitch tools, the Fillet Curve tool, and all three of the Surface Fillet tools. All of these tools will be demonstrated within this and the next chapter.

## Levels of Continuity

Maya offers three levels of continuity, called C0, C1, and C2. Figure 5-13 shows examples of two surfaces that meet using the three levels of continuity.

**FIGURE 5-13** *Surfaces that match edges with C0 (left), C1 (middle), and C2 (right) continuity*

- **C0—Positional**   At this level, the end points of a surface or curve share the exact same place in X, Y, and Z space. In other words, the objects touch but the surface connection is not seamless. C0 continuity is rarely desirable when modeling, because even the sharpest edges have some degree of smoothness to them in real life.

- **C1—Tangent**   This type of continuity offers positional continuity, plus the end tangents between the two curves or surfaces match. This means that not only are the end points used to calculate the continuity, but the adjacent CVs are used as well. This creates a smooth transition between the two objects without leaving a noticeable seam. It is best used when two surfaces need to join smoothly, even though the change in surface is abrupt—as in an edge or corner. An edge with less than C1 continuity will not render a specular highlight. (See Chapter 18 for more information on specular highlights.)

- **C2—Curvature**   C2 continuity has both C0 and C1 levels of continuity, plus matching curvature. This means that three control points from the edge inward on each object are used to calculate the continuity. The resulting surface will not show any changes where the object surfaces join.

# Achieving Continuity Using Tools

As mentioned, several of Maya's modeling tools provide different levels of continuity that result from joining or generating new surfaces:

- **Loft tool**   Creates a new surface between two existing surfaces with C0 continuity
- **Fillet tools**   Create a new surface between two existing surfaces with C1 or C2 continuity
- **Attach tool**   Joins together two surfaces or curves with C2 continuity
- **Stitch tools**   Give edges of objects any level of continuity
- **Align tool**   Repositions any two objects and modifies the position of the CVs to provide any level of continuity

### Exploring Continuity with the Align Command

To understand exactly how continuity is managed, here's a short tutorial that uses two curves to demonstrate achieving the various levels of continuity using the Align Curve command. Keep in mind that there is an analogous command for NURBS surfaces called Align Surfaces.

1. Open a new scene in Maya, and press F4 to display the Surfaces menu set in the menu bar.

2. Choose Create | EP Curve to select the EP Curve tool.

3. In the front view, draw a two-span curve (click three times). Press ENTER (RETURN) to finish drawing the curve.

4. Draw another curve next to that. Select both curves and choose Display | NURBS | CVs. Also, choose Display | NURBS | Hulls. Now you'll be able to monitor the curves' components and watch how they are affected by the different align settings that you are about to apply. They should look similar to the curves shown in Figure 5-14.

5. Select the curve on the left and then, holding down the SHIFT key, select the curve on the right. Choose Edit Curves | Align Curves ❑. In the Align Curves Options window, choose Edit | Reset Settings to return the command to its default settings.

FIGURE 5-14   *Two separate curves displaying their CVs*

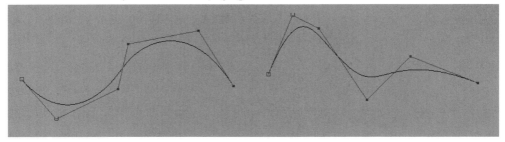

**6.** Choose the Position radio button in the Continuity area. The other settings should match the illustration shown here:

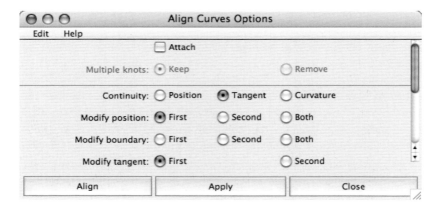

**7.** Click the Apply button. Notice that the first curve you selected (in Step 5) moved. If we had set the Modify Position setting to Second in the Align Curves Options window, the second surface would have been moved. Notice that the positions of the hulls have not changed (see Figure 5-15). This is because we are using a C0 level of continuity.

**8.** Undo the last step by choosing Edit | Undo or by pressing CTRL-Z (COMMAND-Z). In the Align Curves Options window, choose the Tangent Continuity option. Click Apply to see what happens. Look at the hulls and notice that the CVs adjacent to the edges of both curves line up to form a straight line.

FIGURE 5-15    *The two curves after being aligned with positional (C0) continuity*

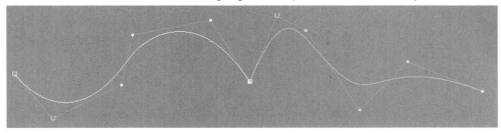

**9.** Use the component marking menu (RMB-click) to select the second CV in from the edge on the first curve. The second CV from the edge is often referred to as the "tangent CV." Choose the Move tool from the toolbox (or press the W key). Try moving the CV upward in the Y direction, and you'll notice that the second CV from the edge on the other curve relocates to maintain tangency with the point that you are modifying, as shown in Figure 5-16.

FIGURE 5-16    *The align nodes continue to evaluate and maintain tangency between the two curves when one is edited.*

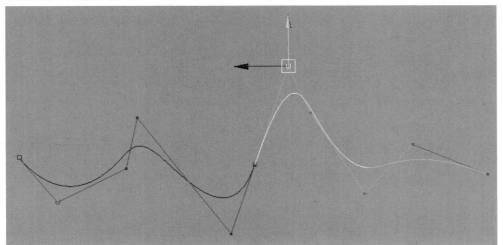

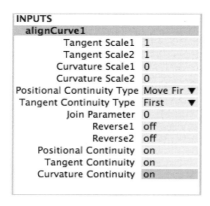

| INPUTS | |
| --- | --- |
| **alignCurve1** | |
| Tangent Scale1 | 1 |
| Tangent Scale2 | 1 |
| Curvature Scale1 | 0 |
| Curvature Scale2 | 0 |
| Positional Continuity Type | Move Fir ▼ |
| Tangent Continuity Type | First ▼ |
| Join Parameter | 0 |
| Reverse1 | off |
| Reverse2 | off |
| Positional Continuity | on |
| Tangent Continuity | on |
| Curvature Continuity | on |

**10.** Now for curvature continuity. Instead of using the Undo command to return the curves to their initial positions, we will edit the attributes of the Align Surface node in the Channel Box. Select the alignCurve1 node under the Inputs section of the Channel Box, as shown in this illustration.

**11.** Click in the field next to the Curvature Continuity attribute and type **one** or the number **1**. Now notice the hulls. Not only did the first row of CVs meet and the second align in tangency, but now the third CV is also taken into the equation. Experiment with this level of continuity by moving the third CV around and seeing the effect the edit has on the other curve.

*NOTE  For on/off attributes in the Channel Box, you can also type 1 to turn on the attribute or type 0 to turn off the attribute.*

### Curvature Continuity with Attach and Detach

Attaching surfaces (Edit NURBS | Attach Surfaces) is a quick way to combine two surfaces together into one. With the Attach Surface command's Attach Method attribute set to Blend, two surfaces will be joined together and the transition between them will be smooth. The Detach Surface (Edit NURBS | Detach Surfaces) command splits a surface into two separate pieces from a selected isoparm. This process of attaching and then detaching produces C2 continuity between any two surfaces.

The result, however, is a little bit different from what it would be if you were to use the Align command. When two surfaces are attached, the tangent row of CVs is removed since the junction is no longer at an edge. This means that geometry is removed and any shape defined by those CVs in the tangent row is lost.

In the example shown on the left in Figure 5-17, each surface has three spans, making for a total of six hulls in each direction. However, when the surfaces are attached together, as shown on the right in Figure 5-17, the total number of hulls in that direction is 9, not 12. Therefore, some detail is lost on the shapes that you are attaching. This can be undesirable, especially for hard surface applications, because you will begin to lose some of the precision of your surfaces. We will practice this process in Chapter 6 when we build the front panel of the car.

*FIGURE 5-17*   *Attaching and detaching is used to obtain C2 continuity between two surfaces. However, the detail in the shape at the edges of the surface is lost during the attach operation.*

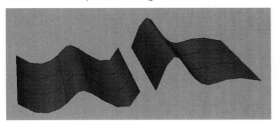
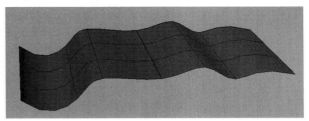

# Curves

The NURBS modeling process usually begins with the creation of curves that match a profile or cross section of the object that you are modeling. Once you establish the curves, you execute commands that use the curves as inputs to generate surfaces. In this section, we will explore many of the curve editing tools and commands and then learn to generate surfaces from them using some of Maya's surface commands.

Curves are often the starting points for models. While they are never rendered, you will probably need to work with them to some degree while building most of your objects. You might use them to trace a bitmap that you use for reference or as the input to some other operation that generates surfaces. Or you might derive a curve from a surface to duplicate and manipulate it to generate a second surface.

When you work with curves, most of the operations that you use involve attaching and detaching curves. A good working knowledge of curves will help you to build them efficiently, which will lead to organized surfaces that you generate from them.

## Creating Curves with the Curve Tools

While we've already been using curves in the last few exercises, we examine curves a bit closer here. Maya has two different methods for creating curves—one that creates them with CVs and another that creates them with edit points (EPs). Regardless of what tool you use, keep in mind that you want to describe your curve by using the least number of points possible. This makes it easy to edit the curves later and build efficient surfaces.

Here's how to create a curve with the CV Curve tool:

**1.** Choose Create | CV Curve from the main menu in the menu bar or from the Hotbox. You can also select this tool from the shelf under the Curves tab.

**2.** Choose Display | UI Elements | Tool Settings ❑ to display the CV Curve Settings window, shown here:

**3.** The default settings are set to draw a third-degree curve with uniform parameterization. Recall from the discussion on curve degree in the "Surface or Curve Degree" section earlier in the chapter, to create one span of the curve, you take *the degree number + 1 CV*. Therefore, for example, you know that you need to click to place four CVs to complete one span of a curve of the third degree.

**4.** Switch to the Front view and click at least four times to place four CVs in the view window. On the fourth click, you will notice that a smooth curve segment is generated. You can continue to click and place as many CVs as you want. When you are ready to complete the curve, press ENTER.

Now we'll draw a second curve; this time, we'll use the EP Curve tool. The EP Curve tool is especially useful for tracing images or snapping and drawing a straight line between two points.

**1.** Choose Create | EP Curve ❑. In the settings window that appears, click the Reset Tool button to set the tool to its default settings.

**2.** Click once in the view window to set the first point. Click again in another place to set the second point. A curve will appear. Because the EP curve is completing an entire span every time a point is placed, you need to set only two points to complete the curve.

Before completing the curve, click again—but this time, LMB-drag around in the view window. You'll see that the curve is updating to follow your mouse movement, showing you how the curve will look if you release the mouse button at various points to set a new point. When you are happy with the placement of the second point, release the LMB.

**3.** Press ENTER (RETURN) to finish the curve.

From these two exercises, you can see that while it takes fewer clicks to create a curve with the EP Curve tool, you don't have as much control of the curvature since the four CVs are being created for you every time you set an edit point. It is therefore advised that you use the CV Curve tool whenever you can. The curves you create will be much more efficient, because you have used all four CVs to control the curvature over each span.

# Curves on Surfaces

In addition to drawing curves on the planes in the view window, you can also draw them directly on a surface, extract them from existing curves or isoparms, create them by projecting another curve in the scene, or create them from the line of intersection between two intersecting surfaces. Most of these methods will be explored at length in the tutorials in Chapter 6.

To give you a taste of what's possible, here are some explanations for the various ways you can create curves.

### Drawing on a Live Surface

A curve can be drawn on any NURBS surface by first making the surface *live*. To do this, first select the surface on which you want to draw. Choose Modify | Make Live, or click the Make Live button on the status line. Once the surface is live, you can use any of the curve tools to draw on it.

### Duplicating Surface Curves

Any curve on a surface or surface isoparm can be duplicated and then edited on its own. To duplicate a curve on a surface or isoparm, first select it and then choose Edit Curves | Duplicate Surface Curves. While the curve is still connected to the surface through history, it becomes a separate object and can be edited accordingly.

### Projecting a Curve

Any curve in a scene can be projected onto a surface and a new curve created. To project a curve, select a curve and a surface and choose Edit NURBS | Project Curve on Surface. In this tool's Options window, you can specify whether the curve should be projected from the *active view* or along the *normal* of the surface.

### Creating Curves Along Intersecting Surfaces

A curve on a surface is created along the line of intersection between any two intersecting surfaces by selecting the surfaces and choosing Edit NURBS | Intersect Surfaces.

### Creating Curves on Surfaces Using Other Tool's Options

Tools such as the Circular Fillet tool have an option that allows you to create a curve on a surface when the fillet operation is performed.

## Attaching and Detaching Curves

Attaching and detaching are probably the most common operations you'll use when working with curves. To attach curves, you connect or blend any two curves into a single curve. The Attach Curves command attaches two curves together based on a few settings in the Attach Curves Options window, shown in Figure 5-18. You can use two methods of attaching: connecting and blending. With the Connect option set, the end of one curve is moved to match the end of the other curve and the curves are attached. This results in a C0 continuity between

**FIGURE 5-18** *The Attach Curves Options window*

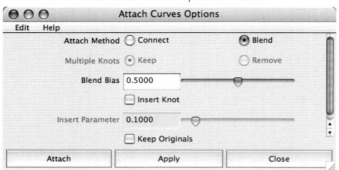

the resulting spans. When the Blend option is selected, the curves are blended together with C2 continuity. Each curve moves a certain amount to blend with the other based on the Blend Bias attribute setting. The default is 0.5000; this means that the ends of both curves move an equal amount to meet and attach. Another option allows you to keep the original curves after the operation.

To attach two curves:

**1.** Draw two curves with either the CV Curve tool or the EP Curve tool.

**2.** Choose Edit Curves | Attach Curves ❑ to bring up the Attach Curves Options window.

**3.** Choose Edit | Reset Settings to restore the tool to its default settings. Then uncheck Keep Originals. Click Apply to attach the curves. The two curves will be attached at the two closest end points and will be blended with a bias of .5, so that each curve is moving the same amount to attach. Figure 5-19 shows the curves before and after attaching.

You can edit the attributes of the resulting attached curve to change which ends were attached. Here's how:

**1.** With the attached curve selected, select the attachCurve1 node in the Channel Box.

**FIGURE 5-19** *Two curves before and after attaching*

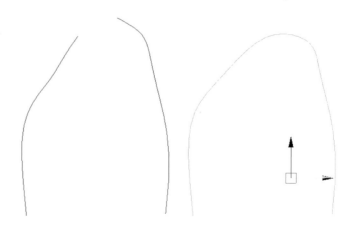

**2.** The first two attributes in the list, Reverse1 and Reverse2, control which points are connected. Select the field for the Reverse1 attribute and type **on**.

**3.** Look at the result in the view window. The attached end of the first curve chosen before the attach operation has been reversed.

**4.** Change the setting for the Reverse2 attribute. This completely reverses the attached ends from their original orientation. The result is shown here:

You can also choose any two points on a curve to attach:

**1.** Select and RMB-click the first curve, and then choose Curve Point from the marking menu. Choose any point on the curve by clicking it.

**2.** SHIFT-click the other curve to select it. SHIFT-clicking keeps the first curve selected as you add the new selection.

**3.** RMB-click the second curve and choose Curve Point. SHIFT-click the second curve to select a curve point.

**4.** With a curve point selected on each curve, choose Edit Curves | Attach Curves, or click Apply in the Attach Curves Options window if it is still open. The curve now attaches at the selected points.

Detaching a curve involves choosing a point on a curve that marks the spot where the curve will be detached into two separate pieces. The only option available in the Detach Curves Options window is one to keep the original curve. To detach the curve, select it, RMB-click it, and choose Curve Point. Choose Edit Curves | Detach Curves. The curve detaches at the selected curve's point. The result is two separate curves.

While this process is basic, selecting points (or isoparms), attaching, and detaching are integral parts of NURBS modeling.

## Cutting and Filleting Curves

When a high degree of precision is needed in creating curves, the attach and detach method may not give you the control you need for creating rounded blends between curves. In these situations, a *fillet* can provide the precision you need. A curve fillet creates a curve between two existing curves and has C1 continuity at both curves.

To use the Fillet Curve command (Edit Curves | Fillet Curve), you first select two intersecting curves. Two different construction methods can be used, Circular and Freeform, which can be set in the Fillet Curve Options window, shown in Figure 5-20. The Circular option builds a circular-shaped fillet according to the size indicated via the Radius setting. When the Freeform option is selected, the point at which the fillet begins on each curve can be set by editing the Curve Parameter attributes on the filletCurve node once the fillet has been created (in other words, you cannot set the Curve Parameter until after the fillet has been created). The Fillet Curve Options window also lets you control the depth of the fillet, or how round the fillet is, and the bias, which sets how the fillet will be weighted toward one curve or the other (similar to the Bias attribute for the Attach command).

**FIGURE 5-20**    *The Fillet Curve Options window*

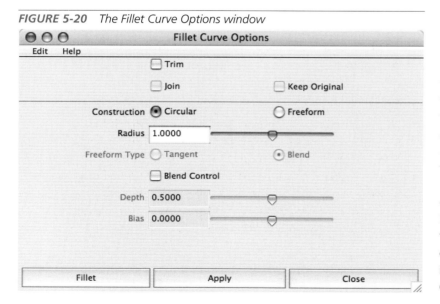

You can also choose to trim the curves at the points where the fillet intersects by choosing the Trim option at the top of the window. If the Join option is enabled, the curves will be attached together after they are trimmed. If the Keep Original option is enabled, the curves will be duplicated before the command is executed. Most of the time, it is best to leave Keep Original disabled.

# Tutorial: Modeling a Teacup with NURBS

In this section we practice creating curves that will be used to generate a teacup surface. In the following lessons, we will build some other props that are related to a table setting.

# Create the Source Curves for a Teacup

As you learn about cutting and filleting curves, you will combine your existing knowledge of drawing, attaching, and detaching curves to create the profile of a teacup. We will use this same curve to generate a surface in the next section. For now, we'll draw the curves.

**1.** Use the CV Curve tool or EP Curve tool to create a curve that resembles the profile of a teacup. Start at the origin and move up. See the adjacent illustration for help:

**2.** Choose Edit Curve | Offset | Offset Curve on Surface. This creates a parallel curve.

**3.** At the top end of the curves, use the EP Curve tool to create another curve that intersects both of the profile curves, as shown in the illustration below.

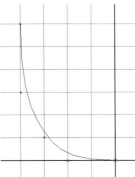

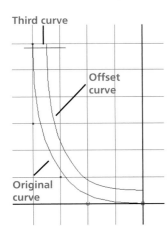

**4.** With both curves selected, choose Edit Curves | Cut Curves. The intersecting curves are cut into multiple pieces. At this point, you could select any two of these curves and use the Attach Curves tool to connect or blend them.

## Fillet the Curves

Most intersecting curves or surfaces have some kind of transition or continuity between them so that we get that nice round edge that will catch the specular highlights when the generated surfaces are rendered. Here, we'll create a simple transition between two of these curves. Using Blend for the Blend Attachment Method attribute in the Attach Curves tool may not give us the amount of fine control we need. It is in these situations that the Fillet Curve tool comes in handy.

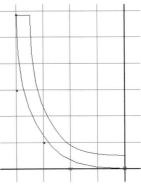

**1.** Choose two of the resulting curve segments and delete the pieces of the new curve that extend beyond the intersections of the existing two curves, so that you have something that resembles this example:

**2.** Before using the Fillet tool to create a round corner where these curves meet, it is important that you check the direction of the curves. Maya will attempt to create this new curve in the angle of the U direction. For this reason, you may need to select one or both curves and choose Edit Curves | Reverse Curve Direction. Figure 5-21 shows the curves before and after the fillet has been created. Notice the direction of the curves in the image on the right.

**FIGURE 5-21**  *Curves with proper curve direction (left) and a circular fillet created between the two intersecting curves (right)*

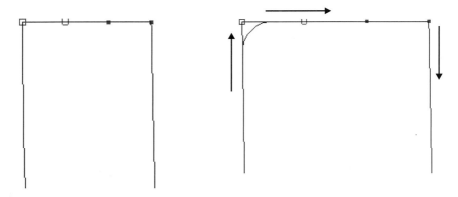

**3.** Select both curves and choose Edit Curves | Fillet Curve ☐.

**4.** In the Fillet Curve Options window, reset the tool's settings to use the defaults and click Apply. You should see a new circular curve that transitions between the two existing curves.

**5.** If the operation failed, it is either because the curves were going in the wrong direction or the Radius attribute in the Fillet Curve Options window was set too large. Undo the fillet operation and try decreasing the size of the radius.

You may often want to attach the fillet to two other curves to make one continuous shape. You can approach this in a few different ways. If you want to continue along the path, you can use the Intersect Curves tool to determine the exact point where the two curves come together:

**1.** Select one of the original curves and the fillet curve. Choose Edit Curves | Intersect Curves. A marker appears at the point where the two curves intersect.

**2.** RMB-click the original curve and choose Curve Point from the marking menu.

**3.** Click the Snap To button on the status line, or press and hold down the V key as you click and drag on the curve until the curve point snaps to the intersection marker.

**4.** Choose Edit Curves | Attach Curves □. From the Attach Curves Options window, set the Attach method to Connect. Click Apply.

**5.** In the Channel Box, adjust the Reverse attributes for the attachCurve1 node, until you see the effect you want, as shown here.

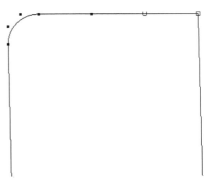

To make this process easier, you can save some steps by using the Trim and Join options in the Fillet Curve Options window before you create the fillet. Here we'll undo the operations we performed on the curves and use this method. The Trim option trims away the parts of the curve not used in the end result, and the Join option attaches the fillet and original curve together.

**1.** Undo all of the preceding operations to just before you used the Fillet Curve tool so that you're back to the two curves intersecting with no fillets.

**2.** Choose Edit Curves | Fillet Curve □ and turn on the Trim and Join options by checking the appropriate boxes in the Options window. Click Apply. The fillet is created, the original curves are trimmed at the intersection points, and the remaining curves are attached.

# Surfaces

While curves play an important role in the modeling process, the surfaces generated from the curves are rendered to create the actual object. Surface generation in Maya begins with one or a series of source curves. These curves are used as the input objects for one of the surface-generating tools. These surface-generating tools, or modeling operations, then use their specific algorithms to generate a surface based on the attribute settings of the tool.

For example, suppose you lofted five curves together. Maya uses the five curves as the input data for a loft operation. The result is a lofted surface. This process can be illustrated by graphing the lofted surface's input connections in the Hypergraph, as shown in Figure 5-22.

In this section, we'll discuss some of the more common surface-generating commands and tools.

**FIGURE 5-22**  *The Hypergraph displaying the dependencies of a surface created by lofting five curves*

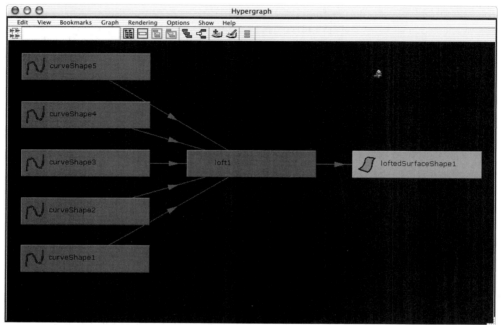

## Revolve

The Revolve command is perhaps one of the simplest of the surface commands in Maya because it needs only one curve to be able to calculate the surface. Any round, symmetrical surface is a perfect candidate for the Revolve command. In this example, we'll use the curve we created in the last section as the input for the revolve operation to make a teacup.

1. Open the scene containing your teacup profile or use the file supplied on the DVD with this book.

2. In the Perspective view, select the curve, and then choose Surfaces | Revolve ❑ to open the Revolve Options window.

3. Choose Edit | Reset Settings to restore the tool to its default settings. Click Apply. The result should look like the surface shown in Figure 5-23.

4. Press the 6 key to display the surface in shaded mode. Press the 3 key to turn up the display smoothness. Note that it's always a good idea to look at a surface with these

settings turned on as soon as it is created. You will usually be able to spot problems right away and know whether you chose the appropriate settings in the surface tool's options to get the desired result.

**FIGURE 5-23**  *The revolved curve produces a teacup.*

5. Because the revolved surface is dependant on the original curve's shape, editing the curve affects the resulting surfaces. To see this better, select the surface and move it along the X axis so that it sits next to its input curve.

6. Use the RMB-click marking menu to go into component mode, select a control vertex on the curve, and then use the Move tool to change its position in the X or Y direction, as shown in Figure 5-24. Notice what happens to the surface as you tweak the CV; it updates to fit the shape of the curve.

**FIGURE 5-24**  *Changing the shape of the input curve changes the shape of the resulting surface.*

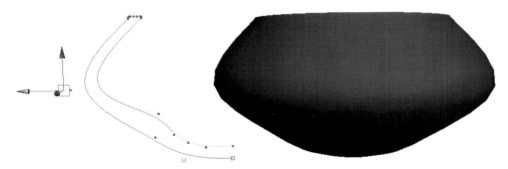

7. Undo any changes that you made to the CVs.

8. Select the revolved surface and choose Edit | Delete by Type | History. This deletes all upstream connections to the surface—including the connection to the original profile curve and the revolve node. If you try to edit the curve again, it will have no effect on the surface. If you select the surface and look at its attributes in the Channel Box, you'll notice that the revolve1 node and all of its attributes are gone.

**9.** Save this scene. Choose File | Save and name it **teacup_surface.mb**.

## Extrude

Another quick way to generate a surface with few input curves is to use the Extrude command. This is a different operation than the Edit Mesh | Extrude command that we have used in previous chapters. With this command, a profile curve can be extruded by any amount along its normal, or it can be extruded along a given path. The method used can be specified in the Style setting in the Extrude command's Options window. Other options can be set to determine whether the object should be extruded from its position or the path's position.

In this example, we will use the Extrude command to create a tube to use as a handle for our teacup. Instead of setting the options in the command's Options window, however, we will extrude the object with the default settings and then edit the Extrude node's attributes in the Channel Box.

**1.** Open the teacup_surface.mb file from the DVD or use the file that you created earlier. Switch to a Side view. Choose the CV Curve tool with the default settings. Draw a curve that resembles a handle on a teacup. It should look something like this:

**2.** Choose Create | NURBS Primitives | Circle. Move the circle out from the center so that you can see it. Press the R key to choose the Scale tool and scale the circle down and make it more of an elliptical shape. Select the ellipse, and then select the path curve. Choose Surfaces | Extrude ▢ to bring up the Extrude command's Options window. Reset these settings to their default by choosing Edit | Reset Settings, and then click Extrude.

**3.** The result may appear strange at first, but we will correct it by editing the attributes in the Channel Box. With the extruded surface selected, go into the Channel Box and select the extrude1 node to view its attributes.

**4.** Set Fixed Path to **on** and Use Component to **Component Pivot**. (This is equivalent to setting Result Position to At Path and Pivot to Component in the Options window prior to the extrude.) The result should look like the surface shown here:

# Loft

If a 3D modeler had to choose one surface-generating tool that he or she could not live without, it would be the Loft command. You select any number of curves to be used as cross sections and then use the Loft command to create a surface between them. The resulting surface will pass through each of the curves in the order they were chosen. Let's use the Loft command to model a napkin to go along with our teacup:

**1.** Use the CV Curve or EP Curve tool to draw a curve with at least three spans.

**2.** While it is not required that the curves have the same number of CVs for a loft operation, having matching parameterization will ensure more predictable results. For this reason, we'll duplicate the curve three times. You may edit each curve individually. In this case, we tucked the end CVs under the curve so that this surface will have the appearance of some thickness.

**3.** The napkin we'll build will be folded over, so duplicate them again and move them down. Also, you can scale them in the negative Y direction so that the folds we modeled into the curves will remain on the inside of the surface once it is generated. Refer to Figure 5-25 to see how the curves are placed.

*FIGURE 5-25    The curves are placed and ready to be lofted.*

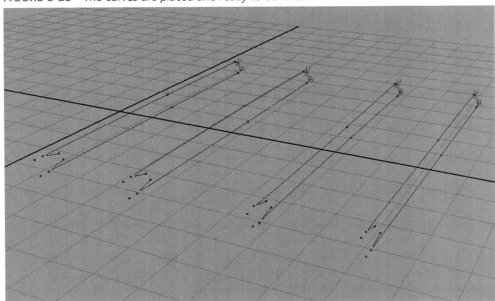

FIGURE 5-26   *A cloth napkin is created by lofting the curves.*

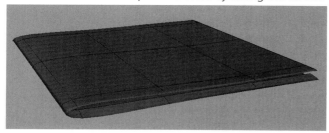

**4.** Select each curve in the order that you want the loft to flow. Choose Surfaces | Loft.

The surface is generated, as shown in Figure 5-26. Again, because the curves are still part of the surface's construction history, any edits made to the curves will be reflected in the surface.

## Birail

Birailing works by blending one or more "profile" curves along two "rail" curves. Because birailing allows you to define all four sides of a surface, as well as any shape along that surface, it is capable of generating the most complex shapes with the fewest number of input curves. (The Loft tool could be used to mimic surfaces generated from the birail tools but might require many source curves, and the results may not be as precise.)

Maya has three different birail tools available in the Surfaces | Birail menu: Birail 1 Tool, Birail 2 Tool, and Birail 3+ Tool. While all of these operations require exactly two rail curves, the difference between them is the number of curves that are used as profiles. Its simplest implementation, the Birail 1 tool, works by sweeping just one profile curve along two rail curves. The following tutorial demonstrates a great workflow for using the birail tools to create very complex surfaces. Let's create a teaspoon using the birail commands:

**1.** In the Top view, use the CV Curve tool or the EP Curve tool to draw a curve that defines the side of a spoon shape.

**2.** Mirror this curve by duplicating it and then setting its Scale X value to −1. These two curves will be used as the rail curves.

**3.** Select the Move tool from the toolbox or press the w key and move the duplicate curve along the X axis. You now have two curves that will be used as the rail curves for the birail operation. The curves are shown here:

**4.** Now to create the profile curve. One of the requirements of all the birail tools is that the profile curves need to intersect both of the rail curves. For this reason, we will snap the first and last edit points of the profile curve to the edit points at

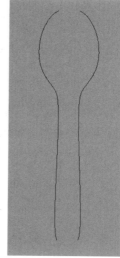

the end of each of the two rail curves. Select both of the rail curves and choose Display | NURBS | Edit Points. This will turn on the display of the edit points and enable you to snap to them when you draw your profile curve.

**5.** Choose the EP Curve tool (Create | EP Curve Tool). Hold down the v key to turn on point snapping and click one of the edit points on one of the rail curves. This snaps the first point of the profile curve to the end of the rail curve. While still holding down the v key, click the edit point at the end of the other rail curve. Press ENTER (RETURN) to complete the curve.

**6.** We now have a one-span NURBS curve whose ends are snapped to the rail curves. Now that the ends are established, we can edit the CVs of the curve to give it more shape. First we'll rebuild the curve so that it has some more CVs for us to work with. Choose Edit Curves | Rebuild Curve ▢. We will use the Rebuild Curve command to rebuild this profile curve to have two spans instead of the one it has now. Set the Rebuild Curve Options window to match the illustration and then click the Rebuild button.

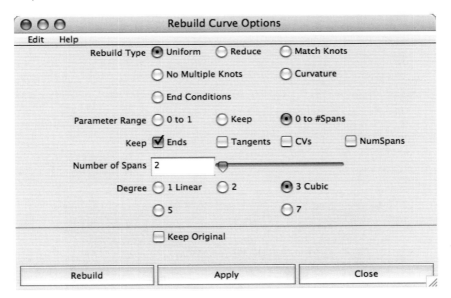

**7.** Shape the curve so that it creates a round end of the spoon.

**8.** Choose Surfaces | Birail | Birail 1 Tool. The birail tools always require that you select the profile curve(s) first and then the rail curves. In the Perspective view, click the profile curve to select it. Now click one of the rail curves and then click the other rail curve.

Press ENTER (RETURN) to complete the birail operation. The illustration shows the resulting surface:

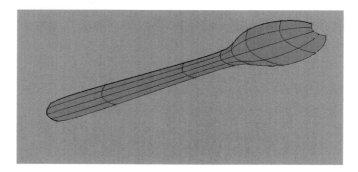

Congratulations! You've successfully generated a surface with the Birail 1 tool. However, it still does not really look like a spoon. The real power of birailing is when multiple profiles are used in the birail operation:

**1.** Select the edge isoparm at the other end of the surface and choose Edit Curve | Duplicate Surface Curve. A curve is created from that isoparm. Delete the birail surface. We can now shape this curve to match the profile at the other end of the spoon. Because this curve was created from an isoparm, we can be sure that it intersects the rail curves. Just be careful not to move the end CVs.

**2.** When using two profile curves, you need to use the Birail 2 tool (Surfaces | Birail | Birail 2 Tool) to generate the surface. Select each profile curve first, and then select the two profile curves and press ENTER (RETURN). The result would look something like this:

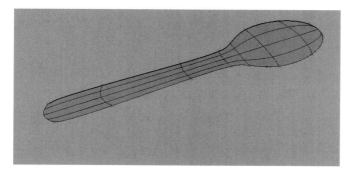

As you can see, the shape of the first profile curve is blended into the second profile curve as it is swept along the two rail curves. While this results in an even interpolation between these two curves, we want to add some depth to the spoon at different points over its length. To fix this, we'll just add some more profile curves by duplicating surface curves on the surfaces and reshaping them.

**3.** Choose Surfaces | Birail | Birail 3+ Tool. Select each of the profile curves. Then select the two rail curves. Press ENTER (RETURN). A new surface will be generated based on the profile curves and should look like this:

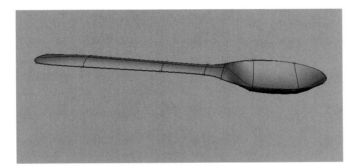

*Note:*
*For Birail 3+ Tool,*
*Select Profile curves,*
*Press Return, then*
*Select the two Rails*

While birailing is by far the most powerful of the surface-generating tools, beginners often have problems with it. Most of the time, any errors are a result of the profile curves not intersecting the rail curves. If you are having trouble with the birail tools, check to make sure that all of the curves intersect.

# Summary

In this chapter, you learned some of the basic principals behind NURBS modeling. We have covered curve creation and editing as well as some of the more common surface-generating commands available in Maya. In Chapter 6, we will demonstrate how to use these same procedures to build some very complex models.

# Advanced NURBS Modeling

**Now that you have some understanding** of what NURBS are, it's time to practice using them by building some models. In this chapter, we will build two hard-surface (nonorganic) models using two different approaches. We'll start off by modeling a wheel rim. The initial surface will be generated using some of the techniques demonstrated in Chapter 5. Details are then added by trimming regions of the surface away and creating fillets to produce nice, round edges.

The second project will lead you through the process of building a car body. This will incorporate all of the techniques that you practiced in the teacup and wheel rim models. However, the focus of the car tutorial will be how to construct the topology of the model with multiple patches that have matching parameterization.

Let's begin by talking about what trimmed surfaces are.

# Modeling with Trimmed Surfaces

Modeling with trimmed surfaces is a quick way to add detail to a NURBS surface. *Trimming* involves creating a curve on a surface by drawing it, projecting it, or using a line of intersection from another intersecting surface. This curve defines the region on the surface that will be kept or deleted once the trim operation is complete. *Surface fillets* are common in these kinds of models and can be used to generate continuous surfaces that blend between these edges, creating nicely rounded transitions.

Figure 6-1 depicts a plane and a cylinder blended together with a circular fillet, with the inside parts trimmed away.

What makes trimmed surfaces so special is that just about any detail can be added to a surface regardless of its topology. This leaves the modeler free to experiment with the form and design of a model rather than wrestle with the technicalities of surface topologies. A hole can be trimmed away, have its edges rounded and its shape changed, and be moved to different locations on the surfaces very interactively. If an idea isn't working, the surface can simply be untrimmed. These advantages make modeling with trimmed surfaces the workflow of choice for industrial designers.

**FIGURE 6-1**   *A circular fillet creates a very precise transitional surface between a plane and cylinder.*

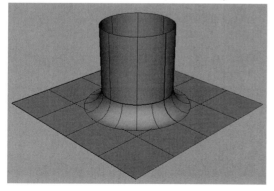

While trims offer a huge advantage in flexibility, they are not without their flaws. For one, the internal calculations that Maya uses to maintain the continuity between the trims and fillets are extremely computationally expensive. For example, if your surface is deformed in any way, trims and the fillets between them will slow down your computer while their new relationship is being calculated and rendered onscreen. Trimmed surfaces are therefore better suited for nondeforming hard-surface models. Another disadvantage is that a trimmed edge will possibly require a lot of geometry in order to tessellate (facet) the model so that the seams between the trims and fillets are not visible when rendered. Since a trimmed detail has no relationship to parameterization of the surface, Maya has no sound method for matching the tessellation at the edges. The solution is to tessellate the model densely around the edges until the artifacts are no longer noticeable. This equates to longer render times.

It is because of these issues that many animation studios choose to use a patch-based approach when it comes to using NURBS. But before we turn that corner, let's use trimmed surfaces to build a wheel rim.

# Tutorial: Modeling a Wheel Rim

In this tutorial, we'll build a model of a wheel rim with trimmed surfaces. We'll build the model based upon a photographic reference that we'll use as an underlay. We'll create a profile curve and revolve it to form the primary surface for the model. We'll project additional curves onto this surface to define regions that we'll then trim away. We'll derive additional surfaces from the trim edges and create round fillets between them. Once the modeling is complete, we'll duplicate the sections to produce the finished model.

### Set the Working Project

As always, the first step in any new project is to *set* the project from the File menu and create a directory structure that will help you organize the project's assets. Use the same procedures demonstrated in Chapters 3 and 4 to set up a project directory. We'll call this project **mcr8_ch06** and name the scene **mcr_mdl_rim_01**.

### Set Up the Underlays

We will use one photographic image as the underlay reference for this model. Place the wheel_rimRef.tga file found on the DVD in the sourceimages directory in the current project directory. Then import this image into an image plane in the Side view using the procedures shown in Chapter 4, in the section "Import Image Planes."

### Generate the Primary Surface

The initial surface will be created by revolving a curve that defines the profile of the rim in the Front view. Since we don't have an underlay available for the Front view, it is suggested that you lay in a polygonal cylinder and scale it to match the underlay and then scale the width (the Z axis) to roughly guess the proportions. This will provide a 3D template for drawing the profile curve.

**1.** Create a polygonal cylinder (Create | Polygon Primitive | Cylinder). Rotate and scale it up so that it matches the underlay in the Side view. This shape will be used only as a blocking reference, so you should also place it in a new display layer and set its property to template. The following illustrations show the cylinder once it's been placed in the Side and Perspective views:

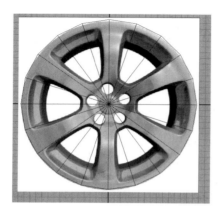
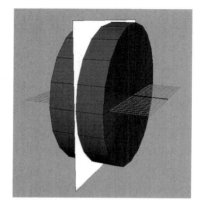

2. In the Front view, use the CV Curve tool (Create | CV Curve Tool) to draw the profile curve. It will be helpful to use the four-view layout so that you line up the details of the underlay in the Side view as you draw it in the Front view. Any hard edge should be defined using three CVs. The last two CVs, the CVs that will exist at the pole of this surface, should be created in a straight line. By holding down the SHIFT key while clicking to set a point, you can constrain the placement of a point to a right angle from the last point drawn. Press ENTER (RETURN) to complete the curve. The resulting curve is shown here:

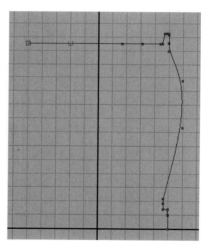

3. Select the curve and choose Surfaces | Revolve ❑ to open the Options window for the Revolve command. Set the Axis Preset attribute to X so that the curve will be revolved around the X axis. Set the End Sweep Angle to 30. This will revolve the profile only 30 degrees. While it's true that the curve should revolve an entire 360 degrees to close the surface, we want to work on just a fraction of the rim and duplicate it. Not only will this be faster to model, but it will ensure that the details are exactly the same throughout the model. Finally, set the Segments attribute to 4 so that there are four spans in the V direction. This should give us enough detail for the surface to tessellate properly at render time. When these options are set, click the Revolve button.

*TIP* **When approaching a model, always look for symmetry. Being able to mirror or duplicate pieces is a great way to make your model faster.**

## Create Trim Curves

In this section we will create the curve that will be projected onto the surface and used to define the region that will be trimmed away. To create curves that flow with the surface, we will derive two of the curves from the actual surface's isoparameters.

**1.** RMB-click the surface and choose Isoparms from the resulting marking menu. Click and drag an existing isoparm to the outer bounds of the cutout of the rim to match the underlay. Hold down the SHIFT key and drag another isoparm to the inner bounds of the cutout. Your selected isoparms will be indicated in yellow and should match this illustration:

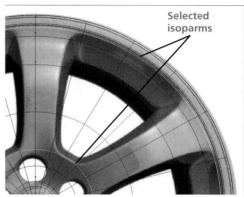

Selected isoparms

**2.** Although we have the selection we need, it is not possible to perform editing operations directly on isoparms. To work around this, we will duplicate these isoparms and convert them into actual curves. Choose Edit Curves | Duplicate Surface Curves. Two new curves will be created in the scene that match the selected isoparms.

**3.** Since we will eventually be projecting these curves, and we still need to connect them, it is best to flatten them before performing any more operations on them. A very useful technique for flattening curves is to scale them to 0 along the axis you wish to flatten them in. In this case, select both curves and set the Scale X values to 0 in the Channel Box. This will flatten them and place them directly on the YZ plane.

**4.** Use the CV Curve tool to draw a curve in the other direction that will follow the image in the underlay and bridge the two curves together. Draw this curve so that it overlaps the other two curves just slightly (shown on the left in Figure 6-2).

**5.** Once the curve has been created, select all three curves and choose Edit Curves | Cut Curve. This will cut all of the curves at their point of intersection. You may now delete the leftover segments of the curves so that you have something that looks like the curves shown on the right in Figure 6-2.

**FIGURE 6-2** *The trim profile curves are drawn (left) and then cut. The extraneous segments are deleted (right).*

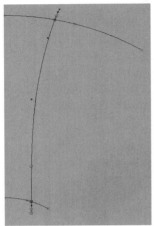
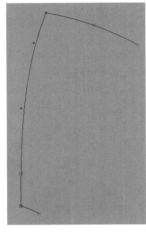

6. Any time a curve (or surface) is cut or detached, it is really important to reparameterize all of its coordinate values so that it starts at zero and ends with a whole number indicating the number of spans. This is done by using the Rebuild Curve command. With all of the curves selected, choose Edit Curves ☐ to open the Curve Options window. Choose Edit | Reset Settings in this window's menu bar to reset the command to its default settings. Set the Parameter Range attribute to 0 to #spans. Enable the NumSpans option for the Keep attribute. Click the Apply button.

Clicking the Apply button instead of the Rebuild button will execute this command with these settings but leave the window so that we can use it again.

7. To create a nice, even, round transition between the curves, we will use the Attach Curves command with the Attach Method set to Blend. Due to the way attaching works (see "Attaching and Detaching Curves" in Chapter 5) we need to increase the number of spans on the middle curve for it to hold its shape and better conform to the underlay. Select the middle curve and, in the Rebuild Curve Options window, disable the NumSpans check box. Now you will be able to enter the desired number of spans in the Number of Spans field. Enter a value of **4**. Click Rebuild to execute the command and close the Options window.

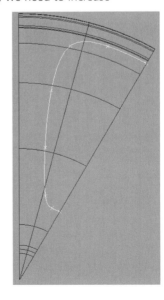

8. Attach the three curves together. This must be done two at a time. Select the outermost curve and the middle curve and choose Edit Curves | Attach ☐. Make sure the Attach Method is set to Blend and Keep Originals is disabled. Click the Attach button. Now select the resulting two curves and press the G key to execute the last command. Choose Edit | Delete by Type | History to delete all dependencies. The resulting curve and underlying surface should look something like the image here.

9. Since the rim surface only has one span in V, use the Rebuild Surface command to rebuild it to have four spans in V.

## Trim the Surfaces

Now we can begin to see the fruits of our labor come together as this curve is projected onto the surface to define the region that will be trimmed away. A surface is trimmed by first selecting the Trim tool (Edit NURBS | Trim Tool). The Trim tool is used to either keep or delete a selected region of a surface. This region is usually defined by an enclosed curve on a surface created by projecting curves, intersecting surfaces, fillets, or curves drawn directly on surfaces. Once the Trim tool is activated, select the surface you wish to trim. A white wireframe display indicates that the surface is in trim mode. At this point, you select the part of the surface you want to discard or keep. Exactly what happens to the selected region can be set in the Trim tool's Options window. If you make a mistake or want to make a change, any trimmed surface can be untrimmed by choosing Edit NURBS | Untrim Surface.

To create the curve-on-surface needed to use the trim operation, we will project the curve we created in the last section onto the surface. Let's begin:

1. By default the Project Curve on Surface command will use the active view panel to determine the direction of the projection. So, to accurately project this curve on the surface, switch the view window to the Side view.

2. Select the curve and the surface together. It does not matter which one you select first. Choose Edit NURBS | Project Curve on Surface. Switch to Perspective view and look at the surface. You should notice that the curve is now lying directly on the surface. With this in place, we are ready to trim.

3. With nothing selected, choose Edit NURBS | Trim Tool ❑. The Trim tool's Trim Settings window (shown in Figure 6-3) appears on the right-side panel of the interface. Choose Set the Selected State attribute to Keep. This tells the Trim tool that the region you select will be kept, and the others will be discarded. You can, of course, change this to be the other way around by choosing Discard.

**FIGURE 6-3**   *The Trim tool's Trim Settings window*

| Trim Tool | Reset Tool | Tool Help |
|---|---|---|

▾ **Trim Settings**
    Selected state: ⦿ Keep      ○ Discard
            ☐ Shrink surface
    Fitting tolerance: 0.0010

▾ **Standard Options**
          ☐ Keep original

4. In the view window, your cursor should take the form of an arrow to indicate that the Trim tool is

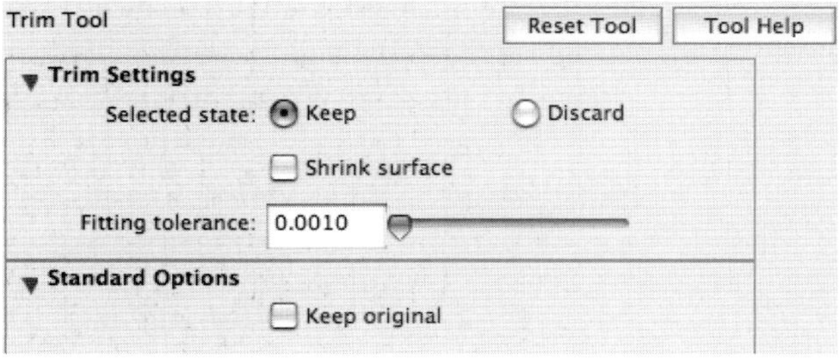

active. Select the surface. It changes its display to a white wireframe with the regions visible. Click the area of the surface that you would like to keep, and a yellow marker appears at that point. Figure 6-4 (left) shows the surface in trim mode with the marker set on the region to keep. If there were multiple regions that you'd want to keep, you could continue to click them, placing the markers. Press ENTER (RETURN) to complete the trim operation.

**FIGURE 6-4**  *The surface on the left is displaying in a white wireframe with the region to keep indicated by the marker. To the right is this surface after the trim operation is complete.*

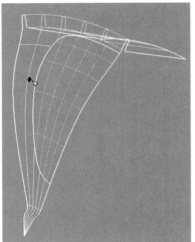
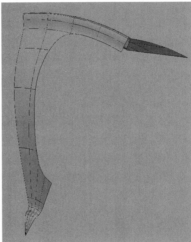

**5.** Now that this is a trimmed surface, a new component type is available when you right-click the surface. To create the inset, right-click the surface and choose Trim Edge from the marking menu. Select the edge on the surface created from the trim operation. It will highlight in yellow.

**NOTE**  *A trim edge is very different from an isoparm. Because it is not related to the surface's parameterization, you cannot attach or align another surface to a trim edge. The only way to calculate continuity at a trim edge is to use a fillet tool.*

**6.** Choose Edit Curve | Duplicate Surface Curve. The trim edge is converted into a curve.

**7.** To accurately place this curve to match the underlay and have continuity with the eventual duplicated pieces, we need to scale this curve down while placing it precisely along the axis of symmetry. We do this by snapping one of the end edit points to a guide curve. In the Side view, select the duplicate surface curve and choose Display | NURBS | Edit Point.

**8.** With the edit points now visible, you are able to use the point-snapping feature. First we'll move the pivot point of the curve to one of its end points. Press the INSERT (HOME) key to enter the pivot editing mode. Now hold down the V key to enable point snapping and drag the pivot point manipulator to the end edit point of the curve.

*TIP* *Holding down the D key lets you edit the pivot point of any transform node. This is a lot faster than toggling the INSERT (HOME) key.*

**9.** Although you can scale this curve down from this new pivot, moving it into place will pose a problem because the axis of symmetry that we are working with does not relate to the world axis. To remedy this, we'll create a guide curve. Choose Create | EP Curve Tool. Hold down the C key to toggle on curve snapping. Click the corner at one end of the trim edge on the surface and then, still holding the C key, click the other end. To make sure your points are exactly at the corner, click along the edge and drag toward the corner while curve snapping.

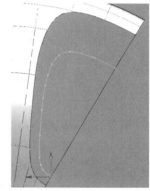

**10.** Select the duplicate curve, move the pivot on top of the guide curve, hold down the C key, and LMB-click and drag until the pivot snaps to the guide curve. Then, continue dragging along the guide curve until the curve is placed to match the underlay. Finish matching the reference by scaling it down. The result should look like the adjacent image.

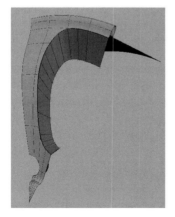

**11.** Move the curve back along the X axis. Now select the curve and then RMB-click the surface, choose Trim Edge, and select the trim edge on the surface while holding down the SHIFT key. You should have two curves selected now.

**12.** To build a surface between these two curves, we'll use a lofting operation. Choose Surface | Loft. The result should look like the image at left.

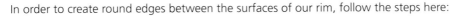

## Round the Edge

We mentioned the importance of beveling or rounding edges back in Chapter 3 when we modeled the church. A round edge adds a high degree of realism to your models. It not only helps in creating a gradation between two surfaces or faces but also catches specular highlights and helps define the form of the model when rendered. The NURBS toolset contains some amazing tools for creating round edges. The two that we will use here are the Round tool and the Circular Fillet command. Both of these tools create a transition surface that maintains C1 continuity between two existing surfaces. These transitional surfaces are known as fillets.

In order to create round edges between the surfaces of our rim, follow the steps here:

**1.** Choose Edit NURBS | Round Tool. LMB-click and drag over the overlapping edges of our trimmed and lofted surfaces. A yellow indicator appears that displays the radius of the surface that the Round tool will create once it has completed. Drag one end of the indicator to shrink the intended size of the radius, as shown here:

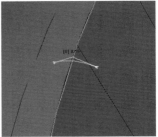

**2.** Press ENTER (RETURN) to complete the operation. A new round edge should now be generated between your two surfaces.

**3.** To better evaluate the edge, add a shiny material. Select one of the surfaces, RMB-click, and choose Materials | Assign New Material | Blinn.

**4.** In the Attribute Editor, make sure the Blinn1 tab is selected. Find the attribute called Specular Color and move the slider all the way to the right to set the color to white. This creates nice, shiny highlights on your model.

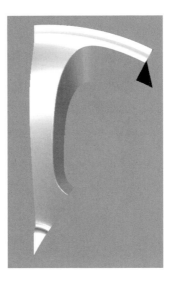

**5.** Add this material to the other objects. You can efficiently do this through the Hypershade or by right-clicking each of the other two surfaces and choosing Materials | Assign Existing Material | Blinn1. The result should look like this:

We could use this same technique to build the holes for the bolts, but instead we'll try a different approach by using the Circular Fillet tool. The Circular Fillet tool creates a nice, round transition surface between two intersecting surfaces in the same way the Curve Fillet tool creates round transition curves between two intersecting curves. One of the Surface Fillet tool's options lets us create curves on the surfaces where the fillet surface begins and ends. We can then use these curves to define the regions that can be trimmed away.

**1.** Create a NURBS cylinder (Create | NURBS Primitive | Cylinder. Rotate it 90 degrees in Z and set its End Sweep attribute to –180.

**2.** In the Side view, move the cylinder up in the Y direction and scale it to match the holes in the underlay.

**3.** In the Perspective view, move the cylinder along the world X axis until it intersects with the rim surface.

**4.** Select the rim surface and then the cylinder and choose Edit NURBS | Surface Fillet | Circular Fillet ▢ to open the Circular Fillet Options window, shown in Figure 6-5. Check the Create Curve on Surface check box, and set the Radius to **.2**. Click Apply to create the fillet between the surfaces.

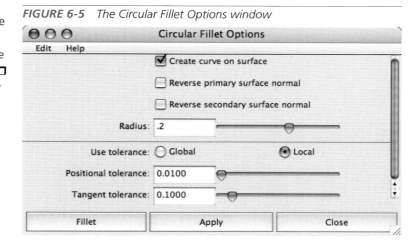

**FIGURE 6-5**   *The Circular Fillet Options window*

**5.** In this case, the fillet is created in the wrong direction. The Circular Fillet tool will create the fillet between the two surface normals' facing direction. If the surface normals are facing a direction other than the intended direction of the fillet, you can use the Circular Fillet tool's options to reverse these normals during the fillet operation. To do this, first undo the fillet operation and click the Reverse Primary Surface Normal check box in the Circular Fillet Options window to reverse the surface normal direction of the fillet calculation (for the first surface selected). The fillet now faces in the desired direction.

***NOTE***   ***Another way to invert the direction of the fillet after it has been created is to edit the Primary and Secondary Radius attributes on the rbfSrf node that is created. Inverting the values to a negative value is the same as using the check box in the Circular Fillet Options window.***

**6.** Now we can trim away the parts we don't want. Choose Edit NURBS | Trim Tool ❑. Select the rim surface and then mark the outer region as the region to keep. Press ENTER (RETURN) to trim the surface. A hole will be created in the surface.

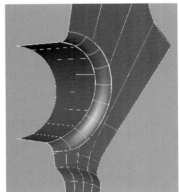

**7.** Repeat Step 6 to trim away the outer portion of the cylinder. The result is shown here:

### Duplicate and Repeat

The modeling part is all done. To finish, we'll duplicate and mirror these surfaces to the other side. This will give us one complete repeatable segment that we can then dupli-cate five more times, with each duplicate offset by a rotation value.

**1.** Select all of the surfaces and delete the history (Edit | Delete by Type | History) to break connections with any dependencies. With the surfaces still selected, press CTRL-G to group them. This group node provides us a single pivot point on which we can scale all three surfaces.

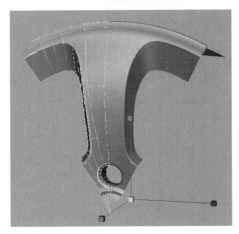

**2.** Choose Edit | Duplicate Special ❑ to open the Options window. In this window choose Edit | Reset to ensure that the Duplicate command is set to its defaults. Set the Scale X attribute to –1 and click Apply. The group will be duplicated and mirrored to produce the image shown at left.

**3.** Select all of the surfaces and group them again. Reset the Duplicate options to their defaults. We need to analyze the problem here to figure out what values we need to set to complete the rim. There are six spokes total in the rim. We have already created one spoke. One-sixth of 360 is 60, so we know that each duplicate needs to be offset 60 degrees from the previous copy. Set the Rotate X attribute to 60. Since there are six spokes total, and we have one spoke, set the Number of Copies to 5. Click Duplicate. You should now have all of the pieces to make your rim complete.

**4.** Clean up this scene by grouping all of the surfaces together in one group, naming the group **rim_grp**, and deleting all of the history. You may also want to look in the outliner and delete any unused nodes or curves. Your completed rim should look like this:

This completes the lesson on modeling with trims and fillets. As you can see, this is a very powerful modeling workflow that produces surfaces with a high degree of precision. In the next lesson, we will model the body of a car but instead of modeling with trims and fillets, we will arrange the four-sided patches to flow in the direction necessary to construct the model's details. However, this does not mean we can't still use the Trim tool. In fact, you'll learn a great technique that allows you to leverage your trimming skills into a patch-modeling workflow.

# Modeling with NURBS Patches

*Patch modeling* involves creating a network of separate, four-sided surfaces, or *patches*, and organizing them so that they flow along and define the form and details of a model. Due to the rectangular, grid-like nature of NURBS surfaces, when a radial detail needs to be introduced (as the Extrude Face command would do on a polygonal surface), the surface has to be split up into multiple patches to allow for the changing structure. The resulting model may possibly be made up of hundreds of separate patches.

This process is best illustrated in Figure 6-6. To create a cylindrical extrusion, or branch from a cylinder, quite a few steps must be taken. Because there is no such thing as extruding a face in NURBS, the structures must be constructed manually.

The initial NURBS plane (Figure 6-6a) is rebuilt and detached into nine equal patches. The middle patch is deleted (Figure 6-6b) to make room for the cylinder (Figure 6-6c). To achieve C2 continuity between the cylinder and the patches on the plane, the cylinder must be split up into four separate patches. In order for all of these pieces to contain enough geometry to maintain their original shape and match the parameterization of their adjacent surfaces, they are all rebuilt to have three spans in the U and V direction (Figure 6-6d).

Since the parameters now match, each patch on the cylinder is attached to the patch on the plane that is adjacent to it. This creates a nice, smooth transition between the pieces (Figure 6-6e). However, the patches at the corners of the plane do not have continuity to their

**FIGURE 6-6** *The process of blending a plane with a cylinder using NURBS patches*

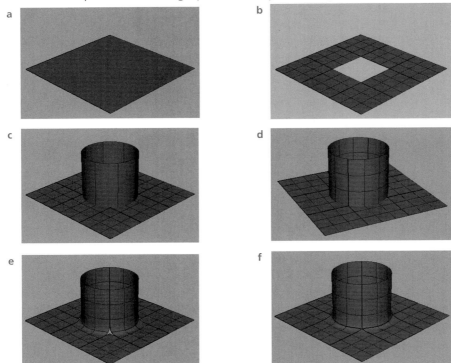

adjacent patches. To solve this, we can detach the cylinder and plane patches that we just attached and use a stitching function (Global Stitch, in this case) so that all of the patches appear continuous even though there are actually 12 separate patches total (Figure 6-6f).

Sure, creating this same shape with a circular fillet would be faster, and creating it in polygons would allow this to exist as a single shape. So what is the advantage of patch modeling when it requires so much work? Well, the advantage over a trimmed surface is that, because the patches all have matching parameterization, it is simple for Maya to match the tessellation of the geometry at render time, eliminating any gaps that frequently plague trimmed surfaces.

The advantage over modeling with polygons is that you can easily add localized detail without affecting the smoothness of a shape. There is no need to reshuffle and resculpt the vertices every time geometry is added, as we had to do in Chapter 4 when we modeled the head. Once you learn how to convert trimmed surfaces into patches, you will find that having the ability to project precise shapes and trim them away is the most ideal workflow for hard-surface modeling in an animation pipeline. Finally, if the surfaces are converted to polygons, the surfaces can be *tessellated* (converted into polygonal faces) and easily merged with the other patched surfaces adjacent to them.

The most important thing to remember when modeling these patches is that they maintain their matching surface parameterization. This allows operations, such as attaching surfaces, to behave in an organized and predictable manner, resulting in well-formed surfaces. Furthermore, surfaces with related parameterization can be stitched together to create and maintain continuity among the parts of an object.

## Tutorial: Modeling a Car Body

Car bodies can pose some of the greatest challenges to a modeler. Most cars, especially those designed in the last 15 years, are made up of some very complex surfaces. Technological advances in manufacturing and the design process have produced car bodies with very round and smooth-curving surfaces. When evaluating what type of geometry to use for a modeling task, it sometimes helps to think about how the object was designed in the first place. These days, almost all cars are designed in Alias Studio, which is really the father of Maya's NURBS modeling toolset. So for this reason alone, using NURBS to re-create a design makes sense.

While the car we build here does not have to meet manufacturing standards, there are many areas on a car that require a certain amount of precision that only NURBS can deliver. For one, the subtle, smooth-flowing surfaces of a car body really show off because they are covered in a very shiny, highly reflective material. The slightest lump or irregularity in a surface is easily visible in the model and, in a visual effects shot, can break the illusion that the car isn't real. When you add the fact that very precise details need to be cut out of the surfaces for the wheel wells, the front grill, and fuel cover, having to sculpt the vertices one by one to match the shape while at the same time not disrupting the surface that these details lie on becomes almost impossible to do without trims.

However, because we are building this model for animation purposes, we would like to avoid the overhead of rendering trimmed surfaces. One of the most powerful modeling features you will learn about in this section is the Trim Convert function of the Rebuild Surface command. As long as it obeys the four-sided nature of a NURBS surface, any trim surface can be converted into a patch. Having a patch model will make tessellating the geometry easier for our rendering engine and will also allow us to produce highly accurate polygonal approximations. We will talk more about this aspect of modeling in Chapter 7. For now, the description of the preceding challenges hopefully has convinced you that NURBS is a good geometry choice for modeling a car.

### Set Up the Workspace and Custom UI

In Chapter 4, we began customizing Maya's user interface by creating a custom shelf for polygonal modeling and created some hot keys so that we could quickly access the tools and commands we needed for our modeling job. In this section, we take the customization a bit farther by creating a custom marking menu that contains some commands. First, let us build a new shelf specifically geared toward hard-surface modeling with NURBS. We'll use this opportunity to review some of the commands we have already used and learn about some new ones.

**Create a Custom Shelf**   Create a new shelf and name it **NURBS**. The first buttons we'll make are for toggling some of the display of components for curves and surfaces:

- **Display Edit Points**   Hold down the CTRL and SHIFT keys while selecting Display | NURBS | Edit Points to create the shelf button for this command. Executing this button will toggle the display of edit points on or off of any selected NURBS curve or surface. You need to have the edit points visible in order to snap to them. Since many of the surface operations require that edit points be snapped, having this available in the shelf is very helpful.

- **Display CVs**   This command is found in Display | NURBS | CVs. Displaying CVs has a similar use as displaying the edit points. By viewing the CVs between objects, you can quickly ascertain the continuity.

- **Display Hulls**   This command is found in Display | NURBS | Hulls. Viewing the hulls gives you a good idea for how the surface is flowing. Since we will eventually convert our NURBS surfaces into polygons at the control vertex level, viewing the hulls will give us a reading on what that polymesh will look like.

We'll add three buttons to the shelf for creating curves:

- **CV Curve Tool**   The CV Curve Tool (Create | CV Curve Tool) allows you to create a curve by placing the CVs. For a third-degree curve, you need a minimum of four CVs to create a single curve segment.

- **EP Curve Tool**   The EP Curve Tool (Create | EP Curve Tool) allows you to create a curve span by span.

- **Arc Tool**   The Arc Tool (Create | Arc Tools | 3 Point Circular Arc) allows you to create a precise arcing curve of any number of spans by positioning three markers that determine the size of the curve and radius of its arc.

The next set of buttons will be used to edit and project curves. For more information about these operations, make sure you've read Chapter 5 and worked through the wheel rim tutorial from earlier in this chapter.

> *NOTE* **All of the following commands can be found in the Surfaces menu set.**

- **Attach Curves**   Choose Edit Curves | Attach Curves ❑ to open the Options window for this command. Disable the check box for Keep Originals. Keep everything else at the default settings. While there are many other useful attributes here that you very well may use in the future, beware of the Connect setting for the Attach Method. Connecting two surfaces produces a flat edge, analogous to C0 continuity. While there may be special cases where you'll need this, it is generally the thing you are trying to avoid in your model. When you have set these options, save them by choosing Edit | Save from the menu bar of the Options window and then click the Close button. Now, when you create a shelf button for this command, the button will store these exact settings.

- **Detach Curves**   The Detach Curves command (Edit Curves | Detach Curves) detaches curves at any selected parameter on a curve. You can use this with its default settings.

- **Project Curve on Surface**   The Project Curve on Surface (Edit NURBS | Project Curve on Surface) projects any curve onto a surface from the current viewing angle. You should be familiar with this command from earlier in this chapter. We will use it a lot in this car tutorial as well.

- **Trim Tool**   The Trim Tool (Edit NURBS | Trim Tool) will keep or remove a region of a surface. This is another tool that you are now familiar with. You definitely want to keep this one handy.

For this tutorial we will use these surface-generating operations:

- **Loft**   Make a button for the Loft command (Surfaces | Loft) that uses its default settings.

- **Birail 3+ Tool**   Setting up the Biral tool (Surfaces | Birail | Birail 3+ Tool) as a shelf button lets you use it as a command. By selecting all of your profile curves, and then your two rail curves last and clicking the button, Maya uses the selected inputs to generate the surface. This is much faster than using it as a tool.

- **Boundary**   When creating a shelf button or the Boundary command (Surfaces | Boundary), it is recommended that you change the default Common End Point attribute to Required. This will ensure that your curve network is connected.

**Attach, Detach, and Insert Isoparms, and Rebuild Surfaces**    Probably the most critical surface-editing commands for patch modeling are the Attach Surface, Detach Surface, and Rebuild Surface commands. Attaching and detaching surfaces is analogous to attaching and detaching curves. Make sure you review the section "Attaching and Detaching Curves" in Chapter 5. Make shelf buttons for both of these commands. You can use the default settings for both with the exception of disabling the Keep Originals setting for the Attach command.

The Insert Isoparm command (Edit NURBS | Insert Isoparm) will also be a frequent operation you'll use. Analogous to the Insert Knot command for curves, Insert Isoparm lets you add geometry to a specific area of a surface. When you select any isoparameter on a surface and execute this command, an isoparm will be created at the selected position.

The Rebuild Surface command is host to many useful functions. Most often, it is used to uniformly rebuild a surface after it has been edited so that the parameterization is uniform with a range from 0 to # of spans. Another common use for Rebuild Surface is to rebuild a surface to a set number of spans so that it matches some other surface. We saw an example of how this command can be used in the "Maintaining Parameterization While Modeling" section in Chapter 5.

As we set up this command in our custom UI, it will be beneficial to have at least two different buttons for the Rebuild Surface command—one that will uniformly rebuild a surface with the existing number of spans and another that will open up the Options window for this command. Choose Edit NURBS | Rebuild Surface ☐ to open the Options window for this command (shown in Figure 6-7).

**FIGURE 6-7**    *The Rebuild Surface Options window set to rebuild with the proper settings for hard-surface modeling applications*

Since we are always going to be using the uniform version of NURBS, the Rebuild Type should always be set to Uniform. The parameter range is 0 to #spans and the Direction is U and V. We want to enable all of the Keep functions. Keeping the number of spans will rebuild the surface so that its parameters will range from 0 to the number of spans that are currently in the surface.

However, when the rebuild function is executed with these settings, the CVs, as well as the isoparms, will be relocated along the surface to comply with a truly uniform B-Spline surface. One exception to this type of rebuilding in a hard-surface modeling application is that we probably want the CVs to stay where they are when a surface is rebuilt. This will help keep geometry in place and make the construction of edges easier. Make sure that your Rebuild Surface Options window matches the one shown in Figure 6-7. Save the settings and create a shelf button. Create another button that will open the Options window every time it is clicked by holding down the CTRL and SHIFT keys and choosing Edit NURBS | Rebuild Surface ❑.

To distinguish between the two Rebuild Surface shelf buttons, you may wish to add an icon name. To do this, click the black triangle to the left of the shelf to see the menu of items to modify the shelf. From this menu, choose Shelf Editor. Select the last item in the Shelf Contents tab, called rebuildSurfaceDialogItem, and type **opt** in the Icon Name field. This label will now appear in the shelf.

The completed shelf up to this point is shown in Figure 6-8.

**FIGURE 6-8**   *A custom shelf is created that contains the commonly used tools and commands that will be used to build the car body.*

> _TIP_   **Make sure that you select Save All Shelves from the list of items that modify the shelves. While Maya saves the shelves and other preferences when you close the application, you will lose any preferences that you set during a session if the program crashes. You can avoid any lost data by manually saving the shelves.**

**Create the Custom Commands**    To maintain uniform parameterization in a model, you must always rebuild a surface after attaching or detaching surfaces or inserting an isoparm. This will ensure that the surface is always uniform. If you do too many operations without rebuilding a surface, the underlying math will start to round numbers as it performs different functions, and the NURBS surface can become unstable. If the surface blows up, there is no undo. It's gone!

It's also a good idea to always delete history after performing any commands on a surface. In that case, create another shelf button for deleting history (Edit | Delete by Type | History).

Now, say you need to add an isoparm on a surface. Currently this takes three separate commands to properly complete the process, Insert Isoparm, Rebuild Surface, and Delete History. Imagine having to go through that every time you need to add isoparms. This can be avoided by stringing together these three commands in the Script Editor and creating a single super command that performs all three.

Open the Script Editor (Window | General Editors | Script Editor) and clear the input window. MMB-drag the Insert Isoparm button from your custom shelf into the input window of the Script Editor. Type a semicolon to end the line. Now drag in the Rebuild Surface button and place a semicolon after that. Repeat this process for the Delete History command. Your Script Editor should look like this:

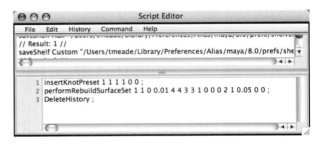

Highlight everything in the input window of the Script Editor and drag it to the shelf. Test it out on a surface by selecting an isoparm and clicking this button. The surface should insert, rebuild, and delete history. Using this same procedure, create another two super commands for attaching and detaching surfaces.

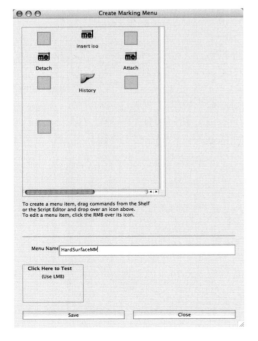

*TIP*  PLEASE do not skip this step! *Take a little time and learn to make these buttons now. It will make the modeling process so much faster and less prone to error. From here on, when the directions instruct you to attach a surface, we are actually attaching, rebuilding, and deleting history.*

**Create a Custom Marking Menu**    Now we will insert these three commands into a marking menu. Choose Window | Settings/Preference | Marking Menu Editor. Click the Create Marking Menu button and name it **HardSurfaceMM**. MMB-drag each super button created in the last section onto one of the blank button spaces in the Create Marking Menu window. You may as well add the Delete History button. Relabel the buttons by right-clicking each one and choosing Edit Menu Item. The window should look like the adjacent screen.

While there is plenty of space left to insert more commands, we suggest that you leave these as empty and spread out as possible. This makes it easier and faster to select a command. We know that if we bring up the marking menu and drag right, we will perform a detach and not have to worry about accidentally selecting another command.

Save this marking menu. Now select it in the list of marking menus in the Marking Menu window. This is where we get to assign how it will be accessed. Set the User Marking Menus in the drop-down menu to Hotkey Editor. Click the Apply Settings button and close this window.

Choose Window | Settings/Preference | Hotkey Editor. In the Categories column of this window, find and select User Marking Menus. The commands will appear in the Commands column. Select hardSurface_Press. You can assign this command to any key on the keyboard, including modified combinations. You may also override an existing hot key if you want to. For the example, we'll use M for the Key setting and choose Assign. Test the hot key in the view window by holding down the M key and LMB-clicking The marking menu appears and should look like this:

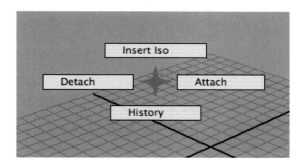

This should give us easy access to most of the tools and commands we will use for this job. Of course, you may find that you are using additional commands a lot. If so, feel free to add them to the shelf or a marking menu, or create a hot key.

## Model the Car

With our interface now properly set up, we can begin modeling the car. The approach we'll use is very similar to the subtractive approach we used for the rim: start with a base surface and cut away the details. Another approach would be an additive approach where you would draw and place every curve in the underlay and build up each patch one by one. While this is good for some models, lots of effort is put into getting C2 continuity between all of the patches. With a subtractive approach, we already have C2 continuity because the patches are detached from a continuous surface.

The modeling process commences by constructing a curve network that defines the primary flow of the body of the car. This curve network is used to birail the top of the car. A second surface is derived from the primary surface to create the side. The topology of the surfaces is reflowed to match some of the details in the car, and then the surfaces are broken down further to handle details such as the wheel wells and front grill.

**Set the Working Project**    As always, the first step in any new project is to *set* the project from the File menu and create a directory structure that will help you organize the project's assets. Instead of creating a whole new project, we'll use the same project directory that we created for the rim.

**1.** Open Maya and choose Edit | Project | Set. Using the file browser that appears, navigate to the directory called **mcr8_ch06** and click Choose. This sets the current project to this directory.

**2.** Save the scene and name it **mcr_mdl_car_01**.

 **Import the Underlays**    For this project we will use three different underlays, one each of the front, side, and top. Place the carRef_Top.tga, carRef_Front.tga, and carRef_Side.tga files (found on the DVD in the sourceimages directory) in the current project directory. Import these images into image planes in each of the corresponding views, using the procedures described in "Import Image Planes" in Chapter 4.

> **NOTE**    *Make certain that the underlays line up by using the Center, Height, and Width attributes on the imageplane node.*

**Create the Main Character Curves**    So, where to begin? The first thing you want to do when approaching a model of this complexity is to identify the main structures. It can be helpful to print your references and draw on them to plan out your structure. Look for the big picture; don't get caught up in the details. The car really has two main structures that define its form—the base of the car, which contains the front hood, side doors, and rear sections, and the canopy, which includes the windows and the top.

Since the base is the largest, we'll start there. Now, study the references and look for any lines that flow the entire length of the car. Just about every car has one line that defines the primary form. We call these defining lines *character curves*. This car is unique in that its character curves not only define the top of the car, but flow all of the way around to define the front and rear of the car as well. Two character curves are shown in Figure 6-9. One of the curves flows around the top and the other around the horizon line. Between them lies the front, top, and rear of the car.

To create curves that are very precise, we want to avoid as much human input as possible. In other words, we want to stay away from drawing these character curves point by point. These curves will be used to generate a surface that the rest of the model will be derived from. Any bumps in this curve will propagate throughout the rest of the model. To avoid this, we'll use the Arc tool.

**FIGURE 6-9** *Identifying the character curves of the car*

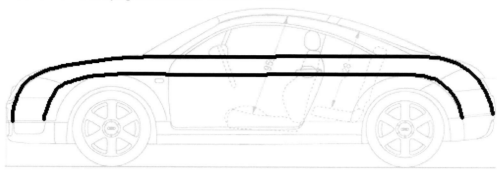

The Arc tool will generate a curve whose number of spans can be set in the tool's settings. The shape of the curve is influenced by three markers. Since it is not possible to define the entire character curve with one arc, we'll break it up into sections:

1. Starting from the Top view, choose the Arc tool from your shelf. Place the first marker at the upper-right corner of the headlight. Place the second marker near the middle where the car is widest. Place the third marker at the front outer corner of the rear taillight. Figure 6-10 shows the placement of the markers from the Top view. You can adjust the markers by clicking and dragging them. When you are finished, press ENTER (RETURN).

2. In the Side view, translate the curve upward to match the horizon line of the car in the underlay. To get the curve to arc along the vertical axis, you can adjust the markers from this view. Select the Show Manipulator tool from the toolbox. In the Channel Box, select the makeThreePointCircularArc1 input node.

**FIGURE 6-10** *The first curve is created with the Arc tool by placing three markers.*

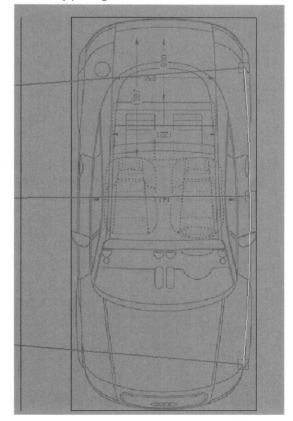

The markers reappear and are ready for editing. Move the front marker down slightly. You won't be able to match the curvature at the front but you can come close.

3. To finish the curve, adjust the two CVs at the front of the curve by moving them down so that the curve now matches the underlay, as shown here:

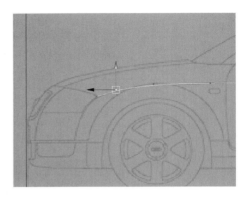

4. Use the Arc tool to create another curve around the rear of the car. Then, use the Top view to align the curve using the Move tool and the Rotate tool. Don't worry about the end of this curve matching the end of the first curve. We'll let the Attach Curve command handle that.

5. Before attaching the curves, we need to assign them with the proper amount of resolution. Choose Edit Curve | Rebuild Curve ☐ to open the Options window for this command. Set the Parameter Range to 0 to #spans and then set the Number of Spans to 4. In the view window, select the second curve and click Apply.

6. Choose both curves and use the Attach Curve command (in your shelf) to attach the curves. They will evenly blend together. Delete the history on the curve. Adjust the CVs around the blend area just slightly so that the curve looks like this:

We could run the same procedure for the front end of the car, but since there is an actual panel seam that separates the front and top surfaces, we will build the front panel on its own. For now, we'll continue to work on this top panel:

1. With the curve selected, toggle on the display of its edit points by using your custom shelf button. Use the INSERT (HOME) key to edit the pivot and snap it to the center of the curve.

2. Duplicate the curve and move it up. In the Channel Box, set the Scale X to 0. This will flatten the curve along the YZ plane. Use the Scale tool to scale it up so that it matches the outer profile of the car from the Side view. It is best to make selections of CVs

around certain areas (the rear arc for example) and scale them by snapping the scale pivot to points on the curve. Make sure you watch the video on the included DVD to get a better idea of how to do this. The character curves are now complete. You should now have two curves that look like this:

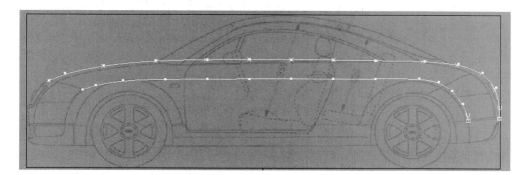

**3.** Invoke the EP Curve tool and hold down the v key to enable point snapping. Snap the first EP to the front end of one character curve and then snap the second point to the front end of the other character curve. Edit the CVs of the curve so that they match the hood seam along the top of the car. Since one span won't be enough to maintain the curvature from the Front view, rebuild the curve to three spans. It should resemble this illustration. We now have everything we need to begin generating a surface.

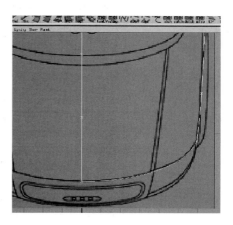

**Generate Primary Surfaces**    By birailing the profile curve along the two character curves, we will get a surface that is too round as it sweeps around the top. However, by selecting isoparms at different parameters along the surface, duplicating them into curves, reshaping those curves, and then re-birailing the surface, we can get it to match the body shape along the length of the car. For more details on this process, review the "Birail" section in Chapter 5.

**1.** Select the profile curve, and then select the two character curves. Click the Birail button on your shelf. A surface will be generated. If not, check and make sure that the edit points of the curves area all snapped.

2. Pick three isoparms at different parameters along the surface, two in the middle and one at the very end. Choose Edit Curves | Duplicate Surface Curves.

3. Delete the birail surface and start modifying these duplicated curves from the Front view so that they are flatter along the top. For the end curve, use the underlays in the Top and Side views to align the shape of the curve with the line in the underlay.

4. Choose all of the profile curves and then the two rail curves and birail the surface. Figure 6-11 shows the curves and the birailed surface in all four views.

5. Just as we did with the rim earlier in this chapter, add a shiny blinn material to the birailed surface and orbit the camera around in the Perspective view. Evaluate the surface by looking for any irregularities. Notice how the highlights on the surface shown in Figure 6-11 are smooth and even.

**FIGURE 6-11**   *The initial surface is successfully birailed and evaluated in all four views.*

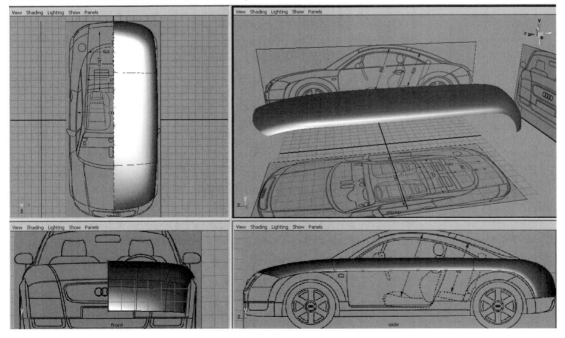

6. Build the front panel of the car. This can be done by creating two arcs, snapping them to the two existing character curves, aligning them to the shapes in the underlays, and

birailing the same profile used to generate the top, along these new arcs for the front. The curves and generated surface are shown in Figure 6-12.

**FIGURE 6-12** *The curves for the front panel are created (left) and birailed to generate a surface (right).*

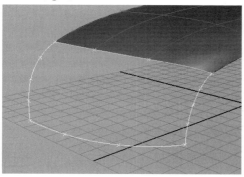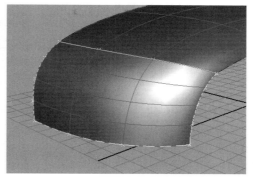

**Reflow the Topology** While the actual shape of the top surface is fine, the isoparms do not conform to any of the details in the reference. The biggest problem is where the surface begins to curve around to form the rear panel (see Figure 6-13). In the reference, there is a panel seam that separates these sections. While creating additional profile curves that match all of the panel seams along the body might be one way to fix this problem, much like we did with the front panel, we are instead going to use a subtractive technique that involves trimming the surface and then converting the surfaces into patches that will then flow to precisely match the details in the reference:

1. In the Side view, use the EP Curve tool to draw a curve that runs along the seam shown in the underlay, as shown in Figure 6-13.

2. Project the curve onto the surface. This surface is now ready to be trimmed.

3. If we trim away one section of the surface, we will lose it. Since we want both surfaces, we need to duplicate the main surface. Press CTRL-D (COMMAND-D) to duplicate the top panel.

**FIGURE 6-13** *The flow of the isoparms does not match the topology of the reference image.*

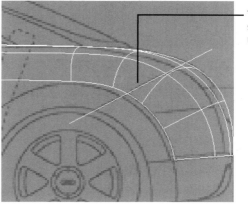

The isoparms should flow to match this line.

**4.** Invoke the Trim tool, select the left region, and press ENTER (RETURN) to trim the right region away. Then choose the Trim tool again and select the right region and press ENTER (RETURN) to trim the left region away. You should now be left with two trimmed surfaces.

**5.** Open the Rebuild Surface Options window. Set the Rebuild Type attribute to Trim Convert. Enable the NumSpans checkbox. Choose the top surface and click Apply. The surface is converted from a trim surface to a normal parameterized surface and the isoparms now transition over the length of this surface to match the seams.

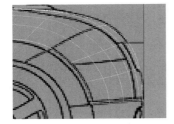

**6.** The rear surface can be rebuilt the same way except that you can uncheck the NumSpans attribute and set the Number of Spans attributes to 4 in U and 3 in V. This should give us enough data to define the curvature of this patch. We will use this technique again and again to reflow the isoparms to match the rest of the panel details. The result should look like this image.

**Boundary the Side Panel**   In this section, we'll create the side panel of the car by using the Boundary command. While we can derive three of the necessary curves (the top and two sides) from the edge isoparms of the existing surfaces, we still need a fourth curve for the lower boundary. We can create this curve by duplicating the top boundary curve and moving it into position. The boundary surface can then be generated.

**1.** Select the edge isoparm on the top surface and use the Edit Curve | Duplicate Surface Curve command to convert the isoparm into an editable curve.

**2.** In the Channel Box, set the Scale Y value of the curve to 0 to flatten it, and then choose Modify | Freeze Transformations.

**3.** Move the curve down to match the base lines that will form the bottom of this patch.

**4.** In the Front view, move the curve inward along its X axis until its position matches the underlay. You'll have to rely on some guesswork, because the line is obstructed by the wheels in the reference.

**5.** Snap the end CVs to the bottom corners of the front and rear surfaces. See Figure 6-14a to get an idea of how the curve is set.

**6.** Select the edge isoparms of each surface, pick the new curve, and execute the Boundary command. A surface will be formed between the selected curves (see Figure 6-14b).

*FIGURE 6-14    Four curves define the sides (a) of a boundary surface (b).*

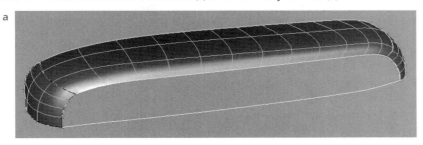

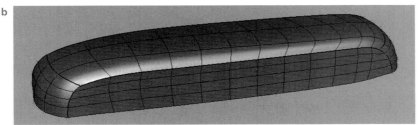

**7.** Reflow the topology to form the doors. To do so, project two curves on the top and side surfaces, and then use the same process demonstrated in the previous section to trim, duplicate, and trim convert each patch until the doors are all created. Figure 6-15a shows the shapes of the door profile curves and Figure 6-15b shows the patches once they are trimmed and then converted.

*FIGURE 6-15    The surfaces before (a) and after (b) reflowing the topology to match the door seams*

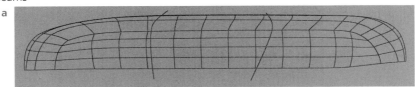

**Create Wheel Wells**    In this section we expand on the trim-converting technique to create the wheel wells. So far, we have been projecting curves, trimming, and trim converting surfaces for the purpose of reflowing their topology, but the actual surface structure has not changed. All of these patches have kept to their square, gridlike layout. When we create the wheel wells, the structure needs to change and flow radially. The effect on the topology will be that five patches will come together at one point.

Planning these topologies is crucial. In most cases, existing patches need to be detached so that the resulting trimmed surfaces maintain four sides. If there are more than four sides on a trimmed surface, the Trim Convert function of the Rebuild Surface command will fail. If we analyze the front wheel well section, the wheel well cuts in across both the side and top panels and is separated by an actual panel seam. This means that we can plan out five-point junctions to sit on these seams. Since the top panel will not be split by this detail, we want to detach it appropriately before trimming so that it results in a four-sided trim surface.

**FIGURE 6-16**  *Once the curve for the wheel well is drawn, the locations of the five-point junctions become apparent.*

1.  Draw the profile for the wheel well. Since we want this to be as precisely circular as possible, we can form the wheel well shape from a NURBS circle. Choose Create | NURBS Primitives | Circle. Rotate it 90 degrees in Z and move and scale it so that it matches the shape of the outer bounds of the wheel well shape in the underlay. Detach the curve, rebuild it to 0 to # of spans, then shape the bottom so that it conforms to the underlay. See Figure 6-16 for the shape of the curve.

2.  The existing circular curve that defines the shape of the wheel well can be projected and converted onto the side surfaces. We can use the same trim-converting process as before to break up the patches and cut an opening. Once we have a nice, precise, circular hole for the wheel well in the side panel, we can work on the top surface. To end up with four-sided trimmed surfaces, we need to detach this surface right where

the projected curve will intersect the edge. Unfortunately, this has to be done by eyeballing it. Before projecting the curve, zoom up very close to this area to select an isoparm and then detach it, as shown here:

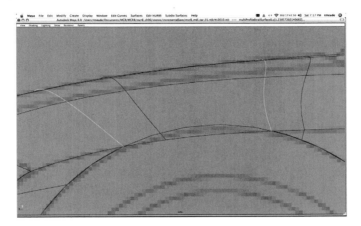

**NOTE** *Instead of eyeballing the isoparm to be chosen, you could snap and project a curve that is precisely positioned to the point where you need the surfaces to separate. While this may take a few extra steps, once you are comfortable with the trim-converting procedure, it can produce more accurate results.*

**3.** Once you have found the intersection point, detach the top surface at these selected isoparms. Project the circular wheel well–shaped curve onto the middle piece and then trim and trim convert.

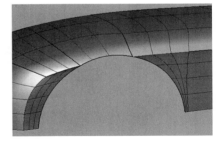

**4.** Go around to all of the patches and rebuild them so that the number of spans in the adjacent patches matches to look something like this:

**5.** The process of modeling the wheel well is very much the same except that the wheel well curve does not intersect the top surface along a seam. It is entirely constructed on the side panel only. This means that you should plan the five-point junction for somewhere close to cutting the circular shape in thirds. Figure 6-17 shows the topology of the rear wheel well.

*FIGURE 6-17*   *The topology for the rear wheel well*

**6.** The final part of the wheel well construction is building the extrusions. To do this, select the edge isoparms around the front wheel well and choose Edit Curves | Duplicate Surface Curves.

**7.** Immediately group the curves (Edit | Group) and center the pivot point for the group (Modify | Center Pivot).

**8.** In the Channel Box, set the Scale X attribute for the group to 0 to flatten the curve in the X axis. Freeze transformations on the group by choosing Modify | Freeze Transformations.

**9.** In the Top view, move the group along the X axis so that it lines up with the outer bound of the wheel well in the reference.

**10.** In the Side view, scale the group down until it matches the profile of the inner wheel well curve in the underlay. You should also move some of the CVs inward toward the body of the car to align the end of the frontmost curve.

**11.** Select each curve segment, and then select its corresponding isoparm on one of the surfaces. Loft these two curves to create a surface. Repeat this process for each curve on both the front and rear wheel wells. The completed wheel wells should look like this:

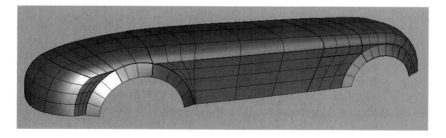

**Create the Canopy Opening**   In this section, we trim and convert the top surface of the car to match the details that flow down the length of the car and the opening for the canopy. Much of this is the same process used for the doors and wheel wells.

**1.** In the Top view, draw a curve that runs along the hood and rear seam lines in the underlay. It is okay to swerve the curve a bit when drawing over the canopy, because that section will be removed anyway. The line drawn is shown here:

2. Project this curve onto all of the patches that make up the front, top, and rear panels of the car. Go through the process of trimming and trim converting the surfaces into normal, parameterized patches. Also, use the Rebuild Surface command to make sure that the parameterization of all of the patches matches. For the most part, sticking to rebuilding 3 spans in U and V will work well at this point. This makes it easier, and faster, to get all of the patches to match. Also, any surface that will be occluded by the canopy can be completely trimmed away. There is no need to leave any surfaces for the inside of the car. At this point, your model should look like this:

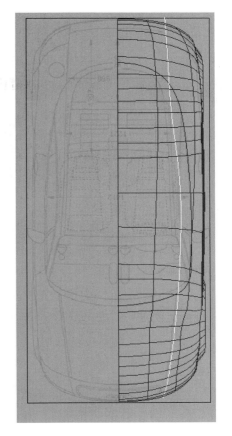

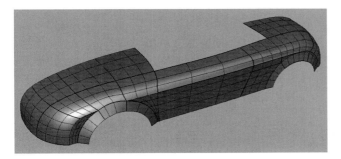

3. In the Top view, draw a curve that matches the shape of the opening of the canopy, as shown here:

4. Just as we did on the wheel well, we need to decide where the structure of the topology will break into a five-point junction, and then detach the surface at those sections. Use the model

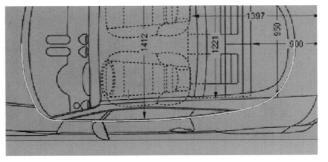

shown in Figure 6-18 to plan your topology and then project the curve, trim the surfaces, and convert them into patches. Once again, make sure that the parameterization matches throughout the model.

*FIGURE 6-18* *The opening for the canopy is created by trimming and trim converting to produce patches with matching parameterization.*

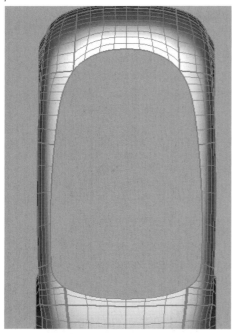

**Model the Canopy**    Modeling the top of the canopy is no different from anything we've already done. Figure 6-19 shows the progression of the modeling. A curve is created that follows the inner boundary of the canopy structure starting from a point snapped to a surface at the rear and continuing to the front. This curve is duplicated and positioned to match the next main character line of the canopy (Figure 6-19a).

One of the edge isoparms from a surface is used as the profile curve to birail over the two character curves (Figure 6-19b). An additional curve is derived from the new surface and flattened along the YZ plane. A third curve is created and birailed over these curves to create the roof (Figure 6-19c). The remaining gap is closed by another birail operation (Figure 6-19d) and the window details are trimmed out and converted to patches (Figure 6-19e).

**Detail the Front Grill**    By this time, you should be getting the hang of the detailing process from all of the work we have done so far. We will use these same techniques to cut the rectangular grill shape out for the front grill. We'll then begin learning how to model nice, round edges between the panel seams. Then we'll look at a technique for modeling the grill geometry. Adding these smaller details to the model helps establish a sense of scale.

*FIGURE 6-19*    *Construction of the canopy*

a
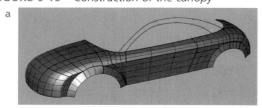

b
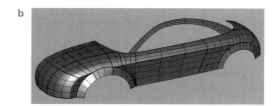

c
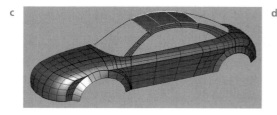

d
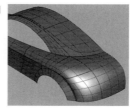

e
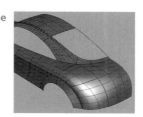

Ideally, you want to be able to constantly breakdown a model from the largest piece to the smallest piece. When the details become too small to model with geometry, they can be added with texture maps. We'll begin this detailing process by adding the ridge that flows down the hood:

**1.** In the Top view, select isoparms on the hood surface patches just right of the edge and detach (Figure 6-20a).

**2.** Group all of these hood patches together and snap the group's pivot point to the inner corner of the hood's surface. Scale the group just a tiny bit from this pivot point so that the center section is now slightly offset (Figure 6-20b).

**3.** Loft the edge isoparms between the hood patches to cover the gap created by the separation.

**4.** The loft operations left us with C0 continuity between these surfaces. This leaves a flat edge that will not catch any highlights. We could attach the surfaces, but then most of the shape would get lost. The Align command is perfect for situations such as this. The Align command calculates any specified level of continuity between any two surfaces. This provides us with a method of achieving continuity without losing any geometry, as we do when we attach and detach (as discussed in the plane and cylinder example earlier in the chapter). Choose Edit NURBS | Align Surfaces □. Set the Continuity to Tangent. Set the Modify Boundary attribute to First. This will move just the edge of the first surface selected instead of the entire surface.

**5.** We will align each of these lofted surfaces to each of their adjacent surfaces. Select one of the lofted surfaces and then a surface adjacent to it and apply the Align command. Select the lofted surface again and then align it to the adjacent surface on the other side. Repeat this process for all of the lofted surfaces. Figure 6-21 shows the hood before and after the alignment. Notice how the aligned surfaces catch the highlights.

*FIGURE 6-20  Detaching and scaling the hood patches to create an offset*

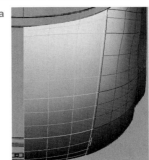

a

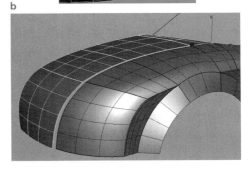

b

**FIGURE 6-21**    *The hood surfaces before (left) and after (right) being aligned*

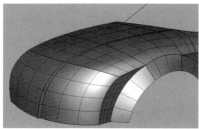 

**6.** Moving on to the grill, in the Front view, draw a curve that matches the shape of the grill's frame in the underlay. Duplicate and scale this curve up so that it is parallel to the first curve.

**FIGURE 6-22**    *The topology for the front of the car is planned out.*

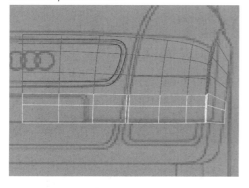

**7.** Based on these curves, analyze the current topology of the model and figure out where to detach the patches. Figure 6-22 shows the curves and where the surface should best be detached for planning the five-point junctions. Remember that when you detach one surface, you want to continue around to all of the adjacent patches and detach those at the same location.

**8.** Project the curves, trim, and trim convert the pieces. But just before you begin this process, duplicate the original patch that is being trimmed and hide it temporarily. Because we are breaking the structure to produce this detail, we have to cut the trimmed surfaces into smaller segments so that they all have four sides. By using the EP Curve tool and snapping the first point to the planned location of the five-point junction, the accuracy of the trim is ensured. One more tip is to detach the surfaces to isolate the corners so that the corner pieces can be rebuilt with a larger amount of span so that they hold their shape. As always, go around the model and rebuild neighboring surfaces so that the parameterization matches between adjacent surfaces. Figure 6-23 shows the curves as they are projected onto the surface (Figure 6-23a) and then rebuilt (Figure 6-23b).

**FIGURE 6-23** *(a) The grill shape is trimmed away and the remainder of the surface is divided into four-sided surfaces by projecting curves. (b) The surfaces are then trim converted and rebuilt.*

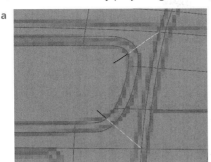

a

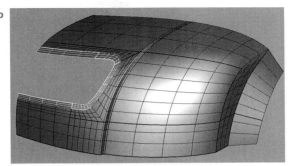

b

9. Once all of the trimmed surfaces have been converted into patches, we can begin to model the round edge between them. Select the edge isoparms of the rectangular grill frame, duplicate the surface curves, and group the selection. Move the curve back about .2 units in Z.

10. Make your way around the grill and loft each curve to the edge isoparm on the surface that it was derived from. Make sure to do this on both the inside and outside of the grill frame. The lofted surfaces are shown here:

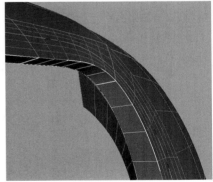

11. Select all of the lofted surfaces and then, in the Channel Box, set the Section Spans attribute to 3. This should give the lofts enough resolution that they hold their shape and produce a nice edge when attached to the frame.

12. Work your way around the frame again, this time attaching the lofted surfaces to the grill frame that was derived from the body. You should also make a loft from the body side of the edge to the duplicate curves and attach these. Once this process is complete, you will have the frame perfectly conforming to the rounded front end of the car with a nice, round edge between it and the body, as shown here. Something like this would be very difficult to do with such precision on a polygon model.

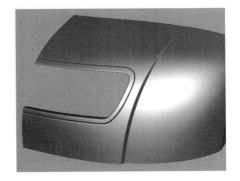

**13.** We will create the actual grill by extruding some simple profile curves along the same curves that conform to the round shape of the front of the car. Begin by making visible the original front patch that we hid before we started trimming. Move this surface back along the Z.

**14.** In the Front view, use the EP Curve tool to draw a line straight across the entire grill shape. Duplicate and move this line three more times and position it so that all four curves are equidistant, as shown here:

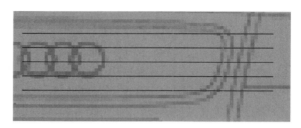

**15.** Select these four curves, select the original surface, and project them onto the surface.

**16.** In the Side view, draw a profile curve that can be used to extrude out the main grill pieces as shown adjacent to Step 18. Center the pivot on the curve.

**17.** Select the new profile curve, then select one of the curves on the surface that was just projected on. Choose Surface | Extrude ❑ to open the Options window for the Extrude command. Set the Style to Tube, the Result Position to At Path, the Pivot to Component, and the Orientation to Profile Normal. Click the Apply button. Repeat this process for the other three curves.

**18.** Draw another set of curves that travel in the vertical direction and project. Draw another profile curve and extrude it along these new curves on the surface. The final grill is shown here:

**Mirroring**    There is still a lot of work left to do on this model. For one, there is an entire base portion that extends below the existing car body. Plus, there are many panel seams that need rounding. However, you can use the exact same procedures that we've used to model the car thus far to finish it yourself. We will revisit this model in Chapter 7 when we show techniques for converting NURBS models to polygons and show some subdivision surface modeling techniques. For now, we will mirror the model over to the other side, attach the surfaces along the axis of symmetry, and evaluate the model.

**1.** Before mirroring, take another pass at the model with your eyes and make sure that you have matching parameterization between all of the patches.

**2.** Select all of the patches and group them (Edit | Group). Duplicate the group. Set the Scale X value of the group to −1. This will create a mirror duplicate of your original surfaces.

**3.** Starting at one end of the model, select the two mirrored surfaces along the axis of symmetry and attach them. Figure 6-24 shows both sides of the completed NURBS model.

**4.** It's a good idea to group everything again and create subgroups that have the different sections of the car named. There are a lot of pieces here, so you want to do everything you can to make this model easy to read when you come back to it later.

**FIGURE 6-24**   *The completed NURBS patch model*

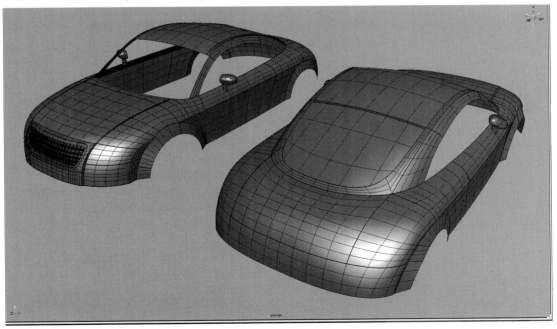

# Summary

This brings us to the end of the NURBS modeling exercises in this book. Throughout the two exercises, we've demonstrated techniques that allow you to deal with just about any modeling problem. While the steps you use are going to be about the same and the process very repetitive, every model you build will present its own unique set of challenges. Good modeling requires a fair amount of experimentation and trial and error until the problem is solved.

# Preparing Models
# for Animation

**Before a model is passed off to other**

parts of the production pipeline, it must meet

certain specifications. The type of geometry used

is one critical factor. While a 3D modeler may

choose NURBS for a specific job, the rendering

pipeline may specify that only polygonal

geometry can be used. Furthermore, that

polygonal geometry may need to be optimized so

that it can be quickly distributed over a network

and rendered in a reasonable amount of time.

In this chapter, we complete the car we were building in Chapter 6 and convert it into a low-resolution polygonal mesh. We will combine and clean up the patches of this polygonal mesh so that we can convert it to a subdivision surface. Working with the model as a subdivision surface will allow us to evaluate the smooth model as we make edits to the geometry. Subdivision creasing will also be discussed.

# Converting Geometry

Maya is capable of directly converting NURBS into polygons, NURBS into subdivision surfaces, subdivision surfaces into NURBS, subdivision surfaces into polygons, and polygons into subdivision surfaces. You can find all the commands to make these conversions in the Modify | Convert menu. The only route that is not available is converting polygons into NURBS, due to the four-sided nature of NURBS. However, it is possible to convert polygons into subdivision surfaces and then convert subdivision surfaces into NURBS, so any combination is actually possible. NURBS geometry can also be converted on-the-fly as surfaces are generated from curves. We'll begin this lesson there.

## Generating Polygonal Surfaces from NURBS Curves

Since NURBS is the most limited in terms of topology, converting NURBS surfaces provides the most options for converting to polygons or subdivision surfaces. Throughout Chapters 5 and 6 we used the Revolve, Loft, and Boundary commands to generate many of the surfaces used in the examples and tutorials. With their default settings active, these commands generated third-degree NURBS surfaces from NURBS curves that were either drawn with the CV or EP Curve tools or duplicated from existing curves on surfaces.

Inside the Options window for any of these commands is an Output Geometry setting that lets you choose what type of geometry will be generated when the command is executed. Figure 7-1 shows the tessellation attributes in the Options window for the Revolve tool when the Output Geometry is set to Polygons. (You have to scroll down the Options window to see all of these attributes.) Notice that many new options appear in the window once this is enabled.

**FIGURE 7-1** *The Revolve Options window with the Output Geometry set to Polygons*

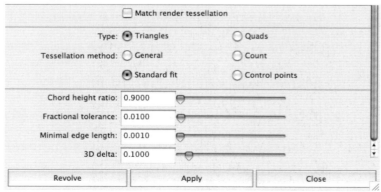

The first option in this new section, Type, lets you specify whether the faces on the resulting surface will be triangles or quads. The next attribute you can choose is the Tessellation Method. When one of these methods is selected, its options appear in the window section below.

Figure 7-2 shows a NURBS curve and the polygonal surface that is generated with each Tessellation Method choice. From left to right, first is the NURBS curve, and then the surfaces generated by selecting General, Standard Fit, Count, and Control Points. These methods are defined next.

- **General**   This option gives the most control for tessellation based on the NURBS object. When the Revolve command was executed, the first tessellated surface shown in Figure 7-2 had the U Type and V Type attributes set to Per Span # of Iso Params with a value of 2 in both directions. Since the isoparms in U were based on the spans of the source curve, the polygonal surface will have two segments for each span on the NURBS surface. Since the Revolve node's Segments attribute was set to 8 in the Channel Box, this means that the resulting polygonal surface will have 16 segments in the V direction.

- **Standard Fit**   This option attempts to match the curvature along the surface as best it can based on the Chord Height Ratio setting. This is the maximum distance from the curve that the face can appear. The result always uses more faces to define curvy areas and fewer faces for flatter areas. While you don't have precise control over the number of segments, as you did with the General setting, the polygonal surface will still be pretty efficient in that polygons are created only where they are needed.

- **Count**   This method allows you to specify the total number of polygons on the surface without any regard for the curvature. The faces will be spread uniformly over the entire surface. While this may be the easiest method for obtaining a desired polygon count, it also leads to the most inefficient surfaces because a curved area will be defined exactly the same as a flat area.

*FIGURE 7-2*   *NURBS curve, General, Standard Fit, Count, and Control Points*

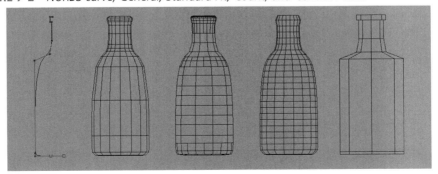

- **Control Points** This tessellation method simply facets the object by using the existing control vertices in the NURBS surface as polygonal vertices. No additional attributes control this setting. This can be a useful method for tessellating an object that you plan to smooth or to convert to a subdivision surface later on.

When the Output Geometry is set to Polygons, Maya actually generates the surface as a NURBS object based on the surface command attributes and then converts it to polygons using one of these methods. Therefore, in the Channel Box, you will notice the appearance of a node called nurbsTessellate1 above the Revolve node. As you experiment with these methods, you can change the Tessellation Method and its attributes in the Channel Box.

## Converting NURBS Surfaces to Polygons

You can convert any existing NURBS surface to a polygonal surface by choosing Modify | Convert | NURBS to Polygons. This gives you the exact same options that you can find by using one of the surface-generating options discussed in the preceding section. In fact, when a surface is generated using one of the commands in the Surfaces menu, Maya internally generates a NURBS surface and then converts it using these same methods; so you end up with the same nurbsTessellate node in the surface's dependency graph.

Yet another way exists for accessing the same tessellation options. You can find the General, Standard Fit, Count, and Control Points tessellation options in the Rebuild Surfaces Options window (choose Edit NURBS | Rebuild ❒). This results in the same nurbsTessellate node being placed above the NURBS surface to tessellate the object into polygons.

### Converting a Multipatch NURBS Model to a Polygonal Model

Often, a NURBS model must be converted into a polygonal model so that it can be exported to another rendering application. Another reason for converting a NURBS model into polygons might be to have control over the UV (texture) coordinates in the UV Texture Editor (discussed further in Chapter 18).

When it comes to NURBS models that are created out of multiple patches, such as our car, it may be desirable to convert the NURBS patches into polygons and then merge the resulting polygons into one complete surface. This is particularly the case when the model will be deformed during animation. Deforming a multipatch model can cause the seams between the surfaces to become visible during the animations. Combining these surfaces into one surface keeps this from being a problem.

You can quickly generate a polygonal approximation of a multipatch NURBS model by choosing certain settings in the NURBS to Polygons Options window. Let's take, for example, the hood of the car. This model was made up of six different NURBS patches. Choose Modify | Convert | NURBS to Polygons ❒ to open the Options window for this command. The options are shown in Figure 7-3.

FIGURE 7-3   *The Convert NURBS to Polygons Options window*

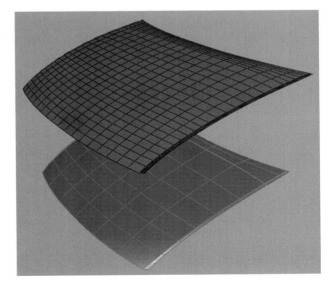

If the Attach Multiple Output Meshes check box is enabled, Maya combines all of the selected NURBS surfaces into one mesh. The overlapping vertices along the edges of the original patches will be merged as long as they fall within a distance of the value set in the Merge Tolerance attribute. The Type attribute should be set to Quads so that the resulting tessellated faces will be converted to four-sided faces wherever possible.

Setting the Tessellation Method to General lets you have precise control over how many faces are created from each NURBS patch, based on the conditions specified in the Initial Tessellation Controls section of this window. If the U Type and V Type attributes are set to Per Span # of Iso Params, you can specify how many polygonal faces will be created for each span in the U and V directions. This illustration shows the NURBS model at the bottom and the converted polygonal mesh above it.

While this method of conversion is fast, the resulting polygonal mesh does not offer the flexibility that we need to finish our model.

Because we want to be able to add surface details to the car and then subdivide it for rendering, converting the mesh using the General method is not going to work for us, as the poly approximation is already based on a smooth, interpolated mesh. Instead we need to convert the NURBS patches into polygonal surfaces at their control vertices. When a surface is converted this way and then subdivided, the result will match the original NURBS surface.

This means that there is going to be some work involved. For one, all of the surfaces must be combined and the vertices merged manually since the Attach Multiple Output Meshes option is not available for the Control Vertices conversion method. Another problem is that all of the edges that are created from a tangent row of CVs in the NURBS surfaces must be removed between all surfaces that were merged. Since the purpose of these vertices was to control the direction of the surfaces' outer edges, they are no longer needed now that they are in the middle of a continuous surface. Leaving them will result in the subdivided model flattening out where that geometry is. Let's work on the front of the car.

### Creating a Low-Resolution Proxy Mesh

In this section, we will convert some of the NURBS patches in our car to polygons at the control vertices. These surfaces will be combined, merged, and cleaned up to produce a geometry that will work as a subdivision surface model.

1. Open the final model of your car or use the mcr_mdl_car_final.ma file found on the DVD.

2. Identify the matches that make up a complete panel of the car. We'll work on the front panel. After examining the reference files, we can determine that the main front panel actually includes the wheel well surfaces, part of the side panel, and some parts that have not been modeled yet. Be sure that the patches all have matching parameterization with their surrounding surfaces. Select them and choose Modify | Convert | NURBS to Polygons ☐. Set the Tessellation Method to Control Points and click Tessellate. The resulting polygonal surfaces are shown here:

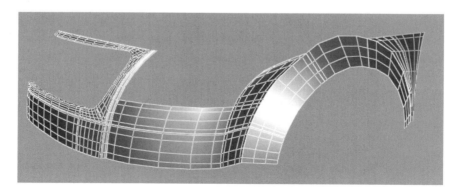

**3.** To successfully merge the edges of these surfaces, the normals must all be facing the same direction. Choose Display | Polygons | Face Normals to display the normals, and look to see which surfaces need to be flipped, as shown here. Select the surfaces that have their normals facing inward. In the Polygons menu set, choose Normals | Reverse, to flip any normals that are facing the wrong direction.

**4.** If you are sure that the surfaces have matching parameterization and are facing the same direction, you can select all of them and choose Polygons | Combine. This combines the surfaces into one single object. However, combining does not merge any overlapping vertices or edges.

**5.** To see any overlapping components, choose Display | Highlight Border Edges. You will be able to see where the multiple surfaces were combined.

**6.** Choose Edit Mesh | Merge. This will merge most of the overlapping components. However, there will be a few remaining nonmerged vertices, particularly in the five corner areas. For these, select the vertices in these areas and use the Merge Vertices command.

The smoothing algorithm used to calculate a subdivision surface (bicubic interpolation) is the same for a NURBS surface. Therefore, if you smooth a polymesh with its components, and that of a corresponding NURBS surface, the surfaces will be identical. However, a polymesh that has been directly converted from NURBS at the CV level will have some extra components that need to be removed.

As we have mentioned before, the second row of CVs on a B-spline NURBS surface control the tangency along the edges. Well, we don't need these, so you will have to go through and remove the second row of edges on all converted polymeshes. You can use the underlying NURBS surfaces as reference and toggle their display by using the Show | NURBS Surfaces in the current view window.

> _**NOTE**_   _**Avoid using the delete key to delete polygonal components, as it deletes only the edges and leaves the vertices, resulting in six-sided polygon faces.**_

To remove an edge loop, select an edge along that loop and then CTRL-RMB-click to display the Polygon Marking menu. In that menu, choose Edge Loop Utilities | To Edge Loop and Delete. (See Chapter 4 for more information on this marking menu.) Executing this command selects the entire edge loop related to the selected edge component and then uses the Delete Edge/Vertex command to delete the edges along that loop and any related vertices. After removing the edges created from the tangent row of CVs, the mesh should look like this:

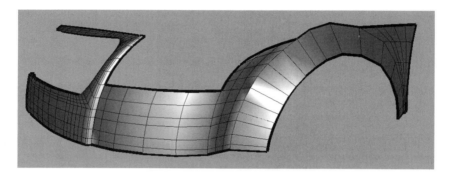

*TIP*   ***Once you have already executed the To Edge Loop and Delete command on one edge, you can select another and press the G key to execute the last command.***

Now you may begin adding any polygonal detail with the polygonal editing tools. Ideally, you'll have planned the topology of the model well enough that you'll have all of the edges in the right place in order to make the necessary extrusions before you convert to polygons. But, there will be cases where something was overlooked, or you decided later to add additional detail.

## Hierarchical Subdivision Surfaces

In Chapter 5, we converted our head model to a subdivision proxy so that we could evaluate the smooth, interpolated version of the geometry. In this section, we look at another implementation of subdivision surfaces.

In Maya, *hierarchical subdivision surface modeling* is called *subdivision surfaces* and is technically considered an entirely separate geometry type on its own. Subdivision surfaces allow you to select certain areas and refine them through hierarchical levels. This lets you add finer details to the mesh without having to add more geometry to the original cage/polygon, ensuring a high level of detail in the resulting surface while maintaining a reasonably simple original (level 0) shape.

You can create a subdivision surface by choosing commands from the Create | Subdivision Primitives menu to create various subdivision surface primitives, or you can convert an existing polygonal or NURBS surface into a subdivision surface by choosing Modify | Convert and then choosing a conversion option. Let's examine some of the features of subdivision surfaces and learn how they work.

## Subdivision Surface Modes and Refinement

Two modes can be used to add detail to a subdivision surface: standard mode and polygon mode. Standard mode gives you access to multiple levels of hierarchical components within the subdivision surface. Polygon mode is similar to the subdivision proxy method. In this mode, you can use any of the polygonal editing tools on the base shape.

**Standard Mode**    By default, a subdivision surface is in standard mode when the surface is created or converted. This mode lets you access polygonal components at various levels of refinement at any area on the surface. If you use the marking menu to select a component (by right-clicking the object and choosing a component), Maya displays the selected component as a number. The sphere on the left in Figure 7-4 shows a subdivision sphere displaying its faces at the base level (level 0). The number 0 indicates the level of refinement with which you are currently interacting.

To start adding detail to the faces, you can switch to a higher level of refinement. The easiest way to do this is to right-click the component and choose Level | 1 from the marking menu. The selected components will *subdivide*. This means that for every one face, four faces will be created for you to work with, as shown in the sphere on the right in Figure 7-4. Notice that the faces are each displaying the number *1* to indicate the current level of refinement.

You can now edit any components at level 1 and then switch back to level 0 at any time by using the marking menu. You can also change levels by changing the Levels attribute in the Channel

FIGURE 7-4    *A subdivision sphere displaying its faces at level 0 (left) and level 1 (right)*

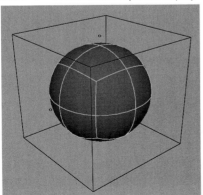
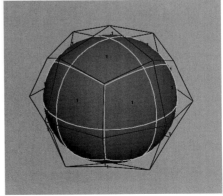

Box. If you've modeled as far as you can at level 1 and still need to add more detail, you can refine the display further by right-clicking the components of the region you wish to refine and choosing Refine from the marking menu. The components will be subdivided another level further.

Probably one of the most useful features of modeling with hierarchical subdivision surfaces is the ability to harden or crease edges without having to add any geometry. By selecting edges or vertices of a subdivision model, you can create either a *full crease* (Subdiv Surfaces | Full Crease Edge/Vertex) or a *partial crease* (Subdiv Surfaces | Partial Crease Edge/Vertex). You can continue to partially crease the selected edge until you are happy with the result. Figure 7-5 demonstrates a sphere whose edges have been fully creased and partially creased twice.

This may seem like the most optimal way of modeling. The ability to add localized detail only where it is needed can save a lot of time. However, it is not possible to change the structure of the surface at different levels of the subdivision hierarchy. Polygonal editing tools such as Extrude Face and Split Polygon cannot be used on subdivision surfaces while in standard mode. Modeling in standard mode is therefore best suited for adding extra details to a model whose structure has already been laid out. Adding wrinkles to skin or creating hard edges are two examples of where this geometry type can be useful. But, to be able to change the structure of a subdivision surface, you must switch to polygon mode.

**Polygon Mode**    You can switch from standard mode to polygon proxy mode at any time by right-clicking the surface and choosing Polygon from the marking menu. This creates a polygonal object that matches the base mesh (level 0) of the subdivision surface. Here, you are free to use any of the modeling tools in the Edit Mesh menu.

The one thing to be aware of here is that every time you switch back to standard mode, the polygon shape is deleted. This means that if you have connected any other nodes, such as a deformer, to the polygon, they will be lost.

**FIGURE 7-5**    *A full crease and partial crease are added to the edges of a subdivision sphere.*

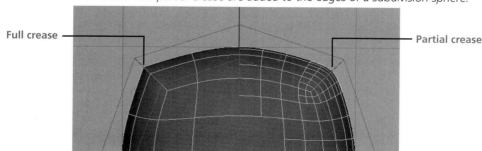

# Tutorial: Modeling with Subdivision Surfaces

In this tutorial, we will use subdivision surfaces to add detail to our car. We'll be using the front panel piece that we converted to polygons earlier in this chapter.

**1.** Open the mcr8_car_frontPoly.ma file on the DVD.

**2.** Select the poly_frontPanel object and then choose Modify | Convert | Polygons to Subdiv ❒ to open the Options window. The main setting to be aware of here is Maximum Base Mesh Faces. By default, this attribute is set to 1000. This means that Maya will not convert the mesh to a subdivision surface if it exceeds 1000 faces. In this case, the front panel is less than 1000 faces, so this setting will work fine, but as you convert some other pieces, you may need to increase this to suit your model. Click the Create button to convert the polygonal mesh to a subdivision surface. Press the 3 key to display the subdivision surface smoothly. Your model should look like this:

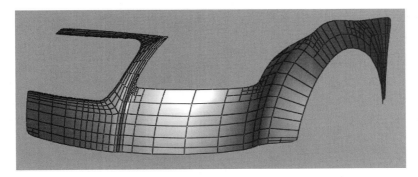

> **NOTE** *If the conversion fails, you need to troubleshoot the model. Start by reading the errors in the Script Editor (Window | General Editor | Script Editor). You can use Mesh | Cleanup to fix any problems with nonmanifold geometry. Usually, it's best to manually inspect the model with the border edges visible. Most of the time, the case will be that normals are reversed or edges are not merged.*

**3.** We will start adding to the existing geometry by extruding the bottom edge to model with missing geometry. To make this type of edit, we must be in polygon mode. RMB-click the mesh and choose Polygon. A wireframe overlay of the original polygonal mesh will appear.

**4.** Select the bottom row of edges and extrude. Move the extruded edges back a bit, then extrude again and move them down to create a nice edge. Extrude one more time so that the surface is aligned to the base of the car in the underlay, as shown here:

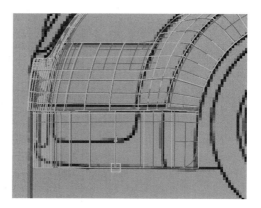

**5.** In the Front view, select some of the vertices on the bottom edge, as shown next, and use the Move Normal tool (Modify | Transform Tools | Move Normal Tool) to move the vertices over a bit so that the edges align to the panel details in the underlay. Using this tool moves the vertices along the surface and ensures that the integrity of the surface is not broken.

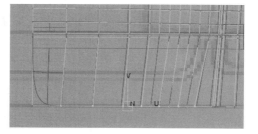

*TIP* *The Move Normal tool is a subset of the Move tool. You can access this setting in the Move Tools Settings window or by holding down the* W *key and RMB-clicking.*

**6.** Use the Insert Edge Loop tool to insert two edge loops, each one aligning to the base of both intakes shown in the underlay. Now we have all of the necessary topology to extrude.

**7.** Select the faces that align to the details in the underlay and extrude them back. Delete the three faces along the axis of symmetry so that the model will mirror properly later on. The result should look like the adjacent image.

**8.** While we could use our trick of extruding three times to create the hard edge, we will instead use a subdivision partial crease to harden the edges. RMB-click the model and choose Standard from the marking menu. The polygonal overlay will be deleted.

**9.** Select the original edges and the extruded edges that flow around the intake details. Choose Subiv Surfaces | Partial Crease. The edges will harden.

**10.** Switch back into object mode and evaluate the surface. It should resemble this:

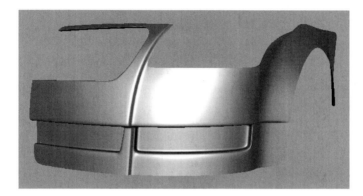

**11.** To give the panel a sense of thickness, that is, that it isn't made of paper, we will *roll* the edges inward. This can be done by extruding the edges inward. Since the edges in this model are running in all different directions, it's best to break up the extrusions into sections. Switch the subdivision surface back into polygon mode. Start with the bottom, front edge, then the wheel well, and up and around to the top of the grill.

**12.** At the end of every extruded series of edges, we want to merge the vertices so that the rolled edge is continuous all of the way around. Figure 7-6 shows the edges between the bottom, front edge and the wheel well before and after merging. Notice how the merged edge is nice and smooth.

**FIGURE 7-6** *The edge before (left) and after (right) merging the corner vertices*

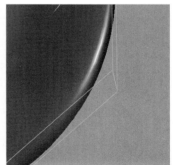

To harden these edges, we could use the creasing technique used on the front intakes. However, some rendering pipelines are not able to read Maya's hierarchical subdivision surface creases. Many productions these days prefer to have the model subdivided at render time. We will demonstrate this rendering feature later, in Chapter 21. Right now, we'll manually insert an additional edge loop around the outside of the edge.

**13.** Choose Edit Mesh | Insert Edge Loop Tool ❐ to open the tool's settings panel. Set the Maintain Position setting to Equal Distance from Edge. This will maintain the offsets from the edges on either side of the edge loop we are inserting. This is useful for creating an edge loop that is close to parallel with the edge we are aligning it with. Start along the front, bottom edge and insert an edge loop. Use the Insert Edge Loop tool to insert a loop along the body side of the front, bottom edge, the wheel well, and the door seam panel.

**14.** The edges along the top of the wheel and around the lightwell are a bit more complicated because the structure changes. For this reason, we won't get too far with the Insert Edge Loop tool. Instead, we'll use the Split Polygon tool and trace along the outside of the edge, being careful to maintain an even distance.

**15.** At every junction where the structure changes, you will end up creating a triangle, shown in Figure 7-7a. This can be cleaned up by selecting the two vertices at the corners of the triangle and merging them (see Figure 7-7b).

*FIGURE 7-7    When inserting edges around a corner structure, a triangle will be created (a) but can be removed by merging the vertices at its two corners (b).*

a

b

**16.** Work your way around the edges and make sure every edge is even and the topology contains all quad faces. After creating the edges for some of the other panels, the front of the car looks like this:

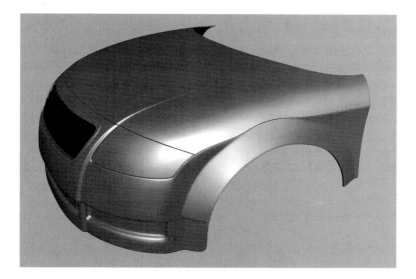

Depending on your preference, you can continue using these techniques to build the rest of the model. If you are using the manual technique, you can convert the subdivision surface back to polygons by choosing Modify | Convert | Subdiv to Polygons. Set the method to Vertices and set the Level to 0.

> **NOTE** *The low-resolution polygonal mesh can be subdivided when rendered in Mental Ray by using the Approximation Editor. The Approximation Editor is discussed in Chapter 21.*

# Summary

In this chapter we concluded our modeling lessons by using the geometry conversion features available in Maya. We learned how to effectively convert NURBS surfaces to polygons and clean them up so that additional detail can be added and subdivided to produce the smoothness of the original NURBS surface.

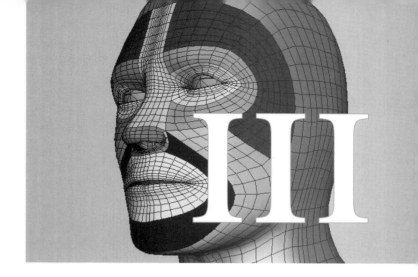

# Character Setup

# Deformers

**Up until this point, we have really only**
dealt with the direct manipulation of geometry
by editing the individual components of a shape.
There are many cases where you need to control
the translations of hundreds, thousands, or even
millions of vertices and have them behave in a
specific manner. Deformers enable relatively
simple control over large amounts of data. Instead
of manipulating each vertex individually, you can
apply a deformer to any piece of geometry,

a selection of vertices, or even a particle system; you can then manipulate the entire shape by editing a few simple controls on the deformer.

Maya offers a variety of deformers you can use to accomplish certain tasks. This chapter provides a reference for using Maya's deformers and includes examples of how they are commonly used. We'll tackle two small projects. First, we'll look at nonlinear deformers and learn to use them in conjunction with one another as we learn to manage the order in which the deformations occur. Then, we'll tour through several of Maya's other deformers and use them to create facial expressions. We will conclude by using these sculpted facial expressions as inputs to a blend shape deformer and set up a head for facial animation. Chapters 9 and 10 introduce the remaining deformers as they can be applied to character setup.

> _**NOTE**_   _**While deformers are often considered an animation/rigging tool, they are just as useful for modeling and particle effects. Any time you need to manipulate large data sets, no matter what type of data it is, consider a deformer.**_

# Nonlinear Deformers

The nonlinear deformers offer a set of simple nonlinear deformation functions. Several deformers can be used on one object or on a group of objects. You create these deformers by selecting an object, some vertices, or a group of objects, and then choosing Deform | Nonlinear and choosing a deformer from the submenu. Once the deformer is created, it can be moved, rotated, and scaled so that it is positioned as you want. The attributes can be controlled using the Show Manipulator tool by selecting the deformer in the Channel Box and then selecting the Show Manipulator tool from the toolbox.

## Types of Nonlinear Deformers

The following list provides brief descriptions of each nonlinear deformer and how its specific attributes affect the object(s) being deformed. Figure 8-1 shows the effect that each deformer has on an object.

- **Bend deformer**   Bends an object. By default, the deformer bends from the center of the object along the Y axis. You can change this default behavior by selecting the deformer and using the Rotate tool to rotate it so that it is deforming in the proper direction. The amount of bend is controlled by the Curvature attribute, which can be interactively edited in the view window by using the Show Manipulator tool and dragging the blue dot at the center of the deformer icon.

*FIGURE 8-1*    *Six nonlinear deformers and their effect on a geometry. Notice that each shows its specific control manipulator.*

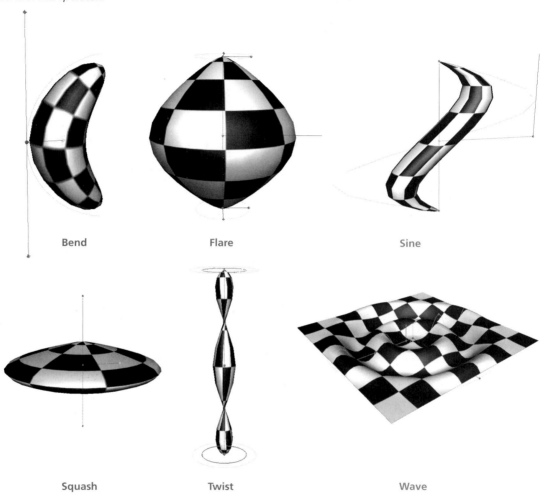

Bend               Flare                Sine

Squash             Twist                Wave

- **Flare deformer**    Lets you narrow or widen the ends of the object being deformed. The attributes for the start and end flares provide nonuniform scale controls for both ends of the deformer. These attributes can be controlled interactively by using the Show Manipulator tool and dragging the blue dots that appear on the circles at each end of the manipulator. The Curve attribute curves the surface between each end of the flare. The Curve attribute can be edited with the Show Manipulator tool by dragging the dot at the center of the manipulator.

- **Sine deformer** Uses a sine function to create a curvy deformation. As in most sine functions, attributes are used to control the amplitude, wavelength, and offset of the sine wave. To control the frequency, select the deformer and use the Scale tool to scale it down. Then use the High Bound and Low Bound attributes to control the length of the deformer.

- **Squash deformer** Squashes and stretches an object. While a similar effect can be achieved with the Scale tool, the squash deformer will keep the volume of an object intact using the Factor attribute. This attribute can be interactively controlled with the Show Manipulator tool by dragging the blue dot that extends from the center of the manipulator.

- **Twist deformer** Introduces a twist into the object to which it is applied. Two attributes control the twisting effect: Start Angle and End Angle. Essentially, these attributes control the rotation at both ends of the deformer.

- **Wave deformer** Simulates a ripple effect into the surface, similar to the way water forms a ripple when a rock is thrown into it. This deformer is similar to a sine deformer in that both share similar attributes, such as Amplitude, Wavelength, and Offset. However, the wave deformer is effective in a radial direction from its center, while the sine deformer works only in a single direction.

## Applying and Using Nonlinear Deformers

In this section, we apply some of these nonlinear deformers to a piece of geometry and learn how they can be set up for use in an animation.

### Set Up Deformers

Multiple deformers can be used on one object. In this section, we will connect three different deformers to a piece of geometry. Let's begin.

**1.** On the accompanying DVD, open the file mcr8_ch08_mushroom_start, which is a model of a mushroom. First, we'll decide what we want the mushroom to be able to do. Let's say that it should be able to bend from side to side, twist along its axis, and squash and stretch. Having the ability to do these things would give an animator plenty to work with.

**2.** We'll start by applying the bend deformer so that the mushroom can rock back and forth. Select the mushroom and choose Deform | Nonlinear | Bend. An object named bend1Handle will be created in the scene and will appear as a vertical line running along the length of the geometry. You may need to view the scene in wireframe to see it.

**3.** To view the deformer's attributes, you can either select the bend1Handle object directly in the view window or Outliner, or you can select the geometry and click the bend1 node in the Input list in the Channel Box. Set the Curvature attribute to 1 and view the result. Notice that the mushroom is bending from the object's center, like this:

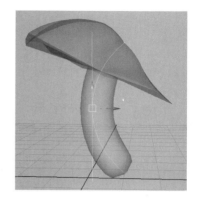

**4.** To have the mushroom bend from its base, the position of the bend1handle must be repositioned. Select the handle and use the Move tool to place the bend handle's pivot point at the base of the mushroom. Now the mushroom is bending from its base but the bend loses its effect when the geometry is no longer in its bounds. To fix this, we'll adjust the Hi and Low Bound attributes on the bend deformer.

**5.** With the bend deformer still selected, use the Channel Box to change the Low Bound to 0 and the Hi Bound to 1.5. Test the deformation by editing the Curvature attribute. The result should look like this:

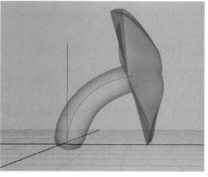

*TIP  You can interactively edit any attribute in the Channel Box by highlighting the name of the attribute and MMB-dragging in the view window.*

**6.** Now that the bend deformer is working correctly, we can add another deformer. Let's add the twist. Reset the Curvature attribute on the bend deformer to 0. Select the geometry and choose Deform | Nonlinear Deformers | Twist.

**7.** Again, reposition the twist handle at the base of the mushroom. Set the Low Bound to 0 and the Hi Bound to 1.5. Test the deformation by editing the End Angle attribute. When you approve, reset the End Angle back to 0.

**8.** Now that we have two deformers controlling one piece of geometry (two input nodes), we will test the deformers together. First, edit the Curvature attribute on the bend deformer to bend the mushroom over. Then, begin to increase the End Angle on the Twist deformer. Notice what is happening. Instead of twisting around its own axis, the mushroom twists around the Y origin, as shown here. This is because the bend is being calculated first, and then the twist is calculated from its original position. This is where the understanding deformation order comes into play.

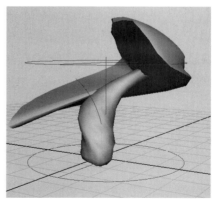

## Deformation Order

By now, you should be becoming familiar with the concept of the dependency graph. We introduced it in Chapter 2 and have been relying on it with everything we have done so far in this book. However, up until now it has only played its role behind the scenes. Now we will learn to control the order in which the nodes are connected and evaluated. Understanding this is crucial to character setup.

**1.** Select the mushroom geometry and then RMB-click and hold on the Inputs to Selected Objects button on the status line. Choose All Inputs from the menu that appears, as shown here:

**2.** The List of Input Operations window opens (see Figure 8-2), which provides us with an editable list of all the nodes that are flowing into the selected object, with the last node evaluated listed at the top. In this window, you can also disable any node by setting the Node State to Has No Effect from the Node State pull-down menu.

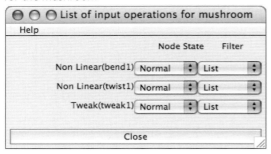

FIGURE 8-2    *The List of Input Operations window for the mushroom*

**3.** To reorder the deformations so that the bend evaluates after the twist, MMB-click and drag on the Non Linear(twist1) entry in the list and drag it below the bend. Figure 8-2 shows the List of Input Operations window with the new order of evaluations.

**4.** Look at the result in the view window. Notice how the mushroom twists along its own axis while it is being bent, as shown here:

**5.** Go through this process again to add the squash deformer. Apply it to the geometry, move the handle, set the Low and High Bounds, and then place the squash deformer before the twist. Test the results.

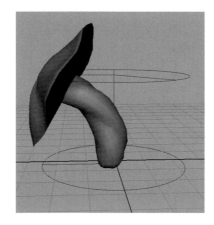

Deformation order is so important when it comes to setting up characters. While we set these up in a specific order here, later we will add multiple blend shape deformers to an object in parallel.

# Specialized Deformers

The deformers that we have explored so far perform somewhat general deformations—that is, they were designed with certain concepts in mind that are common functions in animation (squash and stretch, bend, etc). In this section we look at some of the more advanced deformers available in Maya. These deformers have no one specific function and are usually used in many different ways and combinations to perform very specialized functions. To demonstrate most of these deformers, we will use them to manipulate geometry to sculpt facial expressions. If you wish to follow along, open the file named mcr8_charHead_start.ma.

## Lattice Deformer

A lattice deformer deforms an object through the use of a control cage (see Figure 8-3), similar to the way a low-level subdivision surface model can be used to control a higher-level model. However, the big plus with using the lattice deformer is that you get to choose the resolution of the lattice deformer's cage.

Suppose, for example, that you were working with a polygonal model of a head and had created 5000 faces, and you needed to make the jaw just a little narrower. You could use a lattice deformer and edit just a few of its control points instead of having to select the hundreds of vertices that would be necessary if you chose to edit the model directly. This makes lattice deformers a great tool for modifying the proportions of a model. Figure 8-3 shows a head model surrounded by the control cage of a lattice deformer. The deformer has been used to

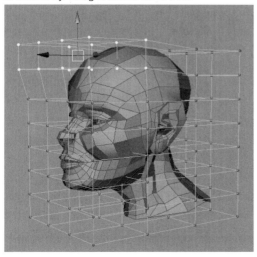

**FIGURE 8-3**   *The entire shape of the head is modified by using a lattice deformer.*

make the top of the head larger and to elongate the nose. In this case, a single lattice point may influence 100 or more vertices on the model. This is why this deformer is great to use for both modeling and animation.

To create a lattice deformer, select an object or a group of vertices and choose Deform | Create Lattice. Once the lattice is created, you will notice that Maya created two lattice nodes in the Channel Box: a lattice and a base, named ffd and ffdBase, respectively (*ffd* stands for free-form deformation). The base lattice is invisible, but you can select it in the Outliner or the Hypergraph. The base is used as the zero point, or origin, for calculating the deform operation.

Lattices use their own local coordinate system, referred to as *STU*. The lattice can be subdivided along the S, T, and U axes by editing the S, T, and U Divisions attributes in the Channel Box or the Attribute Editor. Each point on the lattice can deform the surface it is controlling with an influence that is based on the number of divisions. In other words, if the lattice had four divisions in the S direction, and if the Local Mode box is unchecked in the Lattice Options window (Deform | Create Lattice ▢), as shown in the illustration, the vertices on the lattice have influence over the entire object.

By default, when you create a lattice, its transform node is not connected to the object it surrounds. Therefore, if you move the object, the lattice will not move with it. To remedy this, you can choose options in the Lattice tool's Options window that allow you either to parent the lattice to the object or group the objects together (the Parenting check box and Grouping check box, respectively).

The lattice options can be reset or any tweaks can be removed by choosing Edit Lattice | Reset Lattice or Edit Lattice | Remove Lattice Tweaks, respectively.

When you are finished with the lattice, you can select the geometry and then delete the history. This removes the lattice and keeps any modifications made to the geometry.

## Cluster Deformer

Clusters are essentially a way to assign a transform node to a group of vertices. When the cluster is transformed, the vertices of the affected object will follow. While this is quite useful on its own, the real power of clusters is the ability to adjust the influence of the cluster on each vertex in the cluster's set. Like lattices, using clusters is a great tool for making tweaks to geometry. Figure 8-4 shows a cluster being used to edit the geometry of the upper lip to create a sneer expression; the cluster is being moved upward to create the expression.

Since eventually we want to have several different facial expressions, duplicate the original charHead_baseGeo object and rename it **charHead_LTUpperLip**. This is the object we will modify. To create a cluster, select a group of vertices that you wish to edit. In this case we chose all of the vertices on the left side of the upper lip area. Choose Deform | Create Cluster. A small *C* icon appears in the view window. By selecting this *C* icon, you can use any of the transform tools to edit the cluster. When you first move a cluster right after creating it, it produces a very rigid deformation.

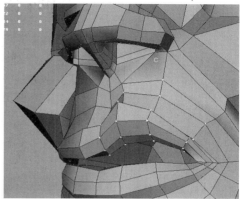

**FIGURE 8-4**   *A cluster is used to sculpt the upper lip section to create a sneer expression.*

In the Channel Box, the default weight (the Envelope attribute) is set to 1. This means that the vertices will be transformed 100 percent with the cluster. If the Envelope attribute were set to 0.5, the vertices would move half as much as the cluster (50 percent). An object can contain any number of clusters, and any vertex can be a member of multiple clusters.

The weight of each vertex can be edited using the Paint Cluster Weights tool, which uses Maya's Artisan brush-based interface to let you assign weight values to the vertices in a cluster by stroking them with a brush. To use the Paint Cluster Weights tool, select a surface that contains a cluster and choose Deform | Paint Cluster Weights Tool ☐ to view the tool's settings, as shown in Figure 8-5.

> *NOTE   If more than one cluster is used on the object, you can choose which cluster you want to modify by LMB-clicking the Cluster1.weights button in the Paint Attributes section. A pop-up menu opens, where you can select the cluster you want to modify.*

When in paint editing mode, the color white on the model represents vertices with a weight of 1, and the black color represents vertices with a weight of 0. You can choose the Replace option to replace the current weight with a value indicated in the Value attribute. Choosing the Add option adds to the current weight by a value indicated in the Value attribute. A quick way to smooth out the weighting of a cluster is to set the operation to Smooth and click the Flood button a few times.

> *NOTE   If you want to edit the weight values of the CVs manually, you can use the Component Editor (choose Window | General Editors | Component Editor). With the Component Editor open, choose the Weighted Deformers tab. Then select a group of CVs in the view window, and click the Load Components button. The Component Editor is discussed in more detail in Chapter 10, when we bind a geometry to a skeleton.*

**FIGURE 8-5** *The Maya UI showing geometry in paint weight mode. Notice the Paint Attributes Tool settings on the right side.*

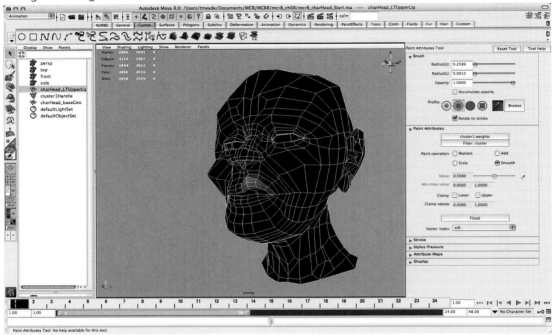

Use the Edit Membership tool (Deform | Edit Membership Tool) to add or remove points from the cluster. With the cluster selected, choose the tool and then SHIFT-drag to add points to the cluster or CTRL-drag to remove them.

## Wire Deformer

The wire deformer uses a NURBS curve to deform an object. You can change the shape of an object by moving a curve or points on the curve. To create a wire deformer, first draw a curve that will be used as the deformer and another object that will be deformed. Choose Deform | Wire Tool. Select the object that will be deformed and press ENTER; then select the curve and press ENTER. The wire deformer will be created.

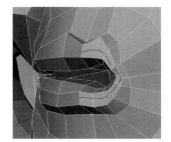

When the shape of the curve is changed, either by transforming the curve or editing its components, it will have a matching effect on the surface that it is controlling. This illustration shows a wire deformer that was drawn to match the shape of the mouth by point snapping the EPs of a curve to the vertices around the lips. This wire deformer can now be used to control and animate the mouth region.

Make another copy of the charHead_baseGeo object and name it **charHead_LTpucker**. Use the EP Curve tool to create the curve and use the previously described procedure to apply it to the head as a wire deformer. If you edit the CVs of the curve as it is, it will have far too much influence over the geometry. To remedy this, lower the Dropoff Distance attribute on the wire1 node to 0.15. Shape the curve as best as you can for one side of the mouth.

> _TIP_   _**It may be difficult to obtain the desired shape precisely with the deformer alone. Use it to make very coarse adjustments to the geometry and then delete history on the geometry and move the vertices one by one or use the Sculpt Geometry tool.**_

## Soft Modification Tool

An excellent tool for making proportional modifications to geometry is the Soft Modification tool. It is very similar to a cluster but editing the effect is much faster when using the Soft Modification tool as a sculpting tool. We will use the Soft Modification tool to widen the mouth.

Duplicate the base model again and name the new shape **charHead_LTwide**. Choose Deform | Soft Modification and click the model near the corner of the mouth. The entire piece of geometry will now be colored with a yellow to red to black gradation radiating from the point where you clicked. This indicates the falloff distance of the effect of the deformer. In the Attribute Editor, select the softMod1 node and lower the Falloff Distance attribute to a value of 0.5. Figure 8-6 shows the model with the falloff distance adjusted. Now you can move and rotate the manipulator to edit the shape of the mouth.

**FIGURE 8-6**   _The Soft Modification tool is used to edit the corner of the mouth._

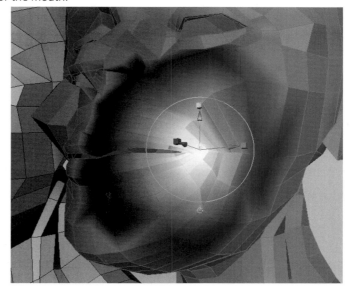

## Blend Shape Deformer

 A blend shape deformer enables you to "morph" the shape of any object into the shape of another object. The original object is known as the *base object*, and the object into which it is being morphed or blended is known as the *target object*. Blend shape deformers are most commonly used for facial animation but we will use them again in Chapter 10 to correct skeletal deformations.

To create a blend shape with what we have created so far, first select the target objects and then select the base object.

> **NOTE**   *Although it is not necessary for the target and base objects to have the same number of edit points, it can be helpful in achieving predictable results.*

Choose Deform | Create Blend Shape to create a blend shape node. Once this node is created, Maya assigns a Weight attribute for each target object you selected. You can set the amount of influence that a target has on the base by editing the Weight attribute for that target in the Channel Box or in the Attribute Editor, which is shown in Figure 8-7.

**FIGURE 8-7**   *The Attribute Editor displaying the Weight sliders for each of the target objects influences*

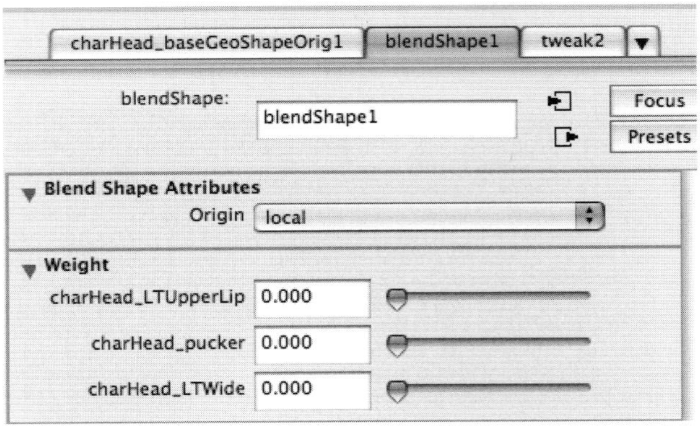

You can add additional target objects to a blend shape by selecting a target object and the base object and choosing Deform | Edit Blend Shape | Add. To remove a target object from a blend shape, select the target object, select the base object, and then choose Deform | Edit Blend Shape | Remove.

Maya offers an additional window where you can edit blend shapes. Choose Window | Animation Editors | Blend Shape, and the Blend Shape window appears, listing all blend shape nodes that are contained in the scene. A slider interface for each target is available to give you better control over your blend shape animation. You'll also see a Key button you can click to set a keyframe for each target's Weight attribute so that you may quickly animate these sliders. Figure 8-8 shows the Blend Shape window in a scene that is set up for facial animation.

# Wrap Deformer

The wrap deformer lets you use NURBS curves or surfaces and polygonal geometry to deform another object. This is a great option for deforming body parts that are built from heavy geometry. In this respect, a wrap deformer is very much like a shapeable lattice. Sometimes, a generic biped geometry is bound and weighted to a skeleton. The final model is then driven by the generic, bound model through the use of wrap deformers. This makes it easy for different people to be working on different parts of a production at the same time.

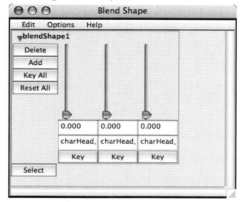

*FIGURE 8-8    The Blend Shape Editor contains many convenient options for animating blend shapes.*

Wrap deformers are also used a lot in simulation. For example, a piece of cloth may be animated with Maya's cloth simulator, but the geometry itself may be too simple for the final rendering (a complex piece of geometry may be too slow to simulate). So, the simulation is carried out on the simple geometry and then applied as a wrap deformer to the complex geometry.

There are so many uses for a wrap deformer. In this section, we are going to use it to address one of Maya's shortcomings—that is, the lack of a shape mirroring function. Since we have sculpted some shapes for the left side of the face, it would be nice to click a button and mirror those shapes over to the other side. We can't use a negative scale on the shapes because blend shaping is based on the unique identification numbers of each vertex. So, we will use the wrap deformer to overcome this hurdle. This may be a little tricky, so follow the steps closely:

1. Select the charHead_baseGeo object and duplicate it. Rename the original **charHead_baseGeoWrap** and then rename the duplicate back to **charHead_baseGeo**.

2. Select the charHead_baseGeoWrap object and set the Scale X attribute to –1 to flip it.

3. Select the charHead_baseGeo object and then the charHead_baseGeoWrap object (they must be selected in this order) and choose Deform | Create Wrap.

4. Select the charHead_baseGeoWrap object and turn the charHead_LTUpperLip attribute on the blend shape node up to 1. Since this geometry is being used as a wrap deformer, the charHead_baseGeo object deforms with it.

**5.** Select the charHead_baseGeo object and duplicate it. Rename this **charHead_ RTUpperLip**.

**6.** Repeat this for the other two shapes. When you are finished, you can zero out all of the blend shape attributes and delete the charHead_baseGeoWrap object. Then select all six target shapes and then the charHead_baseGeo object and choose Deform | Create Blend Shape.

So there you have it! This was an excellent example of how to work around a missing feature. As you become more familiar with Maya, you will learn hundreds of little tricks like this.

## Summary

As you can see, deformers are extremely useful for many applications. Each of them can be utilized for modeling, character setup, and animation. In the next chapter, we'll take a look at a joint deformer and learn how to set up a biped character. Then, in Chapter 10, we will enhance our bound character with even more deformers.

# Joints and Skeletons

**To animate a character effectively,**

you need a structure that allows you to control

the deformations of the geometry based on a

real-life organism. In other words, if the leg needs

to bend at the knee, you need to be able to select

a deformer in the knee region, rotate it, and have

the geometry follow. This type of deformer is of a

special class called a *joint*. In Maya, a hierarchy of

joints, called a skeleton, serves the same function

that a skeleton serves in real life—it forms the internal structure of the character and serves as the framework for performing character movement and deformation.

This chapter not only introduces you to joint deformations, but also commences your journey into the art and science of character setup, or rigging. This aspect of 3D can sometimes seem more technical than, say, modeling or texturing. However, the end result is always to produce something that *looks* right, so, no matter how far on the science side of things you go, always equate it to art. To that end, knowing about how to place joints and control them is a good way for anyone to learn the ins and outs of Maya, be it a modeler or texture artist.

# The Skeleton: Joints and Bones

In Maya, a skeleton is a hierarchy of a unique kind of transform node called a *joint*. Maya skeletons behave like anatomical skeletons, also similar to the armature constructions inside a stop-motion puppet. They allow for easy posing and animating of a character by simply rotating joints.

## Creating Skeletons

Skeletons are created using the Joint tool (Skeletons | Joint Tool). With the Joint tool, you can place a joint into the scene each time you click in a view window. The new joint is automatically parented to a previously placed joint. This hierarchy of joints is visually indicated in the view window by lines, the *bones* that connect the joints. Bones point from a joint down the skeleton hierarchy to the next joint.

Here's how you draw a skeleton:

**1.** Choose Skeleton | Joint Tool, and click in the view window to place the first joint. Click again somewhere else in the view window to create another joint. Notice that the joints are connected with a bone.

**2.** Place a few more joints in the scene and press ENTER (RETURN) to complete the skeleton.

*TIP* **Hold down the X key to snap to the grid or the V key to snap to other joints.**

**3.** Draw another skeleton, but this time, before you press ENTER (RETURN) to finish the skeleton, press the UP ARROW key. This will navigate the selection up the hierarchy to the *parent* of the last joint you created. Click somewhere else in the scene and notice that you have now created a *branch* in the skeleton. A branch occurs when a parent joint has more than one child.

**4.** You can continue to navigate through the hierarchy of the skeleton by using the UP and DOWN ARROW keys.

Figure 9-1 shows a skeleton for a hand. Notice the branches that are created from the hand joint to the fingers and thumb.

## Selecting and Inserting Joints

After you have created the skeleton, you can select a joint by clicking it or by clicking the bone pointing at its child (which is usually easier than clicking the joint). You can then use the transform tools to move, scale, or rotate the selected joint. When you transform a joint with the Move tool, for example, all of the joint's children, or branches, transform as well. This is the expected behavior for any hierarchy in Maya. If you want to edit the position of just one joint in a skeleton and you do not want to affect any of the child joints, select the joint and press the INSERT (HOME) key. This isolates the selection to the single joint.

To insert a joint into a skeleton after it has been created, you can use the Insert Joint tool (Skeleton | Insert Joint). After choosing this tool, click the joint that will be the parent of

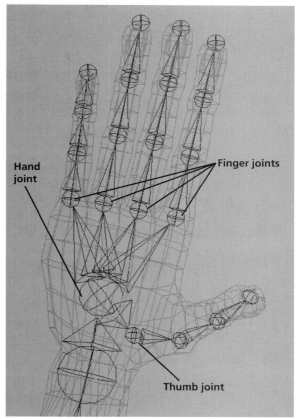

**FIGURE 9-1**    *A skeleton for a hand*

Hand joint

Finger joints

Thumb joint

the new joint you are creating and drag to place the new joint. To remove a joint in the hierarchy, select the joint to be removed and choose Skeleton | Remove Joint.

Suppose you created a skeleton and then decided that you needed to add additional joints—perhaps some joints for ears. You could simply select the Joint tool, click an existing joint in your skeleton's head (if that's where you want the ears to go) to select it, and then continue drawing more joints for the ears. Remember that the skeleton is just a *hierarchy* of *joints*.

Another way to add joints to a hierarchy is to draw the new joints and then use the Edit | Parent command to add the joint as a branch to the skeleton. By selecting the root joint of the new skeleton and then SHIFT-selecting the joint that will be the parent, you can choose Edit | Parent, or simply press the P key, to parent the new skeleton to the intended parent joint.

Definition    ***root joint***    *The topmost joint in the hierarchy.*

# A Closer Look at Joints

As we have mentioned already, a joint is a unique type of transform node. So what makes it so different from a regular transform node, one that controls the placement of a sphere, for example? Well, most of it has to do with how the transformation properties of a joint relate to their parent objects. To better understand this, we will first look at the options on the Joint tool, then discuss joint orientations, and then compare the transformations of a joint's hierarchy to a hierarchy of spheres.

## Joint Tool Options

Let's look at some of the Joint tool's options. Choose Skeleton | Joint Tool ❑. You'll see the tool's options, as shown here:

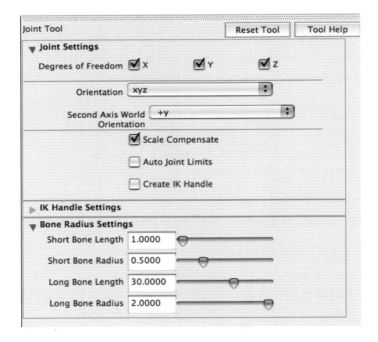

Here's a description of each option:

- **Degrees of Freedom**    Determines on which axis or axes joints are allowed to bend. If all three axes are checked, you will create a *ball* joint that can rotate freely in any direction. If two axes are checked, you will create a *universal* joint that can rotate on two axes only. If only one axis is checked, you will create a *hinge* joint. Editing the degrees of freedom produces the exact same behavior as locking the rotation attributes and is usually the preferred method of isolating rotations.

- **Orientation**   Controls the orientation of a chain of joints when the chain is first created. In other words, it controls which axis of a parent joint will point, or aim, at its child and is determined by the first letter in the list. For example, Maya's default Orientation is XYZ. This means that the X axis will aim toward the child. If you choose YZX, the Y axis will aim at the child. Setting this to None orients the joints to the world instead of aiming them at their children.

- **Second Axis World Orientation**   Controls in which direction the second axis specified in the Orientation attribute will point relative to the world. If we are using Maya's default orientation of xyz, then this attribute controls the direction of the Y axis. You will learn more about how to use this in the next section.

- **Scale Compensate**   Allows you to scale each joint independently. Child joints won't be affected by scaling a parent joint.

- **Auto Joint Limits**   Automatically limits the joint rotation. If the joint is created with a slight bend, for example, this function prevents the joint from bending beyond 180 degrees. You can create much more detailed joint limitations by using the Attribute Editor.

- **Create IK Handle**   Automatically creates an IK chain when the joint creation is complete, although it is recommended that you always set up the IK chain after the joints are created. (We discuss IK in Chapter 12 in "Inverse Kinematics.")

Now that you have an idea of what joints are and how you can create them, let's build a skeleton for a character.

## Orienting Joints

Joint orientation is one of the most important aspects of setting up a skeleton that behaves predictably. In the last section, we talked about the Orientation and Second Axis World Orientation attributes in the Joint tool's settings. This option sets the axis on which the joint will be pointing down the bone. The default setting, XYZ, points the X axis of the joint down the bone. The Orient Joint command can be used to change the orientation of any selected joint or joint hierarchy. Which axis you choose to orient down the bone is up to you entirely. Choose whichever axis best suits your pipeline, but be consistent! Not only do you want to use the same joint orientations in a single character, but you should use the same joint orientations for *all* of the character rigs you build. If you want to drive yourself (and any animators you are working with) crazy, set up some characters to rotate joints on the X axis and then other characters in the scene to rotate along the Y axis. For the examples in this book, we will stick to using Maya's default of XYZ.

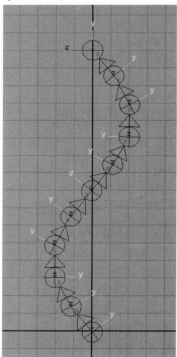

## Local Rotation Axis

To better understand joint orientations and why they are important, we'll study the joint chain shown in Figure 9-2. This joint chain was created with the Joint tool's default setting. It was drawn in the Side view and curves back and forth as it extends along the Y axis of the world.

To view the orientations of each joint, the chain can be selected and the display of the local rotation axis shown (Display | Transform Display | Local Rotation Axis). The local rotation axis is the axis that is relative to the object as opposed to the world. If you were to select one of these joints and translate the joint with the Move tool set to Object, or rotate the joint with the Rotate tool set to Local or Gimbal, the joint would translate along these local axes instead of the world.

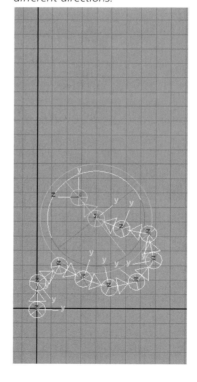

While the joint chain in Figure 9-2 is properly oriented along its X axis, the Y axes are inconsistent as they travel up the chain. If we were to select all of these joints and rotate them around their local Z axis, the joint chain would bend in two different directions, as shown in Figure 9-3. While this may look kind of nice, it is not very predictable. Ideally, when we select a chain of joints and rotate it, we expect all the joints to rotate in the same direction. To fix this, we will use Maya's Orient Joint command.

### Orient Joint Command

Maya's Orient Joint command is used to reorient joint hierarchies based on the settings in the command's Options window, shown in Figure 9-4. These options resemble those found in the Orientation section of the Joint tool's settings. The main difference is that the Orient Joint command can be used after a joint hierarchy is created.

While the joint hierarchy shown in Figure 9-2 is properly oriented along the X axis, the Y and Z axes do not match. When we first placed this joint chain with the Joint tool, the Second Axis World Orientation was set to +Y. Thus, each joint was placed and oriented to point X at the child, the secondary axis, which was Y in this case (based the Orientation setting of XYZ) attempted to point in the world Y direction.

**FIGURE 9-4**   *The Orient Joint Options window*

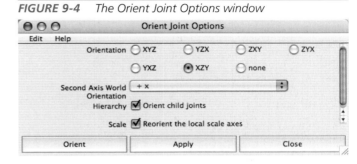

Obviously, since X controls our primary orientation, the Y axis will not point absolutely in the world +Y, but it will do its best. If you reexamine the local rotation axis shown in Figure 9-2, you'll see that the Y axis flips to the other side every time the direction of the joint chain changes. For the first three joints, the Y axis points to the right (which is actually the –Z direction) but when the direction changes at the fourth joint, the Y points to the left. If it were to point right, then it would actually be pointing in the –Y direction. Confused? This concept can be a bit overwhelming at first but it is crucial to understand if you want to properly rig characters. Make sure you spend some time messing around with these settings until you completely understand them.

So to fix this problem, we need to find an appropriate setting for the Orient Joint command so that all of the axes are consistently aiming. For this example, we'll say that we want all of the Y axes to point to the left. Well, in Maya's space, our left is the +Z direction. In the Orient Joint Options window, we can set the Second Axis World Orientation to +Z. With the other options checked, we can select the root joint and click the Apply button. The result is shown next. The Y axes will all point in the +Z direction while the primary orientations remain in +X. If we have two axes properly oriented, then we have the third properly oriented as well.

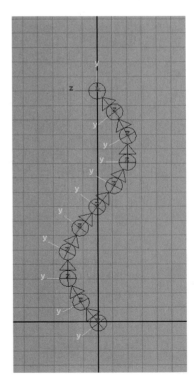

# World, Object, and Local Transformations

A typical transform node in Maya stores information about the location and orientation of an object relative to different spaces. In this section we will examine the behavior of transform nodes as they are parented to different objects and freeze the transforms. We will then look at how joints differ with respect to how they handle their transformations within these different spaces.

## Spatial Relationships of a Transform Node

Three-dimensional positions and transformations exist within different coordinate systems called *spaces*: world space, object space, and local space. *World space* is the coordinate system relative to the scene origin. *Object space* is the coordinate space relative to an object's pivot point. *Local space* is similar to object space but uses the pivot point of a parent object to calculate an offset on which to base transformations.

*FIGURE 9-5    The Attribute Editor showing the object, local, and world space attributes*

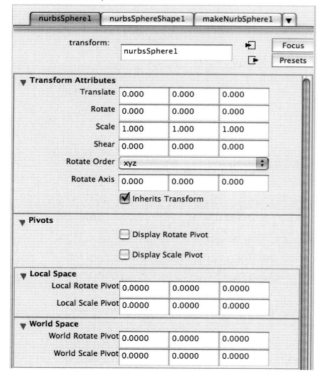

To better understand this, let's take a sphere created at the world origin. Upon its creation, it has Translation and Rotation values of 0. At this point, the world, object, and local spaces all match. To verify this, you can look in the object's transform node in the Attribute Editor and unfold the Pivots folder, shown in Figure 9-5. This section displays the local and world attributes. The object space attributes are the ones listed under the Transform Attributes section and are also the attributes shown by default in the Channel Box.

If the sphere is translated in the X, Y, and Z axes, both the world and object attributes will match. The local space will remain at 0 since there is not yet any offset distance between the object and world pivots. To see a change in the local space, we will change the relationship between the world and object spaces of the sphere's transform node.

One way to do this is to use Modify | Freeze Transformations. When this operation is performed, Maya does not just throw away the translation information. It actually flushes that data into the local space. So, if you had translated the sphere in 2 units in X, Y, and Z, and then performed a Freeze Transformations, the object space attributes will all be set to 0. The local space values will inherit those values of 2 in X, Y, and Z. And, the world space values will remain. As you continue translating the sphere, the object and world spaces will continue updating while the local space values store the offset distance between them.

Another thing to experiment with is to start translating and rotating the sphere and parent it to different objects at different locations in the scene. Create a new sphere and translate it to 5, 5, 5 and rotate it 90 degrees in X. Now duplicate that sphere and translate it to –5, –5, –5 and leave the rotations alone. Now, when you parent the first sphere to the second sphere, the object space attributes of the first sphere change to reflect the offset values relative to the new parent. The object's translate values are 10, 10, and 10, while the rotations are all 0. This is because, relative to its parent, this object has the same orientation, and therefore there is no offset.

All of this is useful information for building animation. You always want to be aware of what space you are in so that you don't get lost! We will make great use of the concepts illustrated in these experiments as we build our control rig in Chapter 12.

## Spatial Relationships of a Joint Node

Alright, let's get back to discussing joints and examine how they differ from basic transform nodes with respect to their coordinate systems. Joints are actually a bit simpler than a basic transform node. A joint's translations are always calculated in local space. That is, a joint's translations are always an offset distance from their parent object, be it the world, another joint, or another transform node. This is where a joint's orientation comes into play.

To experiment with transforming joints, create a small joint chain consisting of three joints. Select the second joint, joint2, open the Attribute Editor, and open the Joint folder to view the joint's attributes. Figure 9-6 shows a joint chain and its attributes in the Attribute Editor. Notice that there are no local or world space attributes. Since the Joint tool was set to orient the joints with X pointing down the bone, the translations of the second and third joints return the offset distance relative to the parent in only one value, X.

If you set the Move tool to translate in local space (set this in the Move tool's settings or hold down the w key to view the Move tool's settings marking menu), the transform manipulator displays the local transform directions. When you grab the red handle and translate the joint, you are moving it along its oriented axis. Doing so allows you to change the position of the joint without breaking the orientation from its parent.

If you move the joint in any other direction than its local X, the parent will no longer be oriented to aim at this joint. Transform values in Y and Z will accumulate, and when this happens,

**FIGURE 9-6** *A joint chain and a selected joint's attributes*

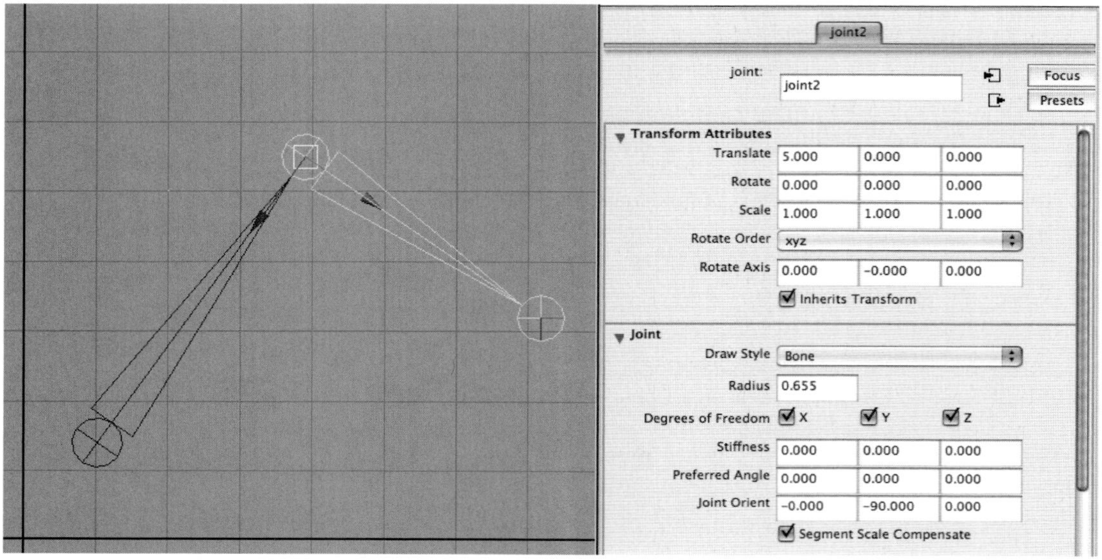

you know that the joint chain needs to be reoriented. By selecting the parent joint and then reorienting the joint chain with the Orient Joint command, the local rotation axes will realign themselves and the child joints will return to having only one offset value given in X (or whatever axis the joint chain is oriented in).

Another thing to point out about joints is that because they do not have other spaces to flush values into, freezing transformations will never affect the translation attributes. The translation will always read the offset distance from the parent. Period! If you set a joint's translations to 0, 0, 0, it will always drop right on top of the parent.

The rotations, on the other hand, can be flushed into another space. If you were to grab a joint and rotate it, then perform a Freeze Transformations, the current rotations would reset to 0 and the offset distance would be calculated relative to the parent joint and flushed into the Joint Orient attribute, also shown in Figure 9-6.

While this information may not be the most fun to learn, understanding it will help you build stable rigs in Maya and enable you to troubleshoot problems when things go wrong. Ninety percent of rigging problems are usually due to joints or controls not being oriented properly. So with that, let's build a skeleton for our character.

# Tutorial: Creating a Biped Skeleton

In this tutorial, we will create a skeleton for a biped character. We'll draw a separate skeleton structure for the spine, leg, arm, and hand. Once the joints are created, we'll edit and fine-tune their placement to match the features in the geometry, reorient them to fix any orientation issues, and name the joints. We'll then parent these separate joint chains together to create a single skeleton. Finally, we'll do some cleanup and save a setup pose so that we can always return to this setup in case things get nudged accidentally.

 To begin, open the file called mcr8_ch09_start.ma. This file contains a character that we will work on over the next several chapters.

> **TIP**  *While we won't step through the process of creating a custom shelf for rigging, we do suggest building one as you work through the rigging of this character.*

## Create the Spine Skeleton

The spine chain will actually include the neck and head. We begin in the Side view:

**1.** Place the first joint near the center of gravity.

**2.** Since most of the rotation of the spine comes from the lower back, we will place more joints in the lower part of the spine. This will give us a nice, smooth deformation in this area when we skin the character later on. Place another four joints evenly from the first joint to the base of the rib cage.

> **NOTE**  *While using a lot of joints in the spine is really nice for deformations, it can make the spine difficult to control. In Chapter 12, you will learn how to use the spline IK to control this spine. If you prefer to animate the skeleton entirely with forward kinematics (rotating singe joints), then you may want to add only one joint in this area.*

**3.** Since we want it to appear as if the rib cage is a solid piece, we only need one more joint in the spine. Place this joint at about where the base of the shoulder blade would be.

**4.** For the neck, place two joints, one at the base of the neck, the other in the middle of the neck. Try to keep these joints near the center of the volume of the neck so that it can twist evenly.

**5.** The head joint is at the base of the skull, behind the ear. Place one more joint. Its precise placement is not so important, as it will not be used to pivot any vertices from. The main reason for having it is so that the joint below it can be oriented.

**6.** When you have finished placing these joints, press ENTER (RETURN) to exit the Joint tool. The entire joint chain is shown in Figure 9-7.

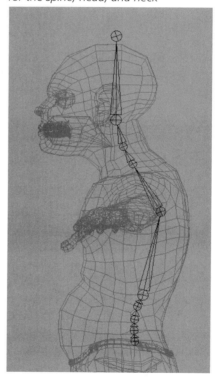

**FIGURE 9-7** *The joint placement for the spine, head, and neck*

**7.** As long as you remain in the Side view, it is safe to edit the positions of the joints however you'd like. You can move them with the Move tool, or press the INSERT (HOME) key to edit the pivot point without affecting the child joints. You may also choose to rotate the joints.

**8.** Now we will use the Orient Joint command to fix any orientation problems that we introduced during the creation or editing processes. Select the first joint and choose Skeleton | Orient Joint ❑ to open the Options window. Reset the settings to their defaults and then set the Second Axis World Orientation attribute to +Z. Click Apply to reorient the entire joint chain.

**9.** Verify that all of the joints are properly oriented. If you choose the first joint and set your Move tool settings to Object, the transform manipulator will show you the object space orientation. Verify that the X is pointing at the child joint and the Y is pointing forward. Then press the UP ARROW key to select the next joint and verify its orientation. You can continue this process all the way up the spine so that it looks like this:

**10.** Another way to verify that the joint chain is properly oriented is to look at the translate values in the Channel Box as you select each joint. If any of the joints (other than the first joint) has any translation values other than 0 for Y and Z, then you know that there is a problem. Go back and reorient the joint chain with the Orient Joint command.

**11.** Eventually, we will have separate controls for the upper and lower body so that we can rotate the hips without affecting the spine, and vice versa. Therefore, we will create another joint chain pointing downward for the pelvis. To snap to the spine joint but not parent the new joint to joint1, select the Joint tool and click somewhere out in space, away from the existing joints. While still holding down the mouse button, hold down the V key and then snap the new joint on top of joint1. Place the next joint below that, and defining the axis that the hips would rotate around. The placement should look something like the image below.

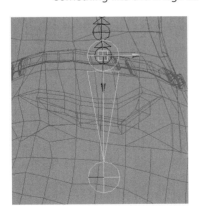

**12.** Use the Orient Joint command to orient this new joint chain. Since this joint is traveling in the opposite direction, set the Second Axis World Orientation attribute to –Z.

**13.** Place one more joint and snap it to the joint1 position. This will serve as the root joint for the entire skeleton.

**14.** Parent the two other joint chains to this new joint. This time, use a hot key. Select pelvis_1, spine_1, and then the root joint and press the P key.

**15.** Name the joints. Give each of these joints a unique name that intuitively describes what it is. Name the new, parent-most joint **root**.

**16.** Name the six spine joints **spine_1**, **spine_2**, and so on. Alternatively, instead of selecting each joint and renaming it separately, you can use Maya's Quick Rename feature. Use the Outliner to select all of the joints in the spine. Select the topmost spine joint in the hierarchy and then, holding down the SHIFT key, select the last spine joint. All of the joints in the hierarchy between these two joints will now be selected.

**17.** On the status line, click and hold the Selection option and chose Quick Rename from the resulting pop-up menu:

**18.** In the field next to Quick Rename, type **spine_**. Press ENTER (RETURN), and all of the selected joints will be named spine_ followed by a number (spine_1, spine_2, and so on).

**19.** You can use the same technique to name the two neck, head, and pelvis joints. The completed joint hierarchy in the Outliner should look like this:

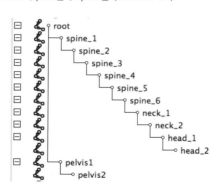

## Create the Leg

The leg consists of the hip, knee, ankle, ball of foot, and toe. We will place the joints in the Side view and then rotate them into position from the front.

**1.** In the Side view, place the hip joint. Use Figure 9-8 as a guide.

**2.** To maintain volume in the knee area when the knee bends, place two joints for the knee, one at the top portion of the knee and the other at the bottom part of the knee. The joints should be placed toward the front of the knee so that the knee does not look rubbery when it bends.

**3.** The ankle joint should go right where the foot rotates from. It is important that it is placed back far enough so that there is a slight bend in the knee from the Side view. This will make the IK chain behave much better when we connect it later.

**4.** Place the ball of the foot along the base of the foot and the toe joint at the very end of the toe.

**5.** Orient the joints so that the Y axis faces forward.

**6.** In the Outliner, name the joints **LT_hip**, **LT_knee_1**, **LT_knee_2**, **LT_ankle**, **LT_ball**, and **LT_toe**.

**7.** In the Front view, select the LT_hip joint and move it along its own Z axis toward the outside of the hip. Since the character more likely will spread its legs outward than inward, this joint placement will provide optimal deformations later on.

**8.** Rotate the hip slightly so that the ankle aligns with the center of the ankle region. Do not bend the knee from this view. When we eventually set up the IK controls in Chapter 12, keeping the knee joint in line with the hip and ankle will help to stabilize the IK solver (the control system that will end up calculating the rotations of the joints).

**9.** Select the ankle joint and rotate it back so that it bends perpendicular to the ground, as shown below.

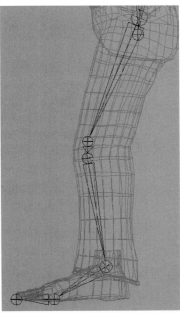

*FIGURE 9-8*  *The leg is placed from the Side view.*

**NOTE**  *Always try to avoid rotating the knee from this angle if possible. While it needs to bend in one axis for the IK solver to work properly, bending it in two axes will also create problems.*

**10.** Freeze transformations on this joint chain so that all of the rotation values are set to 0.

## Create the Arm Skeleton

The arm will contain a clavicle joint, a shoulder joint, an elbow joint, and a wrist joint. Once these are placed, we will add twist joints in the bicep and forearm areas.

**1.** In the Top view, place the clavicle joint toward the front of the base of the neck.

**2.** Align the shoulder, elbow, and wrist joints the best you can from the Top view.

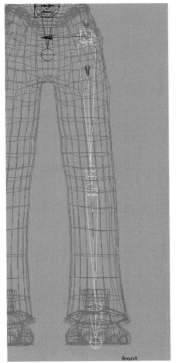

**3.** Before we begin rotating the joints, we need to orient them. Use the Orient Joint command. Again, pointing the Y axis in the +Z will do the job.

**4.** In the Front view, select the clavicle and move it up. Rotate it so that the shoulder sits in place. Then rotate the shoulder down so that the wrist is in position. The arm should look like the illustration:

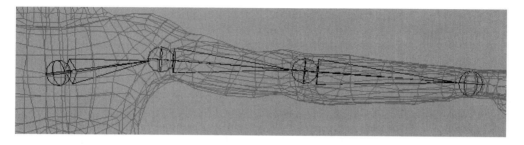

Now for the twist joints. It is important that the twist joints for the bicep and the wrist lie along the oriented axis of the shoulder and elbow. Otherwise, the arm will wobble when the joints twist. To do this, we will unparent the joints in the arm, duplicate the shoulder and elbow, and then move them along their own axis and into place.

**5.** Select the wrist joint and delete it. It's okay! We only placed it in order to orient the elbow joint.

**6.** Select the elbow and shoulder joints and unparent them.

**7.** Select the shoulder joint and duplicate it. Rename the duplicate **LT_bicep**.

**8.** Move it along its X axis to about the middle of the bicep, as shown here:

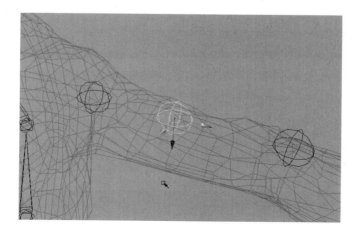

**9.** Select the elbow joint and move it along its own X axis to the forearm area. Name this joint **LT_forearm**.

**10.** Duplicate the LT_forearm joint and move it along its own x axis to the wrist area. Rename this joint to **LT_wrist**.

**11.** Starting with the wrist, parent the arm back together. The final arm is shown in Figure 9-9.

---

*FIGURE 9-9*    *The completed arm with all of its twist joints*

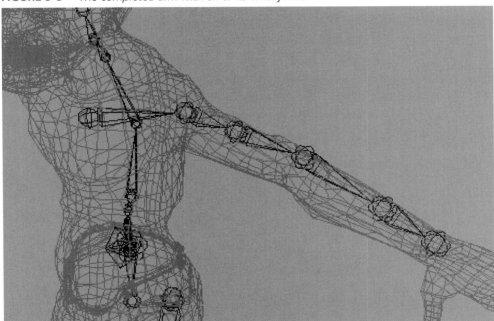

## Create the Hand Skeleton

The hand can take a while to set up because there are so many joints to deal with. Also, the thumb's axis of rotation occurs at an angle and will need to be manually tweaked. As far as placement is concerned, the finger joints should be placed near the very top of the knuckles. While there will be a lot of compression on the inside of the fingers, this is rarely visible, so you don't have to worry about it so much unless it's in a shot.

**1.** Start with the index finger and place the joints from the Top view. Orient them with the Orient Joint command.

**2.** Move the root joint up so that it sits in the first knuckle and then begin rotating the joints down into position. When you are finished, freeze transformations. The finger should look like this:

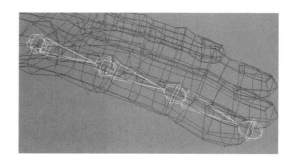

*TIP*  *Because there are so many joints in such close proximity, you may wish to adjust the Radius attribute of the joints to make them display smaller.*

**3.** Duplicate that finger and move it into position to match the first knuckle of the middle finger.

**4.** Set the Move tool to move in local space. Now you can select each joint and change the length of the bone by moving the joint along its local X axis. Rotate the joints if necessary. Freeze transformations when you're done.

**5.** Continue this process for the rest of the fingers.

**6.** For the thumb, duplicate one of the fingers. Move and rotate the finger to match the length of the thumb. The other thing to tackle here is that the thumb joints need to be turned so that the local rotation axis corresponds to the modeled knuckle's axis of rotation.

**7.** Place a hand joint. This can go in the center of the hand. Its orientations are not important, as it will never be animated. Its only function is to hold weight in the hand when the model is bound. Name this joint **LT_hand**.

*NOTE*  *Depending on the needs of your model, you can place additional joints in the hand that will provide the rig with the ability to cup the hand.*

**8.** Select the root joint of each knuckle and then the joint we just created. Press the P key to parent all of the fingers to the hand.

**9.** Renaming each finger with the Quick Rename feature will involve an extra step. The problem stems from the fact that the fingers were duplicated and the Quick Rename feature is based on the absolute, or short, names of objects. While the root joints were renamed when the fingers were duplicated, the children were not. To do this, we will hack around this problem by giving each finger a prefix. Then, we can go through and use the Quick Rename feature.

**10.** Select the root joint of the index finger. Choose Modify | Prefix Hierarchy Names. When the Prefix Hierarchy window opens, you may use the default prefix of **prefix_** as shown here. Click OK. All of the joints in this hierarchy will now have a unique name.

**11.** Now you may use the Quick Rename feature. Select all of the joints in the finger and type **LT_index_** in the Quick Rename field, and all of the fingers will be renamed. Repeat this process for the rest of the fingers, naming them **middle**, **ring**, **pinky**, and **thumb**.

**12.** Parent the hand joint to the LT_wrist joint. Figure 9-10 shows the completed skeleton for the hand.

This completes the creation of the joints.

FIGURE 9-10    *The completed skeleton for the hand*

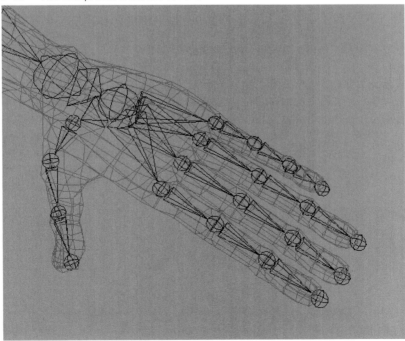

## Mirroring Joints

Now that all of the joints have been placed and named for one half of the skeleton, they can be mirrored to the other side:

**1.** Select the LT_clavicle joint. Choose Skeleton | Mirror Joint ☐.

**2.** Set the Mirror Across attribute to YZ and the Mirror Function to Behavior. To rename the joints for the other side, search for **LT_** and replace it with **RT_** as shown here.

**3.** Click Apply to mirror the joint chain.

**4.** Repeat this process for the leg.

**5.** Parent both hip joints to the pelvis_1 joint. Parent both clavicle joints to the spine_6 joints.

Figure 9-11 shows the completed skeleton.

## Clean Up

Before moving onto the skinning process, there are some things you can do that help keep your skeleton intact while you continue working on it.

**FIGURE 9-11**   *The completed biped skeleton*

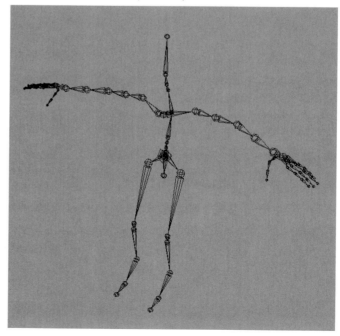

## Check Orientations

If you've made it this far, then you're probably sick of reading about the orientations. However, this is your last chance to fix them before binding. Once the joints are bound, they cannot be reoriented. Walk through the skeleton and look at each joint. Watch the Channel Box to see if there are any translate values other than ones in X (the branch joints, such as the hips and clavicles, are the exception).

## Lock and Make Nonkeyable

When an attribute is locked, its value cannot be changed. None of the joints will be scaled. Since all of the transformations on the joints, with the exception of the root, will be based on rotation, we can lock all of the translations. In some cases, as in the knees and the elbows, we can lock the X and Y rotation attributes. This ensures that nothing gets accidentally moved or rotated.

When an attribute is set to be nonkeyable, it cannot be keyed with the Animate | Set Key command. Usually, this cuts down on the amount of data recorded while animating. In this case, we will be setting keyframes on control objects that operate the joints. However, making attributes nonkeyable removes them from the Channel Box and makes the Channel Box easier to read because there isn't any superfluous information.

**1.** In the Outliner, select all of the joints except for the root.

**2.** In the Channel Box, highlight the Translate and Scale X, Y, and Z attributes, shown here, and RMB-click in the Channel Box to view a marking menu.

| LT_toe . . . | |
|---|---|
| TranslateX | 1.767 |
| TranslateY | 0 |
| TranslateZ | 0 |
| RotateX | 0 |
| RotateY | 0 |
| RotateZ | 0 |
| ScaleX | 1 |
| ScaleY | 1 |
| ScaleZ | 1 |
| Visibility | on |
| Radius | 0.54 |

**3.** Choose Lock and Hide Selected from the marking menu.

**4.** Choose the elbow and knee joints. Lock and Hide the Rotate X and Y attributes.

**5.** Select the bicep and forearm joints. Lock and Hide the Rotate Y and Z attributes.

**6.** Select the Hand joints. Lock and Hide all of the attributes, as we will never edit these.

**7.** Select the root joint. Lock and Hide all of the Scale attributes.

*NOTE  One thing to realize about locking objects in a hierarchy is that if they are unparented, their position will change because the translation values will not be able to update to reflect the offset values for the transforms relative to the world. If you need to make changes to the hierarchy, use the Channel Control window (Windows | General Editors | Channel Control) to unlock the attributes before removing them from their existing hierarchy.*

## Summary

In this chapter we covered quite a few new things. You not only learned how to build a clean skeleton that can be properly controlled, but also learned some things about how Maya uses different coordinate systems to calculate transformations. All of the concepts here will be used again and again as we build on this character rig.

# Skinning and Advanced Deformations

Now that we have created a skeleton that is easy to control, the next step is to make that skeleton control the geometry. This is done through a process known as *skinning*. Skinning assigns how the vertices of the geometry (or lattice) are influenced, or weighted, to a joint. Skinning, or *binding skin*, is the process of assigning how the vertices (control vertices, polygonal vertices, lattice points, and so on) are influenced by the position of a joint. Figure 10-1 shows a character that has been bound to a skeleton, which is then posed by moving the skeleton. This chapter explores various ways to skin the character so that it is ready for animation.

There are two types of skinning in Maya, rigid bind and smooth bind. Each one handles how the vertices are weighted, or influenced by the joints differently.

# Rigid Bind

Rigid binding (Skin | Bind Skin | Rigid Bind) enables only one joint to influence a vertex. This results in a harsh, stiff deformation of the surface surrounding the joints. Many real-time video games use rigid binding for its simplicity and speed, as the deformations can be calculated much faster when weights are not shared between joints.

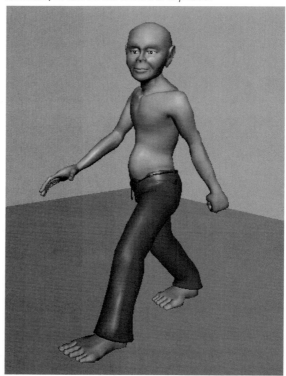

**FIGURE 10-1**   *Once the geometry is bound to the skeleton, it can be deformed and posed.*

## The Rigid Bind Process

This quick tutorial illustrates the basic process of binding using rigid bind. We'll stick to using a cylinder for now.

**1.** Create a polygonal cylinder by choosing Create | Polygon Primitives | Cylinder. Scale it up to 5 units in Y. View the attributes for the polyCylinder1 input node by clicking it in the Channel Box. Change the Subdivision Axis attribute to 8 and the Subdivision Height to 12.

**2.** In the Front view, create three joints by choosing Skeleton | Joint Tool and clicking in the window from bottom to top, as if you were drawing the skeleton for an arm.

**3.** Select joint1 and SHIFT-select the cylinder. Choose Skin | Bind Skin | Rigid Bind ☐. The Rigid Bind Skin Options window opens:

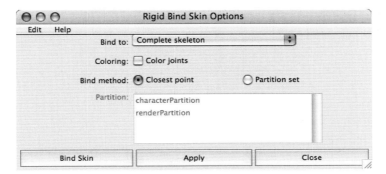

The options are as follows:

- **Bind To**   The Complete Skeleton option allows you to bind an entire skeleton hierarchy while selecting only one of the joints from that hierarchy. The Selected Joints option binds only selected joints to the skin.

- **Coloring**   Allows you to color-code the joints and the skin points with corresponding colors. For example, if the elbow joint is red, all the points that are bound to the elbow joint will also be red. This is a useful option for quickly editing weights.

- **Bind Method**   Determines how points are bound to the joints. The Closest Point option binds points to the nearest joint automatically. The Partition Set option uses premade partitions that you can create before binding. When Partition Set is active, a list of all available partitions will be included in the Partition list.

Set the Bind To option to Complete Skeleton, check Color Joints, and set the Bind Method to Closest Point.

**4.** Select the cylinder, and you now see that the joints are colored individually. Rotate the middle joint and you will see that deformation around the elbow is pretty stiff and collapsing on the inside of the bend. Figure 10-2 shows the result of this rigid deformation.

**5.** Select some of the vertices in the area around the deformation and open the Component Editor (Window | General Editors | Component Editor). You will notice a new tab called Rigid Skins, which shows which joint is controlling the selected vertices, as shown in the following illustration. The weight value for all of these vertices will be

1.000 toward a single joint, since only one joint can influence each skin point. However, this does not mean that the weight values cannot be edited. Indeed, they can.

**6.** In the Component Editor, choose some of the weight values for the CVs that are weighted to joint2Cluster2 and change their values to **0.5**. In the view window, select joint2, rotate it, and note the behavior of the vertices that were just edited. They are now only halfway influenced by the transformation of the elbow joint. However, the remainder of their weight is not redistributed to be shared among other joints as when using smooth bind. Instead, the result is more straightforward. These vertices are simply affected by this one joint only by a factor of their weight. So you can see that it is possible to obtain a smooth deformation using rigid bind alone.

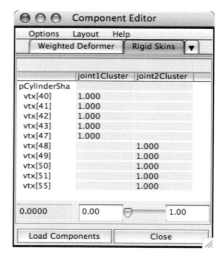

FIGURE 10-2   *The second joint is rotated and the result is this very harsh, rigid deformation.*

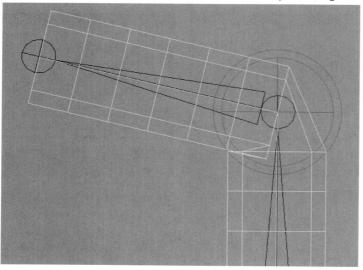

# Edit Membership

The grouping of points for the particular joint is called *set membership*. All the skin points of the rigid skinned surface should belong to at least one set. You can edit skin points' membership in two ways: by using the Edit Membership tool or by using the Paint Set Membership tool.

The Edit Membership tool works like this:

**1.** Select the tool by choosing Deform | Edit Membership Tool.

**2.** Select the elbow joint, and all of the skin points that are members of the elbow joint become highlighted in yellow.

**3.** To add more skin points for this membership, you can hold down the SHIFT key and LMB-click the vertices you wish to add. To remove the points from the membership, hold down the CTRL (COMMAND) key and LMB-click the point. This way, you can quickly add and remove skin points to any joints. Try changing membership and undo back to the original set membership.

*TIP  Removing skin points can lead to "orphaned" points (points that are not members of any joint cluster) and therefore you should avoid this practice. It is better to select the joint that will be added to and use the Add (SHIFT key) function to "pluck" the point from the previously defined membership so points will not be left behind when the skeleton is moved.*

The Paint Set Membership tool works in a similar way, but it uses Artisan tools to edit the membership by a painting method:

**1.** Select the NURBS cylinder and choose Deform | Paint Set Membership Tool ❑.

**2.** Select a set from the Set Membership area of the Tool Settings window.

**3.** LMB-click and drag on the surface to add, transfer, or remove the skin points from the different sets.

# Flexors

A *flexor* is a lattice-like object that enhances the deformation around a joint. While similar to a lattice, the flexor has many built-in attributes that can control a variety of creasing effects. Let's continue using the cylinder from the last section.

**1.** In order to create a flexor, select joint2 and choose Skin | Edit Rigid Skin | Create Flexor. The Create Flexor window opens:

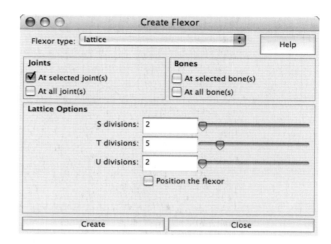

**2.** Make sure the Joints section has At Selected Joint(s) checked. This will create a flexor only at the selected joint. You can also create flexors on the entire length of bones by checking the At All Bone(s) option in the Bones section.

**3.** The Lattice Options determine the number of divisions in the flexor (see Chapter 8 for details on divisions). Let's keep these set at the default: 2, 5, 2. If Position the Flexor is checked, you can change the placement of the flexor. Leave that option unchecked. Click the Create button.

**NOTE** *The Position the Flexor option can be useful in some cases. Many times the automatically generated flexor is at an unusable angle and the placement and dimensions must be adjusted using this option.*

**4.** The flexor appears over the elbow joint. Select the elbow joint and rotate it to bend the cylinder. You see that, currently, too much collapsing is happening around the inner side of the elbow and too much rounding/expansion happens along the outer part of the elbow joint.

**5.** Select the flexor you just created and look in the Channel Box. Click the jointFlexor1 node to view its attributes. To make the elbow stick out more, change the Rounding attribute to 8. To make the inner side of the elbow not collapse on itself, change the Creasing attribute to –2, and then change the Length In and Out attributes to –1.

You can see how the flexor allows you to create better creasing around the joint by easy-to-use preset attributes. Figure 10-3 shows the geometry before and after the flexor attributes are edited.

FIGURE 10-3    *The geometry before (left) and after (right) editing the flexor's attributes*

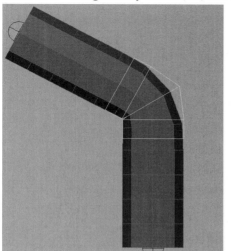
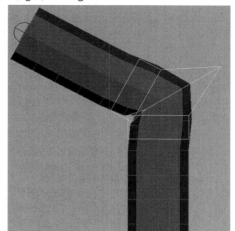

# Smooth Bind

Smooth binding is the other of the two skinning methods available in Maya and is the preferred method of achieving more realistic deformation of the geometry. Smooth binding allows for each skin point to be influenced by multiple joints instead of being influenced by only one joint, as in rigid binding.

## The Smooth Bind Process

Skin Points A is a group of three skin points that lie right in line with the elbow joint. We need to figure out how those skin points should be weighted to the joints around them to achieve the most realistic deformation when the elbow joint is rotated. Let's think about it. If Skin Points A was assigned a weight value of 1 to the elbow joint, it would then rotate around the elbow joint by an influence of 100 percent.

However, this is not very natural. We really want those skin points to stay right in line with the elbow joint. To achieve this, the vertices will have to share their influences between multiple joints. In this case, they will be weighted with a value of 0.5 to the elbow joint and 0.5 to the shoulder joint. This will result in a nice, even deformation.

Realize that this is different from rigid binding. In this case, it is true that you could achieve a similar result by changing the weight of the points toward the elbow joint. However, with rigid binding, the weight is *not* shared between influences. Having multiple influences will be especially useful in areas such as the torso, where you might want to have several of the spine joints influencing each point just a little bit.

## Weight Normalization

Just as the word *sharing* suggests, there is only so much weight available to go around. This means that no matter how many joints are influencing vertices, the total amount of influence between all joints must add up to 1. By default, Maya performs calculations for you through a process called *normalization*.

While this sounds like a good thing, and indeed it is, it can also be the reason for much confusion to a person just learning to use smooth bind. The problem arises when you need to subtract weight from a selected joint. Say you had three joints in a hierarchy and were working on the third. And let's say that the vertices in this area had a weight of 1 to this joint and you wanted them to only be influenced 75 percent. You might expect that if you subtract 0.25 from the third joint, it would go to the second joint. But this is not always the case.

While the smooth skinning tools in Maya have many features that help manage where weights will go when they are subtracted from an influence, issues always arise from the normalization process. When working on a full character, it is possible that by just subtracting an influence of 0.1 from one joint, it will get divided between all of the joints in the entire skeleton. This results in small deformations occurring in, say, the elbow when the toe joint is rotated.

# Tutorial: Smooth Skinning a Character

With a basic understanding of skinning, it is time to practice on a model.

## Prepare the Model

You should always do the following before you begin the skinning process:

- *Finish all the modeling*. While Maya has recently added some features to work around this (Substitute Geometry), it is still always better to maintain a clean model. Unnecessary nodes in the shape's inputs will slow down the rig.

- *Freeze transforms and delete the history*. Strange things will start to happen if you don't begin with a clean geometry. When possible, it is always best to delete all history and freeze the transformations on the geometry. If for some reason your model has dependencies that are needed, consider deleting the history and reapplying or reconnecting the nodes that are needed.

- *Disable IK and zero out joints*. Always return the skeleton to the position it was in when the joints were carefully placed. If you were careful to do this back at that stage of the game, you can type in **0** for all of the rotation attributes for all of the joints in the skeleton. If the IK is currently enabled, you may not be able to edit the rotation values for the joints being controlled by the IK. If you are binding to a skeleton that already has IK or controls set up, you should choose Modify | Evaluate Nodes | Ignore All to disable the controls.

- *Don't bind overly complex models.* The more complex the model, the more time and effort you will spend tweaking weights. Evaluate your model and determine whether it will be best to use a direct or indirect skinning approach.

- *Know your shot.* This goes for any part of the production pipeline. If parts of the model will not be seen deforming, don't spend a lot of time adjusting the skin weights for those regions. Also, add influence objects only where they're needed. Unless the camera will shoot a close-up of a hand, for example, nobody will see or appreciate the hours of work you put into setting up anatomically precise influence objects.

## Skin a Character

Let's apply the smooth skinning to the character. We'll start by binding with smooth skin:

**1.** Open the scene called mcr8_ch10_start.ma from the DVD. This is the same file that was completed at the end of Chapter 9. If you've been following along, you may use your own project.

**2.** To make the most efficient deformations, we only want to bind to the joints that will actually rotate. The joints in question are the tip joints, or the joints that are found at the end of every branch in the skeleton. Realize that the tip of the finger will be deformed by the second-to-last joint, not the last joint. Because these joints were really only created to orient their parent (or used as a pivot for some kind of animation), they do not need to be included as a joint that will have any influence over the geometry. Also, the root joint is currently only serving to drive the translation of the entire skeleton and will not need to be used as an influence object.

**3.** In the Outliner, unfold the joint hierarchy so that all of the joints are visible. Select the spine_1 joint and then, holding down the SHIFT key, select the last joint in the hierarchy, in this case the LT_toe joint.

**4.** Starting from the top of the list and working your way to the bottom, deselect any of the tip joints by holding down the CTRL (COMMAND) key and clicking the joint in the Outliner.

**5.** You can create a selection set that contains these joints by choosing Create | Sets Quick Select Sets. Name the set **skinJoints**. This is a good way to always have record of the joints you are skinning. If they become deselected or you need to unbind your character and rebind later, you can quickly select these joints by choosing Edit | Select | Quick Select Sets | skinJoints.

**6.** Still holding down the CTRL (COMMAND) key, select all of the geometry, excluding the eyes.

**7.** Choose Skin | Bind Skin | Smooth Bind ☐ to open the Smooth Bind Options window, shown here. Set the Bind To attribute to Selected Joints, the Bind Method attribute to

Closest in Hierarchy, and the Max Influence to 1. Deselect the Maintain Max Influences check box and click the Bind Skin button. You may need to wait a bit for all the binding calculation to finish.

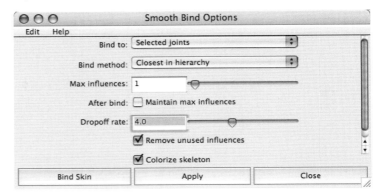

**NOTE**   _The settings suggested here will produce a very harsh initial bind, very similar to what a rigid bind would produce. In many cases, it is easier to smooth out harsh deformations and harden extremely smooth or rubbery deformations. However, this all depends on the model being bound and the number of joints used. As you become more familiar with skinning, you should experiment with other settings._

The character is now bound. Take a few moments and examine the initial deformations. Look for any extreme problem areas. These usually occur in the shoulder region and the hips.

## Paint Skin Weights

Let's fix some of the problems by painting the skin weights. When you paint skin weights, the influence toward a specified joint is indicated by a grayscale color on the model, where black means that there is an influence of 0 and white means that there is an influence of 1. The goal of painting skin weights is to produce smooth, even deformations, _not_ to produce anatomical accuracy. For additional realism, you'll need to use some additional deformers or influence objects, which are discussed later in this chapter. With that goal in mind, let's begin with the wrist:

**1.** Bend the LT_wrist joint and notice how the geometry behaves in the wrist area. If you study the edge loops around the wrist, shown in Figure 10-4, you'll notice portions of them deform while some other parts stay. We want to edit the weights so that the edge loop deforms evenly.

**FIGURE 10-4**    *The wrist area before (a) and after (b) painting the skin weights*

**2.** With the body geometry selected, choose Skin | Edit Smooth Skin | Paint Skin Weights Tool ☐. In the Paint Skin Weights Tool settings window, shown in Figure 10-5, select LT_wrist from the Transform list and change Paint Operation to Add. Set the value to something small, like 0.05. The idea is to gradually add weight to the black vertices around the wrist in an attempt to even out the deformation. Based on the discussion of weight normalization earlier in the chapter, we want to be aware of which influences the weight comes from and where it goes. If we stick to only using the Add operation, we will always be pulling weight from another influence, while using an operation such as Scale or Replace (with a low value) will subtract and therefore flush weights into some unknown location.

**FIGURE 10-5**    *The Paint Skin Weights Tool settings window*

Paint Skin Weights Tool

Reset Tool     Tool Help

▼ **Brush**

Radius(U): 0.2531

Radius(L): 0.0010

Opacity: 1.0000

Profile: ◉ ◉ ● ▣  ╱  Browse

☑ Rotate to stroke

▼ **Influence**

Sort transforms: ○ Alphabetically   ◉ By Hierarchy

Transform: spine_1
spine_2
Hint: spine_3
spine_4
Use the RMB  spine_5
over a joint in  spine_6
the modeling  neck_1
view to select  neck_2
head_1
it for painting  RT_clav

Toggle Hold Weights on Selected

▼ **Paint Weights**

Paint operation: ○ Replace    ◉ Add

○ Scale    ○ Smooth

Value: 0.0500

Min/max value: 0.0000    1.0000

Clamp: ☐ Lower    ☐ Upper

Clamp values: 0.0000    1.0000

Flood

*TIP*  *While the Toggle Hold Weights on Selected button in the Paint Skin Weights Tool settings window will keep the values on a held joint from changing most of the time, we still find it faster to take a pass at the model while only using the Add operation. This keeps the flow of painting weights moving faster, and without the interruption of having to toggle influences on and off.*

**3.** So, if we are only going to use the Add operation on the Paint Skin Weights tool, then what do we do when a joint is influencing the vertices too much? Well, first you decide what joint should be influencing those vertices, then select that joint from the list of influences and add weight to that. Figure 10-4b shows the wrist area after the skin weights have been painted. Notice how much smoother the deformation is.

*TIP*  *If you right-click a joint in the view window with the Paint Skin Weights tool active, a marking menu appears that lets you select that joint for painting influences to instead of having to select it from the list of influences in the Paint Skin Weights Tool settings window.*

**4.** Continue using this technique as you work your way up the arm. Figure 10-6 shows the weighting after taking a pass at painting the weights on the different sections of the arm.

*TIP*  *Don't try to paint weights for the shoulder when it rotates so that it points up. If you do, you'll end up chasing your tail and ruin any chance of it looking right when the arm is down. Understand that as you raise your arm much above the t-pose, the upward rotation of the arm will come from the clavicle.*

**5.** As you paint the body, add a lot of the upper torso's influence to spine_6. You will mostly be pulling weight away from the clavicle. The remainder of the rib cage can be weighted to spine_5. As you are painting, though, concentrate on just the left side of the body, as we will mirror weights later.

**6.** Select spine joints 1 through 5. Rotate them all back 5 degrees in Z and paint weight between joints 1 through 4 so that the vertices in the stomach area receive influence from all of these joints. This will result in a very smooth deformation, without any noticeable single pivot points. Rotate these joints forward and side to side, and twist, fixing the weights for each position. Figure 10-7 shows the deformation of the chest area after the weights have been painted.

**FIGURE 10-6**    *The weights on the forearm, elbow, bicep, shoulder and clavicle*

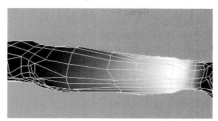

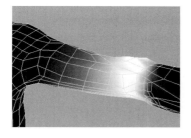
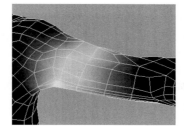
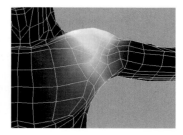

**FIGURE 10-7**    *The body is weighted between the spine joints so that most of the deformation comes from the lower spine and keeps the rib cage intact.*

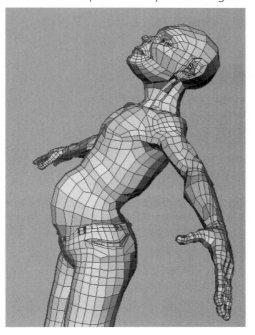
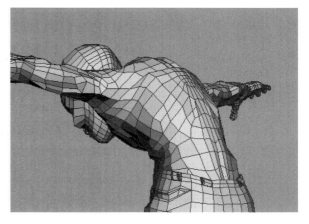

7. Paint all of the weight for the head to the head_1 joint. You may find it easier to select all of the vertices in the head and assign a weight value of 1 to the head_1 joint in the Smooth Skins tab in the Component Editor, very similar to the way we experimented with rigid bound weights earlier in this chapter.

8. The only thing left to paint on the body is the neck. Rotate both of the neck joints in X and then paint each of them until the deformation is smooth from the spine_6 joint to the head_1 joint.

9. Now that we have taken a first pass through the left side of the body, we can further refine the deformations by using the Paint Skin Weights tool's Smooth operation. However, in many cases, this means that we will be subtracting weight from the selected influence. Therefore, we must "hold" them so that any subtracted weight does not get placed in every other joint in the skeleton. Holding the weights keeps the held joint from being added to or subtracted from. To hold the joint, select the joint in the Transform list in the Paint Skin Weights Tool settings window and click Toggle Hold Weights on Selected. The word *Hold* should appear next to the joint's name. Do this for all the joints in this list.

10. Starting from the LT_wrist joint, set the Paint Skin Weights tool's Paint Operation to Smooth. Unhold the weights on the LT_wrist joint and the LT_forearm joint. Smooth out the transition between the two joints. When you are finished, hold the weight on the LT_wrist joint, unhold the weight on the elbow joint, and smooth out the influences between these joints. Continue this process all the way around the rest of the body.

## Mirror Skin Weights

 One of the major advantages of using smooth binding is that you have to paint weights for only one half of the body, because you can mirror your work over to the other side. To mirror skin weights:

1. Select any joint in the body and choose Skin | Go to Bind Pose. This returns the model to the pose it was bound in. It is important that the model is in this pose because it needs to be symmetrical.

2. Unhold all of the joints in the Paint Skin Weights Tool settings window. If any weight is held on any joint on the right side of the body, that joint will not be able to gain or lose the weight necessary to match the other side.

**3.** Choose Skin | Edit Smooth Skin | Mirror Skin Weights ☐ to open the Options window for this command, shown here. Set the Mirror Across to YZ and enable the Direction check box. Click Mirror to mirror the weights.

**4.** Inspect the deformations on both sides. You may have to finesse the weights along the center axis. For this, it is best to use the Smooth operation on the Paint Skin Weights tool.

## Prune Small Weights

No matter how hard you try, it seems that some weight will spill and some of the joints might get only a minuscule amount of the weight. This may lead to a poor deformation or maybe just an unnecessary one. To counter this, you can delete small influences from skin weights:

**1.** Select the body surface and choose Skin | Edit Smooth Skin | Prune Small Weights ☐. The Prune Weights Options window is shown here. Any influences' skin weight values with the Prune Below value will be automatically deleted.

**2.** Try pruning with the default value of 0.01. If you are still seeing a vertex deform by a joint that should not influence it, select that vertex and look up its weight to that joint in the Component Editor. Then, try pruning with the value you find there.

## Save Skin Weight Maps

Another great feature in smooth binding is the ability to save out your weight maps as textures for each joint. The best part about this method is that it is based on the UV layout of the geometry, not the identification number of each vertex. This makes it possible to unbind, edit the geometry, delete history, rebind, and import the skin weight maps. The only bad part for us is that we haven't laid out our UVs yet. (This comes in Chapter 18.)

Also, it should be mentioned that you can use the Detach Skin command (Skin | Detach Skin) to unbind the geometry from the skeleton if you need to. This command has an option to keep the history. This means that as long as you don't edit the geometry in any way that would affect the order or number of the vertices, you could make adjustments to the joint placement and then rebind. Once bound, all of the work you did on the skin weights before unbinding will be there.

## Add an Influence Object

Editing influences of joints over skin points can get you only so far. Even after spending hours tweaking the weights in the Component Editor, the deformation can still appear "rubbery." Animating the compression and expansion of muscle and fat regions cannot be done through skin weighting alone. When this type of anatomical accuracy is required in the deformation of a character's limbs, it is necessary to use an *influence object*.

Influence objects are additional objects (could be geometry, a transform node, or another joint) to which vertices can be weighted. Influence objects can be used to fix areas of your geometry that should lose or gain volume when deformed with the joints. Let's make a bicep muscle and add it as an influence object so that the arm will react to the rotation of the elbow:

1. Any time you add new joints to the skin or add an influence object, you'll need to revert to the bind pose. Select any joint and choose Skin | Go to Bind Pose.

2. Create a NURBS sphere. Name it **LT_bicep_influ**. Place it and scale it in the bicep area of the left arm, as shown here:

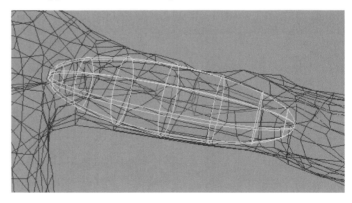

3. Parent this object to the shoulder joint and freeze transformations.

4. Select the body geometry, then the sphere, and choose Skin | Edit Smooth Skin | Add Influence Object □ to open the options window, shown here:

5. Enable the Use Geometry check box and the Lock Weights check box. Leave the Default

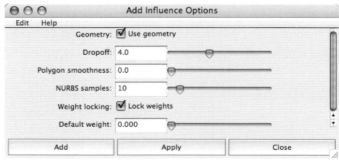

Weight at 0. This is so that when the influence object is added, it will have no initial influence. We will do that manually.

**6.** Return to the Paint Skin Weights Tool settings window and you will find that the LT_bicp_influ object is now listed as an influence. Select it from this list and unhold it using the Toggle Hold Weights on Selected button.

**7.** Use the Paint Skin Weights tool to start adding weight to this object in the bicep region. You may wish to hide the influence object to see what you're doing on the mesh. It should look like this:

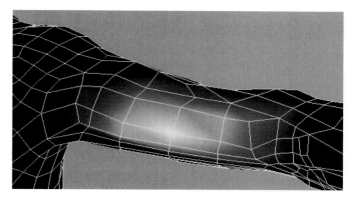

**8.** Duplicate the influence object and move it away from the model so that you can view it easier. Name this object **bicep_bulge**. Sculpt this shape into something that looks like a bulging muscle.

**9.** If you select the original influence object and transform it in any way, you'll see that the geometry deforms along with the transformation. While this is great by itself, we want to take it one step further. To do that, add the bicep_bulge object to the LT_bicp_influ object as a blend shape.

**10.** If you try to increase the bicep_bulge value on the new blend shape node, you will notice that it has no effect on the body geometry. This is because we are modifying the shape node of the LT_bicp_influ object instead of the transform. To fix this, select the body geometry and find the skinCLuster node in the list of inputs in the Channel Box. On that node, set the Use Components attribute to ON, as shown here:

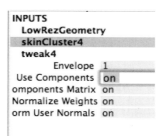

**11.** Try your blend shape. It should affect the body geometry.

The only thing left to do here is to automate the bulging muscle when we rotate the elbow. For this, you need to read the next chapter, in which you learn to connect the attributes of different nodes together.

## Sculpt Deformers

Sculpt deformers are spherical-shaped deformers that can push or pull vertices that they are set to deform. We can use sculpt deformers to solve many of the problems for which we have already used influence objects. However, where they are particularly useful is when you need to have the appearance of skin sliding over muscle or bone. In this section, we'll use a sculpt deformer to create the sliding of the skin over the top of the elbow bone:

**1.** Select the body geometry and choose Deform | Create Sculpt Deformer. Two objects will be created, a sculpt deformer and a stretch origin locator. Move and Snap these to the elbow joint. Group them together and parent them to the bicep joint.

> **NOTE** *Once you learn about constraints in Chapter 11, it is suggested that you parent constrain the sculpt deformer to the bicep joint instead of parenting it.*

**2.** Move the sculpt deformer out toward the elbow area of the geometry. Bend the arm and find the best place for the sculpt node so that it looks like it would represent the bone, as shown here:

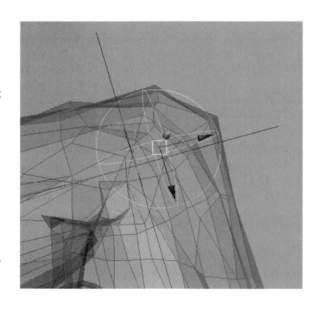

## Add the Facial Blend Shapes

The last thing we'll do here is add the blend shapes that we created in Chapter 8. We need to add them into the geometry so that they evaluate before the skin cluster (joint deformations).

**1.** Import the file that we created in Chapter 8 or use the file on the DVD called mcr8_ch08_finish.ma.

**2.** Select the object called baseGeo and then select our bound mesh.

**3.** Choose Deform | Create Blend Shape ❑. Name the blend shape **faceShapes**.

**4.** Click the Advanced tab in the Create Blend Shape Options window, shown here, and set the Deformation Order to Front of Chain.

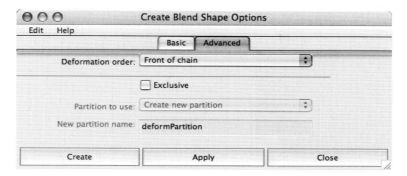

Now, when you bend a joint and invoke a facial expression on the blend shape node, the two work in perfect harmony. The blend shape is calculated before the skin cluster. If it were the other way around, you would bend a joint and then add the smile shape and the model would go back to its t-pose because the blend shape would be overriding the skin cluster.

# Summary

In this chapter, you learned to bind geometry to a skeleton and effectively edit the weights. But we still have a long way to go before the character is fully rigged. In the next few chapters, you will learn how to set up the controls that will operate the skeleton and the other deformations.

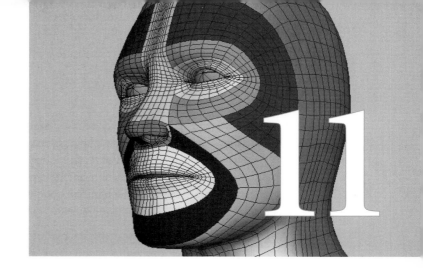

# Connecting
# Attributes

**One of the single most powerful**

features in Maya is the ability to connect

the attributes of any nodes together. These

connections are, after all, the basis of the entire

application. Connections are made every time a

Maya scene is opened. Nodes are connected every

time you set a keyframe on an attribute, assign

a material to a surface, or add an edge loop to a

polygonal mesh. Just about every operation you

carry out by selecting a menu command makes

some type of connection between the nodes in the scene. In this chapter, we are going to begin looking under the hood and making some connections of our own.

# Types of Connections

Maya offers four ways of connecting attributes together: direct connections, expressions, keyed relationships, and constraints.

## Direct Connections

A *direct connection* occurs when the output of one attribute is connected to the attribute of another attribute. For example, if you had two cubes, cubeA and cubeB, and cubeB's attributes were directly connected to the corresponding transform attributes on cubeA, the Rotate X attribute of cubeB would be equal to the Rotate X value of cubeA. When cubeA is rotated on X, then cubeB will also rotate the same amount on X.

The real power of direct connections is that the attributes don't need to be connected to corresponding attributes of another object. In the cube example, we could connect the Rotate X attribute of cubeB to the Translate Y attribute of cubeA, so that when cubeA is translated 30 units in Y, cubeB will rotate 30 degrees in X. Going a step further, we could even connect the Color attribute of the material that is applied to cubeA to the Scale Z attribute of cubeA. This way, when cubeA is scaled down in Z, the color will change. Any attribute can be connected to another attribute. As you can imagine, this provides an enormous amount of control over everything in your scene.

### The Connection Editor

An easy way to make direct connections between attributes in Maya is to use the Connection Editor. The Connection Editor allows you to load the attributes of any two objects and connect them together by clicking and highlighting attributes in each column. Figure 11-1 shows the Connection Editor with the attributes of two cubes loaded. In this example, the output of cubeA's Translate Y attribute is connected to the input of cubeB's Rotate X attribute.

The left column displays the list of output attributes from the loaded node, and the right column displays the list of input attrib-

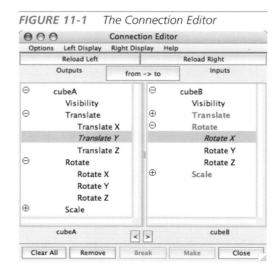

FIGURE 11-1   The Connection Editor

utes on the loaded node. Connections are made from left to right—an attribute is selected from the list of outputs on the left, and then an attribute is selected from the list of inputs on the right. Selected attributes are highlighted. When a connection is made, the attribute's name will show in italic type, as shown in Figure 11-1 for the *Translate Y* output and *Rotate X* input. To break the connection, simply click the highlighted attribute from the list of inputs. The highlight will disappear and the text will return to a normal font, meaning that the connection is broken.

You should also notice that an attribute that has an incoming connection cannot be edited in the Channel Box. A direct connection is indicated in the Channel Box (and Attribute Editor) by a yellow highlight on the attribute's value field. If you were to attempt to change the value of cubeB's Rotate X attribute, for example, Maya would return the following error: "The attribute 'cubeB.rotateX' is locked or connected and cannot be modified."

## Practice Using the Connection Editor

Let's step through a quick exercise so that we can practice using the Connection Editor to set up some direct connections. In this example, we will use the direct connections to animate the curling of a tail that contains six joints. Since our character does not have a tail, we will make one in a new scene.

1. Create a new scene and draw a skeleton that has six joints. Orient these joints so that the Y axis is pointing down the bone and the Z axis is pointing up, as shown here:

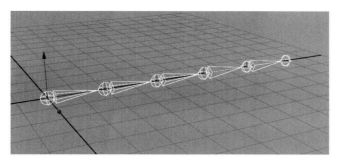

2. Name the joints **tail_1** through **tail_6**.

3. Choose Window | General Editors | Connection Editor to open the Connection Editor.

4. In the Outliner, select tail_1. With tail_1 selected, click the Reload Left button. The attributes for tail_1 will load into the left column.

5. Select tail_2 and click the Reload Right button. The input attributes for tail_2 will load into the right column.

6. Find the Rotate attributes in each column. They will be in a folder called Rotate. Click the plus (+) symbol in front of the folder to see the individual attributes for Rotate X, Rotate Y, and Rotate Z.

**7.** To connect the Rotate X attribute of tail_1 to the Rotate X attribute of tail_2, click Rotate X in the left column, and then click Rotate X in the right column. The attributes will highlight when they are selected. Once the second attribute is selected, a connection will be made and the text will be italicized. The Connection Editor should look like this screen.

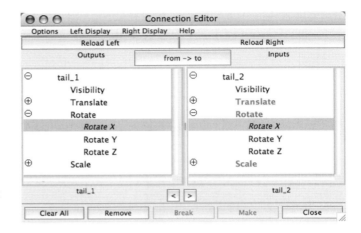

**8.** Select tail_3 and click the Reload Right button in the Connection Editor. Once again, find the Rotate X attribute in the list of attributes in the right column and click it to make the connection.

**9.** Repeat the last step for the remaining joints. Tail_6 can be left alone because it does not have a child joint.

**10.** At this point, all of the Rotate X input attributes for joints tail_2 to tail_5 should be connected to the Rotate X output attributes of tail_1. To test this, select tail_1 and rotate it in X. All of the other joints in the skeleton should rotate the same amount.

**11.** You can verify that they are all rotating the same amount by observing the Rotate X attribute in the Channel Box or Attribute Editor.

## Utility Nodes

Now that you have successfully connected these attributes, you can imagine the possibilities that the power of connections offers and the efficiency that it can provide to a technical director. However, things become even more interesting when the connections are routed through nodes that can process the data. We call these processing nodes *utility* nodes.

Let's say we want each joint to rotate twice as much as its parent joint. This will require some additional calculations. In this case, we can use a type of utility node called a multiplyDivide node.

1. Break all of the connections made to the tail. Select all of the joints in the tail and highlight the Rotate X attribute in the Channel Box, and then RMB-click and choose Break Connections.

2. Choose Window | Hypergraph Input and Output Connections to open the Hypergraph window.

3. The utility nodes are buried in the Create Render Node panel. To access them, select Rendering | Create Render Node from the menu in the Hypergraph. The Create Render Node window, shown in Figure 11-2, contains a list of buttons to create different types of utility nodes.

4. Click the multiplyDivide node to create this node. It will be visible in the Hypergraph.

5. With the multiplyDivide node still selected, open the Connection Editor and click the Reload Right button to load the multiplyDivide node's attributes.

6. Select tail_1 and load it in the left column of the Connection Editor. Click the Rotate attributes on tail_l and connect them to the Input1 attribute of the multiplyDivide node, as shown below.

**FIGURE 11-2** *The Create Render Node window displaying the utility nodes*

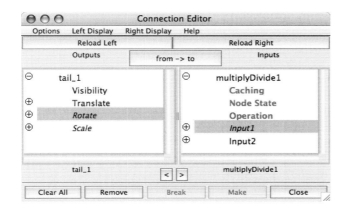

**7.** Select the multiplyDivide node and load it into the left column. Load tail_2 into the right column. Find the Output attribute in the left column and connect it to the Rotate attribute in the right column. We have now finished connecting a multiplyDivide node between the Rotate attributes of tail_1 and tail_2. If we select tail_2 in the Hypergraph and choose Graph | Input Connections, we can see the flow of data in the Hypergraph, shown here:

**8.** Select the multiplyDivide node and, in the Channel Box, set the Input2 X, Y, and Z values to 0.5. This will multiply whatever values are connected to the Input1 attribute by 0.5, resulting in half.

**9.** Now, if you rotate the tail_1 joint 30 degrees in X, tail_2 will rotate 15 degrees in X. You can continue creating these same connections up the tail so that tail_3 gets half of tail_2's rotation, and so on.

Utility nodes provide technical directors with most of the common operations needed to manipulate data. We will use these in many of the chapters to come.

# Expressions

Expressions are a powerful animation tool in Maya. Expressions are script-based instructions that let you control the attributes of objects. This control can be based on a mathematical function, a variable such as time, or the attributes of other objects. In this section, we use expressions to connect the attributes of various objects together.

### The Expression Editor
Choose Window | Animation Editors | Expression Editor to open the Expression Editor, shown in Figure 11-3. The Expression Name field at the top lets you name the expression that you are about to make. The Selection section will show you the name of the selected object and its selected attribute. This is a helpful way to look up or find the names of some attributes that might not be so obvious. In the bottom part of the Expression Editor, you can type in and create new expressions.

One of the simplest expressions you can write makes a direct connection between the attributes of two objects, thus duplicating the behavior of the Connection Editor. Using the tail that we

FIGURE 11-3   The Expression Editor

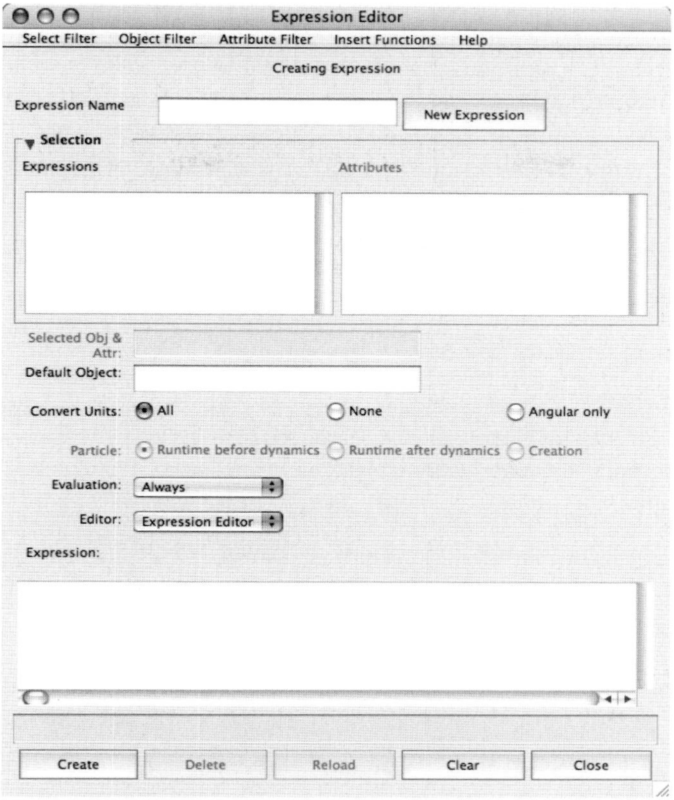

created in the last section, the following expression could be used directly to connect the Rotate X attribute of tail_2 to the Rotate X attribute of tail_1:

```
tail_2.rotateX = tail_1.rotateX ;
```

If we wanted to mimic the tail's behavior that we set up in the Connection Editor, the expression would look like this:

```
tail_2.rotateX = tail_1.rotateX ;
tail_3.rotateX = tail_1.rotateX ;
tail_4.rotateX = tail_1.rotateX ;
tail_5.rotateX = tail_1.rotateX ;
```

Here, we are connecting the Rotate X attribute of each child joint to the root joint, tail_1. The animation that results from rotating tail_1 in X is identical to what we did in the Connection Editor in our first example with the direct connections.

The real power of expressions comes into play when we need to specify exactly how the connections are made.

### Creating Expressions

Let's say, for example, that we want to make each joint in our tail rotate twice as much as its parent joint. To do this, we can write expressions that use mathematical operations, in this case multiplying the rotation of each parent joint by 2:

1. Once again, break the previous connections on the tail skeleton.

2. Instead of using the main menu bar to open the Expression Editor, we will use the marking menu in the Channel Box. Select tail_2 in the Outliner or view window and then highlight its Rotate X attribute in the Channel Box.

3. Right-click in the Channel Box and choose Expressions. The Expression Editor will open with the tail_2.rotateX object and attribute loaded into the Selected Obj & Attr field.

4. Copy and paste the tail_2.rotateX text from the Selected Obj & Attr field into the Expression area at the bottom of the Expression Editor window.

5. Add the following script to complete this expression:

   ```
   tail_2.rotateX = tail_1.rotateX * 2;
   ```

6. Click the Create button in the Expression Editor to create the expression that will connect these attributes in this manner. Once the expression is created, notice that the value field for the Rotate X attribute of tail_2 is purple. This indicates that the value is being controlled by an expression.

7. Test the expression by rotating tail_1 in X and verifying (in the Attribute Editor or Channel Box) that the X rotation value for tail_2 is twice as much.

8. Return to the Expression Editor and write the expressions for the rest of the joints. You do not need to create a new expression for each connection. All of the connections for each joint can be entered into this one expression.

### Using Functions

So far, we have been able to replicate the behavior of the joints with expressions just as we did with direct connections. Another great use for expressions is to insert mathematical (and other types of) functions. For this example we will automate the animation of the tail by controlling the tail_1 joint with a function that is based on time:

**1.** In the Expression Editor, add one more line to the current expression:

```
tail_1 = 30 * sin(time * 20) ;
```

**2.** Click the Edit button in the Expression Editor to make that change and then click Play in the timeline. The sin function will create an oscillating sequence of data so that the tail will rotate back and forth between –30 and 30 degrees. The speed is determined by multiplying the time (Maya's playback clock) by a number, in this case 20.

# Keyed Relationships

A *keyed relationship* occurs when attributes are connected based on some custom input from the user. The user actually specifies, or *keys*, the attributes of an object based on the values of another object. The most common keyed relationship is the basic keyframed animation, in which the attributes of all animated objects are keyed at a certain frame. At another frame, they are keyed with different values. Every time a key is placed, a relationship is made between time and the animated attributes.

What if you wanted to key the attributes of certain objects based on something other than time? For example, if you wanted to pose the tail skeleton at different poses based on the rotation of the first joint instead of time, what would you do? If you were really great at math, you might be able to do this with expressions, but it would take way too long. To handle such a situation, Maya offers a tool called a *Set Driven Key*.

## Set Driven Key

Maya's Set Driven Key creates a keyed relationship between the attribute of one object and one or many attributes of one or more other objects. The output attribute—the attribute to which the other attributes will be keyed—is called the *driver*. In a normal keyframed animation, time is the driver. All of the input attributes that will be keyed based on the value of the driver are called the *driven attributes*. To create an animatable Set Driven Key, you must set at least two keys between the driver attribute and the driven attribute. Once this relationship has been established, all of the values between the keyed values are interpolated by an animation curve.

In this example, we will again use our six-jointed tail as the object for animation. The root joint's Rotate X attribute will be the driver for all of the rotation attributes of the other joints in the tail. Once again, start a project and create six joints. Orient them and name them.

**1.** Choose Animate | Set Driven Key | Set ❑ to open the Set Driven Key window.

**2.** Select the tail_1 joint in the Outliner or the view window.

**3.** In the Set Driven Key window, click the Load Driver button to load the tail_1 object and its attributes. The object will appear in the upper-left column, and the attributes will appear in a list in the upper-right column.

**4.** In the driver's attribute list on the right, find and select the rotateX attribute.

**5.** Use the Outliner to select the rest of the joints in the tail skeleton. Once again, that last joint, tail_6, can be disregarded.

**6.** In the Set Driven Key window, click the Load Driven button to load all of the objects into the list of driven objects in the lower-left Driven pane of the Set Driven Key window.

**7.** Select all of the objects in this list. Their attributes will appear in the lower-right column.

**8.** In the list of attributes, find the rotateX, rotateY, and rotateZ attributes and select all three of them by dragging the cursor over their names or by SHIFT-selecting them. Figure 11-4 shows the Set Driven Key window with all of these objects and attributes selected.

**9.** Click the Key button. A relationship is established between the values of the driven attributes based on the value of the driver attribute.

**10.** Rotate the tail_1 joint to a value of 50 in X. Select each of the joints and rotate them into a pose. They can be rotated in X, Y, and Z all at the same time. The illustration shows a tail in an example pose:

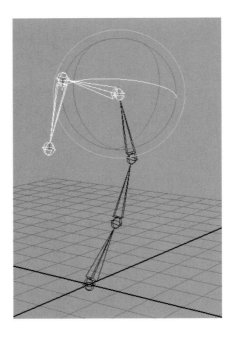

**11.** With the pose set, return to the Set Driven Key window and click the Key button again. This will create another relationship between the selected attributes at their current values.

**12.** Test the Set Driven Key by rotating the tail_1 joint in X. When it is rotated 50 degrees, the rest of the tail will rotate toward that pose.

FIGURE 11-4    *The Set Driven Key window*

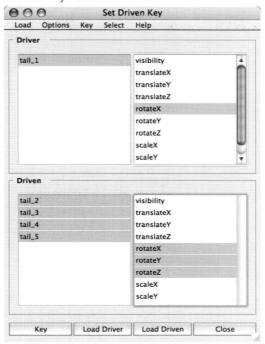

As you can see, keying the relationship instead of connecting it mathematically is an intuitive process. Set Driven Keys are often used to animate fingers, toes, tails, and even the spines of characters. In the next chapter, we will rig the fingers of our character using a Set Driven Key.

# Constraints

Maya's constraints offer another way to connect the transform attributes of one object to the transform attributes of one or several other objects. For example, a point constraint could be used to connect the translations of one object to the other. The object that is being constrained is called the *constrained object*, while the object that it is being constrained *to* is called the *target object*.

Constraints differ from the other types of connections we have used so far in that they are not based on the actual values of attributes. Instead, they use the position and orientation of a transform node's pivot point to base the connections on.

## Types of Constraints

Maya offers nine different types of constraints:

- **Point**   Causes the constrained object to follow the position of the target object(s).
- **Aim**   Orients a constrained object so that it points at the target object along a specified axis.
- **Orient**   Causes the constrained object to follow the rotations of the target object.
- **Scale**   Causes the constrained object to match the scale of the target object(s).
- **Parent**   Enables the constrained object to behave as if it were parented to the target object. For example, a parent constraint can be used for an animation in which a character picks up some object that then behaves as if it were a child of the hand.
- **Geometry**   Causes the constrained object to be restricted to a NURBS surface, curve, or polygonal surface.
- **Normal**   Causes the constrained object to orient itself along the surface normals of the target object. This is especially useful if you want an object to animate over a complex surface.
- **Tangent**   Causes the constrained object to orient itself to the tangent of a NURBS curve.
- **Pole vector**   Constrains the pole vector of an IK handle to a target object. By using this type of constraint, there is no need to control the pole vector of an IK handle with the Show Manipulator tool. A control object, such as a curve or surface, can be used as the target object instead.

## Using Constraints

Creating constraints is a simple process; however, you must keep in mind the order in which the objects are selected. The target object is selected first, and the constrained objects are then selected. When every object has been selected, choose a constraint from the Constraint menu in the Animation menu set.

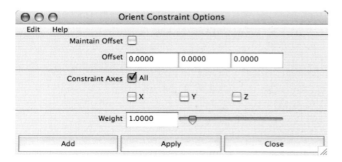

The settings available for the constraints in the Options window are fairly similar for most of the constraints. The point, aim, orient, and parent constraints all have attributes called Maintain Offset, Offset, Constraint Axes, and Weight. The Orient Constraint Options window is shown here.

The following are the options available in this window:

- **Maintain Offset**   Preserves the original, relative value of the constrained object. For example, if an object positioned at 10,10,10 is point constrained to another object at 0,0,0 with the default settings (Maintain Offset is disabled), the constrained object will match the position of the target object once it is constrained. When the Maintain Offset attribute is enabled, the constrained object will maintain its relative distance.

- **Offset**   Specifies an offset value relative to the target point. Note that the target point is the same as the object's pivot point. By default, these offset values are all set to 0.

- **Constraint Axes**   Specifies which axes will be constrained. The default is set to constrain all three axes. However, an object can be constrained to just one or two axes of the target object.

- **Weight**   Specifies how much the values of the constrained object are influenced by the target object. When an object is constrained only to one target, a value of 1 means that the object will follow the target constraint 100 percent. A value of 0 influence will completely disable the constraint. The Weight attribute can be edited or keyframed in the Channel Box or Attribute Editor after the constraint has been created.

An object can be constrained to more than one target. This can be done by selecting multiple target objects, then selecting the constrained object, and then choosing a constraint, or you can select the target and the constrained object and then the constraint can be repeated.

If an object is constrained to two different objects with two different positions, and both of the Weight attributes are set to 1, the actual position of the constrained object will be equal to the average values of the constrained attributes.

While constraints are great tools for animation, as we will see in Chapter 12, they are also great for building control rigs. In that chapter, we will make use of the point, orient, aim, and pole vector constraints.

# Summary

In this chapter, you learned about different methods of connecting attributes together. We will make heavy use of these concepts as we build our control rig in the next chapter.

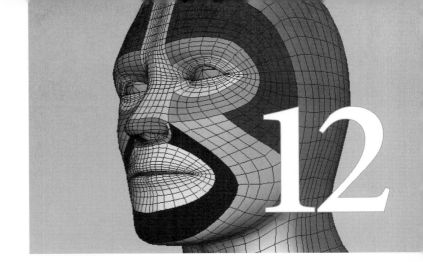

# Character Controls

**In Chapter 9, we created a skeleton**

for our character. In Chapter 10, we bound

the geometry to the skeleton and edited the

deformations. At this point, you *could* begin

posing that skeleton and setting keys directly

on the joints. However, animating a skeleton in

this condition can prove to be less than ideal.

With all of the joints in even this simple skeleton,

there are still too many to control one by one.

The job of the character rigger is to build a

structure on top of the skeleton that will enable an animator to do the job without the complexities of the software getting in the way.

This chapter concentrates on creating controls that operate, or *puppet*, the skeleton. Figure 12-1 shows a completed character control rig with custom attributes on the various control objects. This type of rig is built using the concepts learned in Chapter 11 along with a few new tools that you'll learn about here. The goal for the rig is that an animator will never have to pick a joint to pose the character. Instead, the animator will use elegant and intuitive controls.

# Methods of Posing a Skeleton

There are two primary methods for posing and animating a skeleton: forward kinematics (FK) and inverse kinematics (IK). Each has its advantages and disadvantages and is used to control parts of the rig for which it is best suited. In some cases, rigging the skeleton so that an animator can switch between the two methods is preferable. This setup is discussed later in the chapter. For now, let's discuss these two methods of control.

## Forward Kinematics

Forward kinematics allows you to animate a skeleton so that it behaves as a normal hierarchy. That is, when a parent joint is rotated, all of its children will follow. FK is useful for animating arcing movement, such as swinging or waving an arm or curling the fingers, as shown next. Stop-motion puppets are animated through rotation inside an armature in a way that's similar to FK.

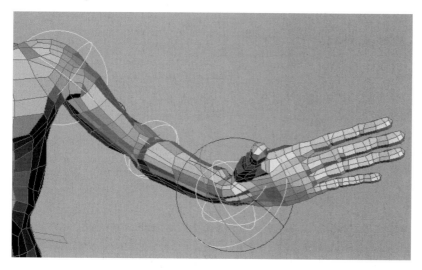

While animating with FK is pretty intuitive to anyone who has been doing basic animation in Maya, it does have some drawbacks for use in character animation. The biggest problem with animating with FK is that you can't anchor the end of a joint chain at a specific location while the parent joints are moved.

FIGURE 12-1    *A real-world production-ready control rig for the short film "Worm"*

CHARACTER CONTROL SYSTEM FOR THE SHORT FILM "WORM"

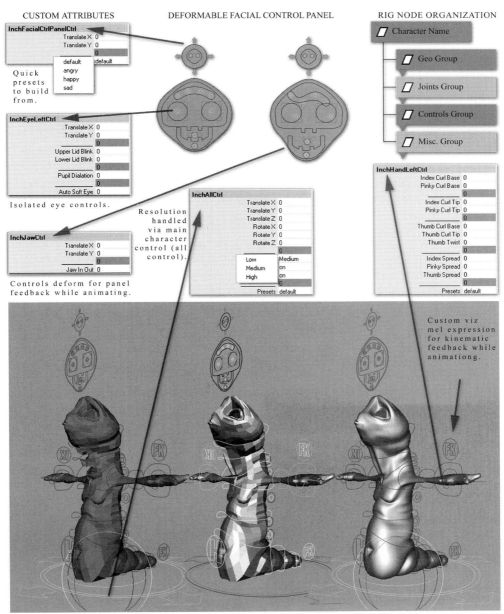

Low Resolution - blocking
(no deformation or facial control)

Medium Resolution - animation
(faceted deformation + facial)

High Resolution - render
(smooth deformation + facial)

For example, if you're animating a character's walk cycle, you want to be able to lock one of the feet to the floor while the body moves and the opposite foot moves forward. This is difficult to accomplish with FK. You would animate the pelvis to move the body forward, but since the feet are children of the pelvis, they will move forward with the body. This means you have to rotate the leg backward at every frame, just to make the foot stay put. To lock down this foot as you move the body, you must use inverse kinematics.

## Inverse Kinematics

With inverse kinematics, a joint's hierarchy is specified to form an IK chain. Once the IK chain is created, two additional nodes are created. The *IK effector* is parented to the joint at the end of the IK chain. The *IK handle* is created and placed outside the joint hierarchy. The IK handle acts as a goal position for the effector. The IK *solver* then evaluates the position of the IK effector and makes the necessary calculations to rotate the joints in the chain so that the joint at the end of the chain will be in the same position as the end effector. This way, the hierarchy is animated by moving the object at the end of the joint hierarchy instead of from the top down—hence the name, inverse kinematics; it's the inverse of FK. Figure 12-2 shows a joint chain with an IK system added.

Like FK, IK has its drawbacks. For many areas on a character, IK can make the simplest movements difficult to control. A character swinging his arm as he walks, for instance, can be difficult to animate with IK. Instead of just grabbing the shoulder and rotating it back and forth as you would with FK, IK would require you to translate the IK handle back and forth. Obtaining a nice arc can be a difficult task in this situation.

**FIGURE 12-2**    *A joint chain with an IK setup that can be used to control the skeleton from the hierarchy's end*

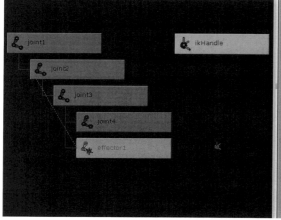
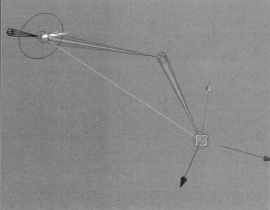

# Tutorial: Building a Control Rig

The most important thing to do before rigging a character is to analyze the shots in which the character will act and figure out what it needs to do. No single character rig will work well in every situation. If a character will be acting in several shots in which its behavior is quite different, you may be better off building different rigs for the separate shots.

For our rig, we will discuss some of the most basic needs of each part of the rig and build a control system based on that specification. Each part of the rig will be operated by one or a few control objects. The purpose of these control objects is to consolidate the animatable attributes of the skeleton for more efficient control over the character. A control object can be any object that will operate a portion of the rig either by transforming the control or by animating custom attributes on the control that are connected to attributes on the underlying mechanism.

By giving an animator access to all of the finger rotations on a single control, they only need to select one object while setting keys. When editing the keys in the Graph Editor, they won't need to reselect all of the joints just to manage their animation curves.

## Create Finger Controls

Animating the finger movements on a hand would be a painstaking task if the animator had to rotate each individual joint in each finger of the hand every time the character had to move its fingers. Unless the shot is heavily focused on the hand and finger movement, we can get away with some pretty simple controls in this area.

For this exercise, we'll create a control object that has attributes to curl and spread each finger. Set Driven Keys provide the perfect solution for connecting the rotations of the fingers to custom control attributes. By keying the position of each finger curl based on the attribute of a driver object, you need to animate only that attribute to pose the finger.

### Create the Controls

One of the underlying philosophies behind a control object is that the appearance of the object should be indicative of its function. In his Maya Masterclass (an Autodesk Maya sponsored educational event), Maya master Jason Schleifer coined the phrase "iconic representation" to describe this design ethic for creating control objects.

With this in mind, we will use a curve drawn in the shape of a hand for our finger controls. As long as your computer has some standard font libraries installed, you can find pretty nice icons by using Maya's text feature. This will allow us to generate NURBS curves from any text. Even better is that the Wingdings font has a hand icon as one of its characters.

**1.** Open the scene called mcr8_ch12_start.ma from the DVD or continue using the file you finished in Chapter 10.

**2.** Choose Create | Text ☐. Select the Wingdings-Regular font and type in **I**. This creates two curves that are grouped under two separate transform nodes. Select one of the curves, unparent it from the group (SHIFT-P), and delete the other elements. Name this curve **LT_handCTRL**.

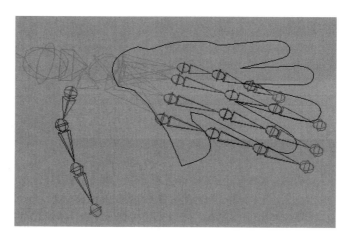

**3.** Position the control somewhere just over the hand, as shown here.

## Add Custom Attributes

Up to this point in the book, every attribute that we have edited has been predefined by the node that it is a member of. Perhaps the most powerful feature in Maya is the ability to create a custom attribute on any node. Connect this custom attribute to other attributes via any of the methods shown in Chapter 11, and the sky is the limit in customizing your scene. We will use these custom attributes to drive the rotations of the finger joints with a Set Driven Key.

Attributes are created by selecting a node and choosing Modify | Add Attribute. This opens the Add Attribute window, shown in Figure 12-3, where you define the attribute name, what type of data it will store, and what values, if any, will define its limits. We'll talk more about data types when we explore MEL scripting in Chapter 13. For now, we'll create attributes of the type

**FIGURE 12-3**   *The Add Attribute window*

float. This means that the attributes can be any number, positive or negative, and contain a decimal. Let's use this to set up an attribute to control the fingers:

**1.** Select the LT_handCTRL object and choose Modify | Add Attribute to open the Add Attribute window shown in Figure 12-3. Name this attribute by typing **indexCurl** in the Attribute Name field. Leave the Data Type at Float (the default).

**2.** Use the Minimum and Maximum fields to set a range for the attribute's value. Enter **–1** for the Minimum, **10** for the Maximum, and **0** for the Default.

**3.** Click the Add button to add the indexCurl attribute to the LT_handCTRL object. Repeat this process for the other three fingers and the thumb. Name the attributes **middleCurl**, **ringCurl**, **pinkyCurl**, and **thumbCurl**. If you select the LT_handCTRL object and look at the attributes in the Channel Box, you will see that the new attributes appear in this list, as shown here:

| LT_handCTRL | |
|---|---|
| TranslateX | 11.481 |
| TranslateY | 6.719 |
| TranslateZ | –3.32 |
| RotateX | –86.412 |
| RotateY | –102.199 |
| RotateZ | –7.837 |
| ScaleX | 0.215 |
| ScaleY | 0.215 |
| ScaleZ | 0.215 |
| Visibility | on |
| Index Curl | 0 |
| Middle Curl | 0 |
| Ring Curl | 0 |
| Pinky Curl | 0 |
| Thumb Curl | 0 |

**4.** At the moment, changing the values for these new attributes won't have any effect, because the attributes are not connected to anything. We will use a Set Driven Key to connect these attributes to the joint rotations of each finger. Choose Animate | Set Driven Key | Set ❒ to open the Set Driven Key window.

**5.** Set up the Driver object first. Select LT_handCTRL in the view window and then, in the Set Driven Key window, click the Load Driver button. The list of keyable attributes will appear in the section on the right. Click the indexCurl attribute in this list.

**6.** Set up the Driven object. In the Outliner, select the joints called LT_Index_1, LT_Index_2, and LT_Index_3. Be sure to select each joint, not just the parent joint. You don't need to select the last joint in the finger because it will not deform anything.

**7.** After you have selected three joints, return to the Set Driven Key window and click the Load Driven button. All of the joints will appear in the list of driven objects.

**8.** Drag-select the joints in the Driven list to select them all, and their keyable attributes will be listed on the right.

**9.** Drag-select all of the Rotate attributes. Figure 12-4 shows the Set Driven Key window as it should look when all of the attributes are ready to be keyed.

**10.** Verify that the rotation values for all of the finger joints are set at 0 by viewing them in the Channel Box. In the Set Driven Key window, click the Key button. This will set a key for the selected joints' current rotation values when the indexCurl attribute is set at 0.

FIGURE 12-4   *The Set Driven Key window with the driver and driven attributes selected and ready to be keyed*

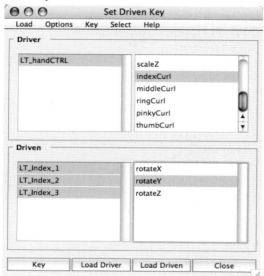

**11.** In the view window, select the LT_handCTRL object and set the indexCurl attribute to **–1**.

**12.** Use the Outliner to select the three joints of the left index finger again. Use the Rotate tool to rotate the joints on their Y axes so that the finger bends back just slightly. Because we took the time to orient the joints when we set up the skeleton, we are able to select multiple joints, rotate them at once, and have the finger curl in one direction. However, the real power of using a Set Driven Key is that the driven objects can have unique values letting you key a specific pose.

*TIP*   **As you rotate your joints, make sure the Rotate tool is set to Local.**

**13.** When you are pleased with the pose of the finger bending backward, return to the Set Driven Key window and click the Key button. Select the LT_handCTRL object and set the indexCurl attribute to **10**. Once again, rotate the joints in the index finger. This time, rotate them so that the finger bends in the other direction, as if the finger were being curled in to make a fist. Figure 12-5 shows the left index finger in a curled pose. When you are pleased with the pose, return to the Set Driven Key window and click the Key button.

**14.** Test the indexCurl attribute by selecting it in the Channel Box and MMB-dragging in the view window to change the value interactively. The finger should curl when the indexCurl value nears 10, and it should bend backward at values from 0 to –1.

**15.** Repeat this entire process, Steps 5 through 14, for the remaining three fingers and the thumb.

**16.** After you have set up all of the fingers to animate with a Set Driven Key, select all of the control attributes at once in the Channel Box, and use the MMB to open and close the entire hand.

**17.** You can use another good control to spread the fingers. Use the Add Attribute command to add an attribute called Spread to the LT_handCTRL object. Give it a

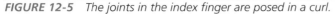
*FIGURE 12-5*   *The joints in the index finger are posed in a curl.*

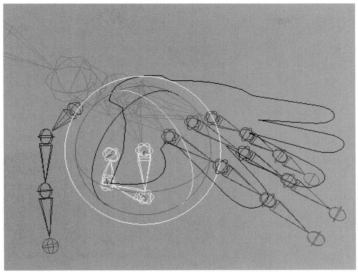

minimum value of **–5** and a maximum value of **5**. Use a Set Driven Key to make this attribute drive the Z rotations of the knuckle joints of the index, ring, and pinky fingers.

**18.** Duplicate the control object and name it **RT_handCTRL**. Position it over the right hand. Unfortunately, Maya does not have any built-in function for mirroring Set Driven Keys. You then need to go through the process of using Set Driven Keys to connect all of the custom attributes to the finger joints on the right hand.

> **NOTE**   *Realize that although Maya may not have a specific feature, its architecture allows users to create their own tools to suit their specific needs through the use of a scripting language (MEL). Highend3d.com has a library of free downloadable scripts that you can install and quickly extend Maya's capabilities without having to write a single line of code yourself. Jason Schleifer, mentioned earlier in the chapter, has an amazing script for copying Set Driven Keys available on his website (www.jonhandhisdog.com).*

## Create the Leg Control

In this example, we will create control objects for the leg. The most important requirement for the leg is that the foot needs to be able to be anchored in place while the body moves. This means that we'll definitely need to use IK. While using one IK system from the hip to the ankle will hold the foot's position, it will not keep the foot flat on the ground as the body moves. Many animators also like the ability to control the rolling of the foot. To address these concerns, we will make IK handles from the ankle to the ball, and another from the ball to the toe.

To control all of these IK handles, we'll build one master foot control and arrange the IK handles in a hierarchy underneath the control. We'll then add a foot roll attribute to the control and connect it to various pivot points, controlling the IKs. Finally, we'll create another control object that let's us point the knee.

### Create the IK Handles

Here we will create three IK handles: one from the hip to the ankle, a second one from the ankle to the ball, and a third from the ball to the toe.

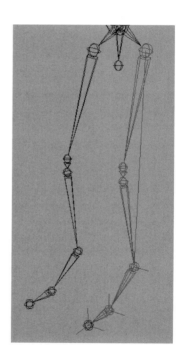

1. Choose Skeleton | IK Handle Tool ❑ to view the IK Handle tool's settings. Set the Current Solver to ikRPSolver. This gives us the ability to control the IK's pole vector.

2. With the tool still active, click the LT_hip joint and then click the LT_ankle joint. This creates the first IK handle. Name it **LT_ankleIK**.

3. Create a second IK chain. Choose Skeleton | IK Handle Tool and click the LT_ankle joint and then the LT_ball joint. In the Outliner, rename this IK handle **LT_ballIK**. Press the Y key to select the last tool used, click the LT_ball joint, and then click the LT_toe joint. In the Outliner, rename this IK handle **LT_toeIK**. Your leg should look like this:

### Create the Control Object

For the foot control, we will use a polygonal cube and scale it so that it covers the bounds of the foot. This will make it easy to select.

1. Choose Create | Polygon Primitives | Cube. Name the cube **LT_footCTRL**. Move and scale it so that it is visible outside of the geometry.

2. Since this is a control object, we don't ever want it showing up in our renders. Therefore, we can adjust some of the attributes on the control's shape node. Select the LT_footCTRL object and open the Attribute Editor. Select the LT_footCTRLShape tab.

Unfold the Object Display section and then the Display Overrides section. In the Display Overrides section, check Enable Overrides and disable Shading by unchecking it, as shown here. You can also set the color of the wireframe with the Color attribute.

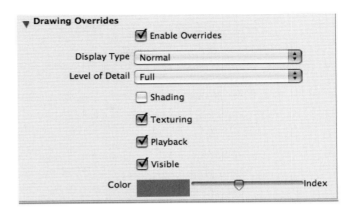

**3.** To make sure the LT_footCTRL object won't render, locate the Render Stats folder in the Attribute Editor and uncheck all of the boxes.

**4.** Press the INSERT (HOME) key, hold down the V key, and move the pivot of the LT_footCTRL object so that it snaps to the LT_ankle joint (you may have to switch to the Perspective view to make sure it did not snap to the RT_ankle joint). Press the INSERT (HOME) key again to exit pivot editing mode.

**5.** Freeze transformations on the LT_footCTRL object.

**6.** Test the LT_footCTRL object by parenting all of the IK handles to it. Then select the LT_footCTRL object and move and rotate it. The foot should move and rotate around its ankle, as shown here. Undo or zero the control back to its default position.

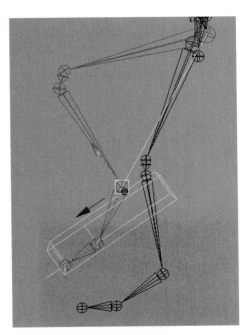

## Build the Control Hierarchy

In order for the foot roll control to work, we need to parent each IK to an additional transform node placed at different positions that the foot will pivot from:

**1.** Create an empty group node by choosing Edit | Group, or by pressing CTRL-G (COMMAND-G). Name the group **LT_ballPivot**.

**2.** Snap the LT_ballPivot group to the LT_ball joint.

**3.** Duplicate the LT_ballPivot joint and rename it **LT_toeLift**. Move/snap the LT_toePivot group to the LT_toe joint. Again, duplicate the LT_ballPivot joint and rename it **LT_ toeWiggle**. Duplicate this once more and move it to the position of the heel. You will need to use the geometry to place this. Name this one **LT_heelPivot**. You now have five transform nodes—the control and the four group nodes.

**4.** When the ball of the foot rolls, as it would during a walk cycle, the toe remains planted as the ankle pivots around the ball. Therefore, if we parent the LT_ballIK and the LT_ankleIK joints to the LT_ballPivot group, the ankle will rotate around the ball when LT_ballPivot is rotated. Since the LT_toeIK joint is not a child of this pivot, the toe will remain planted as the ball of the foot rolls.

**5.** To wiggle the toe, the LT_toeIK joint needs to pivot from the LT_toeWiggle group. Parent the LT_toeIK joint to the LT_toeWiggle group. Select the LT_toeWiggle group and rotate it to test the behavior.

**6.** When the toe lifts or pivots, the entire foot should pivot around the leg. Therefore, we will make the LT_ballPivot group and the LT_toeWiggle group both children of the LT_toePivot group.

**7.** On the other end of the foot is the heel. When the foot contacts the ground, the foot pivots around that contact point. While it is actually a bit more complex than that in real life, we can fake it for now by parenting the LT_toePivot group to the LT_heelPivot group. Now, when you rotate the LT_heelPivot group, the entire foot will pivot from the heel.

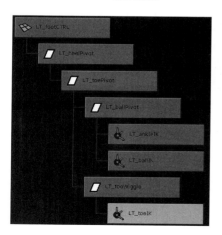

**8.** Parent the LT_heelPivot group to the LT_footCTRL object. The completed hierarchy is shown here.

**9.** Test it! If you select the LT_toePivot group and rotate it, the entire foot will pivot around the toe. If you select the LT_ballPivot group and rotate it, the ball of the foot will roll. Meanwhile, these are all parented under the LT_footCTRL object to move the overall foot. Great!

*NOTE   If the toe is not staying planted, make sure that the Enable IKFK Control option is disabled. Choose Skeleton | Enable IKFK Control to disable this. When disabled, no check mark appears in front of the option in the Skeleton menu.*

Whew! That may seem like a lot of parenting but it's a great example showing the power and functionality that can result from a hierarchy. If your foot does not behave like it should, go back and double-check the hierarchy and try again. You may have to do this several times before actually understanding it.

### Set Up the Foot Roll Control

One problem is that we still have too many objects to select to animate the foot. Therefore, we will create a custom attribute on the LT_footCTRL object that will sweep through the action of the foot contacting the ground and rolling off the toe. This attribute will control the rotations of the different pivot points.

**1.** Select the LT_footCTRL object and choose Modify | Add Attribute. In the Attribute Name field, type **footRoll**. Set the Data Type to Float and set the Minimum and Maximum fields to **–1** and **1**.

**2.** Select the LT_heelPivot, LT_toePivot, LT_toeWiggle, and LT_ballPivot nodes. Open the Set Driven Key window (Animate | Set Driven Key ❒) and load these pivot nodes as the driven items. Select the Rotate Z attribute as the driven attribute.

**3.** Select the LT_footCTRL object and load it as the Driver. Select the footRoll attribute as the driver attribute. The Set Driven Key window should look like this:

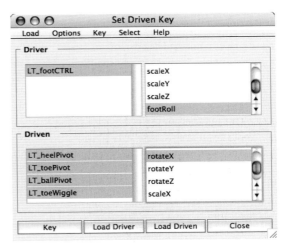

**4.** Since all of the pivot nodes' rotations are zeroed out and the control attribute is set to 0, we will key the current pose as it is since it will be the middle of the roll cycle. In the Set Driven Key window, click Key.

**5.** Now we'll pose the foot for the two extremes. Starting with the foot landing pose, set the footRoll attribute to **–1**. Set the heelPivot node Rotate X attribute to **–50**. Set the LT_toeWiggle node to **–20**. Click the Key button in the Set Driven Key window.

**6.** Now for the other extreme. Set the footRoll attribute to **1**. Set the LT_heelPivot and LT_toeWiggle attributes to **0**. Set the LT_toePivot's Rotate X attribute to **60**. Click the Key button in the Set Driven Key window.

**7.** Now we'll set one more key for when the ball of the foot rolls off the ground. Set the footRoll attribute to **0.5**. Set the LT_toePivot's Rotate X attribute to **0** and the LT_ballPivot's Rotate X attribute to **40**. Click the Key button in the Set Driven Key window.

**8.** Test it. Highlight the footRoll attribute in the Channel Box and MMB-drag in the view window to interactively edit the footRoll attribute. The foot should cycle through the process of rolling along the ground and lifting up.

> **NOTE** *The motion may still look a bit robotic. As you learn more about the Graph Editor, you may wish to go back to this rig and overlap the animation curves so that there is some overlapping of the poses.*

### Create the Knee Control

Our leg control currently has the ability to place and pose the foot but we don't have any control over the direction of the knee. The ikRPSolver that we are using in the leg has the ability to control what is called the *pole vector*. This gives us added control over the twist direction of the joints in the IK system. In this section, we will create another control object that will constrain to the pole vector of the ankle IK.

FIGURE 12-6   *The LT_kneeCTRL object is placed in front of the knee.*

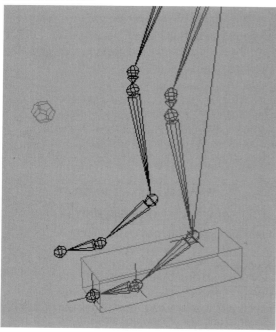

**1.** At this point, we no longer need to have any of the group nodes or IK handles visible, since they are all controlled by the foot control. Select the LT_toeLift node and press CTRL-H (COMMAND-H) to hide all of the IK handles and transform nodes. Now you have a nice, clean leg to animate.

**2.** Choose Create | Polygon Primitives | Platonic Solid. Name the object **LT_kneeCTRL**. Go through the process we used on the foot control to edit the display and rendering attributes of the LT_kneeCTRL object.

**3.** Snap the LT_kneeCTRL object to the LT_knee_1joint so that it is aligned. Then move it out along the Z axis so that it is placed in front of the toe. Figure 12-6 shows the knee control in place.

**4.** In the Outliner, select the LT_kneeCTRL object and then select LT_ankleIK (CTRL-click [COMMAND-click] to select these multiple objects). Choose Constrain | Pole Vector. Now the LT_kneeCTRL object can be used to control the pole vector attributes on the LT_ankleIK node. Parent the LT_kneeCTRL object to the LT_footCTRL object. Freeze the transformations on the LT_kneeCTRL object.

**5.** Test the LT_kneeCTRL object by moving it around. You should also select the LT_footCTRL object and move that around to make sure that the knee control does not go behind the knee, causing the leg to flip. If this is the case, move the knee control further in front of the knee.

## Set the Rotation Orders of the Controls

If there is one major aspect of character setup that is often overlooked by a novice character rigger, it would be rotation order. In fact, shame on us for going this far in the book without mentioning it! The rotation order of a transform node determines the hierarchy of the rotation axis. Just as everything else in a computer must have a logical order of evaluation, the rotation axis of a transform node must evaluate in a defined order of operations.

The default rotation order of any transform node created in Maya is XYZ. To understand how this works, you need to think of the rotations occurring in a hierarchy. In the case of the XYZ rotation order, Z is the parent, Y is a child of Z, and X is the leftmost rotation in the hierarchy.

Setting up your controls to use the proper rotation order is crucial for the rig to be effectively posed and animated. While it is possible to rotate any transform node in world or local space, it is the gimbal control that allows you to view the hierarchy of the rotating node. The Rotate tool will display the gimbal manipulator in the view window.

When determining the rotation order for a control, it is important to anticipate how the rig will be used. You may not always be correct with your decision, but you should be able to come close. In the following steps, we will set the rotation order for the LT_footCTRL object so that it will behave predictably when animated.

**1.** Select the LT_footCTRL object and set the Rotate tool to display its gimbal mode.

**2.** Consider the most frequent behavior that this control will have. Most likely, it will be the rotation along the control's X axis. With a rotation order of XYZ, the Y and Z axes will not follow when the foot is rotated along the X axis. Figure 12-7 shows the foot rotating down in X; notice that the Y and Z rotation axes remain constant. If the animator wanted to twist the foot along the Y axis, the Z axis would also have to be rotated to keep the foot behavior looking natural. And, aside from the appearance of the animation, editing it in the Graph Editor, just to do a simple maneuver might need the control of two, if not three, rotation axes. That is a lot to deal with in the Graph Editor.

**FIGURE 12-7**  *The LT_footCTRL object is rotated using XYZ (left) and ZXY (right) rotation order; note the orientation of the rotate manipulator.*

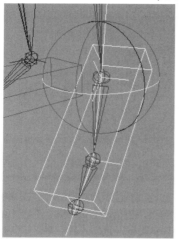
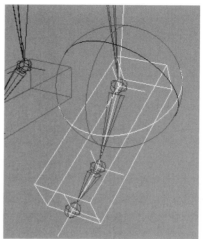

**3.** In the Attribute Editor, find the Rotation Order attribute and set it to ZYX. Now, when you rotate the foot down along the X axis, the Y and Z axes follow, so that you can continue to animate the foot in that space.

For every control that we make that will control the orientation of a joint, we need to set the proper rotation orders.

### Clean Up

At this point, cleanup involves locking and hiding the attributes on the foot and knee controls that will not be edited or keyed. Lock and hide all Scale attributes on all four control objects. For the knee controls, you can lock and hide everything except for the translations.

At this time, the controls for the left leg are complete. You now need to repeat this entire process for the right leg. One shortcut to use would be to duplicate the LT_footCTRL object. The custom attributes we created will also be duplicated. Once again, if you learn to write MEL scripts, you could create a script that would build this entire rig at the click of a button.

## Create the Back Controls

There are many options for rigging the back. For the purposes of this tutorial, we will keep it as simple as we can and aim to fulfill some minimal requirements. We want the ability to bend the character over and be able to twist from the lower back and the shoulders. While an FK back control would give us this functionality, the main issue with our rig is that there are quite a few joints to control.

Sure, maybe some multiplyDivide nodes would work well too, but in this setup we are going to employ the use of an IK Spline Handle tool, which will make the setup process a bit faster. The IK Spline Handle tool bases the IK calculation on a NURBS curve. Instead of moving the IK handle itself, the control vertices of the curve are manipulated and the joints that the IK is controlling rotate to produce some very realistic motion.

The IK Spline Handle tool also provides some additional attributes to control the IK chain. Among them is the Twist attribute and the Advanced Twist features that we will utilize in this rig for twisting the spine. While the spline IK will take care of the twisting and placement of both the root of the spine and the top of the spine, it will still lack the ability to bend the spine from specific points. To counter this, we will build a simple FK control system to provide a rotational pivot point for the spine.

### Set Up the Spline IK

We will begin the setup of the back control by adding a spline IK handle to the spine joints. This will allow all of the joints in the spine to be controlled by a NURBS curve. Let's begin:

**1.** Choose Skeleton | IK Spline Handle Tool ☐ to open the Settings window, shown here:

**2.** Disable the Auto Parent Curve check box so that the IK curve created will not get parented to the skeleton hierarchy. Also, disable the Auto Simplify Curve option. Normally, this option allows you to specify how many spans the IK curve will have. By turning it off, Maya will create a span for each joint segment.

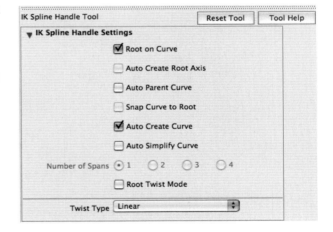

**3.** Click spine_1 and then spine_6. Maya will create an IK handle and a curve. Name the handle **backIKHandle** and name the curve **backIKCurve**.

**4.** Play with this system for a while to get an idea for how it works. Select the CVs of the curve and move them around to get a feel for how the joints react. While having this type of control over the joints is very powerful, actually controlling the CVs one by one can be painful.

### Set Up the FK Control Joints

To control the curve, we will create another set of control joints. The spline IK curve will be bound to these joints. Yes, you heard that right, we are going to smooth bind a curve. This will allow us to deform the curve from three static pivot points. The joints will then be what we use to pose the skeleton.

**1.** Draw four new joints. As you place the first one, remember to click out in empty space, away from the skeleton while holding down the v key and then drag over until you snap to the spine_1/root joint location. This will keep you from creating a branch off of the root.

**2.** Continue creating the joint chain. Snap the next joint to the spine_3 joint, the third joint to the spine_6 joint, and then snap one more joint to neck_1 (we will delete this joint after we orient the skeleton).

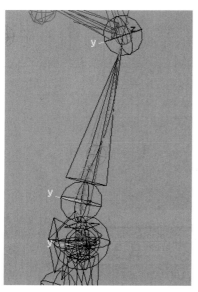

**3.** Orient the joints so that the Y axis is facing in the positive Z direction. Once all of the joints are oriented, delete the fourth joint. Name the remaining joints **lowerBackCTRL**, **middleBackCTRL**, and **upperBackCTRL**. The current rig should look like this.

**4.** Bind the curve to the joints. Select all three control joints and then the backIKCurve curve. Smooth bind the curve with the Max Influences set to 3 and the Dropoff Rate set to 1.

**5.** Now, when you rotate one of the control joints, the curve will deform and drive the IK system in the spine. If you test this out, you will see that it looks pretty nice when you rotate it on Y and Z but the twist in X looks a bit odd.

*TIP*    *You can fine-tune the arc of the spine by editing the weights to the three control joints in the Component Editor.*

### Set Up the Twist Controls

On the spline IK handle, there is an attribute called Twist. This causes the joints in the spline IK system to twist along their oriented axes, starting from the base of the spine to the top of the spine. However, we have three different pivot points along the spine where we would like to influence the twist from. If the hips are turned and the shoulders stay put, for example, the twisting would actually be occurring from the top of the spine downward. Since the Twist attribute will not allow us to have this type of control, Maya provides us with a set of attributes for more complex twisting of a joint chain. In this section we will use the Advanced Twist controls on the backIKHandle handle to add the type of control we require for our rig.

There are many options available in the Advanced Twist attributes. What we want to do is control the twist from either end of the IK chain (since the controls are arranged in a hierarchy, any control will influence the twist). We will define the lowerBackCTRL joint and upperBackCTRL joint as the two World Up Object objects. This means that the IK system will use the orientations of these control objects to determine in what way the joints in the skeleton should aim as they twist around their oriented axis (X). Let's give it a shot:

**1.** Select the backIKHandle handle and find the Advanced Twist Controls folder (under the IK Solver Attributes folder). The attributes are shown in Figure 12-8.

**2.** Enable the twist controls by checking the corresponding check box. Since the World Up Type will be defined by the two control joints, set the World Up Type to Object Rotation Up [Start|End].

**3.** Since our skeleton is oriented in X, we need to pick another local rotation axis that we can use as the Up Axis; in this case, we'll use Y (Z would work too). Set the Up Axis to Positive Y.

**FIGURE 12-8** *The Advanced Twist Controls window with the settings we'll use for this rig*

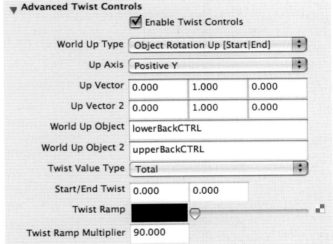

**4.** The Up Vector attributes need to match the Up Axis of the joints to the world orientations of the controls (defined in the next section). Because we were so careful to orient our controls in the same direction as our skinning joints, we can set both Up Vectors to 0, 1, 0. Don't worry about it if your skeleton is flipping out right now…it just needs to know what is defining the World Up axis.

**5.** Copy and paste the name of the lowerBackCTRL into the World Up Object field and the upperBackCTRL into the World Up Object 2 field. The Advanced Twist Control attributes should match those shown in Figure 12-8.

**6.** Test the rig by selecting the lowerBackCTRL and rotating it along its X axis. The spine should behave as expected. The only problem will be that while the upperBackCTRL will control the twist of the spine, its influence over the backIKCurve curve does not affect the rotation of the spine_6 joint. This can be fixed by adding a constraint.

**7.** Select the upperBackCTRL and then the spine_6 joint. Make certain that they are selected in this order. Choose Constraint | Orient ❑ to view the orient constraint's options. The default values will work fine. The key is that the Maintain Offset is enabled. This will make sure that the slave object matches the same orientation of the master constraint. Click Add to create the constraint. If the bind skeleton moves at all, then you know that your controls were not oriented correctly. While enabling Maintain Offset will relieve this problem for the moment, it can come back to haunt you, so it is best to troubleshoot the problem and reorient the rig.

**8.** Test out the upperBackCTRL. It should provide a pretty stable control over the upper part of the body. You may realize that you need to go back and adjust the skin weights so that they work well with your rig.

### Create the Control Objects

While the actual control joints are already set up to control the rotations of the spine, the joints are buried in the middle of the body. We want to give the animator something that is easy to identify as the spine controls. There are many techniques we can use to create this type of control. Orient constraints would be one of them. However, we'll use a little trick that will keep the number of connections to a minimum by simply parenting a control object, a curve in this case, to each joint. But a normal parent-child relationship won't quite get it in this case. What we'll do is parent the shape node, not the transform node (as we usually do when we parent objects) to the joints. Here is how this works:

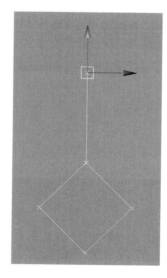

**1.** Create a control object. Anything can be used, but in this case we will use a NURBS curve. In the Front view, use the EP Curve tool set to Linear to draw a curve that resembles this one:

**2.** Duplicate the curve three times. In the Outliner, select Display | Shapes. This will make the geometric objects in the Outliner show their hierarchy of shapes and transforms as separate nodes.

**3.** Unfold the hierarchy of the first curve and select the curveShape1 node. Now CTRL-click (COMMAND-click) to select the lowerBackCTRL joint.

**4.** On the command line, type

```
Parent -r -s ;
```

**5.** The curve's shape node will now be parented to the joint's transform node. This is very different from just using the Parent command, as the Edit | Parent command parents one transform node to another. The relationship that we created with this command is identical to the relation a NURBS sphere's shape node has to its transform node. Selecting the shape in the view window will also select the transform nodes.

**6.** Repeat this process for the other two joints. When you are finished, you can delete the empty transform nodes left over from the curves.

**7.** If you wish to adjust the shape of the curves, do so by selecting the components of the curve. Do not adjust the transform node, as the transform node is the joint and will manipulate your rig.

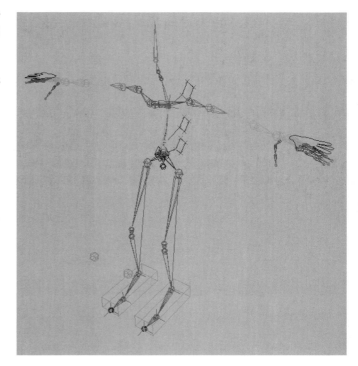

### Set Rotation Orders and Clean Up

Just as we will with every rotation control we create, we need to check the rotation orders of the controls to make sure that they will be usable by an animator. For this rig, ZXY should be a good choice for the spine controls. Lock and hide the Translate and Scale attributes on the three spine controls. The spine is now complete. The current state of the rig is shown here:

# Create the Head and Neck Control

It's completely reasonable to have a separate control for the head and the neck. After all, while the neck and head may twist in the same direction, the head pivots up and down and side to side from the location of the head joint. But, instead of building multiple controls for each, let's cheat a little and create one control that will give us this kind of behavior all in one package. To do this, we will make use of direct connections and multiplyDivide utility nodes.

## Create the Control Object

For the head, we'll create a NURBS circle. The big lesson here is that this circle needs to be oriented to the joint that it is going to control and then have its rotations set to zero without freezing transforms. Otherwise, the control's rotation axis will be out of sync with the joints.

**1.** Choose Create | NURBS Primitives | Circle ❑. Make sure the Axis attribute is set to Y and create the circle. Name it **headCTRL**.

**2.** To orient one object to another, we will utilize some of the concepts discussed in Chapter 9 regarding the spaces that objects are transformed in. To show the importance of this application, we will parent this circle to the head_1 joint. The translations and rotations should all be set to zero. The result will be that, relative to its parent object, the circle will be positioned and oriented directly on top of the head_1 joint. However, we cannot have a control object be a child of an object that we wish to control. That would create a looping, feedback cycle of information.

**3.** Unparent the headCTRL from the head joint. While it is still oriented to the joint, its translation and rotation information is not zeroed. This means that if we tried to use direct connections between the two nodes, the head_1 joint would shift. If we use Freeze Transformations, the headCTRL object will orient itself back to the world. How do we zero out the transformations of an object while maintaining its orientations?

**4.** The headCTRL object must be parented to a transform node that has the exact same transformation attributes as the control. An easy way to do this is to duplicate the object, delete the shape node of the duplicate, and then parent the original transform to the new, empty transform node. Do this by selecting the headCTRL object and pressing CTRL-D (COMMAND-D) to duplicate the headCTRL.

**5.** Immediately press the DOWN ARROW key. This selects the shape node of the currently selected object. Once it is selected, press the BACKSPACE (DELETE) key to delete the shape node. Now you are left with an empty transform node. Rename this transform node **headCTRLgrp**.

**6.** Parent the headCTRL to the headCTRLgrp node. Now look in the Channel Box. The rotation and translation values should all be zero while the actual orientation of the object remains the same. The head control and its attributes are shown here:

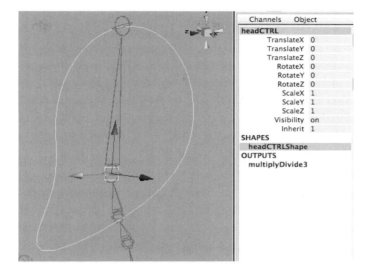

## Make the Connections

The joints will be connected to the control using some direct connections and along with a multiplyDivide node. The head_1 joint's Y and Z rotations will be directly connected to the Y and Z rotations of the headCTRL. The X rotations will be connected from the control to the X rotations of all three joints through a multiplyDivide node that will divide the incoming rotation by 3.

**1.** Open the Connection Editor. Select the headCTRL and load it into the right column. Select the head_1 joint and load it into the left column.

**2.** Find the RotateY and RotateZ attributes of the headCTRL and connect them to the RotateY and RotateZ attributes of the head_1 joint.

**3.** In the Hypergraph, choose Rendering | Create Render Nodes to open the Utility Node window. From here, create a multiplyDivide node. Name this node **headCtrlDivideX**.

**4.** Back in the Connection Editor, load the headCtrlDivideX node in the left column. Connect the RotateX attribute of the headCTRL to the Input1X attribute of the headCtrlDivideX node.

**5.** Load the headCtrlDivideX node into the left column and load the head_1, neck_1, and neck_2 joints into the right column.

**6.** Connect the OutputX attribute of the headCtrlDivideX node to the RotateX attributes of the three joints. Figure 12-9 shows the final connections in the Hypergraph.

*FIGURE 12-9    The connections between the nodes are shown in the Hypergraph.*

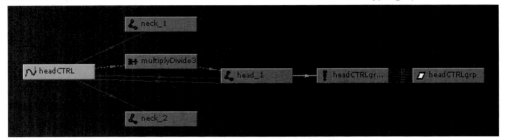

**7.** Test the control by rotating it around to make sure everything works. It should be fine. That is, until you move the skeleton or rotate one of the spine controls. The skeleton will move, but the control will be left behind.

**8.** To fix this, select the head_1 joint and then the headCTRLgrp node (the transform above the headCTRL). Then choose Constraint | Parent. This will parent constrain the headCTRLgrp node to the head joint so that no matter how the skeleton below is rotated, the control will always stay in place. This little trick gets around creating an unwanted cycle error.

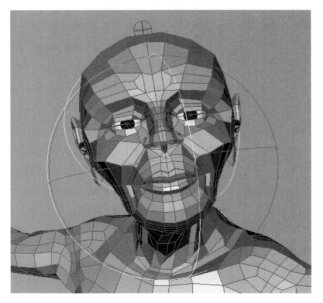

### Set Rotation Order and Clean Up

To finish this control, set the rotation order to ZYX. Then, lock and hide all of the attributes except for the rotations. Test the rig with the geometry visible to verify that all of the deformations in the neck look smooth, as shown here. Next up is the arm.

## Create the Arm Control

The arm control will be the most complex of the entire rig. This has mostly to do with the fact that we have to build a rig that will control not only the rotations of the shoulder, elbow, and wrist, but also the twisting of the bicep and forearm joints. To make it even more complex, most animators prefer that the arm is capable of switching between IK and FK. There are many ways that riggers can handle this. Maya even has a system of its own. Let's look at some of the different options.

## FK/IK Switching

It's often necessary to switch between IK and FK, especially in the arm's motion, so the best possible animation method is utilized for the animation. FK is great for rotationally based animation, such as a waving or swinging arm. It takes much less keyframing to create smooth-swinging animation of arms than does IK. IK is a must for goal-oriented movement of the joint. For example, if the character has to push the box using all its body movement, using FK for this animation requires a lot of counter animation, rotating and keyframing the arm at almost every frame to try to keep the end of the arm planted on the box. It's almost impossible to do.

In some cases, the character must switch between these two types of behaviors over the course of one shot. In other situations, an animator may have a preference or want to experiment with both FK and IK. When necessary, you may need to build a rig that can do both.

Maya provides a method for blending between FK and IK during animation. You can find several commands in the Animate | FK/IK Keys menu for enabling and disabling IK and setting keys on the IK handles and joints that are being animated. A nice feature of using this built-in system is that the animation curves for the FK/IK skeleton will appear dotted or solid in the Graph Editor depending on whether or not the IK handle is enabled. This makes it easy for the animator to know exactly where and when a selected joint is under the control of IK.

However, the system has some shortcomings when the IK handle is constrained to a control object. While a command called Move IK to FK (Animate | FK/IK Keys | Move IK to FK) will move an IK handle to its effector in the FK skeleton, the constraining control object will not be moved. Animating with this system then causes "popping" as the switch is made from FK to IK.

Traditionally, character riggers have relied on a custom three-skeleton system: one skeleton for IK, a second one for FK, and a third one for skinning. The skinning skeleton is orient-constrained to the other two skeletons and custom attributes use Set Driven Keys to blend between the weights on the constraints. Further controls can be set up to move the IK control objects to the FK rig, and vice versa, through the use of some simple MEL scripting.

For our rig, we have the added problem of having to control the twist joints and have the controls blend between FK and IK. In situations like these, you really need to come up with your own solution. But, we're very lucky because we have all of the functionality of Maya right in front of us.

## The IK Arm

Before we begin with the FK to IK switch, we'll build the IK arm. Since the IK arm's orientations must match the skin arm exactly, we will duplicate the joints used to skin the arm. We will then create a control that will control the position of the IK handle and the rotation of the wrist joint.

1. Select the LT_shoulder joint and duplicate it. You can unparent the arm for now so it is easier to work with. If you have locked the translations of the joints, the arm will fly off to somewhere else in the scene. Use the Channel Control window (Window | General Editors | Channel Control). Select all of the joints in the arm and then select the translate X, Y, and Z attributes in the locked column in the Locked tab of the window and click the Move button. The Channel Control window is shown in Figure 12-10.

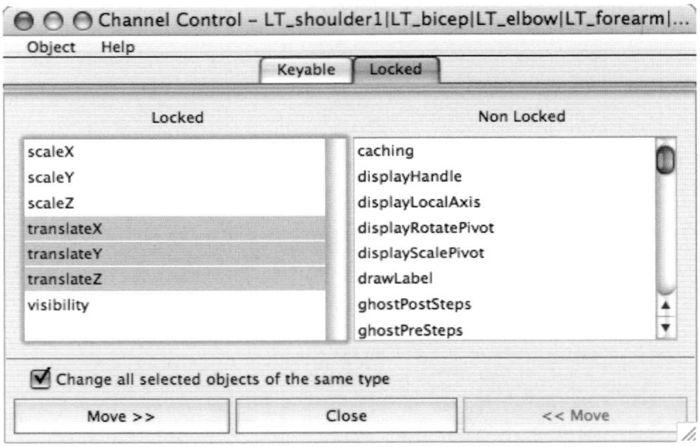

2. Select the root joint and hide it. This will make it easier to work on the IK arm.

**FIGURE 12-10**   *The Channel Control window is used to lock and unlock attributes or make them keyable or nonkeyable.*

3. We will remove all of the joints that we don't need for this control. Since the hands have absolutely nothing to do with the IK, select the LT_hand joint and press the BACKSPACE (DELETE) key to delete the hand joint and all of its children.

4. The ikRPsolver is not going to allow us enough control to deal with the twist joints in this skeleton. Therefore we need to delete them. Select the LT_bicep joint and then choose Skeleton | Remove Joint. This will delete the joint while keeping the hierarchy intact. Do the same for the LT_forearm joint. You will have three joints left over.

5. Select all three joints and choose Modify | Prefix Hierarchy Names. Type in **IK_** and click OK. This will place the IK_ prefix in front of the joint names.

6. Choose the IK Handle tool and click the IK_LT_shoulder joint and then the IK_LT_wrist joint. This will create the IK control for the arm.

7. Now we need a control object. For this we will create another cube. Create a polygon cube and name it **IK_LT_wristCTRL**.

8. Now we must orient the control to the wrist joint. We'll use the same technique as we did for the head control. Parent the IK_LT_wristCTRL object to the LT_wrist joint. Set the translations and rotations to 0, then unparent the joint. Now the control is

oriented, but since its transforms are now reading in world space, the translations and rotations are no longer zero.

**9.** Once again, we'll employ the technique we learned when we set up the head control. Freezing transformations on the control will orient the object back to the world. Instead, we'll parent it to a transform node with the exact same transformation data. Duplicate the IK_LT_wristCTRL object and then immediately press the DOWN ARROW key to select the shape node. Press the BACKSPACE (DELETE) key to delete the shape, leaving only the transform. Rename this transform **IK_LT_wristCTRLgrp**. Parent the IK_LT_wristCTRL object to this group and its transforms will zero out while maintaining their orientation. Figure 12-11 shows the IK_LT_wristCTRL object once it has been oriented to match the LT_wrist joint.

---

**FIGURE 12-11**    *The IK_LT_wristCTRL object is oriented to match the LT_wrist joint.*

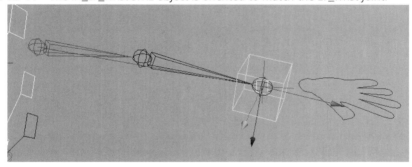

**10.** Select the IK_LT_wristCTRL object and then select the LT_wrist joint. Choose Constraint | Orient. The rotation of the wrist joint will now be controlled by the IK_LT_wristCTRL object.

**11.** Select the IK_LT_wristCTRL object and then choose the LT_wristIK handle. Choose Constraint | Point. The IK handle will now follow the arm wherever the control is moved.

**12.** Test the IK_LT_wristCTRL object. Since the LT_wrist IK handle is constrained to it, the control object will now act as the goal for the IK solver. When the control is rotated, the wrist rotates with it. Another great thing about this setup is that when the wrist is locked down onto an object such as a table or railing, the hand will remain oriented as the body moves. Otherwise, you would have to counter-animate the hand so that it does not go through the table or railing. When you are finished testing, zero out the transforms for the IK_LT_wristCTRL object.

**13.** Now we need a pole vector constraint for the arm. Choose Create | Polygon Primitives | Soccer Ball. Name this object **LT_elbowCTRL**. Snap it to the LT_elbow joint and then move it back along the Z axis so that it rests behind the body. Scale it, freeze transformations, and override the display properties so that only the wireframe is visible.

**14.** Select LT_elbowCTRL and then select LT_wristIK and choose Constrain | Pole Vector. The placement of the pole vector constraint is shown here:

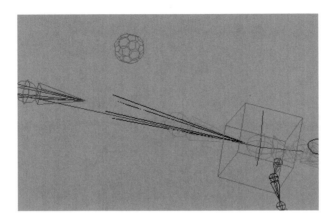

## Create the FK Arm Controls

The FK controls for the arm are easy to set up. Since their orientations need to match the IK, we can simply duplicate the IK skeleton. To make the controls, we'll use the same technique we used for the FK back controls, where we parented shape nodes to the joints.

**1.** Select the IK_LT_shoulder joint and duplicate it. Move it up in Y, just temporarily while we work on it.

**2.** Choose all three new joints and then choose Modify | Search and Replace Names. Search for **IK_** and replace it with **FK_**. The joint chain is now renamed.

**3.** For the control objects, we will use a special kind of object called an *implicit sphere*. These objects make excellent controls because they have no shading properties associated with them. On the command line, type

```
CreateNode implicitSphere ;
```

**4.** Select the implicit sphere, press the DOWN ARROW key, and then select the FK_LT_ shoulder joint. On the command line, type

```
Parent -r -s ;
```

**5.** This parents the shape of the implicit sphere to the joint. Repeat this process for the other two FK joints. The two control arms are now set up and should look like this:

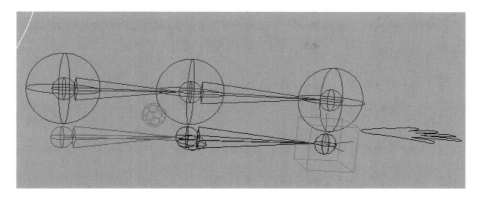

## Make the Connections

We mentioned before, using the more traditional approach of a constraint system may cause the arm rig to become unstable when it comes to handling all of the rotations. We prefer to base the blending on actual rotation values rather than orientations. To overcome any of these issues, we will use some of Maya's utility nodes; specifically, a multiplyDivide node and some type of blend node. There are quite a few different types of blend nodes available in Maya. The one that we will use here is the blendColor node. Sure, it can be used for colors, but what is a color? It's just three values (R, G, and B).

To set up the upper part of the arm (shoulder and bicep), we will take the rotation values from the FK and IK shoulders and connect them into the inputs of a blendColor node. From the blendColor node, the Y and Z attributes will plug directly into the rotation attributes of the skinning shoulder. The OutputX attribute of the blendColor node will run into a multiplyDivide node and divide the input by 2 since there are two joints that need to be controlled. The output from the multiplyDivide operation is then fed into the X rotation of the shoulder and bicep.

When you animate the Blender attribute on the blendColor node, it will blend from the FK to the IK. To make this attribute easier to access, create an FK/IK switch control and connect a new custom attribute to the Blender attribute on the blendColor node:

**1.** Select the LT_shoulder, FK_LT_shoulder, and IK_LT_shoulder joints.

**2.** Choose Window | Hypergraph Input and Output Connections. This will load the selected nodes and their connections into the Hypergraph.

**3.** In the Hypergraph, choose Rendering | Create Render Nodes to open the Utility window. Create a multiplyDivide node and name it **shoulderTwist**. Create a blendColor node and name it **shoulderBlend**.

**4.** In the Connection Editor, load the IK and FK arm joints into the Outputs column (the left) and load the shoulderBlend node into the Inputs column (the right). The Connection Editor should look like this:

**5.** Load the shoulderBlend node into the Output column and then load the LT_shoulder joint in the Inputs column. Connect OutputG to RotateY and OutputB to RotateZ.

**6.** Select the shoulderTwist node (the multiplyDivide node) and load it into the Inputs column. Connect the OutputR of the shoulderBlend node to the Input1X attribute of the shoulderTwist node. Set the shoulderTwist node's Input2X attribute to **0.5**.

**7.** Load the shoulderTwist node into the Outputs column and the LT_shoulder and LT_bicep joints into the Inputs column. Connect the OutputX to the RotateX on both joints. Figure 12-12 shows the flow of the data as it is graphed in the Hypergraph.

**8.** You can test the shoulder to see if things are working. Rotate the FK_shoulder joint in Y. Set the Blender attribute on the shoulderBlend node to 1. The skinning arm should follow.

*FIGURE 12-12    The connections for the control of the shoulder and bicep joints*

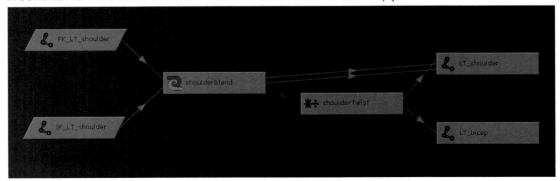

**9.** The elbow is much simpler since there is no twisting. Simply pass all of the rotations of the FK and IK skeletons into a new blendColor node (name it **elbowBlend**) and then pass the output of the elbowBlend node into the LT_elbow joint.

**10.** Use the same process used on the shoulder and bicep for the wrist and forearm.

**11.** You can snap the FK_LT_shoulder joint back down onto the IK and skin shoulder joints so that everything is sitting on top of each other. Select the FK_LT_shoulder and IK_LT_ shoulder joints and group them. Name this group **LT_armCTRLgrp**. Snap the pivot point of this group node to the shoulder.

**12.** Instead of parenting the arm controls back into the skeleton, we will parent constrain them. This will keep our control joints separate from our skinning joints. To stabilize the constraint, we need to create one more transform node that will act as an offset from the location of the clavicle to the shoulder joint.

**13.** With nothing selected, press CTRL-G (COMMAND-G) to create an empty group. Orient this node to the LT_clav joint. Name this group **LT_armConstraint**.

**14.** Parent the LT_armCTRLgrp node to the LT_armConstraint group.

**15.** Select the spine_6 joint and then the LT_armConstraint group. Choose Constraint | Parent. The hierarchy should look like this:

```
LT_armConstraint
  LT_armCTRLgrp
    FK_LT_shoulder
    IK_LT_shoulder
  LT_ArmConstraint_parentConstraint1
```

**16.** When you finish that, it's onto the other side of the body!

**NOTE** *Making all of these connections with the Connection Editor can be a lengthy, tedious process with a lot of room for error. It is these kinds of tasks that lend themselves well to being scripted, so that they can be performed with the click of a button. Once you have a better handle on MEL scripting in Chapter 13, you may want to try scripting this setup.*

## Create a Blend Control

Currently, every time we want to blend from FK to IK we need to select all three blendColor nodes. This is especially difficult since we cannot easily access them in the view window. Instead, we will create another control object. While we can certainly add another object for each side, we are going to try to keep this rig as clean as possible. With a little planning ahead, we know that we eventually want to have a transform node to move the entire rig around with.

Instead of cluttering the scene with little control objects all over the place, we will create this control and add FK/IK blends:

1. Create a NURBS circle. Scale it and place it at the base of the character's feet. Name it **globalXFORM**. Freeze the transformations.

2. Add attributes named **LT_arm_ikToFk** and **RT_arm_ikToFk**.

3. Load the globalXFORM control into the Outputs section of the Connection Editor and load all six blendColor nodes into the Inputs column. Connect the ikToFk attributes to the appropriate Blender attribute.

4. Test the rig. Pose the arm with one set of the controls and then with the other set of controls. Practice blending back and forth between FK and IK.

### Clean Up and Set Rotation Orders

One additional feature you may wish to add is one that will manage the visibility of the appropriate controls as you switch back and forth from FK to IK. When in FK, the IK_LT_wristCTRL and IK_LT_elbowCTRL objects should not be visible. When in IK, the FK joint controls should be hidden. Use Set Driven Key to connect the LT_arm_ikToFk and RT_arm_ikToFk attributes to the Visibility attributes on the controls.

Because of the complexities of this arm rig, and the wide range of motion that the arm is capable of, the rotation orders are especially important. With the joints twisting in X, we know that we want the rotation order for the joints to be either XYZ or XZY. After playing with the IK wrist controls to pose the arm, XZY is the one that will keep us from gimbal lock for most animations. However, no matter how you set the rotation orders, the wrist control is the most prone to gimbal locking. You may find that you need to add a control that handles only the twist of the wrist, or one that handles the waving of the wrist in certain positions. This fix is usually implemented by creating this additional control above the hierarchy in the main controller. Experiment with this concept if you find that the wrist is gimbaling.

## Create the Clavicle Controls

Now for the clavicle control. The setup we will use lets the animator select a control object and move it to point the collar bone. This provides a good amount of control and is simple to set up. We will use a polygonal cube for the collar bone control:

1. Choose Create | Polygon Primitive | Cube. Name this object **LT_clavCTRL**.

2. Move and snap the cube to the LT_shoulder joint. Scale it so that it will not get occluded by the geometry. Freeze transformations.

3. Use the IK handle tool to create an IK chain from the LT_clav joint to the LT_shoulder joint.

4. With the LT_clavCTRL object still selected, select the LT_clavIK handle and choose Constraint | Point.

5. Group the LT_clavCTRL object to itself (CTRL-G [COMMAND-G]) and name the group **LT_clavCTRLgrp**.

6. Just as we did with the arm, we will create an additional group that will act as an offset from the clavicle to the spine. Create an empty group node and name it **clavCTRLRoot**.

7. Parent this LT_clavCTRLgrp node to the clavCTRLRoot node. When you build the rig for the other side, you can parent the RT_clavCTRLgrp node to this node as well.

8. Select the spine_6 joint and then select the clavCTRLRoot node. Choose Constraint | Parent. The completed setup looks like this:

9. Lock and hide the Rotate and Scale attributes on the LT_clavCTRL object.

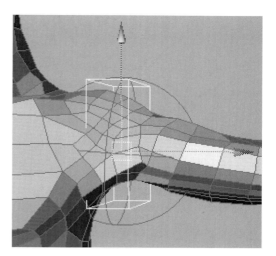

10. Repeat this setup for the other side.

## Create the Root Controls

The last set of controls we'll create will be for the root and pelvis joints. The root control, or COG (center of gravity) as we'll call it, will control the translation and rotation of the entire body. This will be the control that an animator will use to make the character move around in the scene. The pelvis control will control the orientation of the pelvis. Both the pelvis control and the lowerBackCTRL that we created earlier will get parent constrained to the COG.

1. To create the COG, we'll use another icon from one of the font libraries. The Webdings-3 Regular font has a two-way arrow available with the o character. Use the Text command to create this icon. Name the curve **COG**.

2. Unparent the curve from the group nodes that were created with it. Set the Rotate X attribute to 90. Choose Modify | Center Pivot and snap the COG to the root joint.

3. Scale the control so that it is just visible over the geometry. When the placement of the control is complete, choose Modify | Freeze Transformations.

4. Select the COG and then the root joint and choose Constraint | Point and then Constraint | Orient. The root joint's transforms are now controlled by the COG.

**5.** Create another control object. We'll use a NURBS circle this time. Name the control **pelvisCTRL**.

**6.** Orient the pelvisCTRL object to the pelvis joint using the same method we used to orient the headCTRL object to the head_1 joints and the IK_wristCTRL object to the wrist joint.

**7.** Since the pelvis_1 joint and root joint overlap, things can get a little crowded in this area. To fix this, move the pelvisCTRL below the COG and then press the INSERT (HOME) key to edit the pivot point and snap it to the root joint. Press the INSERT key again to exit the pivot editing mode. The controls should look like this:

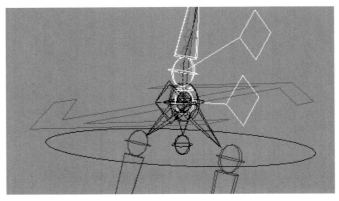

**8.** Since this object will be used as an orient constraint, we do not want to freeze the transforms but instead we want to flush the current transforms into another transform node. Duplicate the pelvisCTRL, press the DOWN ARROW key, and then press BACKSPACE (DELETE) to delete the shape node. Name the transform node **pelvisCTRLgrp**, then parent the pelvisCTRL to it. The pelvisCTRL will now have values of 0 for its transforms and maintain its orientation to the pelvis_1 joint.

**9.** Select the pelvisCTRL and then select the pelvis_1 joint and choose Constraint | Orient.

**10.** Now that the joints are connected to the proper controls, we just need to arrange the controls in a proper hierarchy so that the COG will drive the pelvisCTRL and lowerBackCTRL. So, parent the pelvisCTRLgrp node and the lowerBackCTRL to the COG.

> *TIP*   *To keep things consistent, group the lowerBackCTRL to itself and move the new group's pivot to the root. You can name this group spineCTRLs.*

**11.** Clean up the controls. Lock and hide the Translate, Scale, and Visibility attributes for the pelvisCTRL. For the COG, lock and hide just the Scale and Visibility attributes.

**12.** Set the rotation order for the COG to XZY.

# Clean Up the Scene

If you've looked at your Outliner recently, you'd agree that it is a mess. While we have made a good effort to name everything, the number of objects listed is far too long for anyone to deal with if they open this scene for the first time. We need to arrange all of these objects into one group for the entire character. At least, with that group defined, other characters can be imported in the scene and we could at least decipher what control belonged to what.

But there are additional steps we can take to organize the scene. Underneath the main character group, we will separate the objects into two separate categories that define what space the objects will exist in, local or world. The local group will contain everything that will be transformed. This mainly includes the skeleton and control objects. Because the skeleton is being controlled by the control objects, it is actually never being directly transformed. However, we may want to add a function to the rig that lets us scale the hierarchy. In this case, the skeleton must be scaled along with the controls so that everything matches.

*FIGURE 12-13   The general arrangement of the hierarchy shown in the Outliner*

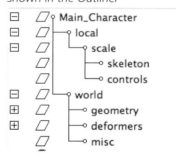

In the world space group, we can add all of the objects that are being driven by the controls and skeleton. This group includes the geometry, whose shape is driven by the skeleton, the deformers, which are also being transformed by the skeleton, and all of the miscellaneous items that are constrained to the controls. This would include any IK handles or IK curves. Figure 12-13 shows the hierarchy and how it should be arranged.

## Local Space Hierarchy

Here we will arrange the local space objects of the rig. We'll begin by creating the group nodes:

**1.** Duplicate the globalXFORM control and delete its shape node. This will give us a new group with a matching pivot. Duplicate this new group node three more times. Name the groups **local**, **scale**, **skeleton**, and **controls**.

**2.** Select all of the control groups in the Outliner. For this rig, that means everything except for the geometry group, the skeleton, the deformer group, the IK handles, and the backIKCurve curve. As you select these controls and groups, make sure to select the globalXFORM control last. Press the P key to parent everything to the globalXFORM control. Parent the globalXFORM control to the controls group.

**3.** Parent the root joint to the skeleton group.

**4.** Select the skeleton group and the controls group and parent them to the scale group.

**5.** Parent the scale group to the local group. The local hierarchy is now complete. Figure 12-14 shows the hierarchy for the local group.

**6.** Lock and hide the transforms for the skeleton and controls groups. For the scale group, lock and hide the Translate, Rotate, and Visibility attributes.

### World Space Hierarchy

Most of the groups for this hierarchy have already been created in previous lessons. The geometry is already grouped to the geometry group, and the deformers are already grouped to the deformers group. We only need to create two more groups:

**FIGURE 12-14**    *The local space hierarchy*

```
□  ◫◦ local
□  ◫  └◦ scale
□  ◫     └◦ skeleton
⊞  ◤       └◦ root
□  ◫     └◦ controls
□  ∿       └◦ globalXFORM
⊞  ∿         └◦ COG
⊞  ◫         └◦ LT_armConstraint
⊞  ◫         └◦ RT_armConstraint
⊞  ◫         └◦ clavCTRLRoot
⊞  ◫         └◦ IK_RT_wristCTRLgrp
⊞  ◫         └◦ IK_LT_wristCTRLgrp
⊞  ◫         └◦ headCTRLgrp
⊞  ◫         └◦ RT_kneeCTRLgrp
⊞  ◫         └◦ LT_kneeCTRLgrp
⊞  ◈         └◦ RT_footCTRL
⊞  ◈         └◦ LT_footCTRL
   ∿         └◦ LT_handCTRL
   ∿         └◦ RT_handCTRL
   ◈         └◦ RT_ElbowCTRL
   ◈         └◦ LT_ElbowCTRL
```

**FIGURE 12-15**    *The world space hierarchy*

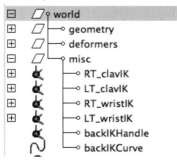

```
□  ◫◦ world
⊞  ◫ └◦ geometry
⊞  ◫ └◦ deformers
□  ◫ └◦ misc
⊞  ⚔    └◦ RT_clavIK
⊞  ⚔    └◦ LT_clavIK
⊞  ⚔    └◦ RT_wristIK
⊞  ⚔    └◦ LT_wristIK
   ⚔    └◦ backIKHandle
   ∿    └◦ backIKCurve
```

**1.** Select all of the IK handles and the backIKCurve curve. Choose Edit | Group to group these objects together. Name the group **misc**.

**2.** Select the geometry, deformers, and misc groups and again choose Edit | Group to group these objects together. Name the group **world**. Figure 12-15 shows the world space hierarchy.

**3.** Select the local and world groups and choose Edit | Group. Name this **MainCharacter**. Actually, this is the perfect opportunity to give your character a name, so name this node anything you want.

## Test the Rig

The entire rig is now complete and is shown in Figure 12-16. However, it is imperative that you test it. Start by attempting a few test poses. Move the character around in the scene with the global transform making sure to rotate it along all axes. If strange things begin to happen, you'll need to troubleshoot the rig.

**FIGURE 12-16**   *The completed control rig as it appears with the geometry in the view window*

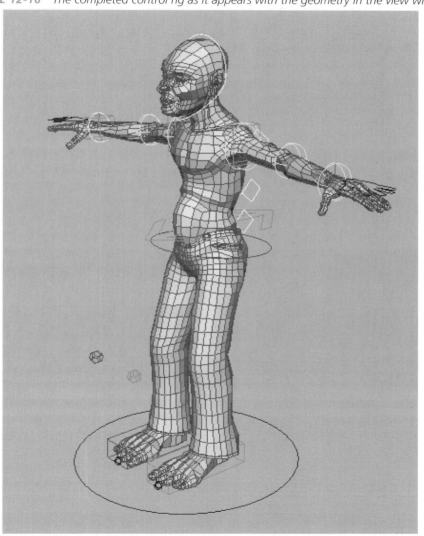

# Summary

If you were able to complete this chapter, then you now have a character that you can begin animating. As you can see, rigging is a very technical process. It's not for everyone, but it is a fascinating puzzle to solve. By understanding the features we explored here, it is really possible to build a rig for anything.

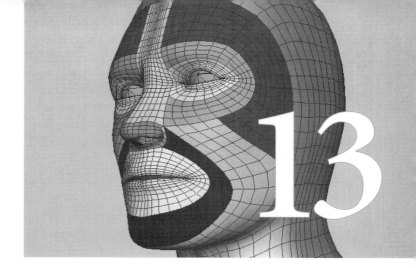

# 13

# MEL Scripting

While the word *scripting* might send many artists away screaming, learning to use MEL is an important aspect of learning Maya. To learn it well, you need to have some basic knowledge and the Maya documentation handy for a description of all of Maya's nodes and commands. However, for many users, just understanding the basics is enough to create short scripts that make working in Maya very efficient. This chapter is meant to demonstrate some of the things you can do in MEL and how to write some simple scripts.

We'll wrap up this chapter by writing a short script that will add some additional functionality to our rig that we built in Chapter 12.

# What Is MEL?

MEL (Maya Embedded Language) is Maya's internal scripting language. Scripting is a simple form of programming. Unlike languages such as C and C++, MEL does not require that the script be *compiled* before it can be run—in other words, the script you write does not have to be translated into machine code to be executed by the computer. In fact, you can think of MEL as a layer between the Maya GUI (graphical user interface) and C++. Any time you click a button on the shelf or choose a command from a menu, that action sends a MEL command to the main Maya program, which then talks to the operating system of your computer that controls the hardware.

## What Can MEL Scripts Do?

MEL can assist you in almost everything you do in Maya. One of the most basic things you can do with MEL is string together several commands and run them in sequence. For instance, the time it takes to convert a multipatch NURBS model into polygons and clean it up could be minimized by creating a script that runs a command to convert the NURBS patches to polygons, combine them into one shape, merge vertices, freeze transformations, and delete history. As an artist, any time you identify a repetitive task, such as the one in this example, you should consider writing a MEL script to handle the task.

Some other uses for MEL are to do the following:

- Redo previous tasks
- Create custom controls, windows, and GUIs
- Control particle systems
- Create, connect, and control multiple attributes
- Align transform attributes of different objects (covered in the tutorial at the end of this chapter)

## MEL Commands

To master MEL scripting completely, you will probably want to research and practice basic programming. Pick up a book on C, C++, Perl, Python, or any other programming language so that you can become more familiar with how programming works and what you can do to optimize and write better scripts.

If you don't want to go that far, a handful of great books on the market are dedicated to MEL scripting. One of the best resources for learning MEL can be found in Maya's online documentation, in the "MEL Command Reference" section. Learning about commands and how to use them is a great starting point to learning MEL. This online section is a dictionary of every MEL command available in Maya, detailing every option for each command and providing examples of how to use it.

> *TIP*   *Every time you open Maya, press* F1 *on the keyboard to open Maya's online documentation, and then go to the "MEL Command Reference" section. Get in the habit of choosing a command of interest, or even at random, and reading about it. Sooner or later, it will all start to make sense.*

Most of the MEL command names are fairly intuitive. The command to add an attribute, for example, is called `addAttr`. In addition to searching the reference alphabetically, you can browse the commands by category. Either way, the names of MEL commands were designed to be as close to English as possible.

## The Script Editor

Perhaps the easiest way to find a MEL command is to execute a command from one of the menus in the main Maya user interface and then look at the Script Editor (Window | General Editors | Script Editor) to see what is journaled. Every time you select an object, open a window, or execute a command from the menus or shelf buttons, these actions fire off a MEL command, which is then journaled, or listed, in the Script Editor. If you have been working in Maya for a few hours, it may surprise you to see that every operation you've done since you launched Maya appears in the Script Editor.

> *TIP*   *Some commands are not journaled in the Script Editor by default. Such commands usually have to do with loading certain parts of the UI. You can see these in the Script Editor by choosing Script | Echo All Commands from the Script Editor menu bar.*

Figure 13-1 shows the Script Editor and its contents after a new scene was created, a polygonal cube was created, and then the Rotate Y attribute was set to 10. The top part of the Script Editor that contains all of the journaled commands is called the History window, and the bottom section is called the Input window, where commands can be typed in and executed.

*FIGURE 13-1*   *The Script Editor*

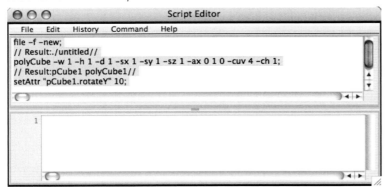

Let's step through the commands journaled so far and figure out what is going on. The first line says

```
file -f -new;
```

The command used here is `file`. This command creates a new Maya scene file and is executed automatically when Maya is launched. The text that follows the dash (-) is called a *flag* (-f, -new). Flags are the options, or settings, for the command. When you set options for a command in its Options window, these options, or attributes, appear as flags in the script. So what do -f and -new do? We could look up the `file` command in the "MEL Command Reference," but while we are in the Script Editor, we can run the `help` command for the `file` command; in the Input window of the Script Editor, type

```
help file ;
```

Then press CTRL-ENTER to execute the command. A synopsis for the `file` command is listed, and all of the available flags are listed with a description of what each stands for. Find the -f flag in this list, and the History window tells you that it stands for *force*. For a more precise definition of this flag, look up the `file` command in the "MEL Command Reference." You'll learn that *force* means to force an action to take place (such as new, open, and save). The -new flag creates a new scene named *untitled*. Finally, the line ends with a semicolon (;), which marks the end of a command.

The next line in the script tells us the result of the `file` command, which in this example is a scene called *untitled*. This result is just feedback from the program telling us what has happened; it is not executing an actual command. Any line preceded by double slashes (//) is a *comment* and is not executed. Comments are often used to document, explain, or comment upon the script.

Next, the command to create the polygonal cube, `polyCube`, is followed by a bunch of flags and their values. Once again, you could use the `help` command in the Script Editor or look up the `polyCube` command in the "MEL Command Reference" section of the documentation. What you will find is that the flags correspond to all of the command's options that you see in the Options window, as shown in Figure 13-2. Compare the information here with that shown in Figure 13-1 and you'll begin to get an idea of how this all links up.

As you can see, while many options are available, MEL commands are quite simple to read. At this point, you could string together a few commands in the Input window and execute them without too much trouble.

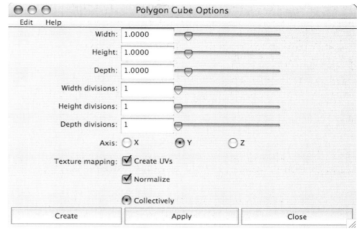

**FIGURE 13-2**   *The Polygon Cube Options window showing all of the attributes, or flags (as they are called in MEL)*

# Using MEL Scripts

One of the most wonderful things about MEL is that additional tools and functionality can be added to Maya without having to wait for Alias to release a new version of the software. However, many of us would rather spend time using tools than making them. The Maya community has generously organized a library of MEL scripts, available at www.highend3d.com, that are available as free downloads. Before you decide that you need to build a new tool, it is a good idea to check this website to see whether you can find a MEL script that does what you need. Chances are, someone has already written it and you can download, install, and use it.

Another good source for MEL scripts is Autodesk's community website (http://area.autodesk .com/). Also, many of the training books and DVDs available from Autodesk contain MEL scripts.

## Where Do Scripts Go?

Okay, so you've just been to Highend3d.com, found the MEL script you were looking for, and downloaded it. So now what? The most common way to use MEL scripts is to place them in one of the scripts directories on your hard drive. Three scripts directories are created when you install Maya.

One of the directories is located in the same directory as the Maya application. Inside this scripts directory, you'll find several subfolders that contain hundreds of scripts. Remember that the entire Maya UI is programmed in MEL. The scripts found in these subdirectories are those that control the Maya UI. Unless you need to modify the existing UI, it is best to stay out of this directory.

Another scripts directory can be found at Documents and Settings\<*your username*>\My Documents\Maya\Scripts (on the Mac, this can be found at /Users/<*your username*>/Library/Preferences/Alias/Maya/Scripts). This scripts directory is a good place to store scripts that are in development or that are loaded in manually.

Scripts that will be frequently executed from Maya should be placed in Documents and Settings\<*your username*>\My Documents\Maya\8.0\scripts (Users/<*your username*>/Library/Preferences/Alias/Maya/8.0/Scripts on the Mac). Any time a script is called, Maya scans this directory, finds the script, and executes the procedures found inside it. Procedures are described in more detail later in the chapter, in the "Procedures" section.

Definition    ***procedure***    *A user-defined function similar to Maya's built-in functions.*

## Executing MEL Scripts

Once you have downloaded a script, it is always a good idea to open it in a text editor and read any instructions that might be included. These instructions will often tell you what the script does, how to use it (such as what things need to be selected before executing), and how to execute it. Sometimes, scripts are written *modularly*. In other words, several scripts may be needed to run a procedure. In the case of the tmBuildFKIKarm.mel script, a separate script called tmFKIKsetup.mel is called on to make the connections between the objects that are created in tmBuildFKIKarm.mel.

If the script is well documented, it will tell you what you need to do to execute it. To execute a script, you can type the name of the command you need either on the command line or in the Input window of the Script Editor. If the script has no documentation, chances are it is a global procedure script.

Most of the scripts that are downloaded are known as *global procedure scripts*. This means that the procedures contained within the scripts can be called from anywhere: from a script, a shelf button, the Script Editor, or elsewhere. A global procedure script needs to be named after the global procedure. If the global procedure is called tmRenderSpec, then the script should be named tmRenderSpec.mel. When tmRenderSpec is typed on the command line or in the Script Editor, Maya searches all of the script paths for a script name that matches the command. If it finds the script, it declares all of the global MEL procedures with that file and executes. Most of the time, it is most helpful if you make a shelf button that contains the name of the command so that you can readily use it without having to look up the name.

# Writing MEL Scripts

Earlier in this chapter, we discussed MEL commands and introduced the idea of stringing several commands together. While this may save you some time in some cases, as you become more proficient with Maya, you will undoubtedly need to take things a bit further. Specific problems or workflow issues will arise and can usually be solved through using MEL. The tmBuildFKIKarm.mel script, for example, was written to automate the process to create those additional skeletons and make the connections between them manually in Maya's interface. This section covers some of the basic elements of MEL scripts.

# Syntax

Every language has its own syntax. In English, every sentence ends with a period, exclamation point, or question mark. In MEL, we conclude a command with a semicolon (;). Learning the punctuation or special characters is one of the first things you'll need to do. If the syntax is incorrect in a MEL script, Maya returns an error. Therefore, it is important that you use the correct syntax as you type in your script. The following table shows some common characters and gives a description of each.

| | |
|---|---|
| // | A double slash is used as a comment indicator. Any characters that follow will not be interpreted by the computer. |
| ; | A semicolon is used to indicate the end of a command. |
| ( ) | Parentheses are used for grouping math equations or strings. |
| [ ] | Square brackets are used to bracket an array index. |
| { } | Curly brackets are used for grouping commands and arrays. |
| " " | Quotation marks are used to indicate that the text contained within is to be treated as a text string for the command line so that any special characters used will not have an effect on the execution of the script. |

# Variables and Data Types

Variables are the cornerstone of any programming language. They are used to store a piece of information temporarily for use anywhere within your script. For Maya to be able to differentiate between a variable and the name of another object or node in a scene, the $ symbol is placed at the beginning of the variable's name.

Variables can hold different types of data. The variable could hold a whole number (integer), a decimal number (float), alphanumeric text (string), or a triple value (vector). Before using a variable, it is important that you declare it in the script by naming it and its data type. Here's a simple example of declaring a variable:

```
float $tmVal = 5 ;
```

This declares that the variable called $tmVal is of a float data type and is equal to 5.

Now that this variable has been defined, we can use it in a math equation. In this line, we use the `print` command to tell Maya to return the result in the History section of the Script Editor and in the feedback line:

```
print ($tmVal + 1) ;
//Result : 6
```

Instead of defining the variable with a constant number, it could be defined using a MEL command:

```
string $tmMaterials[] = `ls -mat` ;
```

In this example, we have defined the `$tmMaterials` variable as a string data type (meaning it can store text). The [] that follows the variable name indicates that the variable can store a list, or array of objects. In this instance, the `$tmMaterials[]` variable is defined by the result of executing another MEL command. The ` at the beginning and end of the `ls -mat` command will return the result of the command run inside of these tick marks. Here, we are using the `ls` command to list and the `-mat` flag to specify what type of objects to list, in this instance, we are listing materials. The result is that the variable `$tmMaterials` will store a list of all materials in the scene.

## Conditional Statements

As you are scripting, you often want to perform a certain function only when a certain condition is met: if "this" do "that" else do "the other." Here is an example:

```
if ($x > 1){
print "X is greater than 1";
} else {
print "X is not greater than 1";
}
```

If we put this to use in a modeling workflow, we can make a script that turns off the display of the wireframe on a model, making it easier to sculpt based on evaluation of the shading rather than on evaluation of the topology. Let's step through this line by line.

First, we want to find out what the display preferences are set to. There is a preference in Window | Settings/Preference | Preferences | Display that sets if or how the wireframe will be displayed on a selected object. To find out the current setting, we will query the display preference. The command for the display preference is `displayPref`. To query a setting, we use the `-q` flag. The attribute we want to query is wireframe on shaded (`-wsa`). Let's get this current setting and store it in a variable called `$wires`:

```
string $wires = `displayPref -q -wsa` ;
```

This query will return either "full," "reduced," or "none." What we want to do is use a conditional statement that says if the setting is currently set to none, then set it to full. Otherwise, set it to none. This lets us toggle back and forth between none and full by using the same script.

```
if ($wires == "none")
     displayPref -wsa "full" ;
else
     displayPref -wsa "none";
```

And there you have it, a very handy modeling tool, especially when you are using the Sculpt Geometry tool. But the real importance is that you can take these examples and formulate them into any script to add some functionality that you need.

## Procedures

Procedures enable you to encapsulate a frequently used piece of code into a single function that can be executed by entering the function's name. With the preceding example we used for the conditional statement, we would have to copy the entire script into the Script Editor and execute it every time. It would be much more efficient to store this function as a global procedure and then save it as a .mel file into our scripts directory. Then we could simply type in the name of the command and execute only that name. What's more, if we were to write another script that needed this functionality, we could simply call the procedure we define instead of having to list the entire script.

First, we will declare the procedure using the global `proc` statement and call it **toggleWireFrame**. After the name of the global procedure, we define any arguments by placing them inside ( ). Arguments are information that is passed to a procedure. A selection of objects, or another user-defined variable are examples of arguments. In this case, we have no arguments, so we leave ( ) empty. The first line of the script should look like this:

```
global proc toggleWireFrame()
```

Now that the procedure is declared, we'll tell it what to do by enclosing the variable declaration and conditional statement inside { and }:

```
{
string $wires = 'displayPref -q -wsa' ;
if ($wires == "none")
     displayPref -wsa "full" ;
else
     displayPref -wsa "none";
}
```

Now, instead of having to use this entire piece of code every time your script needs this information, you can simply call toggleWireFrame from the Script Editor.

Two different types of procedures can be used, local and global. By default, procedures are known only locally. This means that if you have a procedure that is written in MEL, it can be called only from within that MEL script. If you were to type the name of the procedure into the Script Editor, Maya would return an error. If you put the word *global* in front of your procedure, you make it known to all of Maya. This means that it can be called from the Script Editor, a shelf icon, or even another MEL script. You should use local procedures whenever possible and use a global procedure only for the procedure that is called by the user from the Script Editor or Maya UI.

# Tutorial: Using MEL to Enhance IK to FK Switching

In Chapter 12, you learned how to set up a switchable FK/IK arm. While that chapter covers the actual setup of the skeleton and the controls for switching, it leaves out one detail that many animators need if they want to have the position and orientation of the IK skeleton match that of the FK skeleton. Having this functionality will prevent any jumping or popping of the arm when the IK to FK switch is calculated at playback.

## Work Through a Solution

So what is a scriptable solution to this issue? When confronted with a problem like this, the first step is to figure out what the process is—ask "what are the steps involved to accomplish the task manually?" The best way to make this determination is simply to perform all of the necessary actions in the main Maya UI.

How would this work manually? We could move and snap the wrist IK control to the FK wrist control, but this would solve matching only the translations, not the rotations. We need to figure out a way that we can do this in script. Snapping just enables the Move tool to snap to specified components and is therefore not something we can do in script. Copying the translation and rotation values won't work because each object's transforms will differ slightly relative to their parent objects.

A good solution, then, would be to point and orient constrain the IK control to the FK wrist joint and find out what the transform values of the IK control are at that position; then we could copy those values, delete the constraint, and paste those values back in.

Since this will be scripted, a good idea is to make a duplicate of the IK control and constrain it to the FK wrist. This way, we can easily store the target transform values in a variable, and when we are finished setting them to the original IK control, we can delete the entire duplicate. This solves the problem of having to figure out which constraints need to be duplicated.

## Write the Script

Step through the following script. Each line has been commented to document the entire process. Note that some lines have been broken here because of page-width constraints.

```
//Moves IK_LT_wristCTRL to FK_LT_wristCTRL
//This line makes a duplicate of the IK_LT_wristCTRL object and stores
//the object in a variable named $dup. Since selecting $dup actually
//selects two nodes (the transform and the shape) we need to define $dup
//as an array with the "[]" following the declaration of the variable.
string $dup[] = `duplicate "IK_LT_wristCTRL"` ;
//The select command is used to select the FK_LT_wristCTRL object and
//the $dup's transform node in this specific order
select  -r "FK_LT_wristCTRL" $dup[0] ;
//The $dup is point and orient constrained to the FK_LT_wristCTRL
pointConstraint ;
orientConstraint ;
//The translations from the $dup are stored in a variable called
//$pos and the rotations are stored in a variable called $rot
float $pos[] = `getAttr ($dup[0] + ".t")` ;
float $rot[] = `getAttr ($dup[0] + ".r")` ;
//The values in $pos are used to set the translate and rotate
//attributes for the IK_LT_wristCTRL
setAttr ("IK_LT_wristCTRL.t") $pos[0] $pos[1] $pos[2];
setAttr ("IK_LT_wristCTRL.r") $pos[0] $pos[1] $pos[2] ;
//The $dup, along with its constraints, is deleted from the scene.
 delete $dup[0] ;
//We print the result to confirm that it worked
print ("//Result:"+" IK_LT_wristCTRL"+" to " +"FK_LT_wristCTRL" + "\n");
// finished


//Moves IK_LT_elbowCTRL to FK_LT_elbow
string $dup[] = `duplicate "IK_LT_elbowCTRL"`;
select  -r FK_LT_elbow $dup[];
pointConstraint;
$pos = `getAttr ($dup[0] + ".t")`;
setAttr ("IK_LT_elbowCTRL.t") $pos[0] $pos[1] $pos[2];
delete $dup[0];
print ("// Result: " + "IK_LT_elbowCTRL" + " moved to the position
of " + "FK_LT_elbow" + "\n");
// finished
```

Once you have written this script, you can select all of it in the Script Editor and drag the selection onto the shelf along with any other character controls that you may have. This way, you can quickly execute the script while you are animating your character.

## Wrap the Script in a Global Procedure

We now have two functions—one that will match the position of the IK arm to the FK arm, and another that will match the position of the FK arm to the IK arm. Right now, you'd have to execute each script by creating a separate shelf button. While this is fine, you would need to make sure that you save and carry your shelf with you every time you move to a different machine. To make this easier, we can wrap each function into a global procedure and store the entire function in a separate text file that can be stored in the scripts directory.

Once we have verified that the script is working, we can wrap each script inside of a global procedure. This involves defining the global procedure and enclosing it in curly brackets. The result of each procedure should look like this:

```
global proc matchLtIkToFk ()
{
//Moves IK_LT_wristCTRL to FK_LT_wrist_FK
```

This line makes a duplicate of the LT_wristCTRL object and stores the object in a variable named $dup. Since selecting $dup actually selects two nodes (the transform and the shape), we need to define $dup as an array with [] following the declaration of the variable:

```
string $dup[] = `duplicate "LT_wristCTRL"` ;
```

The select command is used to select the FK_LT_wrist object and the $dup variable's transform node in this specific order:

```
select  -r " FK_LT_wrist " $dup[0] ;
```

The $dup variable is point and orient constrained to the FK_LT_wrist object:

```
        pointConstraint ;
        orientConstraint ;
```

The translations from the $dup variable are stored in a variable called $pos and the rotations are stored in a variable called $rot:

```
float $pos[] = `getAttr ($dup[0] + ".t")` ;
        float $rot[] = `getAttr ($dup[0] + ".r")` ;
```

The values in $pos are used to set the Translate and Rotate attributes for the LT_wristCTRL object:

```
setAttr ("LT_wristCTRL.t") $pos[0] $pos[1] $pos[2];
        setAttr ("LT_wristCTRL.r") $pos[0] $pos[1] $pos[2] ;
```

The $dup variable, along with its constraints, is deleted from the scene:

```
delete $dup[0] ;
```

We print the result to confirm that it worked:

```
print ("//Result:"+" IK_LT_wristCTRL"+" to " +" FK_LT_wrist " + "\n");
```

Finally, we just need to add a few more commands that will match the position of the IK_LT_elbowCTRL to the FK_elbowCTRL.

```
//Moves IK Pole vector constraint (IK_LT_elbowCTRL) to the FK_LT_elbow
      string $dup[] = `duplicate "IK_LT_elbowCTRL"`;
      select  -r FK_LT_elbow LT_elbowCTRL;
      pointConstraint;
      float $pos[] = `getAttr ($dup[0] + ".t")`;
      setAttr ("IK_LT_elbowCTRL.t") $pos[0] $pos[1] $pos[2];
      delete $dup[0];
      print ("// Result: " + "IK_LT_elbowCTRL" + " moved to the position
      of " + "FK_LT_elbow" + "\n");
// finished
}
```

Once the script has been tested and is properly functioning, you should create another script that will work for the right arm.

## Store the Scripts in a Script Node

We could now save the script and name it matchLtIkToFk.mel. However, this would bring us back to the problem of the script needing to be carried around with the scene containing the rig that it is related to. We can get around this by storing the scripts inside of a node that exists in our scene, by using something called a *script node*:

1. Choose Window | Animation Editors | Expression Editor.

2. In the Expression Editor, choose Select Filter | Script Node. The Expression Editor is shown in Figure 13-3.

3. Copy and paste your scripts from the Script Editor to the Script input window in the Expression Editor. Click the Create button to create the script node with this information added.

4. Click the Test Script button to run the script and check for any errors. If any errors are returned, go back to your expression and double-check the lines that are returning errors. Usually, the cause of errors is really silly. Perhaps you misspelled an object name or used the wrong case sensitivity. Usually, though, you should not have any problems if you have tested your script before this stage.

FIGURE 13-3   *The Expression Editor showing the script node UI*

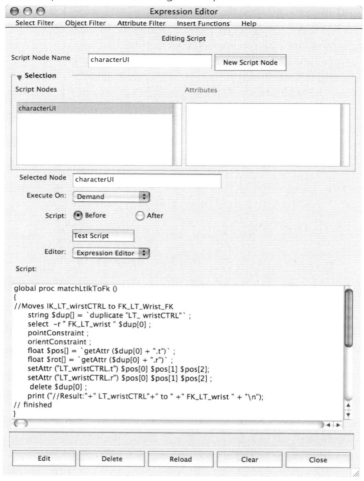

5.   Now, you can type matchLtIkToFk into the command line to execute the matching function. As long as you know the names of the command, you don't have to worry about any external files.

## Build a UI Window

Another wonderful thing about MEL is the ability to create custom windows and panels that contain custom functions. Every single window you see in the entire default interface is a result of MEL building the elements and connecting the attributes to the respective elements.

MEL uses the command called `window` to create a new window. You can then name the window by using the `-t` flag and place a name in quotations. This name will appear at the top of the window, in the title bar. Lastly, you give the window a name that it will be called in script. The entire command to create a window called Character UI Window with a script name of testWin would look like this:

```
window -t "Character UI Window" charUI;
```

To show the window, you need to use the `showWindow` command and then state which window you would like to show:

```
showWindow charUI;
```

Type the preceding two lines in the Script Editor and execute them. A new window will appear in Maya that looks like this:

Before telling Maya to create this window, it is always a good idea to run a check to see if the window already exists. If it does exist, we'll tell Maya to delete it and then use the previous commands to build it again. Also, we can add window dimensions to our testWin script. The entire script should look like this:

```
If ('window -exists charUI ')
deleteUI charUI ;
window -t "Character UI Window" charUI -wh 400 600;
showWindow charUI ;
```

## Add Buttons to the Window

To create a button, you can use the `button` command and then name the button with the `-1` flag (label). Then you can use the `-c` flag to send out a command. For this example, we'll use the script demonstrated in the last section except this time we'll call it buuklaUI. Also, we will use the `rowColumnLayout` command. This will allow us to arrange the buttons in the window in rows and columns so that the buttons will be stacked on top of one another in a column and laid out in a row according to how the panel is defined. We'll create buttons that will execute the matching of the FK to IK and IK to FK. You can use the `-c` flag with the `button` command to execute any command that has already been defined.

```
//check and see if window exists
if ('window -exists charUI')deleteUI charUI ;
//Create window called charUI
window -t "Character UI Window" -wh 400 400 charUI;
rowColumnLayout
      -numberOfColumns 2
-columnWidth 1 200
-columnWidth 2 200;
```

```
//Create Buttons
button -l "Match Left IK to FK" -w 300 -c "matchLtIkToFk " ;
button -l "Match Right IK to FK" -w 300 -c "matchRtIkToFk " ;
showWindow charUI;
```

When you are happy with your window, wrap it in a global procedure called **buildCharUI** and then place it inside the script node we created in the last section. Now, when you execute buildCharUI, the window will appear and give you quick access to the FK to IK matching commands, as shown here:

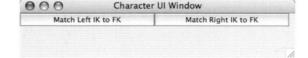

## Summary

In this chapter, you learned a little bit about MEL. Although there is so much more you can do with MEL, with just the information presented here, there are many powerful functions you can create within Maya. Be sure to check out the "MEL Command Reference" section in the Maya online documentation for definitions of every command available in the application.

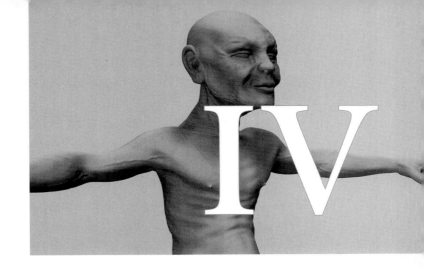

# Animation

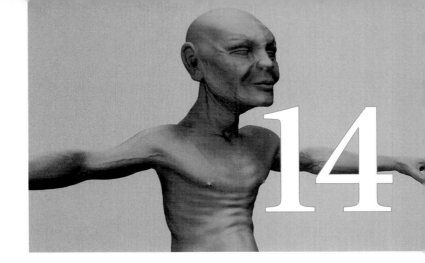

# 14

# Animation Basics

**Animation can be defined as the** process of "bringing art to life"; it is the point at which timing and motion come together to make artwork represent real life. The animation toolsets in Maya provide a wide range of control over timing and motion. But be warned, the realism part doesn't come with the installation alone—you as the artist must practice the art of animation before you'll be able to create realistic characters and objects.

Maya offers dozens of tools and features that let you create animation in many ways. Along with its basic keyframing and motion-editing abilities, Maya lets you animate using path animation, nonlinear animation, dynamics, motion capture, and procedural animation, which will be discussed in the following chapters. In addition, the nature of Maya's dependency graph allows for customization and control over large sets of animated objects.

Throughout the next six chapters, we will examine some of the features that will assist you in bringing your models to life. In Chapters 22 through 26, you will learn to use Maya's Dynamics toolset to create animation by running simulations of natural phenomenon. But let's start with the basics in this chapter. We will focus on learning keyframe animation and using the fundamental animation tools in Maya.

# A Brief History of Animation

Animation has been around since the dawn of humankind—consider the ancient caves of southwestern Europe, which are decorated with paintings that not only contain detailed illustrations of animals from the region but also suggest how those animals moved. One could argue that even though these paintings are still, the artist has brought the creatures contained in them to life; hence, they have been animated, in a way.

However, suggesting motion is not enough. Over time, people have continued to utilize any technology available to make their still pictures move. Flipbook animations were produced by drawing frames along the edge or corner of a stack of paper or a book. As you flipped through the pages in the stack, your drawings would advance from one *frame* to the next. Another early animation technology was the turntable. A series of drawings, or frames, were placed along the inside walls of a cylindrical object that sat on a turntable. The cylinder had little slits between each frame. When a person looked into the cylinder through the slits from the outside while the turntable was spinning, she'd see the drawings advance from one frame to the next.

For both of these methods, animation time was limited and the characters had to be pretty small. In addition, few means were available to deliver such animations to an audience. In today's age of downloadable movies, it is easy for us to forget about all of the fantastic animations that must have been created by amateur artists all over the world. These small glimpses of animation rarely emerged outside someone's home or outside a carnival. For this reason, many of the great early animations have been lost.

What really enabled the assemblage and delivery of animation was the advancements made to the motion picture camera and film stocks in the late 1800s. By the early 1900s, stop-motion animation was accomplished by positioning objects on a set and capturing their positions, one frame at a time. As the film frames advanced, so did the objects. These same basic techniques are still being used today.

The ability to capture a sequence of drawings onto film and play them back through a projector was what eventually brought animation to the masses. However, for animation to be created efficiently for production—that is, for it to become a profitable industry—some procedures needed to be put into place. Disney was responsible for many of these developments in the industry.

*The cycle*, *repeat animation*, and *cross-over* techniques were developed by Disney to minimize the amount of manual labor it took to produce animation, by cutting down on the number of drawings needed for the show. Many of these techniques have carried over into the digital realm and are used in Maya. One of the most basic techniques from the old days is the idea of keyframes and in-between frames. These form the basis for animating in Maya.

> **NOTE** *If you're interested in learning more about animation processes of the early days (or animation in general), be sure to read* **The Illusion of Life: Disney Animation (Hyperion Press, 1995) by Frank Thomas and Ollie Johnson. This book details many of the techniques developed by Disney that enabled the company to produce animated features within a reasonable budget. Later in this chapter, you'll find a list of other books that might be of interest.**

# Keyframe Animation in Maya

In the old days of pencil-drawn animation, large-scale animation production would use a technique known as *keyframing*. Using this technique, a lead animator would draw a character at all of the key poses at different points in time throughout the animation. The key poses usually contained the more extreme positions and gestures for a character or poses that would be held and therefore visible for longer periods of time. An assistant animator would then be responsible for creating the drawings that filled in all of the poses that lie between those key poses. Let's take a moment to look at these processes and how they relate to and are executed in Maya.

## Keyframe and Frame

Animation techniques in Maya are not so different from those developed by Disney some 80 years ago. Once an animator knows what he is going to animate, he will usually go through and *block* out the animation by posing the character at key points in time. (Remember the preproduction process discussed in Chapter 1? *Know* what you have to do *before* you try to do it.) For each of the poses, the animator sets a key for the position of all the different body parts that make up the entire pose at that frame. The values for the attributes being animated are stored at that frame. (Realize that any attribute can be animated in Maya, not just translation, rotation, and scale.) This stored data is called a *keyframe*. In order for any of the animated objects to "move," at least two keyframes are needed, each set at different points in time.

Along with the position of objects in an animation, time is an important variable. When dealing with animation, film, or video, the time variable is broken down into *frames*. How long is a frame, then? That depends on the *frame rate*. The frame rate is measured in frames per second (fps). Different technologies use different frame rates. In general, if you are making animation for NTSC (National Television System Committee standard) video, the rate is 29.97 fps—so three seconds of animation will need about 90 frames. (You can also see it like this: one frame is 1/30th of a second.) For film, the rate is 1/24th of a second, or 24 fps.

Figure 14-1 shows a single ball animated over specified frames. An animation snapshot is created to show the motion.

**FIGURE 14-1** *Examples of keyframes and frames*

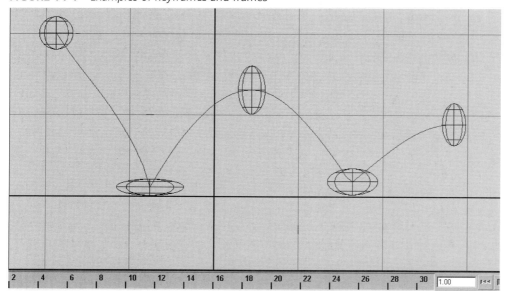

> **NOTE** *The terms keyframe and key are used interchangeably. In this book, the process of animating with keyframes will be called keyframing. The terms keying and editing keys refer to setting and editing keyframes.*

## In-Between and Interpolation

In 2D animation, humans must painstakingly create pictures to fit between keyframes to suggest motion. These frames are called *in-between* frames, and the process of creating them is called *tweening*. The in-between frames of a bouncing ball are shown in Figure 14-2.

**FIGURE 14-2**    *Keyframe and in-between frames of a bouncing ball animation*

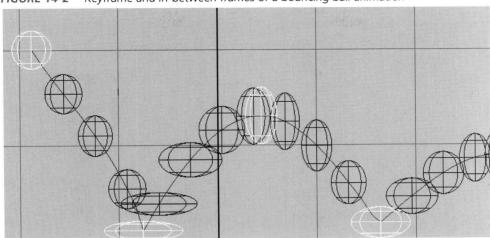

One of the many great advantages that 3D brings to animation production is that the software is able to calculate the positions for in-between frames for you. Since the objects have already been modeled with 3D geometry, the objects in the scene will accurately represent the object at any angle (something that 2D animation in applications such as Adobe After Effects or Macromedia Flash are not capable of doing). The calculation of these in-between frames is called *interpolation*.

The animator can control how the data is interpolated by editing the *animation curves*. An animation curve plots the change in value over a period of frames between two or more keyframes. You can change the interpolation by changing how the tangents behave as they enter and exit a keyframe. This means the curve is completely smooth as it enters and exits a keyframe. These tangents can also be broken and their direction edited independently of one another. The ball shown in Figure 14-2, for example, was animated by breaking the tangents of the keyframes where the ball hits the ground. The tangents were then edited so that they point upward in and out of the existing keyframe. The result is that the ball appears to bounce. In the bouncing ball tutorial later in this chapter, you will go through the steps of editing the curves to create the bouncing ball.

Definition    ***tangent***    *A line or vector that indicates the slope of a curve at a given point.*

In the end, Maya is essentially creating a database, wherein a certain value has been assigned to a certain frame for a certain attribute that is being animated. Every time Maya advances to the next frame, it uses the animation curve to determine the value of the attribute at that frame. This occurs for every attribute that has a keyframe assigned at every frame.

# How to Set Keys

 Maya is all about control and flexibility; it offers a variety of ways to key attributes, but you'll usually key attributes through menu commands and windows in three ways, as shown in Table 14-1.

These methods will create a key at the current frame as set in the Time Slider. The illustration here shows the Channel Box displaying a NURBS sphere's transform node. When a key has been set on a channel, the value fields for that channel will be highlighted in orange.

You can also key transformation quickly by pressing a hot key, as shown in Table 14-2. For example, if you want to key only movement of an object in the X, Y, and Z axes, you can press SHIFT-W.

> **NOTE**   *It's important that you are aware of what attributes you've keyed and limit keying only to necessary attributes to avoid confusion and save time. Maya will evaluate any attribute that has been keyed at every frame. This means that if you have an attribute that has keys set on it but the value is not changing, Maya is evaluating this attribute for no reason. This could slow down your computer's performance.*

The process of setting keys usually goes like this:

**1.** Set the current frame either by moving the Time Slider to the desired frame number or by typing the frame number into the Current Frame field.

**2.** Set the values for the animated attributes. If you are moving a piece of geometry into position, you will use the transform tools to edit the transform attributes.

**3.** After everything is in place, set keys on the desired attributes.

**TABLE 14-1**   *Keying Methods*

| KEYING METHOD | FUNCTION |
|---|---|
| Select the object and choose Animate \| Set Key. | Keyframes all the attributes in the Channel Box for the selected object. |
| Select the object, LMB-click attributes you want to key, RMB-click-and-hold, and choose Key Selected. | Keys only the desired attribute. You can also select and key multiple attributes. |
| Select the object; open the Attribute Editor by pressing CTRL-A (COMMAND-A); RMB-click-and-hold the attribute you want to key, and choose Set Key. | Keys only the specific attribute. The Attribute Editor has more attributes available to key than the Channel Box. |

*TABLE 14-2*  *Transformation Hot Keys*

| HOT KEYS | FUNCTION |
| --- | --- |
| S | Key all the attributes in Channel Box |
| SHIFT-W | Key translation (move) |
| SHIFT-E | Key rotation |
| SHIFT-R | Key scale |

It is important that you set the current frame *before* the object is moved (or before any attributes are changed). For example, if you had keyed a sphere's Translate X attribute at 0 and wanted to animate it to 5, but you set it to 5 without moving the Time Slider, the sphere would jump back to 0 as soon as you touched the Time Slider, because it is reverting to the position of its last key.

> **N O T E**  *If you do set the attributes before setting the Time Slider, you can advance the Time Slider without changing the attribute values for all objects by MMB-dragging the Time Slider.*

## Viewing and Editing Keyframes

Once keys have been set on various attributes, you can view and edit them in a number of ways. As soon as you keyframe, three things will happen immediately to let you see that an object has been animated. First, when an object is selected, red lines, or tick marks, appear in the Time Slider. These indicate that keys are set at those frames. When attributes are viewed in the Channel Box, the fields containing the values will be highlighted in orange to signify that these channels have been keyed. In the Hypergraph, a keyed node will appear skewed instead of as a rectangular shape.

Once you begin to refine your animation, the keys set initially with the Set Key command are rarely used as is. You'll usually choose to edit the values for the attributes or change the frame at which the keys are set. Actually, you'll probably be doing a little of both. Maya offers four main interfaces to use for editing keyframes: the Time Slider, the Graph Editor, the Dope Sheet, and the Channel Box. Let's take a look at them one by one.

### Time Slider

We have already talked about viewing keys in the Time Slider, but what about editing them? The Time Slider offers some basic, yet fast, ways of editing the position of keys over time. After setting the Time Slider to a frame that contains a key, RMB-click to view a marking menu that lets you Cut, Copy, Paste, and Delete keys as well as make other adjustments to the Time Slider.

If, for example, you want to copy a key on frame 1, you set the Time Slider to frame 1, RMB-click, and choose Copy from the marking menu. Then move the Time Slider to the frame where you want to paste it, RMB-click, and choose Paste | Paste.

If the animation is lengthy, editing keys in the Time Slider can be difficult, because the Time Slider may have to squish together all the frames to fit them in the display. You can use the Range Slider to limit how much of the Time Slider's range is displayed. Set the total frame range for your animation in the Playback Start Time and Playback End Time fields at the far left and right ends of the Range Slider (see Figure 14-3). You can slide the Range Slider by LMB-dragging to move through all the ranges. Figure 14-3 shows the Time Slider and Range Slider. The Range Slider has been set up so that the Time Slider displays only frames 1 through 60 in an animation that is 120 frames long.

You can also move and scale (shrink and enlarge) a group of keyframes using the Time Slider. This effectively shortens or lengthens animation. To do this, SHIFT-LMB-drag over the keys. A solid red color appears around your selection. The center arrows in the highlighted area will move the selected keys forward or backward in frames. Arrows at the opposite ends will scale the range of the key. If you scale the selected range down, actions inside the red area will occur in a smaller timeframe, creating a faster animation. If you scale the selected range up, the opposite will happen—animation will slow down. This is the easiest way to change timing of an entire animation. Figure 14-4 shows how timing is changed in the Time Slider.

### Graph Editor

Using the Graph Editor is probably the most powerful way to edit keys. The art of animation involves many subtleties. Animators know that to add those subtleties can be a tedious process, and Maya's Graph Editor is intended to make that task easier. It lets you edit not only the time and value for a key, but also the interpolation between the keys. By being able to edit the curve as it enters and exits a keyframe, the animator has more control over each keyframe, leading to more control over the entire animation and the realism of the shot.

You can display the Graph Editor by choosing Window | Animation Editors | Graph Editor. (You can also open it as a panel in a view window.) As shown in Figure 14-5, the Graph Editor lists all selected objects and their keyable attributes in the Outliner section on the left side. The graph

---

**FIGURE 14-3**    *The Time Slider and Range Slider*

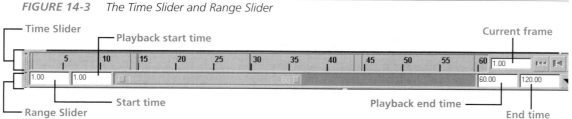

FIGURE 14-4    *You can quickly change timing with the Time Slider.*

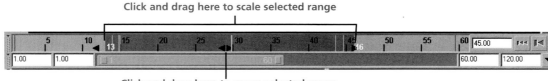

section on the right displays the *animation curves* for each selected attribute. Animation curves graphically represent changes in time (represented on the horizontal axis) and data values (represented on the vertical axis). They contain keys, represented by points, with tangents that control how the curve segments enter and exit a key.

Many commands and tools available in the menus and shelf of the Graph Editor let you control the interpolation of the curves by changing the behavior of their tangents. Keys can be inserted anywhere along the curve by using the Insert Key tool found in the Graph Editor's toolbar along the top. We will cover this editor in more depth in this chapter's tutorial.

FIGURE 14-5    *The Graph Editor shows animation data as a series of curves.*

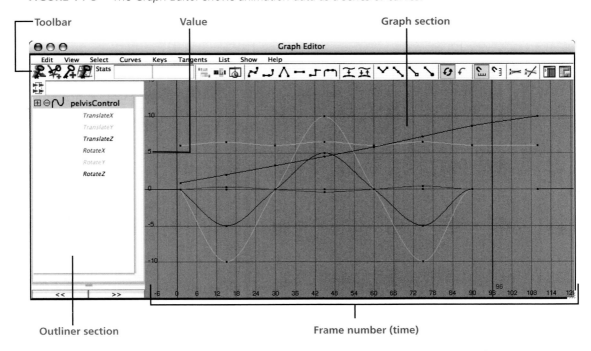

> **TIP** *Opening the Graph Editor in a panel is less cluttering than opening it in a separate window. You can also use Persp/Graph premade panel setup by clicking one of the saved Layout buttons below the toolbox.*

### Dope Sheet

The Dope Sheet can be accessed by choosing Window | Animation Editors | Dope Sheet. Instead of using a curve, the Dope Sheet represents keys as tick marks in frames. It resembles a database, similar to what you'd find in an Excel spreadsheet. Figure 14-6 shows the Dope Sheet.

This editor is most useful for making broad changes to animation to adjust basic timing. It won't let you view or edit the interpolation between the keys, but it does allow you to select and edit the position of keys for specific attributes. This is similar but more powerful than highlighting and moving keys in the Time Slider, because the Time Slider will not let you select specific attributes whose keys you need to edit.

With the Dope Sheet, you can edit all of the frames for multiple objects in a scene or selection or you can edit keys of specific attributes. You can even edit the placement of the keys for all Translate X attributes in the scene.

**FIGURE 14-6** *The Dope Sheet provides a quick way to edit the timing of an animation without worrying about the interpolation.*

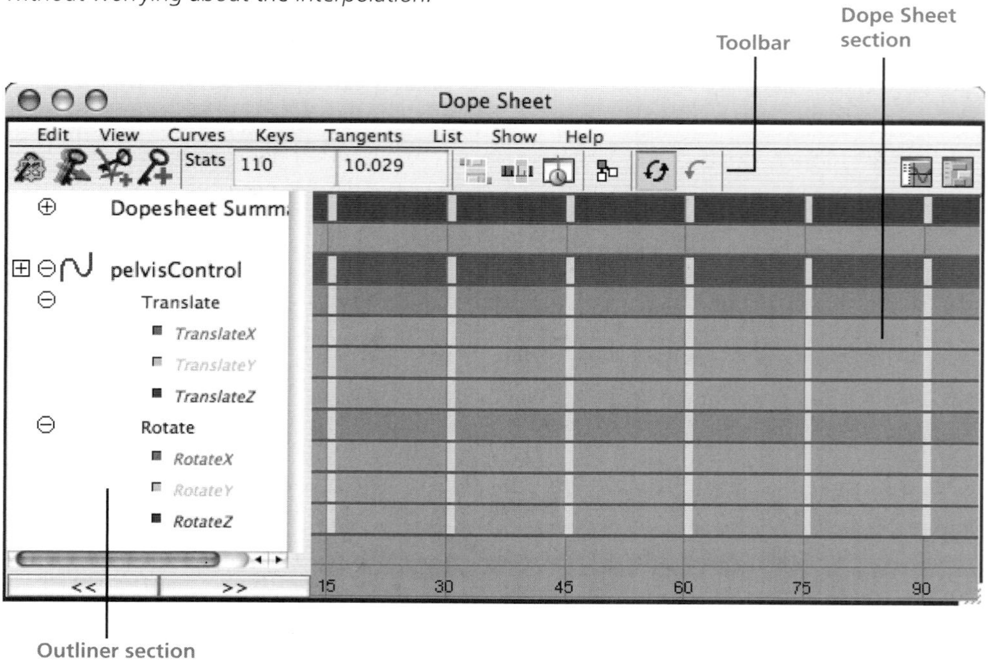

Toolbar

Dope Sheet section

Outliner section

When you open the Dope Sheet, you see a list of selected objects on the left in the Outliner section. On the right side of the window is the Dope Sheet section that shows a green strip with black tick marks. These tick marks are the keys. If multiple attributes of an object were keyed at the same frame, clicking one of these ticks to select it and MMB-dragging with the Move tool will enable you to move all of the keys on that frame to another frame. If you unfold the object in the Outliner section, you can view individual attributes and see what frames have keys on them. From here, you can edit the attributes individually.

At the top of the Outliner is an item called Dopesheet Summary. Clicking this shows you where the keys are placed for all of the selected objects. Using the Dopesheet Summary is an easy way to edit the keys on all selected objects at the same time.

### Channel Box and Attribute Editor

Keys can be set, cut, copied, and pasted in the Channel Box and Attribute Editor by RMB-clicking an attribute and choosing a command from the marking menu. You can delete all animation on any channel by first selecting an attribute in the Channel Box so that it is highlighted (the actual attribute, not the value), RMB-clicking, and choosing Delete Selected from the marking menu. You could also choose Break Connections from the marking menu, which will break the connection to the animation curve.

# Playback Controls

To play the animation, use the playback controls at the bottom right of the Maya Workspace. These controls should look familiar to you. Figure 14-7 shows the playback controls, which are described here:

- **Go to Start of Playback Range, Go to End of Playback Range** Moves the slider, respectively, to the beginning or end of the range of frames shown in the Time Slider.

- **Go to Last Frame, Go to Next Frame** Moves the slider, respectively, backward or forward one frame every time you click.

**FIGURE 14-7** The playback control buttons

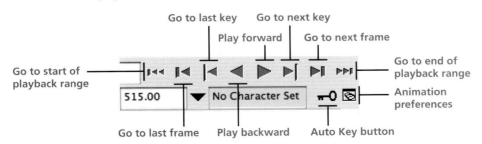

- **Go to Last Key, Go to Next Key**  Moves the slider, respectively, to the last or next keyframe; this is a great way to hop around keyframes.

- **Play Backward, Play Forward**  Plays the animation, respectively, backward or forward.

- **Auto Key**  Click this button, and it will turn red. As long as an object has a previous keyframe, you can simply move to a new frame and change an attribute's values, and the channels will automatically be keyframed. This button is a quick way to keyframe many objects.

- **Animation Preferences**  Opens an Animation Preferences window to let you change settings, such as a time range in the Time Slider, height of the Time Slider, and speed of playback.

*TIP*  *In the Animation Preferences window, Maya's default Playback Speed is set to Play Every Frame. This will play every frame of the animation as fast as the computer is able to play it back. If the scene has many objects or is deforming complex geometry, the scene might play back slowly. If it is a simple animation, such as a bouncing ball, it will play back quickly. When you set the Playback Speed to Real Time, the animation will play back at a constant speed and drop frames to maintain real time if the animation is too complex.*

You can use the hot keys shown in Table 14-3 to move through the animation without touching a mouse. These are great for quickly reviewing an animation.

Let's work on a simple tutorial to learn how to keyframe and edit keys in Maya.

**TABLE 14-3**  *Playback Control Hot Keys*

| HOT KEYS | FUNCTION |
| --- | --- |
| ALT-V (OPTION-V) | Play and stop animation |
| . (period) | Move slider to next keyframe |
| , (comma) | Move slider to previous keyframe |
| ALT-. (OPTION-.) | Move slider to next frame |
| ALT-, (OPTION-,) | Move slider to previous frame |
| K-LMB-drag, K-MMB-drag | Move slider left and right |

# Tutorial: Bouncing Ball

This tutorial will teach you how to keyframe and tweak animation by creating a simple animation of a bouncing ball. First, we will set up the scene for animation, and then we'll create objects, key the ball's animation, and edit and apply character animation techniques by adding more keyframes.

▲ This car model was created using a combination of NURBS and polygon geometry. After applying shaders, it was rendered in Mental Ray using high dynamic range imagery (HDRI) as the light source. Learn to use these techniques in Chapters 6, 7, 18, and 21.

▲ This character was posed using a custom-made control rig to drive the deformations. Displacement texture maps were applied to give the model some additional detail, and then rendered in Mental Ray. Chapters 9, 10, and 11 take you through the character setup process step by step. Chapters 17, 18, and 21 cover the texturing and rendering processes used to create the final image.

▲ *Car from Karmic* by Glen Pitman and Chris Logan. This stylized car model plays a main role in the new animated short film *Karmic*. The car was modeled using subdivision surfaces, and rendered in Pixar's Renderman. More information on the film can be found at www.karmicshort.com.

▲ Caustic reflections from the teapot are cast onto the wall. Learn to create caustics and other rendering effects in Chapter 21.

▲ By rendering a scene in different passes and compositing them in postproduction, you are free to experiment with different surface characteristics without having to re-render the image. Learn to render and composite this image in Chapters 27 and 28.

▲ *William Burroughs* by Shannon Thomas. This polygonal model was sculpted in Skymatter's Mudbox application. Both bump and displacement maps were then extracted and applied to a low-resolution model that was created in Maya. The final image was composited in Photoshop from a series of render passes created in Maya. A detailed tutorial, created for 3D Total, is at http:// 67.15.36.49/ team/Tutorials/chesney _baker/chesney_01.asp.

▲ *Fruit Bowl* by Dario Lopez. Modeled by Dan Wade. This image is part of a lighting challenge series operated by Jeremy Birn. Anyone can download the models from www.3drender.com, add textures and lights, and then post their results and receive critiques from other users of the website.

▲ *Paris* by Dario Lopez. The two helicopters in this image were modeled, textured, and lit to match the background plate. These techniques are explained in Chapters 27 and 28.

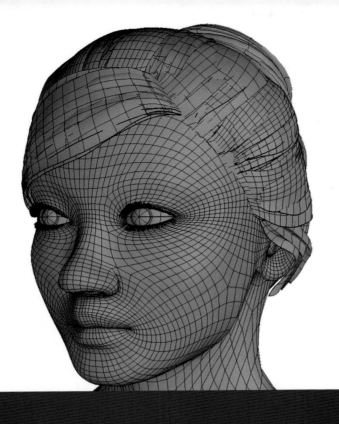

▲ *Young Woman, Wireframe* by Tsun-Hui "Andrea" Pun. This wireframe render is an example of a well-structured polygonal mesh suitable for very articulate facial animation.

▲ *Young Woman* by Tsun-Hui "Andrea" Pun. Realistic skin shaders can be created in the Hypershade by building a network of materials, textures, and utility nodes. Learn all about the Hypershade in Chapters 17 and 18.

▲ *World War II Boeing B-17F* by Timothy Odell. Maya and Photoshop were both used to create, render, and composite these images. Very similar techniques are demonstrated in Chapters 27 and 28.

▲ *World War II Pilot* by Timothy Odell.

▲ *Musa* by Digitrove, Inc. Director: SeRyong Kim, Modeling: HyunJung Shin, Texturing: YunJung Kim, Lighting: JiSun Lee. Images taken from the short film *Duel*. For more information on this project, go to www.digitrove.net.

▲ *Musa* by Digitrove, Inc. (see credits listed above)

▲ *The Beheading* by
Chris Bostjanick. This
piece was initially
modeled in Maya
and then taken into
Pixologic's ZBrush for
additional sculpting.
Displacement texture
maps were then
extracted and applied
to a low-resolution
model in Maya and
rendered, providing
surface details that
would be very difficult
to model with standard
modeling techniques.

▲ *Amazon* by Mark Kobrin. This is another piece that was enhanced by using displacement texture maps that were created and extracted from Pixologic's ZBrush. Rendering with displacement maps is discussed in Chapter 21.

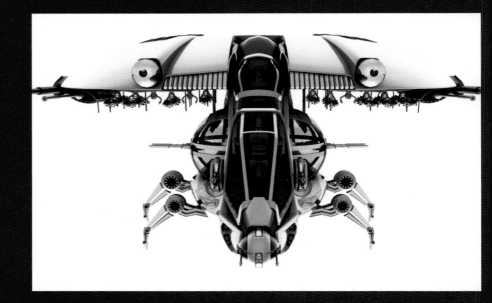

▲ *Stinger* by Chris Bostjanick. This is a NURBS model based on a ship design from *Star Wars: Episode III–Revenge of the Sith*. It was rendered using the Final Gather algorithm in Mental Ray.

▲ *F16* by Roger Ridley. The initial topology of this model was laid out in NURBS and then completed using polygonal modeling. A Mental Ray Ambient Occlusion Shader was used to render the entire image.

▲ *Attack* by Tom Meade. This image was assembled using the same techniques demonstrated in Chapters 27 and 28.

▲ *Domes Scene* by Ping Xie. All shapes on the model were created entirely in NURBS, mostly by revolving complex profile curves. Texture maps and bump maps were then projected onto the surfaces. The image was rendered with Mental Ray.

▲ *Worm* by Landis Fields. This is a still image from the animated short film *Worm*. In this shot, the main character, Inch, contemplates moving his food supply on the long road ahead. More information on this project and many others can be found at www.landisfields.com.

▲ *Worm Environment* by Landis Fields, Roger Ridley, and Shannon Thomas. This is a wide shot of the environment that *Worm* takes place in. This scene made use of high-resolution geometry, texture maps, and Maya Paint Effects.

▲ *Ole Red* by Shannon Thomas. This is a close-up of the truck model featured in *Worm*. Texture and shading techniques were used to create the worn, weathered look of the surfaces.

▲ *BMW M3* by Brian Metzger. This model was created using the process demonstrated in Chapters 6 and 7. It was rendered in Mental Ray.

▲ *Ferrari 430* by Brian Metzger. Another car by the same artist shows off some of the interior pieces that were modeled with polygons.

▲ *RX8* by Shannon Thomas. The modeling technique used on this car employs NURBS to create the precise details, which are then converted to polygons and rendered in Mental Ray.

▲ *Hubcap* by Shannon Thomas. A close-up of some of the fine details of the RX8 model.

▲ *Bats* by Jimmy Wu. Maya's texturing and camera mapping capabilities make it an optimal choice for creating matte paintings.

▲ *Castle* by Jimmy Wu.

*NOTE Even something as simple as a bouncing ball is usually the best and first test of a good animator and a good starting point as a newbie animator to land an internship or job.*

## Set Up the Animation

Before we start animating, let's set up some preferences. First, we need to decide what the final output of this animation will be. If it's NTSC video, we need to set the frame rate to 30 fps. If it's film, we'll use 24 fps. Although you can set the animation to a different frame rate later on, it's much easier to set it at the proper frame rate from the beginning.

*NOTE If you set your animation to 24 fps and output to video, your animation will increase in speed. Therefore, it's important that you figure out your final output before you start.*

**1.** To change the frame rate, choose Window | Settings/Preferences | Preferences to open the Preferences window.

**2.** From the Categories list on the left side, select Settings, and change Time to NTSC (30 fps), as shown in Figure 14-8. Now Maya is set to use 30 fps.

**FIGURE 14-8**  *Setting animation preferences in the Preferences window*

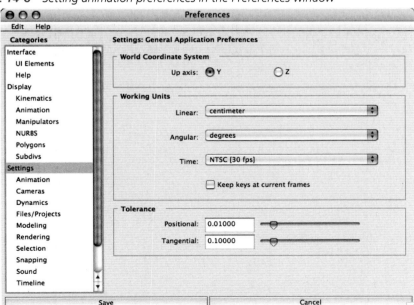

**3.** Change the playback speed to be in real time. From the Categories list, select Timeline. The main Preferences window will update with new information. Under the Playback Speed box, choose Real-time [30 fps]. This way, when you click the Play button, the animation will play in real time.

**4.** Set the range of frames for our animation. In the Range Slider, type **1** in the Playback Start Time field and type **120** in the Playback End Time field (see Figure 14-3). This indicates that you'll use frames 1–120 for your animation and limits the display in the Time Slider to those frames.

**5.** Choose Create | NURBS Primitives | Sphere, and then choose Create | NURBS Primitives | Plane. A NURBS sphere and a flat NURBS plane should appear in your scene.

**6.** Select the NURBS plane surface and scale it in all directions. In the Channel Box, set the Scale X, Scale Y, and Scale Z values to at least 80, so that you leave enough space for the ball to bounce around. The objects should resemble those shown in this illustration.

**7.** Before you start moving and keyframing the sphere, let's move the pivot of the sphere to its base. This will help squash and stretch the sphere's shape toward the bottom of the sphere. Select the sphere, select the Move tool, and press the INSERT (HOME) key on your keyboard. Move the green line down in Front or Side view, so the center of the pivot—the blue square—rests at the bottom of the sphere. Press the INSERT (HOME) key to close the pivot editing mode.

## Set Keys

Now we will begin positioning the ball and setting keys at different frames in time:

**1.** In the Front view, start animating the translation value of the sphere. Make sure you are working in frame 1—if you're not, move the Time Slider to frame 1. Start by moving the sphere about −10 units in the X axis and about 10 units in the Y axis. Press SHIFT-W to keyframe the translation only.

**2.** Move the Time Slider to frame 15.

**3.** Move the sphere so the bottom of it is at the NURBS plane just around the global axis (0,0,0). (See Figure 14-9.)

FIGURE 14-9    *Initial keyframing positions*

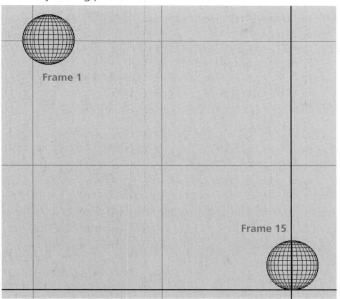

4. Press SHIFT-W to key translation again. Move the Time Slider back and forth between frame 1 and frame 15, and you will see the animation.

   This is how you keyframe animation in Maya. You move the Time Slider to the frame you want to keyframe, adjust values on attributes so it has the desirable values, and then set a key. If you move or scale the object before you change to a different time, the change you made will be discounted as soon as you change a frame. *You should always set the frame first.* Figure 14-9 shows the positions for the initial keyframing of the sphere.

5. Continue keyframing. Move the Time Slider to frame 30 and move the sphere up and to the right side of the view. Since the ball's momentum should diminish a little with each bounce, the bounce lengths should get shorter in the X and Y directions as it progresses through the animation.

6. Move the sphere to around unit 6 in both the X and Y axes.

7. Move the Time Slider to frame 45, and then move the sphere so the bottom of it is at the NURBS plane.

**8.** Keep keyframing every 15 frames, as shown in Figure 14-10, until you reach frame 120, where the sphere rolls to a stop. Figure 14-10 shows the positions of the ball at each frame to which it has been keyed.

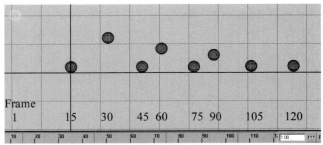

**FIGURE 14-10** *More keyframes are added at 15-frame intervals.*

## Edit the Keys

Now change to the Perspective view. Click the Play button in the playback controls. You'll notice that the animation hardly resembles a bouncing ball. Instead, it looks odd—as though the sphere is swimming through space instead of striking the ground and bouncing. In fact, the ball appears to slide along the plane before moving up.

The reason for its dragging appearance is that we are using *spline tangents,* the default interpolation method. This type of interpolation creates spline curves between the keys, with unified, straight tangents. The tangent of the curve entering the key is the same as the tangent exiting the key. While spline curves are good for fluid motion, they're really not suited for showing the snappy movement of a ball quickly changing direction as it hits the ground.

Definitions   **spline tangent**   *Creates spline curves between the keys, with unified, straight tangents. The tangent of the curve entering the key is the same as the tangent exiting the key.*
**linear tangent**   *Creates straight lines between keys.*
**clamped tangent**   *Makes most of the animation smooth, like spline interpolation, but if two adjacent keys hold similar values, makes the interpolation linear.*

Two other tangents can be used for animation. *Linear tangents* create straight lines between keys, so that movement from one direction to another is abrupt and not smooth. *Clamped tangents* are useful for character animation. They make most of the animation smooth, like spline interpolation, but if two adjacent keys hold similar values, clamped tangents make the interpolation linear. This is a great way to hold movement during those frames. (Chapter 15 examines clamped tangents further.)

You can see how this animation was interpolated by checking the object's animation curve in the Graph Editor.

Select the sphere and open the Graph Editor by choosing Window | Animation Editors | Graph Editor. Figure 14-11 shows the Graph Editor displaying the animation curves for the sphere's translation attributes. These curves are color-coded to match the transform manipulator in the view windows: the X axis translation is in red, the Y axis translation is in green, and the Z axis translation is in blue.

**FIGURE 14-11**   *The Graph Editor displaying the animation curves for our ball*

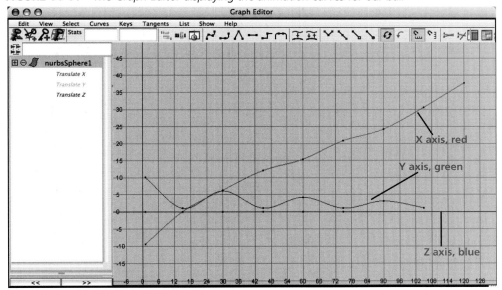

The red animation curve that controls the translation in X is gradually increasing value from frames 1 to 120 because we moved the sphere in the positive X direction every time we keyed it. The green Y axis line appears like a wave, since we moved the sphere up and down every 15 frames. Because you can't translate an object in the Z axis in the Front view, the animation curve for the Translate Z attribute remains flat.

> **TIP**   *In general, it's a good idea to start keying a few attributes at a time and then begin layering more attributes in as you make another pass.*

Now let's edit the animation of the sphere, starting with the Translate Y attribute:

**1.**   In the Graph Editor, select Translate Y in the Outliner section and press the F key to fit the curve in the graph section of this window.

**2.** In the graph section, drag-select all of the points on the curve that have values of 1. These should be the points along the bottom of the arcing curve. These points correspond to the keys you set where the ball hits the ground.

**3.** Notice that keys are now displayed as yellow points with tangent bars. You need to break each tangent bar in half, so you can create a much sharper curve when the ball changes direction. In the Graph Editor, choose Keys | Break Tangent. This will break the tangent bar.

**4.** Select the Move tool (press the w key) and LMB-drag-select the tangent on the left side of the first key. The tangent should now be highlighted in blue.

**5.** MMB-drag the tangent upward. Notice that the tangent is no longer a straight line through the point; instead, it is broken. Try the same thing with the tangent on the right side of the first key.

**6.** Continue editing the other keys by selecting the tangents and pointing them upward. Use Figure 14-12 as a guide. Try previewing the animation in Perspective view. The motion of the ball should now look like it is bouncing after it hits the ground.

**FIGURE 14-12**    *Tangents on the animation curves are broken.*

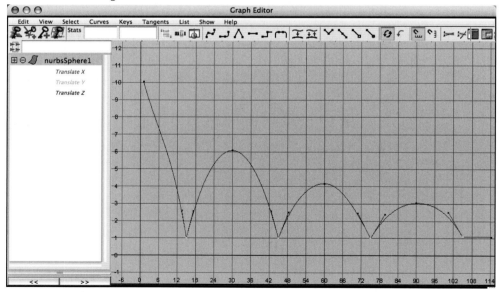

Now let's fix the animation on the Translate X channel:

**1.** Click Translate X in the Outliner section of the Graph Editor.

**2.** Press the F key to fit the curve to the graph.

**3.** Delete all the keys except the first and last keys—LMB-drag-select a marquee box around those keys and press DELETE. This will create a straight line.

**4.** Click the Play button and watch the movement. You should notice that the movement is smoother, but it stops harshly at frame 120. Select the last key and then select its tangent.

**5.** In the Graph Editor, choose Tangents | Flat; the rate of change in the Translate X value around the time the curve enters this key is now very small. The effect is that the ball gradually slows down until it finally stops.

**6.** Adjust the first key so that the tangent exiting the key points the curve so that it is straight or as constant as possible. The finished curve should look like Figure 14-13.

**FIGURE 14-13**    *The completed animation curve for translate X*

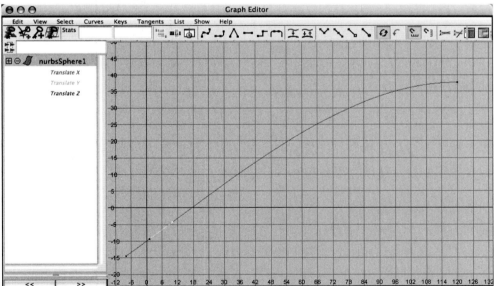

**NOTE**    *If you are still not exactly clear on how this curve is affecting the animation, you should play around with this setup for a little while, moving the keys and tangents and seeing what effect the edits have on the motion.*

## Add Character

Now the ball has motion, but we still need to give it some character—some life! To do this, we will apply a fundamental principle of animation to the ball, which is known as "squash and stretch." Squash and stretch refers to an object's natural flexibility that allows it to absorb inertia by squashing and stretching as it collides with other objects and moves through space. We can add squash and stretch to the ball simply by scaling the sphere. By including this technique, we can add some real character to the ball.

**1.** Select the sphere, and move the Time Slider to frame 1. Press SHIFT-R to keyframe the scale of the sphere.

**2.** Move the slider to frame 14 and press SHIFT-R to keyframe. Continue by placing another key at the frame just before the ball hits the ground, so that the ball maintains its spherical shape until it actually contacts the ground. Otherwise, it would gradually scale down as it approached the ground.

**3.** Click the Auto Key button in the playback controls.

**4.** Move the slider to frame 15, and scale the sphere up in the X and Z axes and scale it down in the Y axis. Better yet, in the Channel Box, type in numbers for the Scale attributes. This will give you more precise control. Type in values of **2** for the X and Z axes and **0.5** for the Y axis. The sphere now appears "well squashed" when it hits the ground. If you scrub the slider, you will notice that the change in attributes is automatically recorded because we have selected the Auto Key.

**5.** Move the Time Slider to frame 30, and stretch the sphere. The scale values should be 0.8, 1.5, and 0.8 for Scale X, Scale Y, and Scale Z, respectively.

**6.** Key these values at frames 44, 45, 60, 74, 75, 90, 104, 105, and 110. You can gradually decrease the values to animate the loss in inertia as the ball comes to a stop. Squash and stretch should diminish as the ball moves forward. Figure 14-14 shows the result of the squash and stretch process.

If you want to change the timing of the ball, the Dope Sheet is the best and quickest way to go. In the Dope Sheet, you can quickly change timing by grabbing keys and scaling them as groups. Open the Dope Sheet, shown in Figure 14-15, by choosing Window | Animation Editors | Dope Sheet. Now let's make the timing of the animation faster:

**1.** Select the Scale tool by pressing the R key.

**2.** Click nurbsSphere1 from the Outliner section in the Dope Sheet.

**FIGURE 14-14**  *Squash and stretch gives the sphere the appearance of weight as it bounces.*

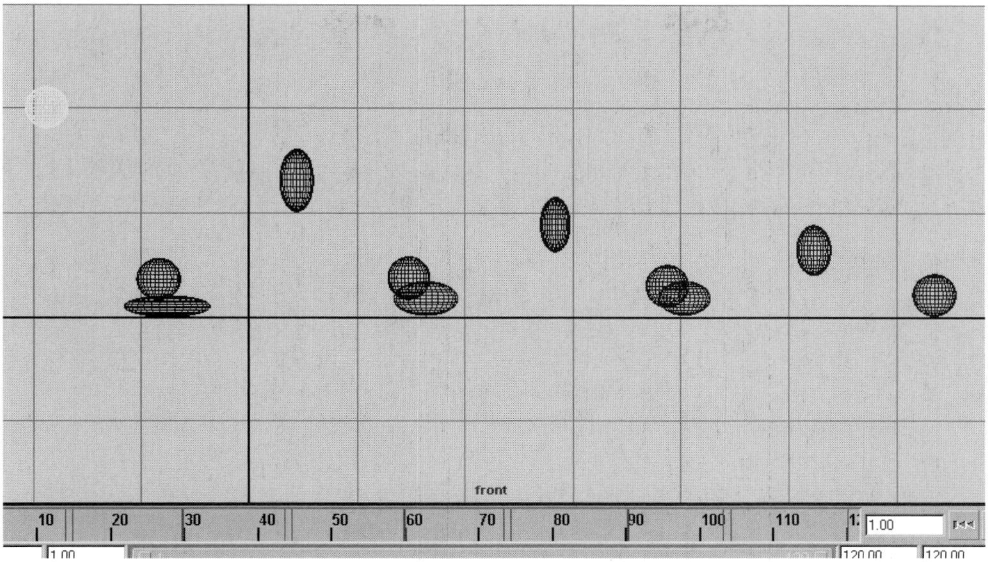

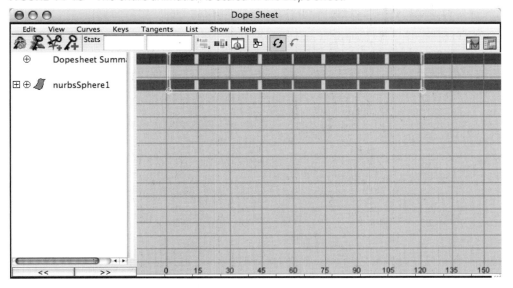

**FIGURE 14-15**  *The entire animation is scaled in the Dope Sheet.*

**3.** Drag-select all of the black tick marks (the keyframes) next to nurbsSphere1. These marks should now light up in yellow, and a white frame will surround all the keys.

**4.** Grab the right edge of the white frame and LMB-drag it to the left. Move it until the edge reaches number 30 (frame 30).

**5.** Play the animation, and you will notice that the animation is now twice as fast. You can move the right edge of the white box to the left to slow down the animation as well.

## Use Playblast

Let's create a movie file directly from Maya to view our bouncing ball animation. Using Playblast is the best way to check the timing of an animation.

**1.** Make the Perspective view active by clicking in it.

**2.** Hold down the right mouse button over the Time Slider until the marking menu is displayed. Choose Playblast □.

**3.** In the Playblast Options window, make sure that View is checked and Show Ornaments is unchecked. This will create a movie from the selected view window, excluding stuff like a manipulator and an axis display. For the Viewer, you can select what player you would like to use to view this animation. For this example, we'll choose Windows Media Player. (Your exact options might be different depending on what platform you are using.)

**4.** Set Display Size to Custom and enter **320** and **240** in the boxes just below Display Size. Set Scale to **1.00**. Check the Save to File check box, and click the Playblast button.

**5.** A window will open asking you where you want to save the movie file. Save it to the best location—save it to the desktop for easy access.

**6.** You should see every frame of the animation played back in the view window, though not necessarily in real time. This is because Maya is rendering each frame onscreen and then capturing it and saving it. After all the frames have been captured, Media Player will show you the movie file automatically. Figure 14-16 shows the Option window for Playblast and Playblast being played in Windows Media Player.

Play the animation and take a look at all the editing you've done to the sphere. This is basically the same workflow you'll use to edit any animation using the Graph Editor and Dope Sheet. No matter how complex your animation job may be, keyframing and editing animation are similar to the process discussed so far.

*FIGURE 14-16*    *Playblast shown in Window's Media Player*

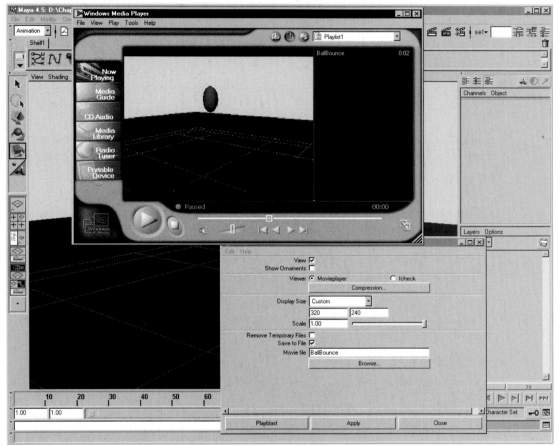

# Basic Animation Principles

Last, but not least, when you are keyframing or editing keyframes in the Graph Editor, you should keep some of the fundamental principles of animation in mind as you work. The following information will help you create more realistic, visually stimulating animations.

Keep in mind that animation is *not* just *copying* natural movement; it's an *exaggeration* of it. In the beginning of this chapter, we talked about how traditional 2D cell animators developed well-established animation techniques to use while animating characters or natural phenomena. Let's take a look at these techniques so you can apply them to your own animations.

## Squash and Stretch

Like a bouncing ball, we humans change shape as we move when we jump up and down: our knees, hip joints, and back all bend straight while we are in the air and bend inward to absorb shock when we land on the ground. The basic theory of squash and stretch gives life to any character or object by adding the appearance of weight to it. Unless you are animating a rock, most things in this world (and some things not in this world) have elasticity of some sort to absorb energy as it is applied to them.

> **NOTE**   *When you add squash and stretch, be careful not to increase the object's size while at the peak of the squash and stretch. The volume of the object should be intact throughout the animation.*

## Anticipation

*Anticipation* is the act of holding energy that is about to be released. In our jumping analogy, you usually swing your arms backward and bend your knees and back inward to compress yourself before you jump. You can jump without this action, but jumping will be stiff and not very elegant. Animation is all about creating and releasing energy or interacting with an external energy source. Anticipation creates the energy in your scene. In addition, viewers can *expect* upcoming action by *feeling* anticipation.

## Follow Through

*Follow through* is the opposite of anticipation. It's a technique used to absorb the shock of an energy by overshooting the action beyond. If we use our jumping analogy again, after a jump is over and you touch down on the ground, your body will not stay at the bending pose of squash. Your body should bend back to its natural posture. This act of continuing motion to reach the object's or character's most natural state is called follow through.

## Secondary Action

*Secondary action* is the animation of extra parts or appendages of an object or a character used to exaggerate motion. To use the jumping analogy again, if you had large ears and a tail and you jumped, your ears and tail wouldn't move at the exact same rate that your body moved, and they wouldn't stop at the same time that your body stopped. Your ears and tail would swing their own ways, pivoting around where they are attached to your body. Adding such detail adds more complexity to your animation and creates believability and realistic qualities in your animations.

## Study Reference

While this entry is not an official principle of animation, studying real-world motion is absolutely necessary when animating. This can be anything from studying filmed reference to studying yourself in the mirror. You can never have too much reference.

# Further Reading and Practice

You might find the following 2D animation books helpful in understanding the finer points of creating realistic animations. Although these books cover 2D animation, the basic theory is the same for 3D animation.

- *The Illusion of Life: Disney Animation*, by Frank Thomas and Ollie Johnston (Hyperion Press, 1995). This book is the best animation resource around. It's fairly thick and expensive, but if you are serious about animation, it's a must-have.

- *Cartoon Animation*, by Preston Blair (Walter Foster Pub., 1995). This is another great book about 2D animation techniques. It has a lot of reference graphics, and it's fun to read.

- *The Animation Book: A Complete Guide to Animated Filmmaking—From Flip-Books to Sound Cartoons to 3-D Animation*, by Kit Laybourne (Three Rivers Press, 1998). This is a great reference book for 2D animators. It also has a lot of information applicable to 3D animation.

- *Animation from Script to Screen*, by Shamus Culhane (St. Martin's Press, 1990). Here's another good animation book from a master of animation. You can use many of the theories discussed in this book.

- *The Animator's Workbook*, by Tony White (Watson-Guptill Publications, 1988). This book offers many exercises for 2D animators.

Use daily exercises to sharpen your eye to get better at animating characters. Always be aware of how your body moves and how people and other characters move. It's a good idea to carry a sketchbook to write about, and draw, your discoveries, or you can videotape yourself or a friend to show a character's specific actions, and then review it carefully.

# Summary

As you may have gathered from reading this chapter, the path to great animation is long and difficult, but it's also extremely rewarding to see your creation move and act like a real-life object or living being. Keyframing is an important part of animating, and, like most other tasks, the more you use this feature, the better you'll get at creating realistic animations.

In the next chapter, we will expand on the basic animation skills taught here as we apply them to a bipedal character.

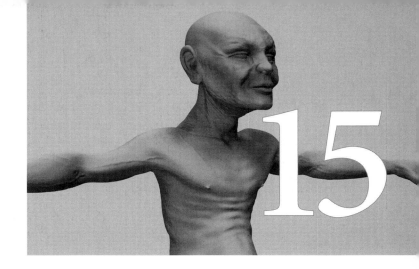

# 15

# Character Animation

**This chapter covers some techniques**

that make the art of character animation fast and

efficient. After some additional setup techniques,

we'll practice setting keys on various control

objects to produce our first character animation.

As you follow the exercises in this chapter,

remember, don't be so concerned with the precise

values that we suggest. Experiment with different

poses and timings that result in a performance

that evokes your own intentions and feelings.

# Tutorial: Walking and Pushing a Box

This tutorial covers how to operate our 3D puppet to make a convincing walking animation. We'll create a fairly complicated setup to ease this animation process. After we make the character walk, we can try layering more interesting animation. Then we'll use Maya's IK blending function to blend between IK and FK to make the character push a box.

## Set Up the Character for Animation

In this section we will prepare our character for animation. The rig is not quite as complex as the rig we built in Chapters 9 through 12, but it is very efficient for this exercise. This rig won't clutter our Workspace with unnecessary controls. It contains only what we need to get these shots done. However, there are some things that we should do before a character is animated that will make it easier and faster to use.

First, we'll create a low-resolution version of the character that is to be animated. This will free us from any computationally intensive operations that can occur when animating and deforming high-resolution geometry. Then, we'll create some shelf buttons that will further ease the selection of certain objects in the scene.

### Set Up the Low-Resolution Geometry

The art of animation is all about timing and motion. To animate successfully, you need to be able to work and view your animation in real time without dropping any frames during playback. While Maya can handle the display and interaction of heavy scenes, at some point, the scene will become too much for your computer hardware to handle.

Creating a low-resolution version of your character animation is one of the best ways to sidestep any possible performance problems that might occur. When making a low-resolution model for animation, two performance problems need to be solved: The first problem is that of the video card's limitation to display a certain number of polygons per frame. Even the newest, most powerful graphics cards will get bogged down when the scene gets complex. This is best avoided by creating a version of the model that has a small number of polygons. The model should contain only enough information to define the overall volume of the character. Being able to view the space occupied by the character will allow you to achieve more accurate placement in the scene and will avoid the character moving through floors, walls, or other characters. There's no specific technique to use when creating a low-resolution version of a character. The technique you use depends on what type of geometry the high-resolution is made of (polygons, NURBS, or subdivision surfaces). For a model created in NURBS, it might be best to convert the model to polygons using one of the tessellation methods discussed in Chapter 7. The same is true for subdivision surfaces. If you have a polygonal model, you might try the Reduce command (Mesh | Reduce). In some cases, it might be just as easy to create a new model quickly from scratch. In any case, the model should have as few polygons as necessary.

The second performance problem to overcome is the one placed on the computer's processor during a deformation. Remember that a deformer, such as the skin cluster used for skeletal deformations, creates a node in the geometry's dependency graph that needs to be evaluated every time the skeleton is transformed. When an indirect skinning technique is used, several nodes may need to be evaluated to result in the final deformation. Even on low-resolution geometry, this could slow down your computer. To remedy this, the low-resolution version of the model needs to be broken up into different pieces and then parented to the corresponding joint. For instance, the face in the thigh region of the model can be extracted (Polygons | Extract) and then parented to the hip joint. The region from the knee to the ankle is extracted and parented to the knee joint. This process continues until each piece of the model is parented to a joint. When the skeleton moves, the geometry parented to it will move as well. Although gaps will exist in the joint regions, and the geometry may overlap in other areas, that is not important. The real gain is in removing any need to evaluate a deformation.

Open AnimationStart.mb from the book's DVD. This scene contains a model that has been rigged and skinned using many of the techniques demonstrated in Chapters 9 through 12. We have already created a low-resolution version for the character used in this scene. Notice that a layer for a low-resolution character is included; you will use this later to animate. Turn off the visibility for the layer named HiRezCharacter and turn on the visibility for the layer named LowRezCharacter to see the low-resolution character. Figure 15-1 shows the low-resolution version of the character.

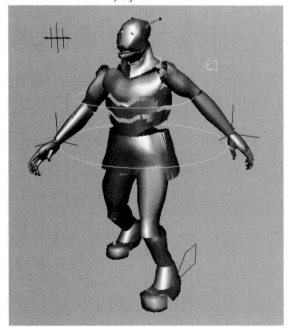

**FIGURE 15-1**   *A chopped-up, nondeforming, low-resolution model will play back in real time.*

### Make Shelf Buttons for Animation Controls

The iconic character animation controls, such as the pelvisCTRL object and the foot controls, were created so that they are highly visible and easy to select in a view window. However, when the scene has many objects obscuring the control, it can still be difficult to select. Therefore, we will create a UI for selecting certain objects by making shelf buttons:

**1.**   Create a new shelf by clicking the down arrow button labeled Menu of Items to Modify the Shelf, at the left side of the shelf. Select New Shelf. The New Shelf window will open. Type **Character_CTRL** for the shelf name and click OK.

**2.** Now you are ready to add shelf buttons. Open the Hypergraph (choose Window | Hypergraph) and the Script Editor (choose Window | General Editors | Script Editor).

**3.** Select pelvisCTRL from the Hypergraph. Now look in the Script Editor. In the top section of the Script Editor (the History window), you should see this command:

```
select -r pelvisCTRL ;
```

**4.** Highlight this by LMB-dragging over it. Then MMB-drag it to the shelf to create a new shelf button. That's it! You made a shelf button to select a pelvisCTRL object.

**5.** Select LT_footCTRL and drag the command to the shelf in the same way. Create shelf buttons for RT_footCTRL, LT_armIK, RT_armIK, spineCTRL, LT_armPoleVector, RT_armPoleVector, neck_a, and eyeCTRL. The order in which you create the buttons is the order in which they'll appear on the shelf. If you create the buttons in the order listed here, the buttons will cover the objects in the character from its foot to its head in a well-organized arrangement. Test the shelf buttons to make sure that each object is selected when its button is clicked.

**6.** It's a good idea to label these buttons to make it obvious which button selects what object. Again, click the down arrow button on the left end of the shelf to view the menu of items that modify the shelf. Choose Shelf Editor, and the Shelf Editor window will open.

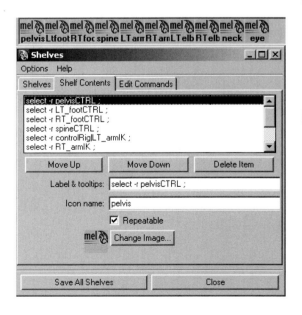

**7.** Click the Shelf Contents tab, and then choose select –r pelvisCTRL from the list. In the Icon Name box at the bottom, type **Pelvis**. Now choose select –r LT_footCTRL, and you'll see that *Pelvis* appears on the bottom of the shelf button.

**8.** Name the rest of the buttons—**LT_foot, RT_foot, LT_arm, RT_arm, Spine, LT_ elbow, RT_elbow, Neck,** and **Eye**. Click the Save All Shelves button. This illustration shows the window and the buttons with their new names. This shelf is probably saved at the default location at C:\Documents and Settings\*Name of your login accounts*\ My Documents\maya\ *x.x*\prefs\shelves. Look inside the folder and you'll find a script named shelf_Character_CTRL.

**NOTE** *While a shelf is quick and easy to set up, you could also create buttons in a UI panel, like the one we created in Chapter 13, that would select each joint.*

### Set the Animation Preferences

Let's set up the scene so that we can animate it correctly. First, make sure you are animating for the proper output. If you are animating for video output, you should be using 30 fps (that is, if you are in the United States).

**1.** To set the frames per second, choose Window | Settings/Preferences | Preferences. Select Settings from the left side of the Preferences window. Set the Time attribute to NTSC [30fps].

**2.** Change the types of tangents that will be used to interpolate the animation data in and out of every keyframe that is set. Select Keys from the list on the left and change the Default In Tangent and Default Out Tangent to Clamped. The Default tangent is set to Spline, which always makes a smooth curve between the keys to simulate smooth acceleration and deceleration. Using Spline as an animation type can cause the object to slide or continue moving after it should have stopped. Using clamped tangents helps this situation by making the animation curve straight, meaning there is no movement, when two adjacent keys have the same value. Figure 15-2 shows how each tangent affects animation.

*FIGURE 15-2* *How spline and clamped tangent settings affect the interpolation of the keyframes*

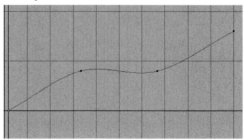

**Spline tangent**

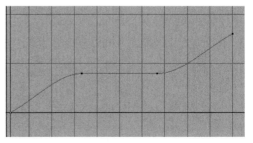

**Clamped tangent**

It's always a good idea to set keys for the bind pose at the beginning of the animation. You can key the bind pose by keyframing all the attributes of all the controls at the bind pose. This is important, because you may want to change skin weight, create blend shapes, or texture the character afterward. It's easy to create a bind pose:

**1.** Make sure the character is at the bind pose. Choose Modify | Evaluate Nodes | Ignore All. This temporarily turns off the IK, constraints, and dynamics.

**2.** Select the spine_root joint and choose Skin | Go to Bind Pose. If you don't see any movements in the joints, you're ready to go. Otherwise, you have to snap each control to the appropriate position, such as spine_CTRL to the spine_root joint, LT_footCTRL to the LT_heel joint, LT_armIK to the LT_wrist joint, and so on. Then choose Modify | Evaluate Nodes | Evaluate All.

**3.** Keyframe all of the control objects at frame –1, so by just scrubbing the Current Time Indicator to –1, you will achieve a bind pose. In the Range Slider, type **–1** in the Playback Start Time field.

**4.** Move the Current Time Indicator to –1 and use the Outliner to select each control object. Press the S key to keyframe all the attributes.

**5.** Try posing a character at frame 1 and then move the Current Time Indicator to –1 to see if the character goes to a proper bind pose.

Now you are ready to animate!

## Create a Walking Animation

We will first attempt to create a walk cycle. In this exercise, you will learn to use Maya's built-in FK/IK switching system called IK blending:

**1.** In the Range Slider, set the Playback Start Time to 1 and the Playback End Time to 100. Frames 1 through 100 should appear in the Time Slider.

 **2.** In the toolbox, click Persp/Graph/Hypergraph.

**3.** Choose Panels | Orthographic | Side from the Hypergraph window. Since the character is walking parallel to the Z axis, we can mainly use the Side view to see the progress. Figure 15-3 shows how the Workspace is laid out.

**4.** It's best to keyframe the main body movement first and then layer the detail of other body parts' animation later. For a walking animation, the first objects to animate are pelvisCTRL, LT_footCTRL, and RT_footCTRL. At this point, be sure that only the low-resolution version of the character is visible.

**5.** Set the Current Time Indicator to frame 1. Select the pelvisCTRL object (either in the view window or using the shelf button) and in the Side view, move it about 2 units forward and 1 unit down.

**6.** Select the LT_footCTRL object and move it forward, so the pelvis of the character is in the middle of both feet.

**FIGURE 15-3** *The panel arrangement is set up for animation.*

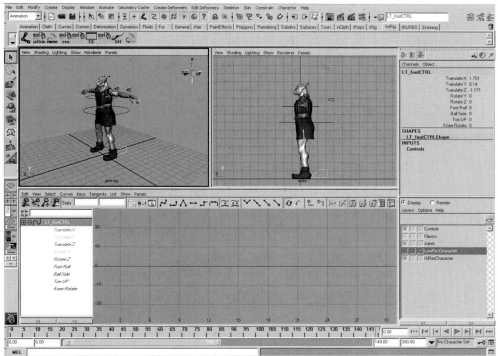

**7.** Key the pelvisCTRL, LT_footCTRL, and RT_footCTRL objects by selecting all three objects and pressing the S key. It's important to key the RT_footCTRL object even though it hasn't moved. Figure 15-4 shows the position of the controls as they are keyed at frame 1.

**8.** Notice that the arms are not following the body. This is because the IK handles for the arms are not children of the hierarchy or constrained to any of the objects in the hierarchy. Select both arm IK handles and change the IK Blender attribute to 0; then press the S key. This turns off the IK solver for arms; they should follow the body movement from now on.

**9.** Drag the Current Time Indicator to frame 30. We will simplify the keyframing by making each step 1 second long. Move the pelvisCTRL object again about 2 units forward and set a key.

**FIGURE 15-4** *The character is keyed in this position.*

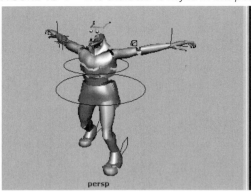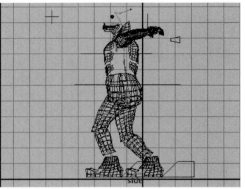

10. Move the RT_footCTRL object so that the pelvisCTRL object is at about the middle of both feet. Key both the right and left foot controls.

11. Scrub the Current Time Indicator back and forth from frames 1 to 30 to see the first step of the walk cycle.

**FIGURE 15-5** *Placement for the pelvisCTRL object and the foot controls every 15 frames*

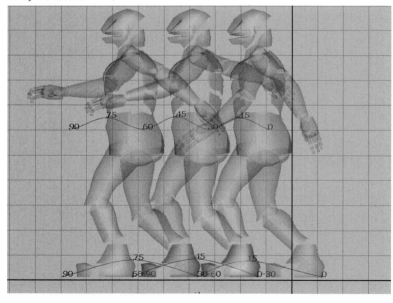

12. Set the Current Time Indicator to 60 and move the pelvisCTRL object forward. Move LT_footCTRL forward, and key all three controls. Repeat this step for frame 90.

13. Activate the Perspective view and click the Play button (or press ALT-V [OPTION-V]) to check the animation. Figure 15-5 shows how the pelvisCTRL object and foot controls are placed every 15 frames.

*TIP*  *If you want to see something similar to Figure 15-5, which shows a snapshot of objects at particular frames, choose Animate | Create Animation Snapshot. Specify from which frame to which frame you want to view and how many frame increments you want to include in the snapshot. You can also show a trail of motion by choosing Animate | Create Motion Trail.*

Now we need to take another pass at this animation to make the feet lift off the ground and the pelvis bob up and down as the character walks. To do this, we need to keyframe the Y translation of the pelvisCTRL object and foot controls. At frames 15, 45, and 75, the pelvisCTRL object should be lifted up a bit to compensate for the foot coming forward that is also lifting up.

**1.** Go to frame 15 and move pelvisCTRL and RT_footCTRL slightly upward. Key them by pressing the s key. Go to frame 45, and this time move pelvisCTRL and LT_footCTRL up and key them. Key the same pattern at frame 75, with pelvisCTRL and RT_footCTRL.

**2.** Play the animation. It should look more convincing now that the feet are lifting off the ground.

**3.** Let's take another pass and add another layer of animation. The next important keyframe will control the shifting of the center of gravity. Because the character is lifting its foot up, it should shift the center of gravity to the left and right. The center of gravity can be controlled by translating the pelvisCTRL object in the X axis and rotating its Z axis. Change from Side view to Front view. If you go to frame 15, you see that the character is doing something impossible: its center of gravity is at the middle of both feet, even though its right foot isn't holding its weight.

**4.** Move the pelvisCTRL object about half a unit toward the left foot and rotate it slightly in the Z axis, about –5 degrees. Key the position for the pelvisCTRL object at this frame.

**5.** Go to frame 45 and move the pelvisCTRL object about half a unit toward the right foot and change the Rotate Z attribute to 5, and set a key. Repeat this step for frame 75. This frame should be similar to frame 15.

**6.** Play the animation, and you can see that the character's movement is much more convincing.

**7.** A big monster like this guy should rotate its pelvis side to side while walking. Go to frame 15 and rotate the pelvisCTRL object about –10 degrees in Y. Set a key. In frame 45, change the Rotate Y attribute to 10 and set a key. In frame 75, change the Rotate Y attribute back to –10 and set a key.

**8.** Play the animation in the Perspective view. The character's lower body now has much more realistic motion.

**9.** Let's finish up the lower body motion by adding some roll to the foot. Select the RT_footCTRL object and go to frame 15. Set the footRoll attribute to 3 and key it by pressing the S key. In frame 25, change the footRoll attribute to −3 and key it again. In frame 30, change the value to 0, and this time key the footRoll attribute individually. You can do this by highlighting the footRoll attribute in the Channel Box, holding down the RMB, and choosing Key Selected. Because this frame doesn't correspond to any other attribute's animation, it's a good idea to key it separately. Again, go to frame 75 and repeat the same process. After finishing with the right foot, select LT_footCTRL and do the same keyframing for frames 45, 55, and 60. The lower body animation should be complete.

**10.** The monster's upper body needs to be animated to correspond to the lower body for its motion to look natural. Also, the arms need to swing to counteract the movement of the pelvis. Start with the spineCTRL object. The spine should rotate slightly in the opposite direction of the pelvis to counteract the movement. Select it and go to frame 1; then key it by pressing S. At frame 20, rotate in Z about 4.5 degrees and set a key. This will counteract the excessive weight shifting of the bottom of the body. In frame 30, key the Rotate Z attribute at 0. At frame 50, key it at −4.5, change it to 0 at frame 60, and key it again. Key it to 4.5 at frame 80 and 0 at frame 90.

**11.** Next we'll work on the spineCTRL object's Y axis rotation. This rotation drives the swinging of the arms. Go to frame 15 and change the Y axis rotation to −10, so that the left shoulder is forward. At frame 45, change it to 10, and then at frame 75 change it back to −10. Make sure to key each time. This movement gives a more animated feel to the upper body, but it can look too exaggerated. Making the neck animation counteract to the spine will compensate for this problem.

**12.** We will add animation for the neck joint in the Rotate X and Rotate Y attributes. Select the neck_a joint and key at frame 1 by pressing the S key. Go to frame 15, and in the Front view, try to make the head point straight ahead by rotating it in the X and Y axes. Key this by pressing the S key. Do the same thing for the neck joint at frames 45 and 75. Set all the rotation to 0 at frame 90 and key. This should make the head face forward throughout the animation.

**13.** The next big step is animation of the arms. We turned off the IK solver earlier on, but since this arm movement will be swinging, the arms are best keyframed using FK animation. We will animate shoulders, elbows, and arm IKs (for wrist and elbowTwist joint's rotation). Before starting the animation of the arms, create shelf buttons that select each shoulder and elbow joint.

**14.** Start at frame 1. Rotate the shoulders and elbows so the arms are closer to the body on the side. Make sure to rotate one axis at a time by highlighting the axis you want to rotate in the Channel Box and MMB-dragging in the view window. If the manipulator is used, you may end up rotating on all three axes. Also remember that the elbow joint is a hinge joint, so make sure that it rotates only in the Z axis. Key all four joints.

**15.** Move to frame 30 and switch the arm rotation, so that the left arm is in front and the right arm is in back. Set a key. You need to highlight only the Y axis and rotate it to swing. Use the same process to swing the arm at frames 60 and 90. Play the animation to check your progress.

**16.** You can make the animation more interesting by slowing the elbow rotation slightly. Make the elbow rotate more, but make it happen about five frames after the shoulders' rotation is at their peak. Key the elbow rotation at frames 35, 65, and 95.

**17.** Delete keys at frames 30, 60, and 90 for the elbow rotation. The easiest way to do this is in the Graph Editor. Select the keys and press DELETE. Figure 15-6 shows the completed animation at intervals of 20 frames.

The basic walking animation is complete. Of course, you should continue to add more secondary movement by continuing to make more passes. Ears can be rotated to simulate swinging motion. Keys can be offset so that corresponding motions are not so mechanical. Any little things that give this monster more character will make it look more realistic.

**FIGURE 15-6**   *The completed walk cycle at 20-frame intervals*

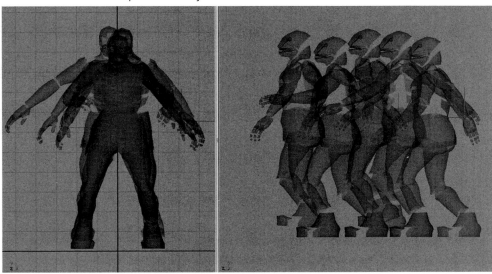

## Push a Box Using IK Blending

Now we will make the character push a box. This will utilize the IK blending function in Maya, which was briefly explained in Chapter 12. IK blending allows you to switch between FK and IK by keying the state of the IK solver so both of the monster's arms can swing freely while walking, and then lock onto the box by switching to IK.

1. Open the scene called pushingBoxStart.ma on the DVD. Play the animation and you will see the character walking and then standing in front of a box. At frame 110, set a key on the IK Blender attribute with its value set to 0. Make sure you create this key by keying just this selected attribute.

2. In the Side view, move the IK handles in front of the body. Key the translation.

3. In frame 140, select both arm IK handles and turn on IK by setting the IK Blender attribute to 1. Highlight this attribute in the Channel Box, right-click it, and choose Key Selected.

4. Scrub the Current Time Indicator between frames 110 and 140, and you should see the arm moving between them smoothly. Adjust the placement of the IK handles and animate pole vectors from frames 110 to 140 to add elbow animation, so the elbows are pointing down more. Also, rotate the IK handles so the character has both hands against the box at frame 140.

5. Animate the pelvis control, foot controls, arm IKs, and box to make the box move forward. Try using different timing, so the box appears to be heavy or light.

As you can see, the IK blending function allows you to switch between IK and FK easily.

# Summary

In this chapter, you learned how a character rig can be animated to produce a variety of actions. As we stated before, you should continue to experiment with these methods to produce your own animation and style. As with just about everything else we do in Maya, references are the best resource for producing good work. Videotape yourself acting out a scene, time it, and then try to re-create it using the rig in Maya. With the basics of keyframe animation understood, you are ready to move to Chapter 16, where you will learn various techniques for processing animation data and interacting with objects.

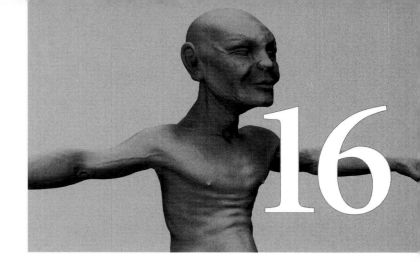

# 16

# Animation Tools

**In Chapters 14 and 15 you learned** how to keyframe and control animated objects and characters. When it comes to working in a production, with a team of people, you may require some additional tools and features. The ability to reference files, work with clips, and transfer animation data is part of every animation pipeline. In this chapter, we will explore some of Maya's built-in features for dealing with these issues.

# File Referencing

Let's examine a common issue in any character animation pipeline. Imagine a situation where there are 60 different shots that are divided between ten animators. Toward the end of the production, the director decides that a character's shirt—a character that is used in all 60 shots—needs to be a different color. If the animation pipeline was set up based on each animator just importing (File | Import) the character into their Maya scene for each shot, someone would have to go into each file and change the color of the shirt.

While many studios handle this problem differently, Maya offers the ability to reference files as opposed to importing them. When an asset such as a character rig or modeled environment is referenced, it points to the files that are the source of this data every time the scene is loaded. Therefore, if a character model, rig, material, or light changes, the reference to those assets in the file can easily be updated.

## Creating References

Creating a reference is no more difficult than importing a file. To create a reference:

1. Start a new Maya scene.

2. Choose File | Create Reference. The file browser opens and lets you browse for a Maya scene to reference. Choose the mcr8_ch16_rig_01.ma file from the accompanying DVD. The character will show up in the scene all ready to be animated.

3. Add another reference, this one for the environment. Choose File | Create Reference and find the mcr8_ch16_env_01.ma file. The environment now appears in the scene.

## Managing References

Maya lets us manage referenced scenes in the Reference Editor (File | Reference Editor), shown in Figure 16-1. The Reference Editor provides many functions to efficiently manage a complex project. Some of the most common uses of the Reference Editor are for replacing old assets with new ones and loading and unloading of referenced assets in the existing scene.

Let's say that the model was changed and that the scene needed to be updated. This can easily be accomplished by selecting the referenced file from the list in the Reference Editor and choosing Reference | Replace Reference. You can then browse to the new file. In this case, you can reference the mcr8_ch16_env_02.ma file. The environment will be updated with the new reference.

The real power of file referencing comes in when the referenced objects are animated. Any animation created on this rig will remain intact if the rig file is replaced by a new, updated version.

The one rule is that you cannot change the name of anything in the scene that is animated. You can add all of the objects you want, but renaming objects, including renaming parent objects in a hierarchy, will break the referencing.

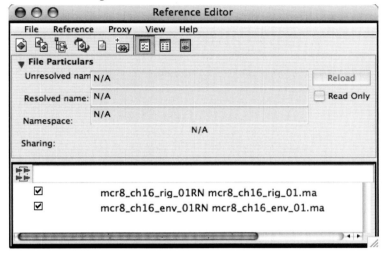

FIGURE 16-1  *The Reference Editor currently displaying the references to the character rig and the environment*

Another great use for the Reference Editor is to quickly load and unload referenced files from the current scene. If the animator doesn't really need the environment loaded to animate the rig, the animator could unload the environment by unchecking the box next to the environment's filename in the list. This would reduce the added processing and memory overhead that the environment may add to the scene. When the environment needs to be loaded again, the check box in the Reference Editor can simply be re-enabled. Any transformations that you may have made to the environment will remain intact.

As you can imagine, file referencing is an amazing feature that can save immense amounts of time managing even the smallest projects.

# Tutorial: Creating Nonlinear Animation

Nonlinear animation, an industry buzzword for quite a while, is similar to nonlinear film or video editing. Before digitization of film or video footage became widely available, all the footage had to be recorded on tapes. A tape is linear; you can't easily cut and insert footage, because all the other sequences have to be pushed back, too. But nonlinear editing allows you to cut and paste footage anywhere in the recording of time, because you are arranging clips, and entire sequences will conform to your operation. Nonlinear animation works on the same principle.

Nonlinear animation can treat each animation as a clip of footage, and you can insert that clip anywhere in the timeline. You can create a library of motions and simply drop the motions in wherever they belong. For long sequences with many repetitive motions, you and your team members can pull out motion clips and add, mix, and blend them together, resulting in a shorter and more efficient animation workflow. Nonlinear animation is used often in games and TV series, where many team members can access a bank of animation clips and poses to aid in creating a lengthy animation quickly.

## Create Character Sets

Before you start converting animation into clips, you should group objects' keyable attributes into sets to make clip handling more efficient. You can group keyable attributes into special groups called *character sets.* Character sets can hold keyable attributes from any objects. When a character set is selected, you can view the list of all animatable attributes in one list in either the Outliner or the Channel Box.

Making a character set is easy:

**1.** Open the scene called TraxWalkCycle.mb from the DVD. This scene contains a walk cycle animation of a character. The character is walking in place. The values for all of the animated attributes are the same at frame 1 and frame 120, making the animation seamlessly cycled.

**2.** Let's organize this animation's character sets into upper body and lower body sets, so we can add more motion to these two sets separately. Open the Hypergraph and select pelvisCTRL, LT_footCTRL, and RT_footCTRL.

**3.** Choose Character | Create Character Set ❑ to open the Create Character Set Options window, shown in Figure 16-2. Type **LowerBody** in the Name field at the top. Several options appear here:

**FIGURE 16-2**   *Create Character Set Options window*

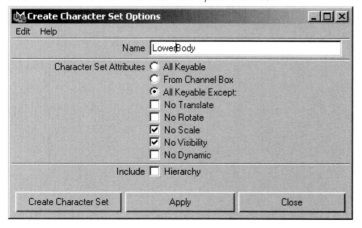

- **All Keyable**   Makes all keyable attributes of the selected object part of the character sets.

- **From Channel Box**   Lets you select specific attributes in the Channel Box, and creates character sets only for those selected attributes.

- **All Keyable Except**   Lets you check a number of boxes below to be part of the character set. Because it says *except*, any box with a check mark *won't* be a part of the character set. The default is a check mark for No Scale and No Visibility. These attributes won't be a part of character sets.

- **Include Hierarchy**   If this is checked, all the children below the selected object's attributes will be part of the character sets.

**4.** Leave these options at their default settings and click Apply. The LowerBody character set is created.

**5.** Take a look at the Channel Box. All of the attributes in this set are listed here. Also at the bottom-right side of the interface, just to the left of the Auto Key button, "LowerBody" appears as the active character set in the Character Set Selector, as shown in the illustration. While this is active, every time the Key All command is used (the S key), all attributes in this set will be keyed. This keeps you from accidentally not keying something because it was not selected.

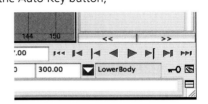

**6.** In the Hypergraph (or Outliner), select LT_armIK, RT_armIK, LT_armPoleVector, RT_armPoleVector, spineCTRL, neck_a, LT_shoulder, LT_elbow, RT_shoulder, and RT_elbow. Choose Character | Create Character Sets ❒ and name this set **UpperBody**.

**7.** Click the Create Character Set button. The UpperBody character set is created.

### Edit Character Sets

Because we used the Channel Control window earlier to clean up keyable attributes in the Channel Box, all of the attributes in the character sets should now be pretty well organized. No unneeded attributes should appear in the list. If you do want to delete or add attributes to any particular character set, it is best to use the Relationship Editor (see Figure 16-3):

**1.** Choose Window | Relationship Editors | Character Sets. Click a character set you want to edit at the left side of the Relationship Editor window.

**2.** Click the plus mark next to the character set on the right side to see all of the attributes for the character set.

**3.** To remove an attribute from the Relationship Editor, select it and choose Edit | Remove Highlighted Attributes.

**4.** To add attributes, select the character set to which you want to add attributes from the left side of the Relationship Editor, and then find an object whose attributes you want to add from the right side of the Relationship Editor. Click the attribute and it is automatically added to the selected character set.

**FIGURE 16-3**  *Relationship Editor for character sets*

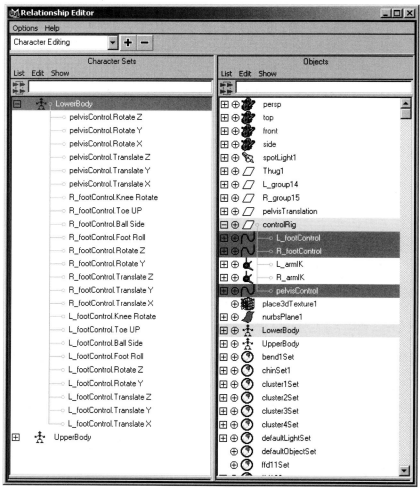

## Use the Trax Editor

The Trax Editor lets you mix animation clips in much the same way that you might assemble a movie in video editing software. You can use the Trax Editor to arrange a bunch of clips in any order and blend between them. Each clip can be mixed, sped up, slowed down, cycled, and split so that you can quickly assemble a complex animation.

## Create the Clips

Before you can begin editing the clips, you need to convert an animation or pose into clips so that it can be used in the Trax Editor. A *clip* is just a chunk, or block, of animation data for a specific set of attributes. Let's make some clips from the walk cycle animation:

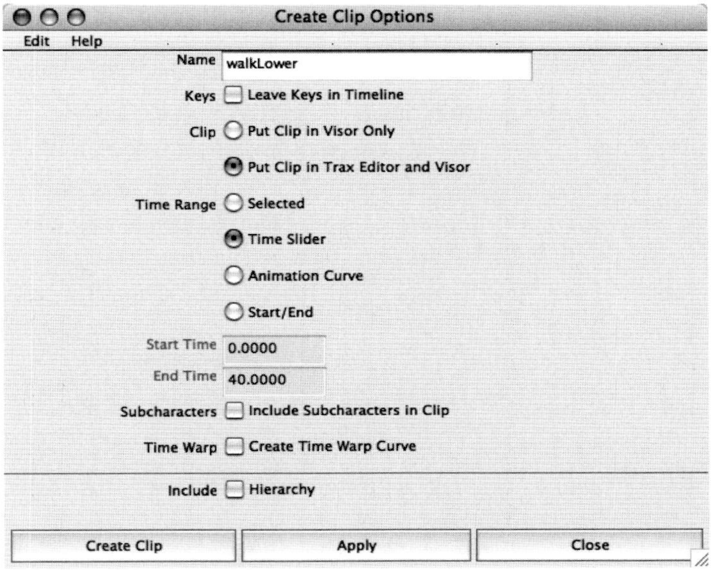

**1.** Use the Character Set Selector drop-down menu to make the LowerBody character set active. Choose Create | Clip ☐ to open the Create Clip Options window, shown in Figure 16-4. The following options appear in this window:

*FIGURE 16-4    Create Clip Options window*

- **Name**   Type in a name for your clip.

- **Keys __ Leave Keys in Timeline**   If this box is checked, the animation data remains in the Graph Editor and Time Slider. Usually, since you are trying to create a nonlinear editing environment, you want to leave this unchecked.

- **Clip**   You have two choices for where to put the clip:

  - **Put Clip in Visor Only**   Removes the keyed animation and creates a clip, but the new clip will be stored only in the Visor (another window used for selecting nodes in Maya).

  - **Put Clip in Trax Editor and Visor**   Creates a clip and immediately adds the clip to the Trax Editor and the Visor. It's the default option and the most logical one if you want to use the Trax Editor right away.

- **Time Range**   This determines the range of frames from which the clip is generated:

  - **Selected**   Uses the selected time range. To select a time range, SHIFT-drag in the Time Slider. A range will be highlighted in red; this is the selected range.

  - **Time Slider**   Uses the active range in the Time Slider.

- **Animation Curve**   Uses the start and end of the animation curve of animating attributes.

- **Start/End**   Lets you specify the clip's time range by typing in the Start Time and End Time of the range.

- **Subcharacters __ Include Subcharacters in Clip**   This is a subset of a character set. By keying the character set, the subcharacter set under it will be automatically keyed as well. But the subcharacter set can also be independently adjusted. By checking this box, the clip will contain subcharacter sets.

- **Time Warp __ Create Time Warp Curve**   Lets you adjust the speed of the clip with a curve without changing the animation curve of the original clip. This slows down the workflow, so it should be used carefully.

- **Include __ Hierarchy**   All the objects under the selected object's hierarchy will also be included in the clip. If it's unchecked, only the selected object will be included in the clip.

**2.** Set the options to match those shown in Figure 16-4. Click the Apply button.

**3.** Choose Window | Animation Editors | Trax Editor.

**4.** Load the clip into the Trax Editor by choosing List | Load Selected Characters from the Trax Editor's menu bar. The Trax Editor is shown in Figure 16-5 with the walkLower clip loaded.

**FIGURE 16-5**   *The walkLower clip is loaded in the Trax Editor.*

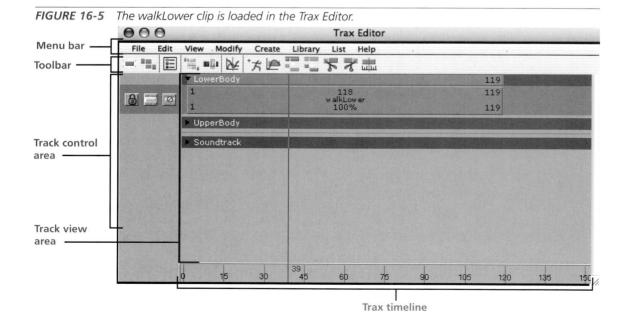

**5.** You'll see that a clip called walkLower is generated and placed in the Trax Editor. If you click the Play button, it doesn't look like it has changed. But select the pelvisCTRL object and open the Graph Editor. There is no animation in the Graph Editor, because the animation of the pelvisCTRL object is now totally controlled by a clip in the Trax Editor instead of in keyframes.

**6.** Let's try making a clip for the UpperBody set. This time we'll create the clip from the Trax Editor. Select the UpperBody set in the Outliner, and then in the Trax Editor choose Create | Clip □.

**7.** Name the clip **walkUpper** and choose the Put Clip in Visor Only radio button in the Create Clip Options window. Click the Create Clip button.

**8.** Play the animation. Since no clip appears in the Trax Editor and keyframes have been removed, there is no animation on UpperBody.

**9.** You can insert the generated walking clip easily by choosing Library | Insert Clip | walkUpperSource. You should see that the walkUpper clip is now loaded into the Trax Editor, and the body is properly animating.

### Animate in the Trax Editor
By simply editing the clips in the Trax Editor, we can change the timing and cycle the clips so that the clip can be repeated. Let's change the timing of the animation by manipulating the clip. It's easy to do this in the Trax Editor:

**1.** Move your cursor to the lower-right corner of the walkLower clip in the track view area, where it says "119." The cursor should turn into a slider. Click and drag it to shrink the clip to frame 60. Do the same thing for the walkUpper clip.

**2.** Play the animation, and the character is now walking twice as fast as before and stops at frame 60. To slow down the animation, simply lengthen the clips. For now, keep the clips at 1 to 60 frames.

**3.** You can cycle this walk easily using the Trax Editor. Hold down the SHIFT key and move the cursor to the bottom-right corner of the walkLower clip. When you see the cursor change to an arrow in a circle, LMB-drag it to frame 119.

**4.** Do the same for the walkUpper clip. Now the character is walking twice as fast for 120 frames. That's how you can cycle the clips. Figure 16-6 shows the shortened and cycled clips.

FIGURE 16-6   *The clips have been scaled to 60 frames and then cycled to 120 frames.*

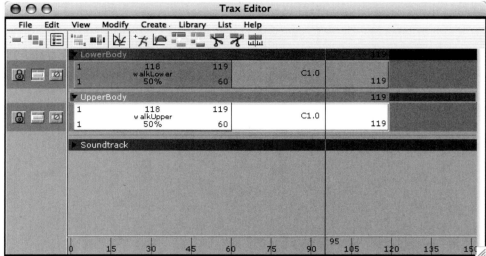

5.   Double-click the walkLower clip in the Trax Editor, and look at the attributes for the clip in the Channel Box:

   • **Weight**   The influence of the clips affecting the character set. A setting of 1 will influence the clip 100 percent. If this attribute's value is 0, no animation appears in a clip.

   • **Enable**   Turns on/off the clip.

   • **Offset**   Lets you offset the animation.

   • **Absolute**   Uses exact values from the clip—for example, if the character is walking forward and you made a clip out of it, the Absolute value will play the animation until the end of the clip, and the character will move back to its original position at the first frame of the clip when it's cycled.

   • **Relative**   Makes the character move forward and keeps it moving.

   • **Start Frame**   Lets you manually assign when the clip starts.

   • **Cycle**   Lets you numerically input how many times the clip should cycle.

   • **Scale**   Lets you shorten and lengthen the clip. This controls the speed of the animation. The smaller the number, the faster the animation will become.

   • **Start and Duration**   Shows the current starting frame and length of the clip.

**6.** If you are happy with the results in the Trax Editor, you can merge the walkLower clip and its cycle into a new, combined clip. This lets you treat it as a totally new single clip. In the Trax Editor, select walkLower and choose Edit | Merge ☐.

**7.** Type **fastWalkLower** in the Name field and check Add to Trax. This will immediately replace walkLower with a newly merged clip. Checking the Add to Visor option will not replace the original clip, and the merged clip will be added to the Visor.

**8.** You can export these clips to their own folder so that you can reuse them later for other projects. In the Trax Editor, select fastWalkLower and choose File | Export Clip.

**9.** Name the clip **fastWalkLower** and click the Export button. It should be exported to the clips folder of your current project.

**10.** To import a clip, select a character set to import and choose Edit | Import Clip. Use the same procedure to create and export a fastWalkUpper clip.

## Combine and Blend Clips

Now we will create a new clip and blend it with the clips we just exported:

**1.** Change the Range Slider's range from frame 1 to frame 90. Go to frame 1 and delete both clips from the Trax Editor. There should be no more animation.

**2.** At frame 1, pose the character so it is in a neutral pose, before jumping. Select the LowerBody character set using the Character Set Selector and press the S key to key the pose for the lower body.

**3.** Select the UpperBody character set from the Character Set Selector and key by pressing S. Keying various poses using character sets is easy to do in this way.

**4.** Go to frame 30 and move the character down, so it is anticipating a jump. You can move the character down by translating the pelvisCTRL object.

**5.** Change the knee rotate attributes for both foot controls, so the legs are spread apart. Put the character's arm behind the body.

**6.** Rotate both the pelvisCTRL and the spineCTRL objects, so that the body is curled forward a little. Key both character sets.

**7.** In the Time Slider, MMB-drag from frame 30 to frame 40. This moves the Current Time Indicator but leaves all animation channels constant. Key both character sets. The purpose of this key is to hold the anticipation pose for ten frames.

**8.** Go to frame 50 and make the character jump up by translating the pelvisCTRL object and both foot controls upward. Rotate the pelvisCTRL and spineCTRL objects so the spine appears to be bent backward. Rotate the shoulder and elbow so that they're moving forward. Key both character sets.

**9.** Create an extreme pose for the mid-air jump. Go to frame 60 and extend the back a bit more. Rotate the arm a bit more forward. Move the pelvisCTRL object up just a bit more as well. Key the character sets.

**10.** MMB-drag from frame 40 to frame 70 in the Time Slider, and set keys for both character sets to hold this pose. MMB-drag from frame 1 to frame 90 and key both character sets. This key will set the character back to the default position so that the animation is now cycled.

The jumping animation is done! Keying using character sets is a great way to make animation based on specific poses. Look at Figure 16-7 for the four poses you want to make for the jumping move.

*FIGURE 16-7*    *Poses for the jumping animation*

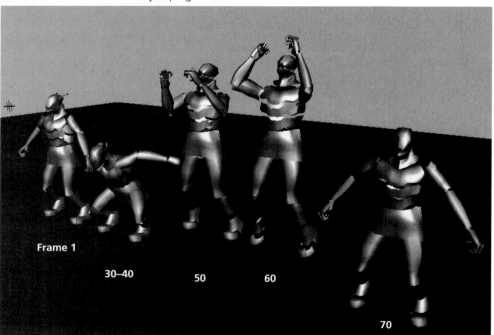

Now let's create a clip out of the animation you just made:

**1.** Select the LowerBody character set from the Character Set Selector and choose Create | Clip ▢ in the Trax Editor.

**2.** In the Create Clip Options window, name the clip **jumpLower** and enable the Put Clip in Trax Editor and Visor option.

**3.** Choose the Time Slider for the Time Range setting. Do the same procedure you did for the UpperBody character set, but name the clip **jumpUpper**. Set the Range Slider's range from 1 to 300.

**4.** Move the jumpLower clip to frame 120 by LMB-dragging so the jump starts at frame 120. Also, move the jumpUpper clip to frame 125. Now a little delay occurs in the arm motion, which makes the jump look more natural.

**5.** Add the merged fast walk clips to both character sets. Choose File | Import Clip, and then select fastWalkLower and click the Import button. Do the same for fastWalkUpper.

**6.** LMB-drag to move both clips to run from frames 2 to 120. Figure 16-8 shows how the clips should be laid out.

**7.** Play the animation. The motion is abrupt between the two clips, causing them to snap. You can fix this problem by creating a blend between the clips. In the Trax Editor, select fastWalkLower and SHIFT-select the jumpLower clip.

**FIGURE 16-8**   *The clips are laid out in the Trax Editor.*

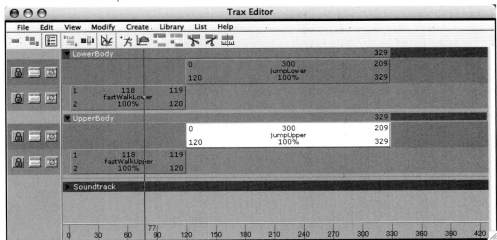

**8.** Choose Create | Blend. A blend is created between the two clips in the Trax Editor. Apply a blend to the UpperBody clips, too. Move jumpLower and jumpUpper about 20 frames to the left. Play the animation again. The transition between the clips should look much smoother now.

At some point, you may wish to edit the keys in an individual clip. As mentioned, the keyframe data was removed when the clips were created. However, you can edit the keys by selecting a clip and choosing View | Graph Anim Curves. This will show the animation curves in the Graph Editor for the selected clip. Edit this in the Graph Editor to fine-tune the motion. Figure 16-9 shows the Graph Editor displaying the animation curves for the clip selected in the Trax Editor.

Animating in the Trax Editor takes a little more work to set up and get used to, but the benefits of using it are incredible for any production environment. The ability to have several people working simultaneously on a motion library for a character is an invaluable asset. You can even use clips on different characters, as long as the rigs are similar.

**FIGURE 16-9** *The Graph Editor displaying the animation curves for the selected clip in the Trax Editor*

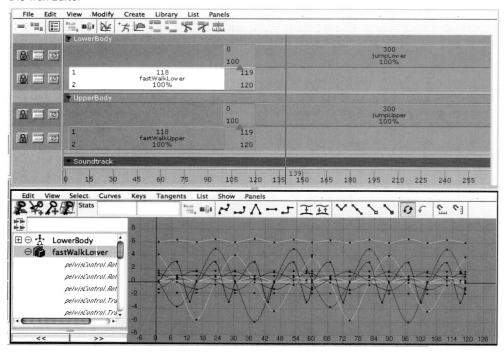

# Animation Retargeting

Animation retargeting lets you transfer the animation data from one skeleton to another, even if the skeletons are proportioned differently. This means that once you have spent time building a rig for a character, as we did in Chapter 12, any animation created from that rig can be transferred to another character with the use of retargeting. This is also a great way to import animation data that was created by a motion capture system.

Definition     **Motion capture**    *The process of capturing the performance of a real-life actor and recording the data in a way that can be translated to a 3D character.*

## Retargeting Workflow

The workflow involves setting a neutral pose for both the source (the animated) skeleton and the target skeleton. The skeleton is usually in the neutral pose when the joints are placed to match the model. Any animation will be transferred using this neutral pose as the base reference.

Once the neutral pose is set, the joints in both the source and target skeletons must be labeled. Joint labeling is different from just naming joints—to label joints, you choose Skeleton | Retargeting | Joint Labeling to label each joint in a hierarchy. This process tells Maya where similar joints are located in different skeletons. For example, the right elbow joint must be labeled the same for both the source and target skeletons, even though the actual names of each of the joints might be something more specific.

Once the joints in both skeletons are labeled, the retargeting options are set and the retargeting of the animation is performed. Once completed, the target skeleton will animate to match the source skeleton, but at the same time the target will retain all of its unique proportions and characteristics.

## Tutorial: Retargeting the Animation

In this section, we will practice retargeting the animation between a biped skeleton that has been animated using motion capture and retarget it to the skeleton we built in Chapter 12.

> **NOTE**   **Since the animation will be created by transferring it, we do not need any controls.**

### Import FBX Data

The source file with the animation data is in the FBX file format. This is the native file format for files created in another Autodesk product called Motion Builder. Any FBX file can be read by or written from Maya. Because so many other 3D packages support the FBX file format, it can be used to export entire Maya scenes and import them into other packages. While this heads outside the scope of this book, it is important to know about the FBX import and export features in Maya.

**1.** Open the scene on the accompanying DVD called mcr8_ ch16_boundMesh.ma.

**2.** Choose File | Import ☐. In the Import Options window, set the File Type to Fbx, as shown here:

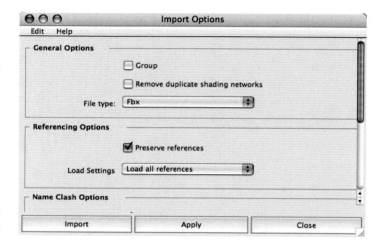

**NOTE**  *If the FBX file format does not show up in the list of available file types, make sure the fbx plug-in is loaded in Window | Settings/Preferences | Plugin Manager.*

**3.** Click the Import button and browse to the mcr8_ch16_moacap.fbx file. Opening mcr8_ch16_moacap.fbx imports a skeleton that has motion capture data on all of the joints. Click Play to watch it.

## Retarget the Data

Now we will retarget the animation data on the motion capture skeleton to the bound character:

**1.** Select the root joint on the mocap skeleton. In this case, the root joint is named mocap_pelvis. Choose Skeleton | Retargeting | Set Neutral Pose.

**2.** Select the root joint on the *target* skeleton. This joint is named char2_pelvis. Choose Skeleton | Retargeting | Set Neutral Pose to set the neutral pose for this skeleton.

**3.** Select both skeletons, and choose Skeleton | Retargeting | Show All Labels. This will show the labels for all of the joints. By default, they are all labeled None.

**4.** Because the names of the joints in both skeletons differ only in their prefix, we can quickly label the joints by using the Label Based on Joint Names command. Select the root joint of the source skeleton, and then choose Skeleton | Retargeting | Label Based on Joint Names. The joints will be labeled. Repeat this process for the target skeleton.

**5.** Inspect the labels. Because the target skeleton did not have any prefixes specifying whether the joints were on the left or right side of the body, you must use the Joint Labeling menu to specify this. Select one of the joints on the right side of the target skeleton and choose Skeleton | Retargeting | Joint Labeling | Right Side. Choose a joint on the left side and choose Skeleton | Retargeting | Joint Labeling | Left Side.

**6.** You must set a custom label for some joints, such as the ball joints in the foot. Select the left ball joint on the source skeleton and look in the object's Attribute Editor. In the Joint Labeling section, set the Type to Other and label the joint in the otherType field **ball**, as shown in the illustration.

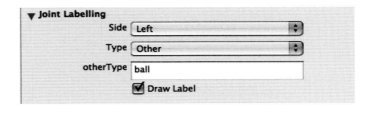

**7.** Continue using these techniques until all of the joints are labeled and each joint in the source skeleton has a matching joint in the target skeleton. Figure 16-10 shows both skeletons with joint labeling turned on.

**8.** Select the root joint on the mocap skeleton (the mocap_pelvis joint). Hold down the SHIFT key and select the root joint on the target skeleton.

*FIGURE 16-10*   *The two skeletons with joint labeling turned on*

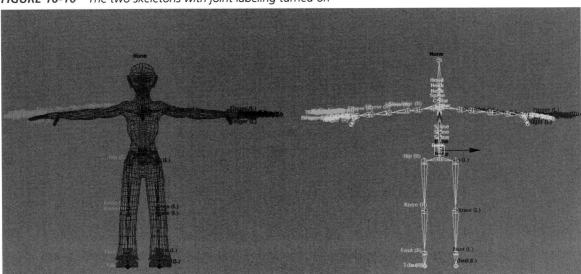

**9.** Choose Skeleton | Retarget ❑. Set the Time Range to Start/End and set the End Time to **40**. Click the Retarget button. Maya will transfer the animation data from the source skeleton to the target skeleton.

**10.** Play back the animation and see that the bound character is now animating.

At this point, you can throw away the mocap skeleton. This animation is now complete.

# Object Interaction

How would you animate objects that are being "picked up" and "put down"? Animating the visibility of several duplicates in different locations might be one solution. Counter animation, or keying the carried object at every frame so it remains in the same place, might be another. However, both of these options require lengthy setups, and making changes to either of these setups could require hours of work. In other words, neither is a good option in a fast-paced production environment.

## Constraints

In Chapter 12, we used constraints to help rig our skeleton, but constraints are also a valuable animation tool. Constraints offer an excellent method for solving the "pick up/put down" scenario. Constraints can be keyed on and off, making the relationship between the constrained object and the target object temporary. While point and orient constraints may be used together for object interaction, Maya's parent constraint allows for constrained objects to behave as if they were temporarily parented into the hierarchy. The constrained object will move and rotate around its target object.

## Using the Parent Constraint

In this example, an arm will pick up a ball and put it down on top of a cube. For this to work properly, the ball must be parent constrained to both its original position and the hand. However, the constraint to the hand should be added with the hand in position to pick up the ball. This will enable us to take advantage of the Maintain Offset option in the parent constraint's settings. The constraint Weight attribute to the original position is keyed at 0 (off) and the hand lifts the ball to the cube. Then the ball is put down and set to its resting position on the top of the cube. The Weight attribute to the hand can now be keyed at 0 (off) and the hand can move away.

**1.** Create a new scene and draw three joints to create an arm.

**2.**   Use the IK Handle tool to create an IK chain that controls the joint hierarchy. Either parent or point constrain the IK handle to a control object. For this example, create a cube to use as a control object.

**3.**   Create a sphere and another cube to use as props in the scene. Scale and position them so that they look similar to those in Figure 16-11.

*FIGURE 16-11    Position the sphere and cubes in the scene.*

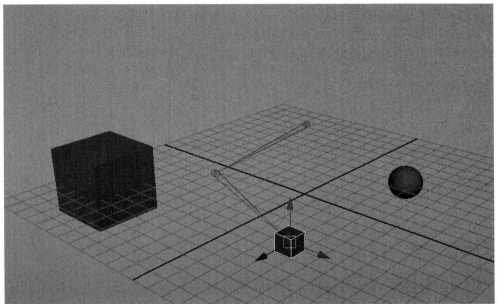

**4.**   Create a locator (choose Create | Locator) and parent constrain the ball to this locator. Select the locator, SHIFT-select the sphere, and then choose Constrain | Parent ❑.

**5.**   In the Parent Constraint Options window, make sure that Maintain Offset is checked.

**6.**   Select the constraint in the Outliner, and then set a key on the constraint weight at a value of 1 by clicking the Locator1 W1 attribute in the Channel Box; then right-click and choose Key Selected from the marking menu.

**7.**   Key the arm's IK in its original position. Then, at frame 30, key the IK so that it is in position to pick up the ball.

**8.** Select the hand (the control cube), then select the ball and choose Constrain | Parent ☐. Make sure that Maintain Offset is checked. Also, you can set the Weight attribute to **0** before constraining the object. Key the weight at this frame.

**9.** Advance one frame, set the Locator1 W1 attribute to **0**, and set the P Cube1 W0 attribute to **1**. Key these values. Your scene should look like this:

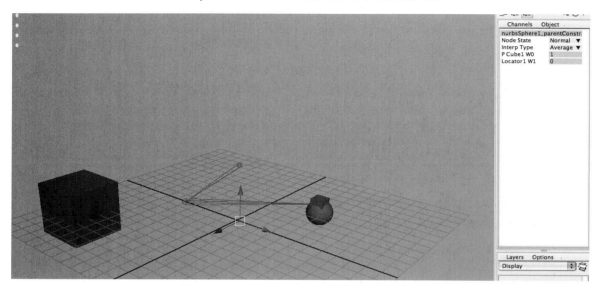

**NOTE** *You may have to set your tangents to Stepped so that no interpolation occurs between the keys set on the constraint's Weight attributes. You can quickly do this by selecting the constraint and then holding down the* SHIFT *key and drag-selecting all the keys in the Time Slider. Then right-click and choose Tangents | Stepped from the marking menu.*

**10.** Thirty frames later, the IK handle should be placed and keyed in position to set the sphere on the large cube, as shown in Figure 16-12.

**11.** Select the sphere, and then choose Constrain | Set Rest Position. This will keep the ball from snapping back to its origin when all constraints are set to 0.

**12.** Select the constraint and key the P Cube1 W0 attribute to **0**.

FIGURE 16-12    *The sphere is positioned on the large cube.*

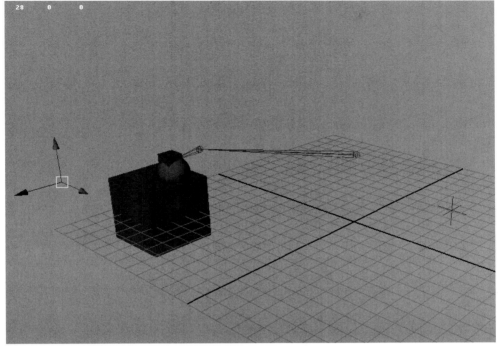

**13.** Move and key the position of the control cube so that the arm is away from the ball.

**14.** Play back the animation to see the result.

# Summary

Character animation can take a lifetime to master. To create good animation in a reasonable timeframe, the most successful approach requires a combination of your technical and creative abilities. We hope that the information provided over the last four chapters will help lay a foundation for your future animations.

Next, we will switch gears and explore the world of surfaces and textures.

# V

# Texturing, Lighting, and Rendering

# Texture Basics

**One of the fundamental concepts in**
color theory is that color cannot exist without
light, nor can light exist without color. In this
chapter, we explore Maya's tools for adding light
and color to your scene. *Texturing* generally refers
to the process of building a surface's detail and
indicating how that surface will react to light. In
the real world, all surfaces can be described by
their color, shininess, smoothness, and opacity. In
Maya, a surface's material contains attributes to
control all of these characteristics. By assigning
different values to these attributes, you can make

a surface appear to be made out of a variety of materials, including plastic, metal, glass, or wood. These values can be controlled by a texture—either a bitmap file or a procedural texture—giving the material attributes some detail, irregularities, or pattern.

As is usually the case in any visual art, your greatest asset in the texturing process is your own eyes—use them to study the world around you. Being able to recognize and break down a surface and how it reacts to light is the first step in this process. Learning about the different materials and how to control their attributes in Maya is the next step. Therefore, it is always important to do research. Find photographs of surfaces similar to the ones you are trying to emulate. If possible, keep something made of this material at your desk so you can play with it and move it around in the light.

## Hypershade: Maya's Texturing Interface

The Hypershade window (Window | Rendering Editors | Hypershade) is shown in Figure 17-1. This is where you build and edit materials for all the objects in your scene. Like the Hypergraph, the Hypershade displays a dependency graph of the different materials in your scene. In fact, these materials display the exact same nodes, and you could do all the same editing and connecting by using the Hypergraph and other menu commands. The main difference between the two is that, when dealing with material-related nodes (also known as *rendering nodes*), the Hypershade displays its node icons, known as *swatches* in Maya, in more detail. This means that a file texture node's icon displays a thumbnail version of the actual texture. A material node's swatch will display everything from the falloff of the specularity to the peaks and troughs of its bumps. This, in addition to all the material-related menu sets readily available with a right-click, makes the Hypershade a much more artist-friendly interface than the Hypergraph.

 _TIP_   **When working with the Hypershade, it can be helpful to use one of the saved panel layouts. A button that changes the layout to the Hypershade/ Perspective layout is available in the toolbar.**

The basic workflow of building a material is to start with a rendering node, such as a basic material, and edit its attributes. Other nodes, such as textures, can be connected to these attributes by dragging and dropping them onto a material and choosing the desired connection. Placement nodes are then edited to direct how a texture is positioned on a surface. For more advanced materials, additional *utility nodes* are connected to tweak the attributes of the other nodes in the network. As you'll see, even a relatively simple exercise in the Material Editor can lead to a few dozen separate nodes and connections. The network of connected nodes that feeds into a material is called a *shading network*. Therefore, a completed material is often called a *shader*.

**FIGURE 17-1** *The default Hypershade window*

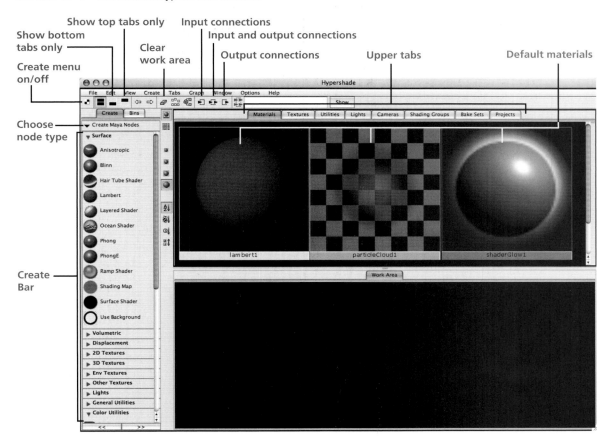

## Hypershade Sections

The Hypershade is divided into three main sections: the Create Bar, the top tabs section, and the bottom tabs section. In the Create Bar, you can create any of the rendering nodes by clicking them. Click and hold the Choose Node Type arrow to view the different types of rendering nodes. You may choose the Create Maya Node (the default), Create Mental Ray Nodes, or Create All Nodes. Once a selection is made, the relevant nodes will appear in the Create Bar. Items in the Create Bar are also available by right-clicking in the other sections of the Hypershade window, so to create more space to work, you may want to hide the Create Bar. To do so, click the Create Bar On/Off button in the toolbar, shown in Figure 17-1. The top tabs section contains all of the rendering nodes in the current scene; they are organized by category: Materials, Textures,

Utilities, Lights, Cameras, Bake Sets, and Projects. Here's how this works. If, for example, you click a Blinn material in the Create Bar, a new Blinn material icon will appear in the Materials tab. Likewise, any texture that is created will be displayed in the Textures tab. This feature organizes the different kinds of nodes in your scene.

The third section of the Hypershade, the bottom tabs section, is where you can view, select, and edit the connections of the rendering nodes. To view a node in the work area, you can MMB-drag the node's icon from one of the top tabs down into the work area. To view its connections, you can use one of the three connection buttons in the Hypershade toolbar. This will expand that node to show its input connections, output connections, or both, in the work area. To clear this space, use the Clear Work Area button in the toolbar. Selecting one of the rendering nodes from the scene library tabs and clicking one of the display connections buttons will automatically clear the work area and display the connections for the selected node.

These two sections of the Hypershade can be hidden by using the three buttons in the upper-right corner of the Hypershade toolbar. Clicking the first button hides the lower tabs and expands the space for the upper tabs. Clicking the second button hides the upper tabs and shows only the lower section, and clicking the third (the default view) displays both tab sections.

To move around in the Hypershade window, you use the same pan and zoom controls used in most of the other windows in Maya. To zoom in and out, hold down ALT (OPTION) and then RMB-MMB-drag. To pan, hold down ALT (OPTION) and the MMB-button and drag. Pressing the F key fits all the swatches in the active window. All the items available from the Hypershade menu bar are available with a right-click. When you start a new project, Maya loads three materials by default, as shown in Figure 17-1. These are the Lambert, a particle cloud, and a shader glow. Every time you create a new piece of geometry in the scene, it will use this Lambert material. Any particle cloud or glow will use the other defaults, respectively.

## Working with the Hypershade

Let's practice working with the Hypershade by creating a Blinn material and connecting a texture to the Color attribute:

**1.** The Create Bar should be turned on and displaying its Create Maya Nodes option; click the Choose Node Type arrow and select Create Maya Nodes from the resulting menu, if necessary.

**2.** Click the sphere icon labeled Blinn. A Blinn material will be added to your scene and will appear in both the Materials tab and in the Work Area tab. If this is the first Blinn material you have created in the scene, it will be named blinn1. The next time you create a Blinn material, it will be named blinn2.

3. Scroll down in the Create Bar and find the Textures section. Unfold this section to see the texture nodes. Click the Checker icon (see Figure 17-2). A checker texture node is added to the scene and appears on the Textures tab and the Work Area tab. Depending on where your Blinn material was positioned in the Workspace, your checker texture node might overlap it. If this is the case, simply select one of the nodes in the work area and drag it out of the way so that all the nodes are visible. The work area should look like that shown in Figure 17-2. In addition, notice that a 2D placement node has also been created and connected to the checker texture node. This controls how that texture fits into the texture space. We'll look at this feature in more depth in the "Placement Nodes" section later in the chapter.

4. We want to have this checker texture control the Color attribute of the Blinn material. We can make this connection directly in the Hypershade by dragging and dropping the checker node's icon onto the material node's icon. In the work area, select the checker

**FIGURE 17-2** *2D placement node, checker texture, and Blinn material in the work area*

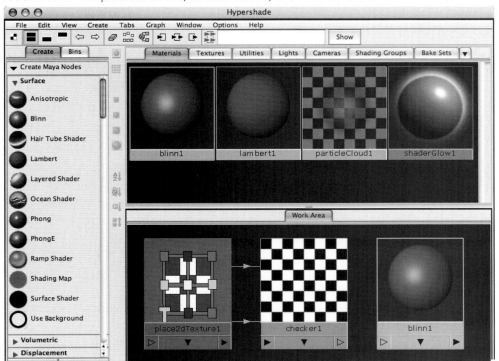

texture node and MMB-drag it onto the Blinn material node. When you release the mouse button, a marking menu appears with a list of attributes that the checker texture

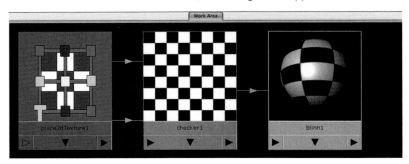

can be mapped to by default. Since we want to map this to the Color attribute, choose Color from the list. A connection is made, and the material node icon is updated to display the checker texture on the material. Your work area should look like the illustration.

**5.** Lastly, we assign the material to an object in the scene. Create a NURBS sphere (Create | NURBS Primitives | Sphere). You can assign the material to the object simply by MMB-dragging the material swatch onto the object in the view window or the Outliner. Another way would be to select the object, right-click a material in the Hypergraph, and choose Assign Material to Selection from the marking menu. To view the texture on the object in a view window, make sure you turn on hardware texturing by pressing the 6 key.

*TIP*   *You can quickly assign a texture to a group of objects by first selecting them all and then using the marking menu option to assign materials.*

## Using the Attribute Editor for Editing Materials

The Attribute Editor (shown in Figure 17-3) also plays an important role when you are working with materials. While connections can easily be made in the Hypershade window alone, the actual editing of attribute values is done in the Attribute Editor. Connections to other textures can also be made in the Attribute Editor. In fact, this tool offers access to more of the possible connections than the marking menu that is displayed when a node is dragged onto a material in the Hypershade.

To get the most out of material editing, it is best to use the Hypershade and Attribute Editor together. Generally, you use the Hypershade to view connections and select particular nodes. Once selected, that node's attributes are displayed in the Attribute Editor, where they can be edited. Connections to textures are probably best made through the Attribute Editor, and then additional utilities and optimizations can be made back in the Hypershade.

*NOTE*   *When connecting nodes in the Hypershade, you might also use the Connection Editor. You will see examples of this workflow in Chapter 18.*

Let's try using the Attribute Editor to make connections to textures. We are going to continue working with the material we created earlier. If you no longer have that scene available, follow Steps 1–5 in the preceding section to create a Blinn material with a checker texture connected to the Color attribute.

**1.** Open the Attribute Editor, either by pressing CTRL-A, clicking the Show Attribute Editor button in the status line, or double-clicking a material node in the Hypershade.

**2.** Click the different nodes in the Workspace and notice how the Attribute Editor loads the specific attributes for the selected node. The Blinn material node attributes describe the basic surface characteristics, the checker texture attributes contain values that edit the colors in that checker pattern, and the placement node attributes have values for translations, rotation, and other spatial attributes.

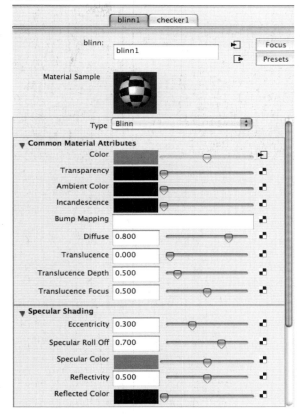

**FIGURE 17-3** *The Attribute Editor with a material node loaded*

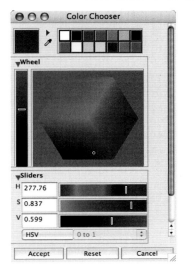

**3.** Let's change the color of the checker pattern. Select the checker texture node in the Hypershade. In the Attribute Editor, click the color swatch in the Color1 attribute. The Color Chooser, shown here, will open. Use the Color Chooser to change the white to red. Click the Accept button to close the Color Chooser.

**4.** In the Attribute Editor, you can use the slider to the right of the color swatch to change the brightness value of the red. Change the Color2 attribute to blue.

**FIGURE 17-4** *Changing the placement values in the Attribute Editor*

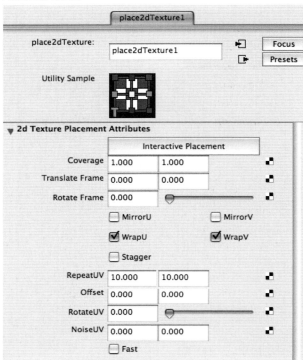

5. Now we'll change the tiling of the checker pattern. Select the 2D placement node (named place2dTexture1) in the Hypershade to view its attributes in the Attribute Editor. Change the value of RepeatUV to **10** for both U and V, as shown in Figure 17-4. The checker pattern on the checker node icon in the Hypershade window will change to show more tiles. Moving downstream, the Blinn material node also updates to show the new checker pattern mapped to it.

*TIP* **You can zoom up and pan in the Hypershade window just as you can in an orthographic view window by using ALT (OPTION) in combination with the left and middle mouse buttons and the middle mouse button alone.**

6. Next, we'll edit the Bump attribute of the Blinn material node by connecting a texture to it. Select the blinn1 material in the Hypershade window to display its attributes in the Attribute Editor, as shown in Figure 17-3. Find Bump Mapping in the Attribute Editor (under Common Material Attributes). Click the checker button at the right to add a connection to this attribute. This will bring up the Create Render Node window.

7. The Create Render Node window, shown next, is really the same as the Create Render Node menu in the Hypershade window. Make sure that 2D Texture creation setting is set to Normal at the top of this window. Then select the Checker texture. You'll notice that in the Hypershade, a new checker texture node is connected to a bump2d node, which is in turn connected to the Bump attribute of the Blinn material node. A bump map creates raised and indented areas in a surface when it is rendered. Lighter grayscale values raise the surface, while darker values create indentations. (We'll talk more about this attribute in the "Material Nodes" section, later in this chapter.) Of

course, this can be reversed by assigning a negative value to the Bump Depth attribute or by using a reverse utility node.

*TIP* **As you continue to add nodes, the work area in the Hypershade will start to become cluttered and difficult to read. For this reason, it is a good idea to select the material node, in this case the Blinn1 material, and click the Input Connections button in the Hypershade window's toolbar (see Figure 17-1). This will regraph the entire connection network and clean up the work area.**

**8.** We want the new checker pattern to control the bump value of the surface, but we also want it to align with the checker pattern connected to the Color attribute. We could go in and make sure that the RepeatUV values of the placement node of the color map match the values of the bump map, but this would be time-consuming and would likely result in errors if more attributes in that node needed to be duplicated. Instead, we'll connect the entire placement node that is controlling the checker pattern on the Color attribute to the new checker pattern that is connected to the bump map.

**9.** Select the 2D placement node that is connected to the checker texture node controlling the color channel. Then MMB-drag it onto the checker texture we created in Step 7. Choose Default from the resulting marking menu to complete the connections. Now if you roll over the new connection lines, you'll see that the OutUV and OutUVFilterSize output attributes are connected to the UVCoord and UVFilterSize input attributes, respectively. The old placement node is no longer connected and can be deleted by selecting it and pressing the DELETE key. The work area should look like the example shown in Figure 17-5.

**10.** Lastly, we can edit the amount of effect that the bump map has on the surface by editing the Bump Depth attribute in the bump2d node. Select the bump2d node in the Hypershade window to view its attributes in the Attribute Editor. You will see two attributes displayed: Bump Value and Bump Depth. Bump Depth is the attribute that the checker texture is controlling. The Bump Depth attribute controls the amount of the effect that the map will have in creating the bumps on the surface.

**11.** Change the Bump Depth value with the slider or by entering a value numerically. Watch the effect this has by viewing the material node in the Hypershade window.

*FIGURE 17-5*   *Blinn material with one checker texture node controlling the color and the other checker node controlling the bump; both checker textures share the same placement.*

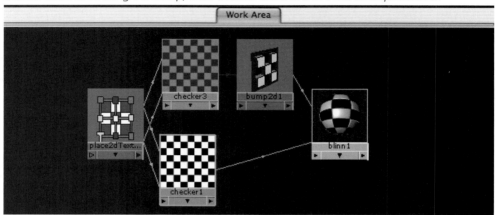

The illustration shows some material applied to a sphere. The material on the right uses a positive Bump Depth value, while the material on the left uses a negative value. The effect is that the bumps are inverted.

# Rendering Nodes and Their Attributes

In the last section, you saw some of the basic rendering nodes. Rendering nodes can be divided into five basic categories: materials, textures, utilities, lights, and cameras. The Blinn and Lambert materials are examples of material nodes, the checker node is an example of a texture node, and the placement node is a type of utility node. While we have already used these nodes, connected them, and edited their attributes, we have only touched the surface (so to speak). In the next few sections we will dig deeper and examine some of the individual attributes of the material, texture, and utility nodes. Then, in Chapter 18, two tutorials will allow you to put these basics to use in building several materials to texture map surfaces of different shapes.

## Material Nodes

Material nodes are responsible for controlling the characteristics of a surface. The main material nodes in Maya are Anisotropic, Blinn, Lambert, Phong, Phong E, and Ramp Shader. While the texture nodes might display a pattern hinting at the material that a surface could be made of, the material node is primarily responsible for communicating this effect. Even without a texture containing pictorial details, the type of material node and the settings of its attributes will largely tell you the material is made of plastic, metal, glass, rock, or other material. The way material interacts with its environment— the objects, colors, and lights that surround it—provides most of the visual clues about its substance.

Look at the candle pictured here. What is it about its surface that tells us it is made of wax? The reflection of the light source on the object is very soft, because the light is being absorbed into all the little pores on the surface instead of reflecting back. This characteristic is controlled by the *specularity attributes* of the material. Notice that the top of the candle is illuminated from the flame inside it. You can see the light coming through the wax. This is controlled by the *translucency attribute* of the material.

Take a moment to analyze the surface characteristics of a mirror. What color is a mirror? It is actually black, but it does not give off any of its own color. Instead, its surface is almost completely reflective, so it just reflects what is around it. To create a material attribute for a mirror surface, you would turn down the *diffuse value* all the way, so that the color of the object is not reflected back into the camera.

All the materials in Maya share the same set of attributes. These are known as the *common material attributes*. Where these materials differ is in the way they calculate their specular and reflection attributes, if they have any. The Lambert material does not have any specular attributes. It is best used for matte surfaces, such as chalk or certain types of paper or fabric.

## Common Material Attributes

The common material attributes for all of Maya's materials are as follows:

- **Color** The color of the material. Assigning an RGB or HSV (Hue, Saturation, Value) value using the Color Chooser sets this attribute. Its brightness value can be controlled with the slider.

- **Transparency** Controls the transparency/opacity of the surface. A value of 0 (black) renders the image 100 percent opaque. A value of 1 (white) renders the object totally transparent. A color value can also be used; it will act as a color filter over objects behind it.

- **Ambient Color** Sets the color of the darkened or black areas. Think of it as bringing up the black contrast levels of an image in Photoshop. When used together with ambient lights, the color and brightness of the lights control the amount of ambient color specified in this attribute, which will contribute to the final render.

- **Incandescence**   Controls the self-illumination, or luminance, of the object. This attribute creates the illusion that the object is emitting light. The material will illuminate the color set using the Color Chooser with the brightness amount set with the slider. Note, however, that this will not cause the object to illuminate any objects around it.

- **Bump Mapping**   Alters the surface normals to create the illusion of raised areas and indentations on a surface. This attribute uses an 8-bit grayscale image to assign transparency values: 0 (black) is the lowest or deepest, while 1 (white) appears to be raised or highest.

- **Diffuse**   Sometimes called falloff, this attribute controls the amount of light reflected back from the surface. At a value of 1, the surface will reflect back 100 percent of the color specified in the color channel. The default value is 0.8.

- **Translucence**   Controls the amount of light that can pass through an object. Think of the way light travels through wax, leaves, and certain plastics and rubbers. At a setting of 0.0 (the default), no light shows through. At 1.0, all light shows through.

- **Translucence Depth**   Controls the amount of decay that will occur as the light moves through the object. When the value is set to 0, no decay occurs and the light emitted through the other side will be just as bright. In most cases, the light will lose some of its brightness and decay before it reaches the other side of the object.

- **Translucence Focus**   Sets the amount of blurring on the surface of backlit objects. The light scattering on a thin surface, such as a leaf or skin, will cause this effect. At 0.0, light is scattered in all directions.

### Specular Shading Attributes

As mentioned, specularity controls the reflectivity of a surface. Even a surface that doesn't necessarily reflect objects around it will reflect the light sources that illuminate it. This is known as the *specular reflection*. The specular attributes differ slightly depending on the material you are using. Usually some type of control can be used for the size of the highlight and another control for how it falls off; this is illustrated in Figure 17-6. Some materials, such as the Phong material, use composite algorithms to calculate the specular shading; others, such as Phong E, Blinn, and Anisotropic, break up the specular attributes into several attributes for finer control. The Anisotropic material even lets you control the shape of the specular reflections.

**FIGURE 17-6**   *The specular attributes give you control over the highlights and how they fall off.*

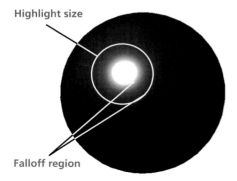

Highlight size

Falloff region

The reflectivity of an object determines how much of the environment or other objects will be reflected from the surface. The color of the reflection is determined by the reflected Color attribute. For a surface to reflect objects around it, raytracing must be turned on in the Render Settings window. (Render Settings will be discussed in detail in Chapter 21.) Because raytracing is very processor-intensive, it can take a long time to render. Consequently, reflections are sometimes faked by applying an environmental reflection map to the reflected Color attribute. We will play with this in the next exercise.

# Texture Nodes

Texture nodes contain the picture or pattern that will be mapped onto a material. In Maya, you can use a procedural texture or a file texture.

### Procedural Textures

Procedural textures are completely computer generated from mathematical equations. Checker, bulge, fractal noise, and cloth are all examples of procedural textures. Each has its own specific parameters to achieve its unique results. One of the biggest advantages of using procedural texture maps is that they are resolution-independent. Whereas bitmapped file textures start to soften and break apart as you zoom closer, procedural textures remain infinitely sharp no matter how close you get.

A subset of procedural textures is called *3D textures*. Because a procedural texture is not limited to 2D pixel information, the computer is able to calculate the depth of a texture. Marble or any kind of rock is a good example from the real world. The patterns in marble are continuous through its mineral layers, changing slightly from slice to slice. A 3D texture in Maya behaves the same way. This texture is especially useful if the surface is semitransparent.

The placement of 3D textures also has some advantages. Because you are not limited to a 2D projection method, distortions and discontinuities are less of an issue with 3D textures. We'll discuss this further in the section, "Placement Nodes."

### File Textures

File texture nodes contain a bitmapped image that is imported into the scene. This image can be loaded through the Image Name attribute in the file texture node's Attribute Editor. Further adjustments can be made to this image in the Color Balance section of the file texture's Attribute Editor. Color Gain and Color Offset control the brightness and contrast of the texture's RGB channels, and Alpha Gain and Offset control the levels of the file texture's alpha channel.

# Placement Nodes

A *placement node* is a type of utility node. These nodes are automatically created with and connected to a texture node. They can also be created through the Create Utility menu. Placement nodes control how a texture is mapped to a surface. Two basic kinds of placement nodes can be used: 2D and 3D. To understand the difference between these nodes and how to use them, we first need to talk about how a texture is mapped to a surface.

### Implicit UVs and Projection Nodes

Maya uses one of two methods of mapping textures: normal and as a projection. The normal method uses the existing UV coordinates in a surface (known as *implicit UVs*) to decide how the texture is mapped across it. The projection method uses a primitive shape—such as a plane, sphere, cube, or cylinder—and projects the texture from that shape onto a surface. Whenever you create a texture node from the Create Render menu, you specify whether it will be mapped normally (as in implicitly) or as a projection. Normal and projection modes can be used on both NURBS and polygonal objects. However, while NURBS surfaces already have UVs due to their nature, polygonal objects might require a fair bit of work to assign them properly.

Recall the discussion from Chapter 5 about surface parameterization. Any point on a NURBS surface can be described by a parameter with values for U and V between 0 and 1. The surface origin is at 0,0 and the opposite corner could have a parameter value of 1,1; 2,2; 5,5; or 6.3,8.9. A 2D texture has a coordinate system as well. Think of a texture, a bitmapped file created in Photoshop, laid out on a grid. Texture coordinates have U and V coordinates from 0

to 1, with 0,0 at the origin (bottom-left corner) and 1,1 at the end (upper-right corner). When the texture is applied to a surface, Maya internally parameterizes the surface from 0 to 1 if it isn't already done, and then it can easily match the corresponding coordinate from the texture. No matter how curvy or twisted the surface, the texture borders of the texture map will match all edges of the surface, as shown in the NURBS surface here. The 2D texture node can then be used to manipulate the position of this texture on the surface, control how much of the surface it will cover, or control how many times it will repeat.

A polygonal object, however, does not have these implicit UVs. UV sets can be created and assigned to any polygonal object by using one of the texturing tools found by choosing Edit Polygons | Texture. With these texture tools, the UVs can be assigned planarly, cylindrically, or spherically. Individual faces, or groups of faces, can be selected and assigned with a different type. If the surface is irregular or has areas obstructed by other polygonal faces, such as a humanoid face, the UV texture tools can be used to unwrap the UVs and lay them out flat so the UVs can be moved around to coincide with specific points on the texture. Once an object has a UV set assigned to it, a texture can be mapped to it in normal mode.

Whenever a texture is projected, Maya creates a *projection node* (a type of utility node) to specify how this texture will be projected. The texture's output is fed into the projection node, and then the projection node is fed into the desired attribute of the material. This projection node is capable of projecting that texture in a number of different ways: planarly, cubically, cylindrically, or spherically. These types of projections can all be set in the projection node's Attribute Editor by changing the Proj Type attribute.

The projection node actually maps the texture onto an imaginary surface with implicit UVs and then projects it onto the object from that position, regardless of the surface's UVs. The result is an even distribution of the texture on the faces or parameters that are perpendicular to the projection plane. However, as the angle of the projection to surface increases, the texture begins to stretch, as shown here:

The 3D placement node determines the position of this projection node in world space. The exact position of the texture map on the surface can be edited by grabbing this placement node, moving it, and rotating it in the view window to achieve the desired placement. The Scale tool can be used to change the tiling.

Projections are most useful for mapping a single texture across multiple surfaces. (A great example of this can be found in Chapter 27. It could also be useful to apply textures to the multipatch NURBS car model that we built in Chapter 6.) Another good use for projections is when you want to map a texture to a surface so that it specifically does not use the surface's implicit UVs. Sometimes, the parameterization or the layout of the UVs flows in an undesirable direction for texture mapping, or it is uneven. This illustration shows two identical NURBS surfaces with checker textures applied to them. The poles of these surfaces (the point where all of the isoparms intersect) are at their centers. The surface on the left has the map applied in normal mode so that it follows the surface's parameterization. The surface on the right has the texture projected from a planar projection type.

## 2D Placement Node

Whenever you are dealing with a 2D texture—whether it's a procedural or a bitmap image—you will use at least the 2D placement node. This node controls how the texture map is placed within the UV space from 0 to 1. Let's take a look at some of the attributes of the 2D placement node and the ways they affect how the texture is mapped. The first three attributes listed in the

Attribute Editor deal with how the texture frame is laid out. The rest of the attributes deal with how the texture behaves within that frame.

FIGURE 17-7    *Interactively editing the Coverage attribute with the Texture Placement tool*

- **Coverage**    Controls the amount of surface the texture covers. A value of 0.5 for the U value covers 50 percent of the surface in the U direction. This value can be interactively changed in the view window by clicking the Interactive Placement button in the 2D placement node's Attribute Editor and MMB-dragging the edge of the 2D placement frame on the surface in the view window (see Figure 17-7).

- **Translate Frame**    Sets the position of the covered area. This value can be interactively edited in the view window using the placement tool: MMB-drag in the middle of the frame and position the texture in the desired position.

- **Rotate Frame**    Orients the covered area. To rotate the frame interactively with the placement tool, MMB-drag one of the corners of the frame.

- **Mirror U and V**    Mirrors the tiles next to one another to hide possible tiling artifacts, such as seams. This attribute requires that the Repeat values be set to more than 1. Figure 17-8 shows an example.

FIGURE 17-8    *Two surfaces with the same texture repeated, but the texture on the right has Mirroring turned on*

- **Wrap U and V**    Sets whether a map is repeated in U and V over the entire surface.

- **RepeatUV**    Specifies how many copies of the texture map are mapped within the coverage area along either the U or V direction. This is sometimes called *tiling*.

- **Offset**    Offsets the pattern of the texture map.

- **RotateUV**    Rotates the texture itself. Realize that this differs from Rotate Frame in that the actual transforms of the frame remain. It is just the texture map within that frame that rotates.

- **Noise UV**    Randomly displaces the colors of the texture. This can be helpful in breaking up recognizable repetitions or patterns.

*TIP    All of the attributes in this second set can also be edited interactively with the Interactive Placement tool. When you select this tool, it is set to label mapping mode by default. This mode lets you edit the coverage/frame attributes as we did previously. If you double-click the Interactive Placement tool icon in the toolbar and change its setting to Surface Placement, you can edit the other tile-based attributes. MMB-drag the UV frame and place it on the surface. Select the edges and MMB-drag to change the coverage. If a point on the corner is selected and MMB-dragged, the rotation of the frame can be edited.*

### 3D Textures and 3D Placement Nodes

As mentioned, the 3D placement node is used mainly to position a texture or projection node in 3D space. While you have already seen how it is used to position a projection node that is projecting a 2D texture, you've not yet seen it used for a 3D texture. A 3D texture is a special kind of procedural texture map. Instead of being a flat surface, the 3D texture has depth to it. This means that you could map a 3D texture to a sphere and not worry about it stretching anywhere. All points in the surface in 3D space are evenly mapped. The place3dTexture node positions the 3D texture. This illustration shows a 3D marble texture applied to a head. The box around the head is the placement node used to position the texture.

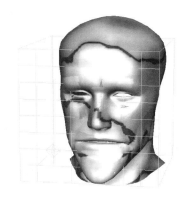

The 3D placement node is really not much more than a basic transform node. In this situation, the placement node is positioned in world space. The transform tools are used to move, rotate, and scale the texture. Since the placement of a 3D texture is based on the place3dTexture node's world space coordinates, realize that any object with a projection or texture that is mapped with a 3D placement node will "move through" the 3D texture space when the object is transformed in any way. One way to avoid this would be to group the place3dTexture node to the object to which it is being mapped. However, while this will work fine for rigid objects such as spaceships or cars, you'll have problems if the mapped geometry deforms. Therefore, when using 3D textures that need to stay put, it is a good idea to "bake" them onto a material. This can be done by selecting the material node and the surface it is mapped onto and choosing Edit | Convert to File Texture in the Hypershade. This will save a texture map corresponding to how the material is mapped to the surface and will create a duplicate of the material with the texture nodes mapped as normal textures.

## Summary

By now, you should have a pretty good idea of how to create and edit materials and be somewhat comfortable with using the Hypershade to make connections between nodes. We will practice using all of this information in the tutorials in the next chapter.

# Texturing in Practice

**To gain a full understanding of all the**

nodes you learned about in the last chapter, it is

best to practice the process of creating materials,

editing their shading attributes, and applying the

materials to objects in the scene. In this chapter,

we set up and work through two different

projects. The first project focuses on building

materials for different elements that can be

applied to the church that was modeled in

Chapter 3. The second tutorial is geared toward

laying out UV coordinates and building materials for more organic models, such as the head that we built in Chapter 4.

Before you go on, make sure that you have carefully studied the information in Chapter 17. You should have a good understanding of how to make connections in both the Hypershade and the Attribute Editor using the techniques demonstrated in that chapter. The instructions in this chapter assume that you know what to do when we say "connect a 2D placement node to the texture," as steps such as these will not be defined in great detail.

# Tutorial: Building Basic Shading Networks

In this first exercise we will build shading networks for two different types of material: stone and metal. By understanding how these shading networks are managed, you can create shading networks for any material imaginable. Here, several attributes on each material will be connected to texture maps and, in some cases, other utility nodes to enhance their appearance. Each material will then be applied to an object in the scene and adjusted so that the textures properly fit and tile correctly.

## Texture the Stone Wall Material

We will apply a stone wall texture to the main geometry that makes up the church model. We will build the shading network starting from a Lambert material and then connect textures to control its Color and Bump attributes. Once we set up the material, we will apply it to the geometry and adjust it so that it fits properly.

Let's get started.

### Create the Material
Setting up a shading network begins with creating a material node that will hold all of the shading and texture information. As we begin to build this shading network, it is crucial that we keep our scene organized by naming all of the nodes as we create them.

1. Open the file called mcr8_ch18_church_start.ma from the DVD. Set the window layout to use the Hypershade/Perspective layout. It is strongly recommended that you copy the entire project folder over and set your current project to this directory.

2. In the Hypershade window, create a material for stone walls. Use a Lambert material. Select the new material and rename it **mStone** (the *m* before the name is the naming convention we'll use for all material nodes) in the Attribute Editor.

3. In the Attribute Editor, click the checkered button in the color channel of the Attribute Editor to open the Create Render Node window. Because the church does not yet have

any texture coordinates (UVs) set up, we will apply the
textures as a projection. Choose the As Projection
option, shown here, and then click the File button.
This creates a file texture node and connects it to
the Color attribute. Since we chose As Projection, a
projection and 3D placement node were also created.

4. In the Hypershade window, select the mStone material
and click the Input Connections button in the toolbar.
This displays the input connections to the material and
cleans up the organization of the Workspace. Figure 18-1 shows the nodes and how
they are connected.

*FIGURE 18-1*   *The shading network after connecting the color texture*

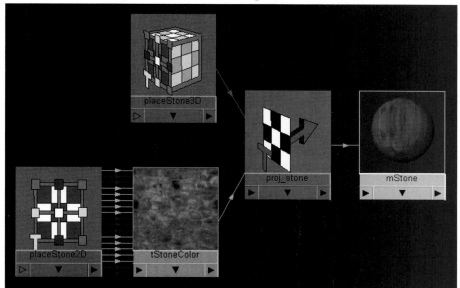

TIP  **You may also wish to hide the upper tabs section of the Hypershade
window so that you have more room to work with this material.**

5. Because we'll be creating quite a few nodes in this material, it is essential that you keep
the project organized by naming the nodes as you create them. Select the file texture
node and rename it **tStoneColor**. Then select the 2D placement node connected to
that, and name it **placeStone2D**. Name the 3D placement node **placeStone3D**.

**6.** Select the tStoneColor node to view its attributes in the Attribute Editor. Use the File Browser button in the Image attribute's field to browse and open the mcr8_stone_color.tiff image. The tStoneColor file texture now references this external TIFF file.

**7.** Use the same procedure demonstrated in Steps 3 through 6 to create and import a texture for the Bump attribute. This time, import the file called mcr8_stone_bump.tiff.

### Apply the Material to the Geometry

Now that the material has been created, we will apply it to the geometry. This involves making some adjustments to the nodes' attributes and connections.

**1.** In the view window, select the three main objects in the church model: main_base, RT_tower, and LT_tower.

**2.** In the Hypershade, RMB-click the mStone material and choose Apply Material to Selection from the resulting marking menu.

**3.** In the view window, press the 6 key to see the texture maps. You should notice that while the appearance has changed, the texture is difficult to recognize. To fix this, we will set the projection type and edit the placement nodes.

**4.** Select the proj_stone projection node and view its attributes in the Attribute Editor. The projection type is currently set to Planar, which means that the texture is projecting along one direction. While this may look correct for the front of the church, the textures on the sides will smear. To fix this, set the Proj Type to TriPlanar and then click the Fit To Bbox button, shown here. This changes the projection type and then moves and scales the projection to fit the selection. This placement information is stored in the placeStone3D node.

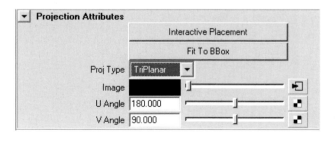

**5.** Do a test render to evaluate the placement of the current textures. While it is starting to look better, the placement of the bump maps is out of sync with the color map that we just added. To fix this, we'll connect the placement nodes that are currently connected to the color texture and projection and connect them to the nodes controlling the placement of the bump. In the Hypershade, locate the placeStone2D node and MMB-drag it on top of the tStoneBump node. When you let go of the MMB, a marking menu appears. Choose Default. This makes all of the connections that we need between the placeStone2D and the tStoneBump nodes.

**6.** Connecting the placeStone3D node to the proj_bump_stone node requires a little more work. In this case, choosing Default will not make any connections for us but will instead prompt us to make our own connections by opening up the Connection Editor. With the Connection Editor, you want to connect the World Inverse Matrix on the placeStone3D node to the Placement Matrix on the proj_bump_stone node, as shown here:

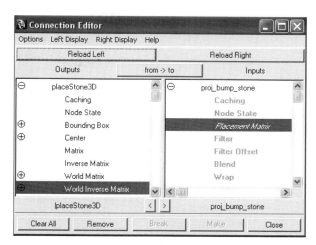

**7.** Now that these connections are made, you can delete the two placement nodes that are no longer connected to anything. Figure 18-2 shows the completed shading network for the stone material. Since the textures and projections use the same placement nodes, we only need to edit the placement in one location. This is just one of the big advantages offered by node-based shading networks.

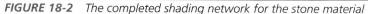

*FIGURE 18-2* *The completed shading network for the stone material*

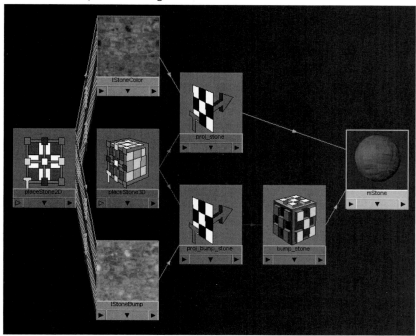

**8.** Set the proj_bump_stone Proj Type attribute to TriPlanar. Do a test render of the scene to check your placement. It should look something like the scene shown in Figure 18-3.

**FIGURE 18-3** *A test render of the scene*

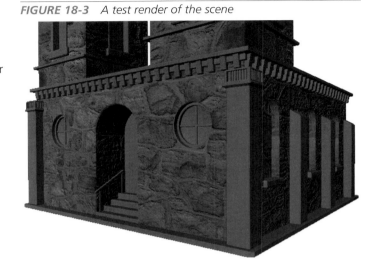

**9.** While the texture is projecting properly, the scale of the stones in the texture is too big. Texture maps should communicate a clear sense of scale to your geometry. With stones that large, the church would have to be the size of a toy playset. Since this is not our intention, we will increase the tiling. Select the placeStone2D node and set the RepeatUV attributes to **5** and **5**. Do a test render. It should look like the image shown in Figure 18-4. Compare this to the image shown in Figure 18-3 and notice how the texture looks like it meets the proper scale that we need for a large building.

**FIGURE 18-4** *The scene is re-rendered after editing the RepeatUV attribute.*

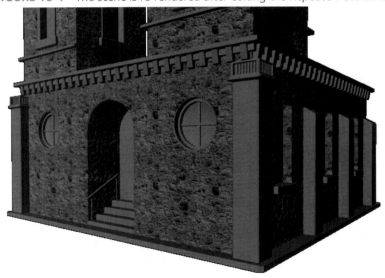

*TIP  Sometimes the texture maps appear blurry in hardware shaded mode. You can increase the quality by selecting the material and editing its Texture Resolution attribute in the material's Attribute Editor under the Hardware Texturing. Use this pull-down menu to change the display resolution from Default to Highest [256×256].*

### Apply a Layered Texture

In many situations, you'll want to use multiple texture maps to control a single material attribute. This not only increases the detail in a texture, but can also aid in hiding any noticeable tiling artifacts caused by a repeating texture.

A *layered* texture node is a type of rendering node that allows multiple texture files to be combined together in a singe node. This is similar to compositing an image using Photoshop layers. The textures can be arranged in any order, their opacity edited, and the blend type chosen to control how each layer will affect the layer underneath. Masks can also be applied to any layer by connecting other textures to the alpha channel of the layer you wish to mask.

**1.** Right-click in the Workspace area and choose Create | Layered Texture from the marking menu. The new node appears in the Workspace. Name it **tStoneColor**. Notice that the node is green. This is because the layered texture has a default green layer already in it.

**2.** Create another file texture from the Create Bar in the Hypershade. Make sure that the Create options are set to Normal, instead of As Projection. (Because we will use the existing projection node, proj_stone, to project the tStoneColor node through, there is no need to create an additional projection node). Name this texture to **tGrunge**.

**3.** Reroute the connection into the proj_ stone node through the layered texture node. MMB-drag the tStoneColor node onto the proj_stone node and choose Image from the marking menu. This replaces the previous connection to the tStoneColor file texture node.

**4.** Route both color textures into the tStoneColor node. Select the tStoneColor node and look in the Attribute Editor. You should see a big, white space with a green block inside it. This is where you can arrange your textures into layers. The leftmost texture will be the one on top. Figure 18-5 shows the Attribute Editor with the tStoneColor node loaded.

**FIGURE 18-5** *The layered texture, named tStoneColor, is loaded into the Attribute Editor and has two layers loaded.*

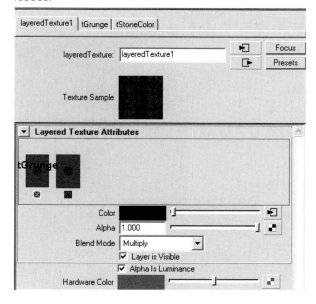

**5.** MMB-drag the tStoneColor texture from the Workspace window into this white space in the Attribute Editor, shown in Figure 18-5. Do the same thing with the tGrunge node.

**6.** You should now have three blocks in the Layer Editor: two textures and the default green layer. Click the check box under the green block to delete that layer. Arrange the remaining two layers so that the grid texture is on top or to the left of the checker texture in the Attribute Editor. When you are finished with these steps, the layered texture's Attribute Editor should match Figure 18-5.

**7.** Set the Blend Mode to Multiply. This multiplies the pixel values between the two layers and, hence, darkens the pixels.

**8.** Select the placement node that is connected to the tGrunge file texture. Name this **place2DGrunge**.

**9.** Select the place2DGrunge file texture and set the RepeatUV attributes to **2** and **2**. Since 2 is not a multiple of 5 (the Repeat Value used on the place2DStone texture), the tGrunge texture will not tile at the same frequency as the tStoneColor texture, thus breaking up any noticeable tiles.

**10.** Do a test render to see the results. This render is shown in Figure 18-6. Compare this to the render shown in Figure 18-4. Notice how the two textures repeating at different rates create unique pixel values across the surfaces.

**FIGURE 18-6**   *The church is rendered with a layered texture as its color.*

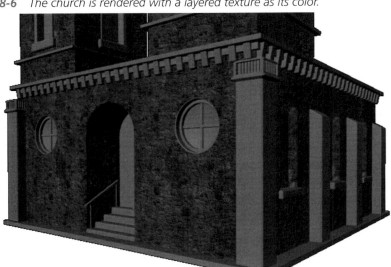

This completes the setup for the stone material. However, you can continue to add more layers if you'd like. This is an excellent way to build up very complex patterns without having to use an external imaging application. Next, we'll look at how to deal with metallic surfaces.

> _TIP_  *Layering textures can increase rendering time due to the fact that more files need to be loaded into RAM. You can flatten or "bake" a layered texture into one file by selecting it and choosing Edit | Convert to File Texture (Maya Software) from the menu in the Hypershade.*

## Create the Bronze Material

Bronze is a type of metal. Some of its general characteristics are that it is somewhat shiny (that is, specular and reflective) and has a medium diffuse (dull) color. When aged, this bronze color turns greenish. In this section we will create a weathered bronze material that will be applied to the spires of the church. Although we are focusing on bronze in this section, similar metallic materials can be created from very similar settings.

### Create the Material

When trying to emulate the characteristics of a metallic surface, it is usually best to start with a Blinn material. So, let's begin with that:

1. In the Hypershade window, create a new Blinn material. Name it **mBronze**.

2. Set the Color attribute to something approximating a pale green. The settings used for this example are H = 93, S = 22, and V = 8.

3. While metal is usually not very porous, this material does have some wear, resulting in the decay or minor rust of the surface. This wearing will have an effect on the specular attributes. Set the Eccentricity to 0.5 and the Specular Roll Off to 0.7.

4. MMB-drag the material onto one of the spires. It is important to only use one of the spires for now, as any more than one would slow the renderer down.

The remainder of the relevant attributes will be controlled by a texture map. While we are suggesting values for many of these settings, it is important to keep in mind that these are only suggestions. Feel free to experiment with other values. Depending on the lighting conditions in the scene, these settings will always need to be tweaked to get the desired effect.

## Create the Specular, Diffuse, and Reflectivity Maps

We need something to break up the diffuse, specular highlights, and the environment's reflections over the entire surface. All of these attributes relate to each other. With a few tricks, we can optimize this shading network and use only one file texture to control all three characteristics. We'll begin by attenuating the specularity:

1. Select the mBronze material in the Hypershade. Click the checkered box in the right column of the Specular Color attribute to open the Create Render Node window. Make sure the As Projection radio button is enabled and create a file texture node. Name this node, **tMetalGrunge**. Name the projection node connected to the tMetalGrunge node **proj_metalGrunge**.

2. Select the proj_metalGrunge node and set its Proj Type attribute to Cylindrical. Click the Fit To Bbox button to place it. Select the placement node and set the Repeat UV values to **4**.

3. MMB-drag the tMetalGrunge texture on top of the mBronze material. Choose Other from the resulting marking menu to open the Connection Editor.

4. We need to connect the output of the tMetalGrunge texture to the Diffuse and Reflectivity attributes of the mBronze material. However, we must realize that both of these attributes only accept a single value as their input, as opposed to Color, which is three values, R, G, and B. Therefore, we will use the alpha output from the file texture node and connect that to the material. In the Channel Box, find the Out Alpha attribute in the left column and then find the Reflectivity attribute in the right column and connect them.

5. For the Diffuse attribute, we need to add another node in the chain of connections. Generally, when an object is more reflective, it is less diffuse and vice versa. In the example of a mirror, the mirror has little to no diffuse, as its color is predominantly a reflection of the environment around it. With this inverted relationship, we need to add a Reverse node between the proj_grunge node's Out Alpha attribute and the mBronze material's Diffuse attribute. Use the Create menu in the Hypershade to create a Reverse node. This is located in the General Utilities folder.

6. Use the Connection Editor to connect the Out Alpha attribute from the proj_grunge node to the Input X attribute of the Reverse node. Then connect the Output X attribute of the Reverse node to the Diffuse attribute of the mBronze material.

## Create a Reflection Map

The last connection we'll make to this material will be to the Reflection Color attribute. The Reflection Color attribute controls what the surface will reflect with the use of a texture map. This helps us to evaluate the reflectivity of the surface without having to wait for a raytrace calculation. Any image can be used as a reflection map, but ideally you want to use some

360-degree panorama. This way, you can move your object through the scene and have it reflect back an entire environment. If you were to use a regular photograph, you might notice the seams where the environment textures wrap around.

A variety of special cameras and software let you create perfect 360-degree environment maps. dvGarage, a company dedicated to providing tools and training for digital artists, offers a product called the Reflection Toolkit (www.dvgarage.com), which contains a collection of 30 reflection maps that have all been shot and corrected for seamless projections. We have included on the DVD one of these images as a sample, which is shown in Figure 18-7.

***FIGURE 18-7*** *A perfectly seamless spherical environment texture map (image courtesy of dvGarage)*

Even though your actual environment might not look anything like the one photographed in the reflection, any reflection map that is seamless will work for most situations. It really helps to integrate all of the objects in a scene when they are all using the same reflection map. For objects that are close together and need more accurate reflections, raytracing can be added. You can make additional render passes with raytracing and composite these with the reflection pass in postproduction. We will look at ways to handle this in Chapters 27 and 28. For now, let's take a look at the effect that adding a reflection map has on the scene:

**1.** Use the Attribute Editor to connect a texture to the Reflected Color channel of the mBronze material. When the Create Render Node window appears, scroll down and reveal the Environment Textures section. Choose Environment Sphere from the list. This will add the envSphere1 node to the scene.

**2.** Connect a file texture to the Image attribute of the envSphere1 node and name this node **envReflect**.

**3.** With the envReflect node selected, click the File Browser button next to the Image Name attribute. This brings up the file browser, where you can navigate to the location of the mcr8_reflection_outside.tiff file and open it. This will load into the envReflect node.

**4.** Do another test render to make sure all of the attributes are being attenuated. While the connections are fine, we need to fine-tune some of the attributes.

## Use IPR to Fine-Tune the Material Attributes

The texture maps we've added so far control the basic attributes for this material. From here on, we will be making some adjustments to these textures to fine-tune the rendered image. In this process, you can determine whether the material needs to be more or less reflective, or whether the bump map shows up fine or is too deep. To adjust the strength of the texture maps, we will be adjusting the texture's Color Balance attributes. These essentially control the color levels and contrast of the bitmap images within Maya. If the reflection is too bright on the surface, you know that the file texture is too light and needs to be darkened. This tweaking process is best handled with interactive photorealistic rendering (IPR).

**1.** Frame the spire in the view window and click the IPR button in the Render view's toolbar. Drag a box to define the region that you want to see updated with the changes you make. It is best to find an area of the surface that contains sharp specular highlights. as shown here:

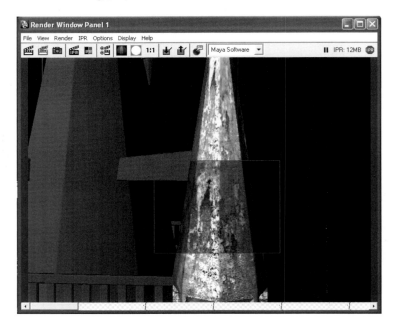

**2.** In the Hypershade, select the tMetalGrunge node and view its attributes in the Attribute Editor. First, we'll adjust the specularity by adjusting the Color Gain attribute. Scroll down through the attributes on the tMetalGrunge node and reveal the Color Balance section. The Color Gain and Color Offset attributes can be used to control the outgoing connection, which is the Specular Color. Set the Color Gain to about **0.7** and leave the Color Offset to 0.

**3.** To adjust the reflectivity and diffuse of the object, select the tMetalGrunge node in the Hypershade. Reflectivity attributes use the alpha value of the texture map to control reflectivity. This is convenient because the reflectivity maps are often the same as the specular maps. However, you usually want to be able to control the reflectivity and specular maps separately. This means that you can use the Alpha Gain and Alpha Offset attributes to control the levels of the texture, and control the overall reflectivity as a result. Use the Alpha Gain attribute to turn the reflection down with a value of about **0.6**. Turn the Alpha Offset attribute up to about **0.2**.

# Tutorial: UV Texture Mapping

So far, all the textures that we have applied to geometry have used the geometry's existing UV coordinate or a projection node for the textures' placement. In this section, you will learn how to assign UV coordinates to a polygonal mesh. A NURBS surface, by its nature, already contains this information. Polygonal geometry, on the other hand, does not contain any UV information. How, then, would we map a texture to a polygonal surface?

One solution is to use a projection node, as we have been doing so far in this chapter. Projections are a quick way to apply a texture to any surface that does not have any UV coordinates assigned—or, in some cases, where the existing UV information is undesirable for texture mapping. A problem with projections is that severe stretching or warping can occur when the surface being mapped has complex curvature or overlapping regions. Also, problems can occur with projections on an animated or deforming mesh. The workarounds to address these issues can result in lengthy render times. In these situations, it might be best to assign UV coordinates to the surface and then apply the texture normally.

UVs can be assigned to any polygonal surface by using one of the four methods available by choosing Create UVs | Planar Mapping, Cylindrical Mapping, Spherical Mapping, and Automatic Mapping. Once the UVs are assigned, they can be manually edited in the UV Texture Editor (Window | UV Texture Editor). The goal is to have the UVs evenly laid out, without any overlapping pieces, to fit within the UV coordinates of 0 and 1.

The process begins by a careful evaluation of the geometry. Does the shape of the model conform to any primitive shapes available in the Create UVs menu options? Can the model be broken down into separate segments that can then be individually assigned UV coordinates? What about this creature's head? How will you overcome all of these overlapping regions from the nose? Does the mesh have to be textured from a single map, or can you use multiple maps for different parts of the body? How much time do you have? These are the types of questions you must ask yourself and the problems you need to solve.

## Create UVs

This tutorial takes you through the steps of creating UVs to a polygonal model of a human head. However, you can use this same procedure to map any object.

Open the mcr8_ch18_charGeo_start.ma file on the DVD. This model does not have any UVs assigned to it. If you were to create a material with a texture map, it could not be mapped on the surface and therefore would not show up when the surface is rendered. Analyze the model and try to determine how it might be divided up into separate sections, or *shells*. In this tutorial, we will concentrate on the head and try to find a UV mapping method that can be used. Cylindrical mapping works well for objects such as this, so that is what we will use to assign UVs to it. To evaluate the evenness of the UV layout, we will use a special texture map designed just for this task.

### Create the UV Evaluation Material

First, we'll create a Lambert material with a file texture map that will be used to evaluate our UV layout as we work:

1. In the Hypershade, create a new Lambert material and name it **mUVeval**.

2. Connect a file texture node to its Color attribute. Make sure that you choose Normal instead of As Projection when you create this node, because we want to use the UVs of the mesh, not a projected texture.

3. On the file texture node, load the mcr8_ch18_landisUVmap.jpg file into the Image attribute.

4. Apply the material to the body_lowRez geometry. It won't look like anything just yet because we don't yet have any UV coordinates to map it with.

### Assign UVs to the Head

Here we will concentrate on evenly laying out UVs for the head of the character. Later, we will decide exactly how they will reside in the UV texture space with the other geometry.

1. Select the faces in the head and neck. Deselect the faces that make up the inner cavity of the mouth—these will only get in the way at this stage in the process. Also, it isn't possible to map the very top of the head with the cylindrical projection oriented around the face, so deselect the 18 faces on the top of the head.

2. Choose Create UVs | Cylindrical Mapping.

3. The default settings for the Cylindrical Mapping command show the Projection Horizontal Sweep attribute set to 180. This means that the cylinder will wrap around the head only 180 degrees, and therefore the UVs that are not within that cylinder will fall outside the 0 to 1 range. The Projection Horizontal Sweep attribute can be interactively edited in the view window by clicking and dragging the red square on the edge of the cylinder. Drag it around until the cylinder wraps a complete 360 degrees around the head, as shown here:

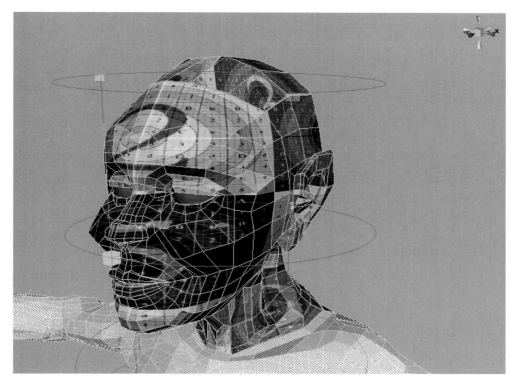

If a texture map were assigned to the model at this point, it would appear to wrap around the head as if it were being cylindrically projected. However, the top of the head and the area around the nose would display texture stretching. To fix this, we will manually edit the UVs in the UV Texture Editor.

## Edit UVs in the UV Texture Editor

Figure 18-8 shows the UV Texture Editor displaying the UVs from the head model. This is where all of the UV editing will be done. Look at the grid in the UV Texture Editor, and you will notice that it is marked to show the coordinates in the U and V directions, both positive and negative. The texture map will occupy the region that falls within positive 0 to 1 space. Any UVs outside this area will either not be covered by the texture or will get the repeated texture (if the Repeat attributes are enabled in the Place 2D texture node). Therefore, it is desirable that all UVs are assigned to this region.

**FIGURE 18-8**  *The UV Texture Editor displaying the UVs that were assigned to the selected faces*

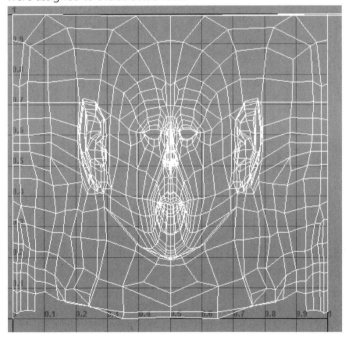

By setting the cylindrical mapping's Projection Horizontal Sweep attribute to 360 degrees, Maya was able to keep most of the UVs within the 0 to 1 space. What we need to do now is look for areas that are overlapping. The chin is a good example. If you look at this region in the UV Texture Editor, you will notice that some of the UVs are overlapping the UVs in the neck. This means that if you want to texture the chin so that it has a pimple, for example, that same pimple will appear on the neck. Let's fix this.

### Relax UVs

As you begin editing the UVs from their initial assignment, you will find that there are many areas where the UVs overlap. The ears, nose, and under the chin are some of the most common areas that need fixing when using the cylindrical technique shown earlier. The Polygons | Relax command helps unfold overlapping UVs. Once the UVs are unfolded, you can more easily edit them manually. In this section, we will use the Relax UVs command to fix the overlapping UVs in the chin area.

1. Select the head in the view window and choose Window | UV Texture Editor. The UV Texture Editor opens and shows the current layout of UVs for the head model.

2. In the UV Texture Editor, right-click the UVs and choose UVs from the marking menu. Drag-select the chin area with overlapping UVs. The selected UVs will be highlighted as shown in the illustration.

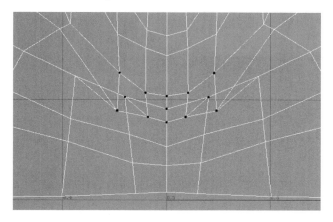

*NOTE  Once the UVs have been selected, you can use the transform tools to move, rotate, or scale them in the UV Texture Editor. Realize that while the UVs will be highlighted in the view window, they cannot be transformed there. That is because the view windows let us work in 3D space while the UV Texture Editor considers only two dimensions (U and V).*

3. An easy way to fix these UVs is to use the Relax UVs command. In the UV Texture Editor, choose Polygons | Relax UVs □. We can leave all of the options set at their defaults except for the Pin Unselected UVs option. Make sure this option is enabled. This will hold all of the unselected UVs so that they are not affected by the Relax UVs command. Click the Relax button. Notice that the UVs in the chin are no longer overlapping.

4. You can continue selecting UVs and using the Relax UVs command for other areas that have overlapping UVs—such as the nose and the eye areas. If the Relax UVs command is unable to fix the problems, select individual UVs and use the Move tool to lay them out manually.

5. Once the overlapping UVs have been taken care of, the UVs will need some manual work to make sure the texture is fitting evenly onto the mesh. This is done by selecting the UV in either the view window or the UV Texture Editor and moving the UVs in the UV Texture Editor while watching the model in the view window. Figure 18-9 shows the UVs after some relaxing and some manual editing.

*FIGURE 18-9*    *The UVs after relaxing the overlapping areas*

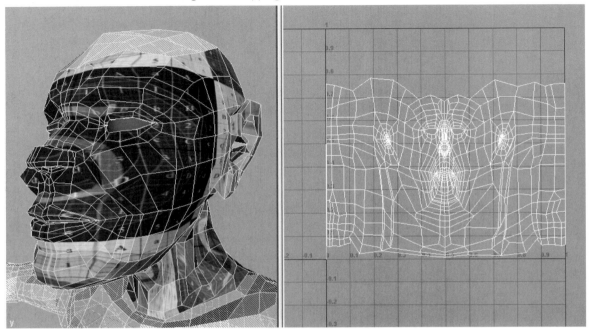

### Move and Sew UVs

Earlier we talked about multiple UV shells. A UV shell is a section of the geometry where all of the UVs are merged together. Sometimes, you will want to keep different parts of UV layout in separate shells. The head, arms, and torso are examples of parts that are better left in separate shells. Other times, having too many small shells makes it difficult to texture later on. In this section, we will assign UVs to the top of the head and use the Move and Sew UVs command to merge the two shells together:

**1.** Select the faces on the top of the head. These should be the faces that currently do not have any UVs.

**2.** Choose Create UVs | Planar. This applies a planar projection to the faces.

**3.** Use the manipulator to rotate and scale the projection so that the faces get the cleanest projection with the least distortion, and scale the manipulator so that the texture pattern is square.

**4.** In the UV Texture Editor, move the entire shell above the face layout so that it is out of the way.

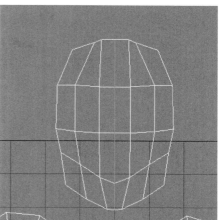

**5.** We will cut this shell in half. Select the edges along the middle of the shell, shown here, and choose Polygons | Cut UV Edges.

**6.** Now we have two shells for the top of the head. You can easily select a shell by selecting one of the UVs on the shell that you wish to select, RMB-click, and choose To Shell from the marking menu.

**7.** Figure out where this piece should be integrated back into the face shell. With one of the shells selected, scale it so that the edges in the shell match the scale of the edges in the head shell.

**8.** Select the edges that you would like to merge. Notice that for each edge that you select on one shell, its corresponding edge highlights on the other shell, as shown below.

**9.** Choose Polygons | Move and Sew UV Edges. The two UV shells merge into one.

### Create a UV Template

Once the UVs have been assigned and laid out so that no pieces overlap, you can move on to texture mapping the object. This might include importing premade texture maps or using Maya's 3D Paint tool, discussed in Chapter 19. One of the best ways to create an accurate texture for this object is to use the UV layout as a template that you can paint on in a paint package such as Adobe Photoshop. A great new feature in Maya lets you create a layered Photoshop document (PSD) from right within the Maya UI. The layers in this PSD file can be connected to any material attribute desired, plus, you have the option of including the UV layout in a layer, so you know exactly what part of the geometry you are painting on in Photoshop. Using the PSD file for your textures can also help you better organize your files since you will not need a separate file for each attribute.

Let's create the Photoshop document and connect it to a material that is assigned to the model:

**1.** Right-click the head geometry and choose Material | Assign New Material | Blinn.

*FIGURE 18-10*   *Create PSD Network Options window*

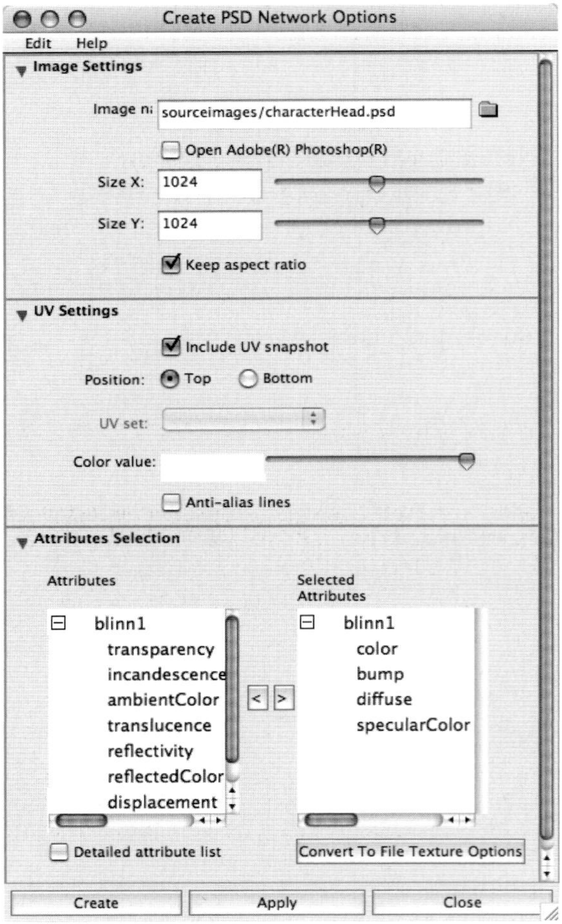

**2.** Select the head geometry in the view window, and then choose Texturing | Create PSD File. This opens the Create PSD Network Options window, shown in Figure 18-10. Here we can create and connect a unique layer from the Photoshop document to a specific attribute on the current material.

**3.** In the Image Name field, set the path and filename for the new PSD file that is about to be created.

**4.** Set the resolution of the file using the Size X and Size Y attributes. It isn't so important that you set the correct resolution here, as the PSD file can easily be resized in Photoshop before you begin painting. However, the texture should be created large enough with enough resolution to draw the UV template.

**5.** In this window, we need to make a selection of attributes that we want to create and to which a layer will be connected. Select the Color attribute from the Attributes list and click the right arrow (>) button to move the attribute into the list of Selected Attributes. Do this for the Bump, Diffuse, and specularColor attributes as well.

**6.** Click the Create button. A Photoshop file will be created and written to the location specified in the Image Name field. This PSD file will contain five layers—four layers for the attributes and one layer that contains the UV snapshot.

**7.** In the Hypershade, graph the input connections to the Blinn material and check that four file textures are connected to the material, all mapped to the appropriate channels.

**8.** Now you can open the PSD file in Photoshop. You will see that a separate layer has been created for each material attribute specified in the Create PSD Network Options window.

**9.** With everything now set up, you can begin painting in the different layers. As soon as you save the PSD file, you can immediately see the results in Maya by selecting the PSD file texture node in the Hypershade and clicking the Reload File Texture button in the Attribute Editor.

### Transfer the UVs onto the Bound Mesh

The geometry that we have been using in this chapter has not been bound or rigged. Chances are, we want to take the work we've done here and transfer it over to our rigged character. Do so as follows:

**1.** Select the UV'd mesh and delete its history. Save this file.

**2.** Open the rig you completed in Chapter 12 and Import (File | Import) the UV'd version of the model.

**3.** Select the UV'd model and then select the rigged model. Choose Mesh | Copy Mesh Attributes.

**4.** From the Copy Mesh Attributes Options window, shown here, enable the UV Sets option and click Copy. The UV information will be transferred onto the bound mesh. You can now apply the material that is on the UV mesh to the bound mesh and it will fit exactly the same.

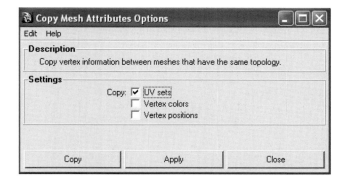

## Summary

In this chapter, we experimented with several different material attributes and learned how the textures can be accurately mapped onto a surface. While the tutorials had different focuses, the lessons learned can be combined to create high-quality rendered images. Remember, though, the real ability to create good surfaces comes from careful study in the real world.

In the next chapter, we will look at the Paint Effects toolset in Maya and discuss ways to paint directly on objects in the view window.

# Painting in Maya

**19**

The paint tools in Maya offer some of the most unique and innovative features and solutions in the program. You can use these tools to paint onto a standard 2D canvas, paint directly onto 3D models, or even create 3D objects such as plants and hair with the stroke of a brush. This chapter will show you how to use the paint tools for all of these applications.

Maya's paint tools are divided into three main categories: the Artisan tools, the Paint Effects tools, and the 3D Paint tools. We have already used the Artisan tools, such as the Sculpt Surfaces tool and the Paint Skin Weights tool, to manipulate geometry and manage our smooth skinning data. In this chapter we will focus on the Paint Effects tools and the 3D Paint tools.

# Paint Effects

You can use Paint Effects to create entire objects just by dragging on a canvas or in one of the view windows. Paint Effects is more like a particle-based technology than a geometry- or pixel-based one. This means that instead of being made up of polygons or pixels, objects created with Paint Effects are made up of tiny particle-like matter. Maya 8 ships with hundreds of Paint Effects brushes that let you paint all kinds of objects. Brush strokes can be as simple as a pastel or oil paint, or as complex as trees, flowers, or rain. Once they are created, objects made with Paint Effects can be animated to appear to grow over time or be affected by forces such as the wind.

## Strokes, Brushes, and Tubes

Paint Effects works by applying paint, or a pattern that is defined by a brush, to a stroke that is created when you click and drag your cursor in a canvas or view window with the Paint Effects tool (Paint Effects | Paint Effects Tool). A *stroke* is a curve attached to a hidden NURBS curve that instructs brushes on how the paint should be applied. In the case of an airbrush, for example, the paint is applied to the curve and rendered as a simple airbrush stroke. However, some of the brushes available will "grow" *tubes* as the stroke is created.

Take a look at Figure 19-1. The image on the left shows a stroke that was created with the birchMedium.mel brush. The tubes—branches and leaves—are extending from the stroke. The image on the right shows what the stroke looks like when rendered.

Tubes are used to simulate organic growth or branching. As you drag the cursor in a canvas or view window, tubes will extend from the curve. The tubes are created by Maya sampling the curve made by the stroke. If the points on the curve are far apart, meaning that the stroke was made quickly, fewer tubes are created. If a tree brush was selected, for example, the resulting branch may not have as many little branches or leaves as a branch that was created from a stroke that was denser or drawn over a longer period of time. At some point, the maximum number of tubes for each brush will be reached and the object will stop growing. In Paint Effects terms, this object will have reached its *lifespan*.

You can select a brush from the brush library in the Paint Effects tab of the Visor window. Open the Visor window either by choosing Window | General Editors | Visor or by choosing

**FIGURE 19-1**  *The view on the left shows a Paint Effects stroke and tubes; the image is rendered on the right.*

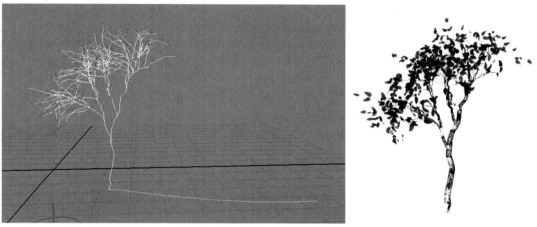

Paint Effects | Get Brush. The Visor window is shown in Figure 19-2 with the Paint Effects tab selected and displaying the available brushes. The Paint Effects brushes are organized into separate folders on the left side of the window. Clicking a folder will display its contents on the right. Clicking a brush will make the Paint Effects tool active with the selected brush loaded.

**FIGURE 19-2**  *The Paint Effects brush library shown in the Visor*

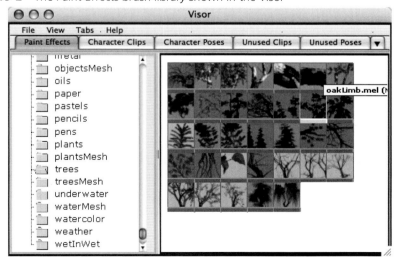

The brush that you select from the library is known as a *template brush*. The idea is that you use a brush's settings as a starting point to enhance and customize the look of the Paint Effects objects that are created. We will get into editing the template brushes in the tutorials in this chapter.

## Paint Effects Limitations

Because Paint Effects is such a unique technology, it does have some limitations when used with other parts of Maya. The most significant limitations have to do with rendering. Paint Effects is supported only in the Maya software renderer. Mental Ray will not render objects created in Paint Effects. Even when you're using the Maya software renderer, render times can be quite long—almost unmanageable—if the scene contains a lot of strokes that contain many segments. Be mindful of this as you use the Paint Effects tool. Using fewer strokes to cover an area will render much faster than using many strokes to cover the area.

Paint Effects does not work well with raytracing. Objects created with Paint Effects won't reflect or refract. (See Chapter 21 for more information on raytracing reflections and refractions.) To make the Paint Effects reflect and refract, you must convert the Paint Effects into a 2D image and apply it as a texture or choose Modify | Convert | Paint Effects to Polygons (more on this step later in the chapter). Also, since it does not work with raytracing, Paint Effects can't cast raytraced shadows; use Depth Map Shadows instead. You can also render Paint Effects objects separately and composite them with the rest of the image in a compositing application.

One of the most amazing aspects of Paint Effects is its ability to add animation. Turbulence can be added to simulate objects blowing in the wind. However, you cannot directly apply Maya's dynamic fields to the Paint Effects objects. All animation done to the Paint Effects object must be created by editing attributes on the Paint Effects' *tube node*. This may make it difficult to match any dynamic simulation on other objects in the scene that use fields or have the Paint Effects objects collide with other objects in the scene. While converting the Paint Effects objects to polygons and using soft and rigid body dynamics (explained in Chapter 24) is a workaround, you should be aware of these limitations if you plan to use Paint Effects in your productions.

## Tutorial: Creating Paint Effects on a 2D Canvas

While Maya is generally considered a 3D application, it can be used as a traditional 2D paint application to paint on a 2D canvas. However, the brushes available in the Paint Effects toolset let you go beyond just painting color, as you would with a brush in Photoshop or Painter. Simply by clicking and dragging across the canvas, you can use Paint Effects to draw tree branches, clouds, hair, or even flesh.

Using Paint Effects on a 2D canvas is great for painting texture maps or creating matte paintings that you will eventually reuse in your Maya scenes. However, it is also a fast way to spice up a web page or page layout for print design.

Let's get started by learning some of the basic interface options and terminology used in Paint Effects:

**1.** Create a new scene. Choose Panel | Paint Effects or press the 8 key in the Perspective view. This opens the Paint Effects window. The default for the window is the 2D mode panel. Within this panel, you can paint as you would with any 2D paint program. Figure 19-3 shows the Paint Effects window.

*FIGURE 19-3    The Paint Effects window*

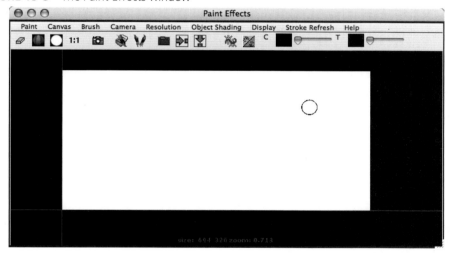

> **NOTE**    *While you may use the Paint Effects tool in 3D by painting in the view windows, you can also paint in the Paint Effects window by choosing Paint | Scene. This loads one of the existing camera views in the scene. You can then use the Camera menu to select a specific view. The advantage to painting in this window is that some of the brushes do not display details while working with them in the regular view windows. The Paint Effects window was designed to allow you to see the results of your brushes without having to do test renders.*

**2.** Load template brush settings by selecting a brush from the Visor. Open the Visor by choosing Window | General Editors | Visor, and click the Paint Effects tab.

**3.** When loaded, the left side of the Visor shows folders containing Paint Effects brushes. Select any brush and paint in the Paint Effects window to experiment. Some of the brushes, such as oil paint and pastels, behave like more common 2D paint brushes, while others, such as the flower and tree brushes, create branching structures.

**4.** Click the Clear Canvas/Delete All Strokes button on the toolbar. This erases whatever you have done so far. Now let's create a leaf texture to apply to a plane geometry.

**5.** Choose Canvas | Set Size. Set the X and Y sizes to **512** and click the Set Size button.

**6.** To make the texture tileable, you can use the Wrap Canvas functions by clicking the Wrap Canvas Horizontally and Wrap Canvas Vertically buttons. This makes the Paint Effects texture wrap around vertically and horizontally so seams will be invisible when the texture is repeated over a surface.

**7.** Open the Visor and select the mapleCluster.mel brush from the trees folder.

**8.** Paint two or three strokes from bottom to top in the Paint Effects window. LMB-click and drag only very little to make the stroke fill the top of the panel. You should see several limbs with a lot of leaf details. It may take several tries to get the look you want. The illustration shows the result of the strokes so far:

_NOTE_  **When you use the Paint Effects tools here in 2D, Maya immediately applies the paint to the stroke and then deletes the stroked curve, leaving just the paint. Therefore, you cannot go back and manipulate that curve or edit the brush attributes once the object has been painted, as you would be able to do in 3D.**

**9.** You can now export this image into a tileable, seamless texture by first saving the image. Choose Canvas | Save.

**10.** In the Save dialog box, name the file and click Save.

_NOTE_  **You can save the image in a variety of formats. If you choose Save As □, you can also check or uncheck the option to Save Alpha. Having an alpha channel allows you to composite your Paint Effects creations easily in other applications.**

The texture you have created can now be applied as a texture map to a square plane to simulate the leaves and branches on a tree or cast shadow for light passing through the leaves and branches. Paint Effects is perfect for helping you quickly create organic textures.

## Tutorial: Creating Paint Effects in a 3D Environment

The most exciting aspect of Paint Effects is its ability to create 3D objects. A Paint Effects stroke can be created simply by painting in the view window or directly onto geometry. In this section, we will use Paint Effects to create an underwater scene with seaweed, coral, sea anemones, and bubbles. To create the initial geometry, we will use one of the Artisan brush tools—the Sculpt Surfaces tool. Let's get started.

We'll start by creating the ocean floor geometry:

**1.** Open a new scene and choose Create | NURBS Primitives | Plane.

**2.** In the Channel Box, set the Scale X, Y, and Z attributes to **30** and change Patches U and V to **15**.

**3.** Use the Sculpt Surfaces tool (Edit NURBS | Sculpt Surfaces Tool) to add bumps to the surface to resemble the ocean floor. Save and name this plane **oceanFloor**.

**4.** You can deform a surface like this grid by using the Map function of the Sculpt Surfaces tool, which uses the luminance values or alpha channel of a texture map to push or pull the CVs on the surface. To use this function, open the Sculpt Surfaces Tool Settings window. Choose Push as the Operation, and change Opacity and Max Displacement to **1**.

**5.** Click the Map tab at the top of the Tool Settings window and then click the Browse button and find the OceanFloor.iff file in the Sourceimages folder of the Chapter 19 project folder on the DVD. This texture map is used to displace the oceanFloor surface. Look in the view window to verify that this has happened.

**6.** Click the Reload button to add more displacement. Figure 19-4 shows the Map tab and the deformed surface.

**7.** Open the Hypershade and create a Lambert material. Apply it to the oceanFloor surface.

**8.** Map OceanFloor.iff to the Color attribute of the material. Using this map to displace the surface as well as using it as a color map will make the sunken areas darker.

*FIGURE 19-4*    *The Map tab and the deformed ocean floor*

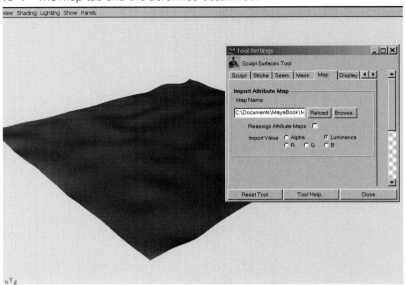

Next we'll create Paint Effects strokes in 3D. If we were to start painting in the view window right now, the strokes would actually be created on the view panel's grid, not the geometry. To paint directly onto geometry, we need to make the surface paintable.

**1.** Select the oceanFloor surface and choose Paint Effects | Make Paintable. Now this surface is a paintable surface and the Paint Effects' strokes will be created on the surface.

**2.** Choose Window | General Editors | Visor to open the Visor. Find the underwater folder on the left side and open it.

**3.** Click kelp.mel and LMB-drag to start painting on the oceanFloor surface. You should see the curve generated on the surface and green kelp tubes growing out of the surface.

*TIP*    *The speed at which you drag a brush over the surface affects the complexity of the tubes that grow out of the strokes. A slower stroke will create more sample points and hence create more tubes that grow. Experiment with creating strokes to get the results you want.*

**4.** Try using other brushes, such as fanCorals.mel and seaUrchins.mel, to populate the ocean floor. You can control the size of the objects you are creating by adjusting the size of the brush. To change the size of the brush, hold down the B key and drag left to right in the view window. The brush cursor gets larger and smaller, depending on which direction you drag.

**5.** Render the scene by clicking the Render the Current Frame button on the status line. Figure 19-5 shows the rendered image. It looks pretty good, but each element is still too uniform looking, and we may want to make the kelp a bit taller.

*NOTE   You can paint in either the view windows or the Paint Effects window by choosing Paint | Paint Scene (instead of Paint | Paint Canvas). Painting in the Paint Effects window will let you view your strokes in greater detail than you can in the view windows. However, viewing all of these details can put a strain on your computer. It's more efficient to paint in the view windows and do test renders to see detail.*

*FIGURE 19-5   Rendered ocean floor*

**FIGURE 19-6** *The Attribute Editor displaying the attributes for the kelp stroke's shape*

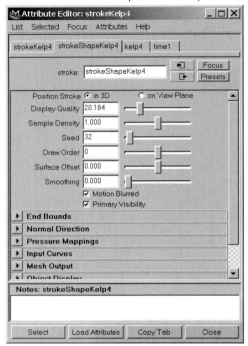

Now let's edit the attributes for the strokes and brushes. If you want to change the size and other aspects of the kelp, you can open the Attribute Editor to access a stroke and edit the brush's attributes for individual strokes. You can also change how complex the tubes are displayed or the number of seaweeds emerging from the strokes.

Select one of the strokes and open the Attribute Editor. Click the strokeShapeKelp tab that contains the stroke's attributes. Figure 19-6 shows the attributes for the stroke.

*TIP* ***You'll find it easier to choose strokes by using the Outliner window than to click in the view window.***

A few sliders here are very important:

- **Display Quality**   Changes the complexity of how tubes are displayed. A higher number displays more accurate tube shapes, but will slow down the interactivity.

- **Sample Density**   Determines how many sample points the stroke has. The more sample points, the more places tubes will grow.

- **Seed**   Sets how each tube is randomized. Set various Seed values on strokes so that the objects look different from one another.

By manipulating these three values in particular, you should be able to see tubes more clearly, have more tubes grow, and add or decrease randomization.

The strokeKelp tab contains attributes for the kelp brush. Here, you can edit the shape of the tubes, and control how the tubes will appear once they are rendered. You can control such aspects as color (texture) of the brush and the number and complexity of the tubes, turn on shadow casting, and apply animation. In fact, so many attributes are available here that it can be daunting. By experimenting with all of the attributes that are discussed in the following section, changing values, and viewing the results of those changes, you will gain a good understanding of how Paint Effects works.

*NOTE* *While this example explores brush attributes by editing attributes of a brush after it has been applied to a stroke, you can edit the brush settings before creating a stroke by choosing Paint Effects | Edit Template Brush Settings. A window opens that contains all of the brush attributes discussed here.*

Figure 19-7 shows the kelp's brush attributes in the Attribute Editor, described here:

- **Brush Type** Determines what kind of brush this is. This is the most defining attribute in any brush, as it determines its general function: Paint, Smear, Blur, or Erase. Any time you need to create a Paint Effects object, set this to Paint.

- **Global Scale** A very important attribute that sets how large the Paint Effects object will be. By adjusting this attribute, you can make the kelp bigger or smaller per stroke. Realize that this attribute is controlled by the brush size at the time of creating the stroke.

- **Tubes** Contains many important attributes for controlling the behavior of the tubes. Remember that the tubes control the general shape and complexity of the Paint Effects brush. A lot of these attributes are related to how the tubes will branch out from other tubes as the object grows. The following are the two most important attributes:

  - **Tubes** The Tubes check box enables tubes on the stroke. If this is not selected, no branching occurs. Brushes such as the airbrush do not use any tubes. If you were to disable this check box now, the kelp plant would disappear, because without any tubes, the brush will be able to apply only a texture to the stroke.

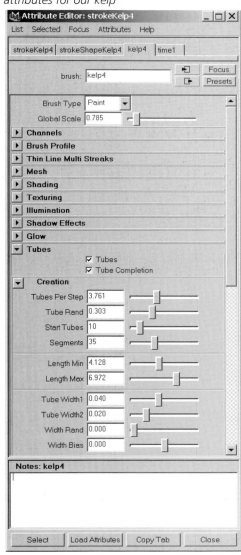

*FIGURE 19-7* *The long list of brush attributes for our kelp*

- **Tube Completion**   As you paint a stroke, a tube is "planted" at the various sample points. As you continue the stroke, the tubes grow until they reach the end of their lifespan. This means that the tubes from the last sample point may not be as fully grown or complex as the tubes from the first sample point of the stroke. When Tube Completion is disabled, the tubes remain at their current growth when you release the mouse button. However, if Tube Completion is checked, all of the tubes grow to reach their maximum lifespan.

- **Creation: Tubes Per Step**   Specifies the number of tubes planted per the number of sample points in the stroke.

- **Creation: Start Tubes**   Specifies the number of tubes created from the first sample point of the stroke. Many of the Paint Effects brushes, such as the kelp brush in this example, have this attribute set to 0. This means that only a single object will be created from this stroke.

- **Creation: Segments**   Changes the length of each segment. As the value increases, the kelp becomes smoother and straighter.

- **Creation: Length Min/Max**   Sets the minimum and maximum length of the tubes. Paint Effects creates tubes with random lengths within the Length Min and Length Max values.

- **Creation: Tube Width 1 and Tube Width 2**   Sets the width of the tubes as they grow from the base (Width 1) to the tip (Width 2).

- **Growth**   Other elements, such as branches and flowers, can be added by checking the appropriate box (shown in the illustration) and tweaking the attributes relating to each growth. Checking one of the boxes will activate the corresponding controls for branches, twigs, leaves, flowers, and buds.

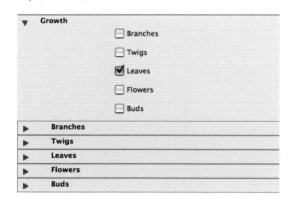

Adjust attributes for each stroke curve to create a variety of different Paint Effects objects. Slightly altering each one allows you to create a more natural, less uniform scene.

## Animate the Brushes

Another powerful feature of Paint Effects is its ability to add animation to the tubes. You can, of course, set keys on any of the attributes available in the Attribute Editor and animate them over time. On the kelp, for example, you can also keyframe Azimuth Min and Azimuth Max attributes found in the Width Scale section of the brush's Attribute Editor. This adjusts the tilting of the tubes and simulates them drifting in water.

However, Paint Effects brushes were designed with specific attributes that let you animate Paint Effects objects so that they can appear to grow over time, react to forces or turbulence, spiral, bend, or twist. Many of the brush presets—such as those for kelp, for instance—have some of these animation attributes already enabled. If you click the Play button next to the Time Slider, you will notice that some animation has already been applied to the kelp—it sways and bends a bit to simulate being affected by underwater currents. This is due to the kelp having Turbulence applied to it. Turbulence applies noise to the motion of the tubes.

Let's adjust this animation to make the current appear rougher:

1. Use the Outliner to select one of the kelp plants in your scene. In the Attribute Editor, select the brush node and expand the Tubes | Behavior | Turbulence section so that you can see the attributes shown in this illustration:

2. Notice that Turbulence Type is set to World Force. This means that the turbulence is added in "world" space. This setting is more useful than the Local Force option, because the World Force appears more uniform when applied to different strokes. If you wanted to disable the turbulence from the kelp, you would set the Turbulence Type to None.

3. To make the kelp animate faster, increase the Turbulence Speed attribute. Click the Play button to see the result.

*TIP  You don't have to stop the animation to see the results of editing the animation attributes. You can adjust the attributes while the animation is playing.*

4. Now let's add some bubbles to the scene that animate flowing upward. In the Visor, select the bubbles.mel brush in the underwater folder.

**5.** In the view window, draw a few strokes on the oceanFloor object. The bubbles will be created. If you open the Attribute Editor and look at the Turbulence settings (Tubes | Behavior | Turbulence), you'll notice that Turbulence has already been enabled.

**6.** Do a test render of a few frames to see how the bubbles look. The bubbles flow upward because a negative gravity force has been applied to them. To slow down the animation a bit, we can adjust this gravity to lessen its effect on the bubbles. Make sure that the bubbles stroke is selected, and look in the Attribute Editor. Find the Forces section (Tubes | Behavior | Forces). In this section, you'll see an attribute called Gravity. Change it to **–.8** to slow down the effect.

### Render the Scene

When you are finished adding Paint Effects objects to the scene, you can begin to set up the scene for the final render. To make this scene look believable, we need to add lights, a background, and shadows. First we'll add the lights:

**1.** Create a directional light (Create | Light | Directional Light) and aim it at the ground. Set the color of this light to light blue.

**2.** Add another light. This time, create a spot light. Position this light above the scene. This light will be used to cast the shadows and water caustic pattern.

**3.** Map a file texture to the Color attribute of the spot light. Import the caustic.jpg file found on the DVD.

**4.** Enable Depth Map Shadows on the spot light. Do a test render.

**5.** While the caustics are visible on the ground, the Paint Effects objects are not casting any shadows. For Paint Effect objects to cast shadows onto other objects, you must turn on the Cast Shadow attribute for each brush in the scene. Select all of the Paint Effect brushes in the scene from the Outliner, and then find the Cast Shadow attribute in the Channel Box. Set this attribute to **1** (on).

*TIP* *You can set attribute values for an entire selection of objects by editing the attribute in the Channel Box or the Attribute Spreadsheet Editor (Window | General Editor | Attribute Spreadsheet Editor). This will not work, however, in the Attribute Editor.*

**6.** Add a background. To do this, create a NURBS sphere and scale it up so that the entire scene, including the perspective camera, is inside the sphere.

**7.** In the Hypershade, create a Surface Shader material and set the Out Color attribute to a dark-blue color. Apply this material to the NURBS sphere.

**8.** We need to make sure that this background object will not cast or receive any shadows from the spot light. Select the NURBS sphere, and in the Attribute Editor, select the nurbsSphereShape1 node. In the Render Stats section, disable the check boxes next to Cast Shadows and Receive Shadows.

Figure 19-8 shows the final render of this underwater scene.

*FIGURE 19-8    The final render of the underwater scene*

By this point, you should have a good understanding of how to use Paint Effects to create organic forms, edit brushes, and add animation to the brushes. These same techniques can be applied to create a variety of objects quickly and populate your scene to create forests, hair, or outer space star fields and gases.

# 3D Paint Tool

The 3D Paint tool in Maya allows you to paint textures directly onto a 3D model in the view windows. This feature has quite a few applications in Maya. You could use the 3D Paint tool to create textures that control any attribute on the material of the target surface. This means that you would not have to leave Maya and revert to a painting application such as Photoshop to create texture maps for your models. Instead, all of the painting could be done in Maya, directly on the model. Furthermore, you can use 2D Paint Effects brushes to paint on the models, giving you access to and control over hundreds of patterns that are not available in other paint programs.

Unfortunately, painting directly on a model does not offer the precision or speed that you might find in a dedicated painting package. The ability to apply filters and work with an unlimited number of layers is also lacking. (Although a layering system *could* be set up using Maya 8's new PSD texture node and layered texture nodes, it would behave slowly.) If you want to be able to 3D paint with all of these features, you are better off using Maxon's BodyPaint 3D (www.maxon.net) or Right Hemisphere's Deep Paint 3D (www.righthemisphere.com).

If the final output for your objects will not require precise texturing, such as SWF images for the Web rendered with the Maya vector renderer, using the 3D Paint tool may be all that you need. However, by using the techniques demonstrated here in combination with the UV layout techniques shown in Chapter 18, you will be able to create great texture work.

## 3D Paint Workflow

This section demonstrates a step-by-step approach to using the 3D Paint tool to paint a model. We cover laying out the UVs on a model before painting, setting up a material that is paintable, and then using some of the 3D Paint tool's main functions to paint the texture. This tutorial should provide all you need to get started painting with the 3D Paint tool. The rest is just practicing the art of painting.

### Prepare the Surface for Painting

When you paint on a surface, Maya samples the UV coordinates of your brush stroke and writes the stroke to a bitmap texture file. Therefore, before you can paint on a surface, you must ensure that no UVs overlap on the model. If any overlapping UVs exist, for example, a brush stroke on the lip of the model may also end up on its ear.

If you will be importing the file that we are about to paint into another paint program, you would probably want to spend some time laying out the UVs using the techniques you learned in Chapter 18. However, if you are going to use only the 3D Paint tool and you do not require a organized UV layout, you can use the Automatic Mapping command to lay out the UVs automatically so that they do not overlap.

Here's how it's done:

**1.** Open the mcr8_ch19_painthead.ma file found on the DVD. (This is the same head that we modeled in Chapter 4.)

**2.** Select the head and choose Window | UV Texture Editor. The UV Texture Editor displays the current layout of the UVs for this model. As you can see, it is a mess right now!

**3.** To paint cleanly on this surface, we need to organize these UVs. Choose Edit Polygons | Texture | Automatic Mapping. This lays out the UVs based on the projection of six planes. Figure 19-9 shows the UVs in the UV Texture Editor before and after the UVs have been laid out. While the UVs at the right may not be as organized as, say, the UVs shown in Chapter 18 in Figure 18-9, we can rest assured that there are no overlapping UVs. The model is now ready to paint.

*FIGURE 19-9* *The UVs displayed in the UV Texture Editor before (left) and after (right) the Automatic Mapping command is applied*

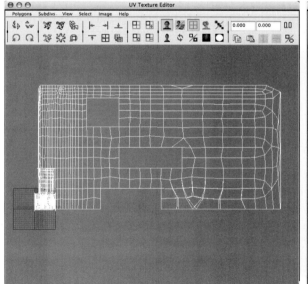 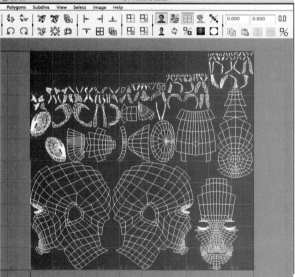

### Create a File Texture

Neither the geometry nor the material is capable of storing texture information. As you should know by now, texture information is stored in a texture node, such as a file texture. If we were to try to use the 3D Paint tool on the model at this time, an error would result, telling us that no file texture has been assigned to the current attribute on the selected model.

Remember that when you paint on a model, you are actually just painting onto a texture map that is connected to the material that has been applied to the model. Therefore, you must assign a file texture to the attribute of the area on which you want to paint. To make things easy, Maya lets you assign a file texture to a specific material attribute right in the 3D Paint tool's settings panel. Here's how its done:

1. Right-click the head geometry and choose Material | Assign New Material | Blinn.

2. Select the head geometry and then open the 3D Paint tool's settings window by choosing Texturing | 3D Paint Tool □. Scroll down to the File Textures section. The options for this section are shown in Figure 19-10.

**FIGURE 19-10** *The File Textures settings in the 3D Paint tool's settings window*

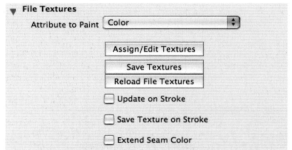

*NOTE* **To open a file in this format from within Photoshop, you need to install the iff plug-in that comes with the Photoshop installation CD. If you do not have this CD, it is best to choose another file format, such as Targa (tga).**

3. Click the Assign/Edit Textures button. A window opens that allows you to specify the resolution for the file that is about to be created. Set the resolution of the file using the Size X and Size Y attributes. When setting the size, you must be concerned about using textures that are too big or too small. A texture that is too large will hinder performance, while a texture that is too small will break up and look "jaggy" in the renders. For now, set both of these attributes to **1024**; note that it is always a good idea to do a few tests before committing to a resolution. Also, choose an image format. It is recommended that you use the Maya IFF format as it will perform faster. Click the Assign/Edit Textures button. A file texture will be created and connected to the Color attribute of the Blinn material.

4. Change the paintable attribute to Bump by choosing Bump from the Attribute to Paint pull-down menu. Once again, click the Assign/Edit Textures button to set the resolution size and create a file texture connected to the Bump attribute of the Blinn material.

5. Continue assigning a file texture to the Specular Color and Diffuse attributes of this Blinn material.

**6.** In the Hypershade, graph the
input connections to the Blinn
material and see that four
file textures are connected to
the material, all mapped to the
appropriate channels, as shown
in the illustration:

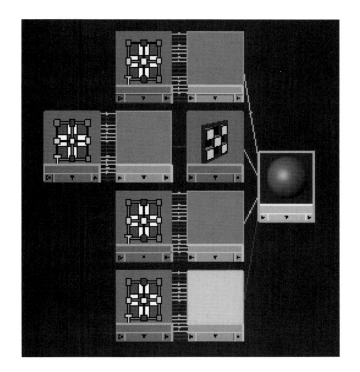

## Paint on Surfaces

Now we are ready to begin painting.
For this part, we will use many of the
other settings found in the 3D Paint
tool's settings panel, which is shown in
Figure 19-11. We'll begin by painting
in the color channel:

**1.** In the 3D Paint tool's settings
panel, under File Textures, set
the Attribute to Paint back to
Color. This allows us to paint
onto the file texture that is
connected to the Color attribute.

**2.** At this point, the default color of the texture is gray. We want to change this to a flesh
color. We can fill the entire texture to one color by using the Flood option. In the Flood
section, click the color swatch next to the Color attribute. Find a color that works well
for skin in the Color Chooser and click Accept. Click the Flood Paint button and the
entire model will fill in that color.

**3.** Let's paint the lip area a reddish color. In the Brush section, select an Artisan brush with
a soft edge (the first one on the left).

**4.** In the Color section of the Attribute Editor, set the Color attribute to a reddish color.

**5.** In the Perspective window, click and drag over the lips to paint that area red.
Remember that you can hold down the B key to change the brush size interactively.

**6.** Continue painting any other detail that you want—perhaps redden the cheeks and
darken the areas around the eyes. When attempting to paint subtle colors, use a low
Opacity setting on the brush, such as 0.2, and stroke the geometry a few times until
the paint becomes visible, to produce the best results.

*FIGURE 19-11*  *The 3D Paint tool settings*

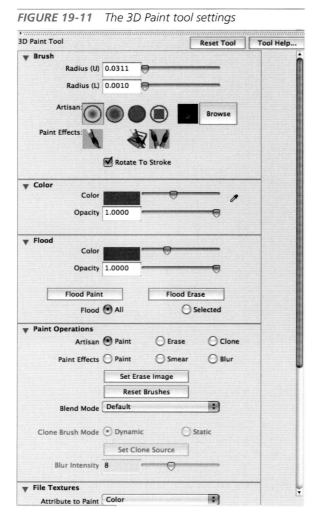

**7.** When you are happy with the paint, save the strokes to the connected PSD file. In the 3D Paint tool's settings window, scroll down to the File Textures section and click the Save Textures button. You can use the Save Texture on Stroke and Update on Stroke options to write the data directly to the PSD file and save that file as soon as you are finished painting a stroke. However, while this may seem convenient, it can slow things down if you are painting to a large texture. Therefore, it is better to manually save the strokes when you are ready.

**8.** Another important option in the File Textures section is the Extend Seam Color check box. This is especially helpful when the UVs were laid out with the Automatic Mapping command, as was used here. Because you may be painting across multiple UV shells, it is possible that you will start to see seams in your strokes, where one UV shell ends and another begins. This option extends the stroke a bit past the actual seam in the UV shell, making your seams invisible.

**9.** You can also use Paint Effects brushes to paint on your textures. Because these are actual file textures that you are painting on, the resulting Paint Effects brushes will be in 2D. If you wanted to add 3D Paint Effects objects, such as hair, you would need to make the surface paintable by choosing Paint Effects | Make Paintable, selecting a brush from the Visor, and then painting. For now, though, we will add some eyebrows with the Paint Effects brush. To select a brush from the 3D Paint tool settings, click the Get Brush button. This opens the Visor.

**10.** In the Visor, look for the hair folder; inside is an eyebrow.mel brush. Select this brush and drag over the eyebrow area on the geometry. An eyebrow appears. Unfortunately, because of the way that paint is applied to the tubes in a Paint Effects brush, Maya is unable to extend the seam color, resulting in visible seams, as shown here:

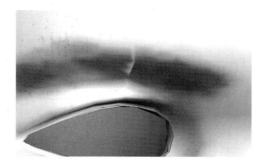

**11.** We'll just have to touch it up. Select the soft-edged Artisan brush that we used to paint the lips. Click the Eyedropper button next to the Color attribute to activate the sample color function.

**12.** Click over the eyebrow where the Paint Effects stroke is visible and sample its color. Make the brush size very small, and paint the areas where the seam is visible. This should clean up the area pretty well.

**13.** Once you are pleased with the color map, you can select another attribute to paint onto. In the 3D Paint tool settings, scroll down to the File Textures section and select another attribute from the Attribute to Paint pull-down menu.

**14.** If you would rather continue working on the other channels in Photoshop, just make sure that you click the Save Texture button. You may then open the file in Photoshop, or any other image editing application. Once you make any changes in the Photoshop file, you can go back to Maya and click the Reload Texture button in the 3D Paint tool settings to see the updated texture.

## Other Applications

If you've gone through the previous tutorial and have found that the 3D Paint tool will not be useful for your production needs, perhaps you'll find it useful in other ways. This section describes a few other uses for the 3D Paint tool: marking areas on a surface and cleaning up existing textures.

### Marking Areas on a Surface

One of the most useful applications for Maya's 3D Paint tool is to use it simply as a marking tool to direct painters where to paint in a 2D paint program. Even when the UVs have been laid out in an orderly and recognizable fashion, the scale of the UV layout may not be proportional to the actual texture that you need to paint. In other words, if you paint a circle on a texture map in Photoshop and then apply it to a material in Maya, the circle might appear stretched due to the layout of the UVs. In such a case, you can use the 3D Paint tool to mark areas that need to be painted on the model.

Another excellent use for this tool is for use with dynamics. This topic is covered heavily in Chapters 22, 23, and 24, where you will learn that many particle attributes can be controlled with a texture—Surface Emissions are one example. You could easily paint onto an area of a model that will control the point at which the particles are emitted.

### Cleaning Up Existing Textures

Earlier, we used the Eyedropper tool to sample a color, and painted out the seam with the Brush tool to clean up artifacts that resulted from painting over multiple UV shells. While that was a simple example, you may find that at times the detail in the texture map is too complex to use the Eyedropper tool. In such cases, the Clone feature of the 3D Paint tool is just the solution.

Figure 19-12, left, shows a surface whose two sides do not have corresponding UVs, and an obvious seam exists in the texture. By setting the Paint Operation to Clone and then sampling an area with the Set Clone Source button, the brush clones the sampled area when a stroke is painted. The image on the right shows what the surface looks like after being touched up with the Clone operation.

**FIGURE 19-12**    *The Clone operation is used to fix visible seams in a texture map.*

## Summary

This chapter explored some of the features available to you with Maya's paint tools. You can see that no shortage of options exists when it comes to painting in Maya. The trick is finding out which tools are right for your particular job.

In the next chapter, we will explore rendering methods that will give your images extra shine.

# Lights and Cameras

**Once a scene has been modeled,**

textured, and animated, it is time to light it and

shoot it through the eyes of the camera. The art

of cinematography plays a large role in any story.

The creative use of lighting can provoke a mood

while the framing of a shot decides what the

audience can actually see at any given time. In

this chapter, we will explore light and camera

nodes in Maya. Each of these nodes functions

very similarly to its real-world counterpart.

# Light Nodes

Maya's light nodes are designed to work just like lighting in the real world. This allows a traditional lighting designer to begin using Maya quickly, because the tools have familiar names and behaviors.

Traditionally, lights serve two main purposes in film and still photography. The first is, obviously, to illuminate the scene being photographed. Without light there would be no color—the printed film would be nothing but black. The other purpose for lighting is to add a dramatic effect or mood to a scene. In fact, lighting may be one of the first visual effects ever used. Heroic characters are traditionally well lit, even shining, whereas villains lurk in shadows or are backlit. Just as it is important to do research and collect references on your models and textures, it is important to take notice of lighting, both in the natural world and in the movies.

## Types of Lights

Maya has six different types of lights: ambient, area, directional, point, spot, and volume. All these lights can be created from the main Create | Lights menu or in the Create Lights menu in the Hypershade. The different types of lights (shown in Figure 20-1) create the following effects:

- Ambient light simulates diffused lighting in all directions. Use this light as a fill light to brighten the dark areas of the scene or to even out the lighting entirely.

**FIGURE 20-1**    *Different types of lights and how they illuminate*

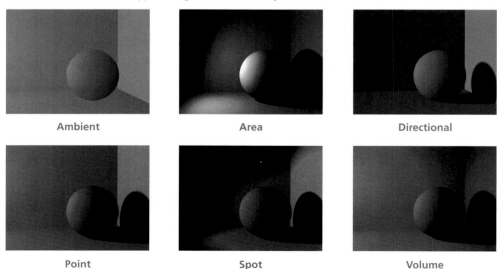

| Ambient | Area | Directional |
| Point | Spot | Volume |

- An area light is a 2D rectangular light source. The size of the light, which can be determined using the standard scale tools, has a direct effect on the distribution of the light in the scene. Sunlight coming through a window is a common example: the light can be scaled to fit the window shape while the reflection of the light on, say, the floor is brightest closest to the window and falls off farther away from the window.

- Directional light shines evenly in one direction only, such that the rays are emitted parallel to each other from an infinitely large plane. Use this light to simulate a light source such as the sun. Maya uses a directional light as the default light in the scene.

- A point light shines evenly in all directions from the location of the light. Use a point light to simulate an incandescent light bulb.

- A spot light shines evenly within a narrow range of directions (defined by a cone) from the location of the light. Use a spot light to create a beam of light that gradually becomes wider—for example, a flashlight or car headlight.

- A volume light illuminates objects within a finite region defined by a primitive shape: sphere, box, cylinder, or cone. The direction of the light can be set to outward, inward, or down, making a volume light capable of simulating many of the other light types in Maya. Another advantage of using a volume light is that its total decay distance is indicated by its icon in the view window, so precise placement is easily achieved without having to do numerous test renders. The color range is also controllable through a ramp interface. This allows you to control the color of the light at any point along its volume decay. It can also be used to control the falloff.

## Light Attributes

Lights in Maya can be translated, rotated, and scaled just like any other piece of geometry in the scene. After adding a light to a scene, the best way to place it is to select it and choose the Show Manipulator tool from the toolbar. This lets you position the light as well as its reference or target. Another way to place a light is to look through it and use the standard camera tools—track, dolly, and zoom—to place and point the light precisely. Just select the light and choose Panels | Look Through Selected.

You can edit a light's attributes in the Attribute Editor, shown in Figure 20-2. Attributes common to all lights are Color and Intensity, which are pretty straightforward. The color of the light is used to set the color the objects in the scene will receive from their illumination. It is even possible to map a texture to this attribute. The intensity of the light controls its brightness. You can get a rough approximation of the general effect a certain Intensity value will produce by monitoring the Intensity Sample swatch (shown in the figure) in the light's Attribute Editor.

It is important that you realize that lighting is an additive/subtractive process. If you have two lights with exactly the same attributes and positions in space and both have an Intensity value

FIGURE 20-2   *The Intensity Sample swatch in the Attribute Editor shows the general effect of the light intensity.*

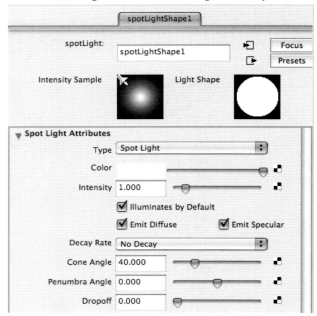

of 1, they will produce the same amount of light that one light with an Intensity value of 2 would produce. It is also possible to use a negative Intensity value to subtract or dim light that is emitted from other lights in the scene.

Area, point, and spot lights all have an attribute called *Decay Rate*. The Decay Rate attribute controls how quickly the light's intensity decreases over distance. The default is set to No Decay. This means that the light's intensity is continuous, and therefore it can reach every object in the scene regardless of how far the object may be from the light. To make the decay fall off faster, use Linear (slowest), Quadratic, or Cubic (fastest). The falloff of a spot light with the four different decay settings is shown in Figure 20-3.

## Spot Light Attributes

FIGURE 20-3   *From left to right, a spot light with No Decay, Linear, Quadratic, and Cubic Decay settings*

A spot light contains three main attributes for controlling its beam. The *cone angle* is simple. It is the angle in degrees from one edge of the light's beam to the opposite edge. The *penumbra angle* controls the falloff, in degrees, of the edges of the beam. This can be a positive or negative value. If it's positive, the falloff will occur from the edge of the light, determined by the cone angle, outward by the penumbra angle. Therefore, if the Cone Angle is set to 30 and the Penumbra Angle is set to 5, the light has a total angle of 40 (30 + 5 + 5—that is, 5 degrees on each side). The beam's falloff would begin at 30 degrees and would fall off to an intensity of 0 at 40 degrees.

The overall shape can be previewed and edited in real time by monitoring the Light Shape swatch found in the spot light's Attribute Editor. Figure 20-4 shows a spot light's beam projected onto a surface. The spot light has a Cone Angle value of 40 (the default), but the left half has a Penumbra Angle value of –5 and the right side has a Penumbra Angle value of +5.

The *Dropoff attribute* is also unique to the spot light. It is similar to a penumbra with a negative value, in that it controls the falloff from the beam's cone angle inward. But it behaves more like the decay rate, in that it is based on the distance from the center of the beam instead of from the edges. The Light Shape swatch in the Attribute Editor should give a good indication of what effect the Dropoff setting will produce.

*FIGURE 20-4    A spot light beam with a negative (left) and positive (right) penumbra angle*

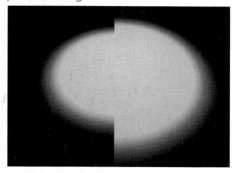

## Light Effects

The point, spot, area, and volume lights are all capable of producing various lighting effects, such as Light Fog, Light Glow, or both. These effect settings are located in the Light Effects folder of the Attribute Editor. To apply these light effects, click the map button next to the desired attribute. This creates the respective node in the scene and connects it to the attribute. For example, if you click the Light Fog map button, a lightfog node is created, which contains the attributes for editing the color and density of the fog. The material editing tutorial in the next chapter gives more detail on the use of Light Fog.

When Light Glow is activated (by clicking its map button), an opticalFX node is created in the scene and connected to the light. This node contains all the attributes used to edit any property of an optical glow effect. For a light to be visible to the camera, Light Glow must be enabled. This produces a sort of star-shaped glowing region that simulates the effect of a light viewed through a lens (see Figure 20-5).

*FIGURE 20-5    A rendered image of a glow effect with a lens flare*

**FIGURE 20-6** *Attributes and folders in an opticalFX node's Attribute Editor*

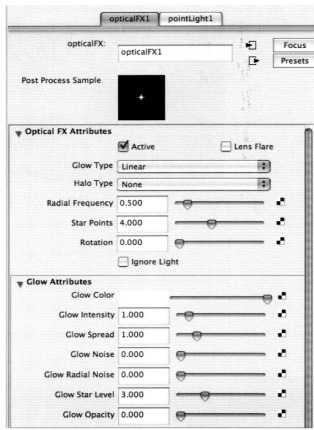

Five different glow types are available, and their specific attributes can be edited in the Glow Attributes folder of the opticalFX node's Attribute Editor. You can also add a halo around the glow. Its attributes are then edited in the Halo Attributes folder shown in Figure 20-6. Finally, a lens flare effect can be added by selecting the Lens Flare check box and editing the respective attributes.

# Shadows

Lights not only illuminate objects in a scene, but also produce shadows. Maya offers two different methods for calculating shadows: depth map and raytraced. In this section, we focus on depth map shadows. Raytraced shadows will be discussed in Chapter 21, when we deal with rendering.

Before you can create shadows, the light must be capable of casting shadows. All light types except for ambient lights are capable of producing depth map shadows. These shadows are calculated using an image map that is rendered from the light's perspective relative to the surface it illuminates. The renderer then uses this depth map to determine which objects are lit and which ones fall into shadow.

To turn on shadows, select the light and find the Shadows folder in the light's Attribute Editor. Then check the box next to Use Depth Map Shadows. If you render your image with the default shadow settings, chances are that the shadows produced will have somewhat jagged edges (see the leftmost image in Figure 20-7). To fine-tune the shadows, you can use some of the depth map shadow attributes.

First, edit the Dmap Resolution setting. This sets the resolution of the depth map that Maya uses to calculate which objects are illuminated. It is best to use the same resolution as the output resolution of the image. (For more information on rendering and output resolution, see Chapter 21.) If your output resolution is set to 640×480, set Dmap Resolution to 640. From

*FIGURE 20-7*  *Two images lit by a directional light with Use Depth Map Shadows turned on: the image at left uses the default shadow attributes, and the image at right has increased Dmap Resolution and Dmap Filter Size attributes.*

there, you can begin increasing the Dmap Filter Size setting in small increments until the edge of the shadow is smooth enough (see the rightmost image in Figure 20-7).

> **NOTE**  *Be very careful with these attribute settings, because increasing them too much can increase render times.*

Occasionally, when the shadow is filtered, the blurring of the shadow will cause it to cast on the light side of the object. To compensate for this error, you can adjust the Bias setting of the Depth Map Shadow. Figure 20-8 show an object casting a shadow before and after adjusting the Bias attribute. The default setting is 0.001. Usually, this attribute only needs to be slightly adjusted. The second image in Figure 20-8 was adjusted to 0.005.

*FIGURE 20-8*  *The shadow biasing*

 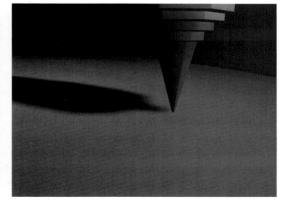

# Camera Nodes

Though you might not have been aware of it, we have been using cameras throughout this book. All the view windows are seen through cameras, and additional cameras can be created with the Create | Cameras command from the main menu bar. Cameras can be transformed and animated just like any other object in the scene. While they can be selected in the Outliner or Hypershade, probably the easiest way to select a camera is to choose View | Select Camera from the view window's menu bar. At that moment, you could press the s key to key its transforms or view its attributes in the Attribute Editor or Channel Box.

Many camera attributes simulate the workings of a real-world camera, so it is important that you understand how they work and how these attributes relate to and affect one another. Take, for example, a common 35mm camera; *35mm* is a measurement of the width of the film. The piece of film ready for exposure is held in place by the *film gate*. While the film gate is 35mm wide in this case, it could have various heights and, as a result, produce images with different *aspect ratios*. If you wanted to choose a lens for a 35mm camera that would most closely approximate normal vision, you would need a 50mm lens—50mm is the distance between the lens and the piece of film in the film gate, which is known as the *focal length*.

With this in mind, look at the camera attributes in Maya. You will find the Film Gate Aspect Ratio attributes in the Film Back folder and the Focal Length attribute in the Camera Attributes folder. Now suppose that we want to do a close-up of a character's face. In the real world, you have two options: move the camera closer or change the focal length. Changing the focal length would require a lens change or the use of a zoom lens. In Maya, you can simply change the Focal Length setting. A higher value zooms in closer to the scene, simulating a telephoto lens; a lower value gives you a wide view. Notice as you change the Focal Length setting that the angle of view also changes. These attributes are inversely proportional to one another—that is, as the focal length gets longer, the angle of view becomes narrower, and vice versa.

Another way you could zoom in for your close-up is by changing the Film Gate setting. In the real world, this would require a change of cameras. But in Maya, if you change to a 16mm film gate, the 50mm lens that gave you a normal view with the 35mm film gate will behave like a telephoto lens—the 100mm lens with the 35mm film gate.

Note that switching the film gate is not really a great option, like switching to a different camera. For continuity's sake, it is best to use the same camera all the time. Switching the film gate can change your aspect ratio—the width and height ratio of the frame—which is not desirable. Usually, the aspect ratio of your view window does not reflect the aspect ratio of your film gate. This could make it difficult for you to frame your shot. Luckily, you can view the film gate's aspect ratio in the Cameras view window either by turning on Display Film Gate in the Display

Options folder in the Attribute Editor or by choosing View | Camera Settings | Film Gate. You should see a frame in your window. To set how much space is around that gate, find the Overscan attribute in the camera's Attribute Editor and change its value. Your view should look something like Figure 20-9.

*FIGURE 20-9    The view window displaying the film gate*

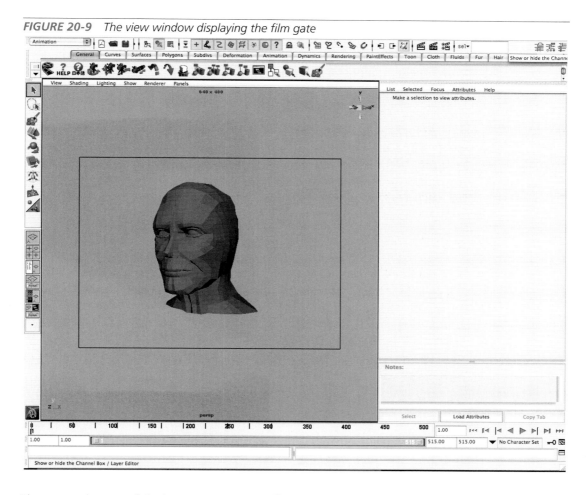

These are only some of the important camera attributes. We discussed a camera's image plane attributes in Chapter 4, when you set up underlay images on image planes in the camera views. We will discuss other camera attributes throughout the book. Of particular interest is the tutorial in Chapter 27, which deals with setting up a camera projection.

# Tutorial: Indoor Lighting

In this section we will exercise our knowledge of lighting as we practice on a small indoor scene. The main focus of this tutorial is to establish the key light, a practical light, set up shadows, and learn some shortcuts and timesavers for faking shadows.

1. Open the scene called mcr8_ch20_indoor_start.ma from the DVD. It contains a basic hallway scene with a table, lamp, and cup. The hallway contains a window that will act as one of the main light sources in this scene. The other main light source will be the lamp. Therefore, we will have two key lights. We'll begin with the sun.

2. Create a spot light and name it **lSun**. In the view window, choose Panel | Look Through Selected. Use the camera orbiting tools to position the light so that it points downward through the window in the hallway, emulating a mid- to late-afternoon sun, as shown here:

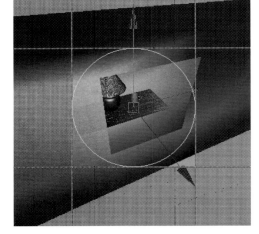

3. Select the lSun node and load the lSunShape node into the Attribute Editor. Set the color of the light to a warm color. A yellow-orange color is suitable for this afternoon quality of light. Turn on Enable Depth Map Shadow. Render the image and evaluate the scene. It should look like this:

4. Let's soften the shadow just a bit. Set the Filter Size on the lSunShape node to 6. This softens the edges of the shadow and removes the aliasing artifacts. We can leave this light alone now and move on to the lamp light.

5. The lighting for the lamp will consist of a few tricks made up of several lights. The first light will be a point light placed inside the light and act as a practical light source emitting from the incandescent bulb. Choose Create | Light | Point Light and place the light in the center of the lampshade where the bulb should be. Name it **lBulb**.

**6.** While the logical reason for this light would be to illuminate the indoor scene, we will instead use it to fake the color spill from the lamp shade onto the wall and surrounding objects. Set the color to something reddish that matches the lampshade color.

**7.** If you do a test render, you'll notice that the scene is much too red. We want the effects of lBulb to be subtle. We can fix this, not by lowering the intensity, but by changing the way that the light falls off from its point of emission. Set the Decay Rate of lBulb to Quadratic. When you increase the Falloff attribute, you will need to increase the light's Intensity setting. Set the Intensity to 50. Disable the Emit Specular attribute on this light. A test render should look like this:

**8.** Now we need something to cast shadows from the light bulb. Using the existing lBulb would not be very efficient. First of all, it is the wrong color to emit from the gaps of the lampshade. Secondly, point lights make poor shadow casters because they create a total of six shadow maps. We can instead fake the light and shadow casting from each end of the lampshade with a spot light. Create two spot lights and place them inside of the lampshade. One should point upward and the other downward. Try to match their cone angle to the size of the lampshade, as shown here:

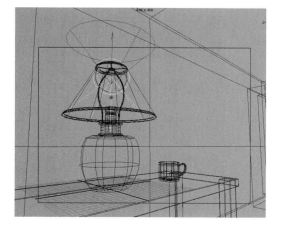

**9.** Set the Intensity of the lights to something low, around 0.4. Enable Depth Map Shadows, set the Filter Size to **8**, and do another test render. Notice how the spot lights inside the lamp begin to cast shadows from the lamp's interior pieces, adding a greater depth to the overall lighting in the scene by breaking up any regularity.

**10.** Now we'll add some fill lights. Again, we will use spot lights. Create a spot light and name it **lFill_01**. Set the Intensity to 0.3, and change the color to something earthy, like a greenish blue. Place the light so that it projects from underneath the scene. Increase the Dropoff attribute so that we could never make out the cone shape. Disable the Emit Specular attribute on this light.

**11.**  Add another fill light with similar settings but change the Color attribute slightly. Cast this light from underneath the scene in the other direction from lFill_01. When you render the scene now, you should start to notice that the light is uneven everywhere—that is, that every point in the scene should be in a state of gradation from its surrounding pixels. The final render should look like this image.

Remember, this tutorial was really just designed to give you a clear understanding of a lighting workflow and introduce you to a few tricks. With the help of some additional reference and research, you should be able to push this image further toward reality.

## Tutorial: Outdoor Environment Lighting

In this tutorial, we will light the outdoor scene to re-create outdoor environment lighting. Compared to artificial indoor light, outdoor environment lighting requires a very different approach. There are several things to consider when trying to mimic natural outdoor light, as listed next. Figure 20-10 illustrates some of these lighting phenomena.

*FIGURE 20-10*   *The directions of different light rays in a typical outdoor environment*

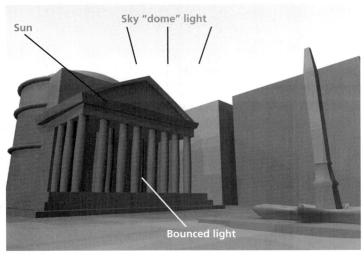

- The sun is the main source of outdoor light during the day. It is one of the strongest light sources and it alone creates high-contrast, harsh lighting. However, most of the time (except for clear, midday), the outdoor light isn't very harsh because the sunlight is scattered through many materials before reaching the ground. Also, at sunrise and sunset, the sun gets a warmer and weaker reddish color because the scattering of the sun's light rays increases when the sun is at the lower angle.

- The sky "dome" light accounts for the sunlight that gets scattered as it passes through the atmosphere. During the day, the sun is high up and the light is scattered moderately, so the atmosphere casts blue light rays toward the ground. This blue light is most obviously found in shadows of objects. Sunlight is also scattered by cloud cover. Clouds further diffuse sunlight, so the light is much softer and weaker.

- The bounced light compensates for the light that is reflected off of the ground or other objects. Large expanses of green lawns create a slightly greenish bounce light. White snow cover creates very strong white light, which is almost blinding. Including these bouncing colors is very important to make the scene real looking.

## Set Up the Scene

The basic environment for this tutorial is already prepared. Your job is to create various complicated lighting components to create a realistic lighting effect.

> **NOTE** *This tutorial uses Maya's Software renderer. Mental Ray renderer's Final Gathering algorithm can produce similar results but will render much more slowly.*

1. Open the scene called mcr8_ch20_outdoor_start.ma from the DVD. The scene has the simple Pantheon Italian courtyard environment geometry. Also, there is a camera called render_cam. There is not a single light in the scene yet.

2. Render the scene with the default light, as shown here. We will add necessary light one by one to make this image fuller and more realistic.

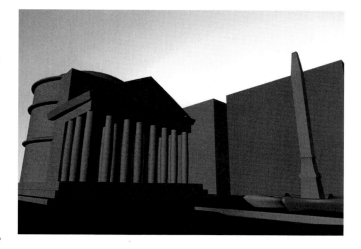

3. Add a key light that imitates the sun. Create a spot light by choosing Create | Lights | Spotlight and name it **lSun**. Next, aim the light by looking through it. Do this by choosing Panels | Look Through Selected. Since there are no buildings on the left side of the

Pantheon, we will aim the light in that direction. In the Attribute Editor for the key light, set the Color attribute to be slightly yellow and set the Intensity to 0.7. Render again and see how it looks.

**4.** Once you are satisfied with how the light falls, turn on the shadow to see the key light's shadowing effect. For this tutorial, we will use raytraced shadow. Select the key light and open the Attribute Editor. Open the Shadows folder on the lSunShape node and enable Use Ray Trace Shadows in Raytrace Shadow Attributes section. Test render the scene and notice where the shadow falls.

**NOTE**  *To see the effects of raytracing, you must enable Raytracing in the Render Settings window under the Raytracing Quality folder in the Maya Software tab.*

**5.** The shadow is a bit too harsh for the key light. In the Attribute Editor for the key light, set the Shadow Color attribute to 50% gray with a tint of blue. Now the color of the shadow reflects the color of the sky. But the shadow looks a bit too sharp still.

**6.** Still in the lSun's Attribute Editor, in the Raytrace Shadow attributes section, change Light Radius to 1 and Shadow Rays to 6. This creates a more diffused, smooth shadow edge. That's the initial setup, the result of which is shown here:

# Create the Dome Light

The next step is to create a dome light to imitate the scattering light from the sky. Dome light consists of 17 spot lights in the shape of a dome. Each one casts slightly bluish light and casts shadow.

**1.** Create a spot light and name it **domeLight01**. Press the T key to activate the Manipulator tool. Move the light in the positive Z axis until Z axis translation is about 50 units. Press the W key to activate the Move tool. Press the INSERT key to activate the edit pivot mode. Hold the X key to use grid snap and move the light's pivot to the global center (0,0,0). The top view makes this easier to see. Press INSERT again to exit pivot mode.

**2.** In the Attribute Editor for the domeLight01 spot light, change the Color attribute to be slightly blue, the Intensity attribute to 0.08, and turn on Raytrace Shadows. Choose Edit | Duplicate Special and set Rotate X to −30 and Number of Copies to 6. Geometry Type should be Copy and Group Under should be set to Parent. Click the Duplicate button. This creates the basis for the dome light.

**3.** Select the six lights at the bottom; don't select the topmost light or the key light. Choose Edit | Duplicate Special again, and this time choose 0 for Rotate X, 45 for Rotate Y, and 3 for Number of Copies; the rest should be the same. Click Duplicate. This completes the dome-shaped light group consisting of 25 lights. We don't need the eight bottommost lights (lights at the ground plane level), so delete them.

**4.** In the Outliner, select all the domeLights and choose Edit | Group. Name the group **domeLightGroup**. Rotate the domeLightGroup group 20 degrees in the Y axis, so the lights don't exactly line up with the environment. Attributes of these lights can be quickly accessed and edited by choosing all the dome lights in the Outliner and choosing Window | General Editors | Attribute Spread Sheet. Test render to see how these lights will give more realistic treatment to the scene. It takes much longer time to render as well.

**5.** The last thing to add are lights pointing up from the bottom to imitate light bouncing back from the ground. Select the domeLightGroup group, choose Edit | Duplicate Special, and set Rotation Y to 0 and Number of Copies to 1. Click Duplicate and change Rotate X of the newly created group to −180. It should now flip to the bottom so that the lights are pointing up. Rename the new group **domeLightGroup_ground**.

**6.** Select all the lights in domeLightGroup_ground and, using Attribute Spread Sheet, change the color of light to a slightly reddish color. An easy way to do this is to select one of the lights and open the Attribute Editor to change color interactively and note the RGB or HSV color values. Then, select all of the lights, open the Attribute Spread Sheet, and copy the same RGB value into the RGB values for all of the lights. While you are in Attribute Spread Sheet, turn off Use Ray Trace Shadow; we don't want any shadow from these lights. Also change Intensity to 0.03. Test render the scene again. It should look like the following image:

# Summary

In this chapter, you learned about the different types of lights and their properties. The techniques shown in the tutorials outlined some very useful procedures you can use to create some very realistic lighting effects without using any physically accurate rendering schemes. As you will see in the next chapter, while these new rendering schemes produce very realistic lighting conditions, they come at a cost of rendering time. Therefore, it is always good to know some tricks when it comes to lighting, some of which are explained in Chapter 21.

# Rendering

**3D rendering is the process of**
translating all of the information in a 3D
scene to a bitmapped image. Maya's software
renderer uses your computer's CPU and Maya's
software rendering code to convert all of
the 3D elements of your scene—geometry,
textures, lighting, and effects—into a pixel-based
image that can be opened in an image-editing
application, a video-editing application, or an
Internet browser. In this rendered image, none
of these elements can be transformed in 3D space
or have their materials edited. They are fixed, and

any changes that need to be made will require that the image be re-rendered or the pixel information modified in an image-editing program.

The software renderer is also capable of making many additional calculations that are not possible in real time and, hence, not displayed in your view window. In software, rendered image anti-aliasing and texture filtering can be calculated to give your image smooth edges and textures. Fog from lights can be rendered. Reflections can be raytraced and refracted. Motion blur can be added. Maya ships with four renderers. The Mental Ray renderer includes some enhanced features to add caustic effects, Global Illumination, and HDRI (high dynamic range imagery, discussed later in the chapter in the section "HDR Images with Final Gather"). This chapter examines all of these cases and shows examples of each. We will start by looking at some rendering options in the Render Settings window. Then we'll talk about some techniques for improving smoothness and eliminating seams in NURBS geometry. After these basic concepts are explained, we'll move through a tutorial that shows some additional rendering effects and introduces a workflow for working with the Mental Ray renderer.

# Rendering in Maya

Maya offers both software and hardware renderers. The hardware renderer uses the real-time, OpenGL (Open Graphics Library) enabled rendering engine of your computer's video processing unit. This type of rendering is used to view geometry, textures, particles, lights, and shadows in the view windows. All of these can be viewed in real time, or near real time, depending on the power of your video card. While this type of rendering is great for fast interaction, the image quality and limited display of certain effects prohibit it from being used as a final image for television or film production. Hardware renderers similar to Maya's are sufficient for real-time games that you might find on a video game console such as a PlayStation or Xbox. Some elements in Maya, such as several of the particle types, can be rendered only in hardware. For these, you can use the hardware renderer to render each frame of an animation as a file or sequence of files to be composited together with the software renderer to create the final image or animation. We will discuss the hardware renderer in detail in Chapter 22.

This chapter focuses on Maya's software renderer and Mental Ray. In this kind of rendering, all elements in the scene for a given frame are gathered. Each object in the scene is positioned according to its keyframed translation values at a given frame. If the object is being driven by inverse kinematics (IK), a dynamic simulation, or expression, these are calculated first and the object is translated to match. If the object is a piece of NURBS geometry, it is tessellated into polygonal facets. Every object is combined with its material into a shading group node. This shading group is then calculated along with the light to produce shadows, bumps (from the bump map attributes), reflections, highlights, and any other effects such as fog.

With all of this information gathered, the renderer divides the image into chunks based on the amount of RAM available and begins calculating each pixel, one at a time, for the final image.

All of the chunks are combined and written into a file as one frame of the animation. Before they are written, the renderer does some additional work. Anti-aliasing and texture filtering can be performed to eliminate the jagged edges between objects and any noise that might arise from the detail in the texture map. Other effects, such as motion blur, are also added to moving objects.

This intense process can take minutes, hours, or even days, depending on the complexity of the scene. Understanding the different attributes and their settings can help you minimize these render times and give you precise control over the final look of the image.

## Preparing to Render

The process of rendering usually involves these steps:

1. Set up the scene (modeling, texturing, animation, lighting, and so on).

2. Select a rendering engine.

3. Set up render settings.

4. Render the scene by clicking the Render Current Frame button on the status line, or render the entire range of frames specified in the Render Settings window by choosing Render | Batch Render.

## Render Settings Window

The Render Settings window (shown in Figure 21-1) is the main interface for editing the resolution, quality, and complexity of an image or animation. It can be accessed by choosing Window | Rendering Editors | Render Settings or by clicking the Display Render Settings Window button on the status line.

In this section, we look at some of the most commonly used attributes in the Render Settings window. These attributes, such as File Format and Resolution, are native to this window and do not require any tweaking on a per-object basis. Attributes that merit a lengthier discussion and examples will be discussed in their own sections later in the chapter. Finally, we will review this information when we set up the rendering of our final scene in Chapter 27. For now, we'll concentrate on the basics.

## Select a Rendering Engine

Currently, Maya ships with four different rendering engines: the Maya software renderer, the Maya hardware renderer, the Mental Ray renderer, and the vector renderer. Other rendering engines, such as Renderman, Turtle, and Gelato, are also available for purchase as plug-ins. In many situations, you may need to render different elements of your scene using different rendering

*FIGURE 21-1*    *The Render Settings window*

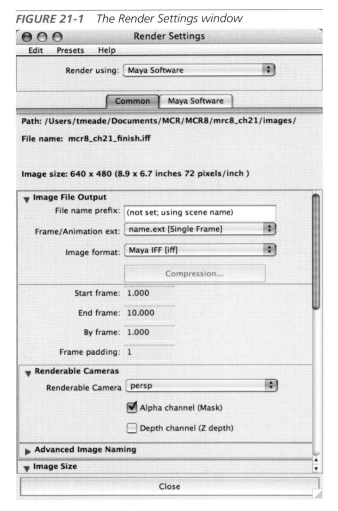

engines. To make this more manageable, the Render Settings window has been organized into two main sections, each grouped under a specific tab. Figure 21-1 shows the Render Settings window with the Maya software renderer loaded.

The Common tab displays the attributes that are common to all four rendering engines. These settings control such things as output resolution and range of frames—things that are not so specific to the "look" of the render.

The other tab displays attributes that are specific to the rendering engine that is selected from the Render Using pull-down menu at the top of the window. These attributes contain settings for quality (anti-aliasing and texture filtering) and specific rendering algorithms (raytracing and Global Illumination).

## Common Rendering Attributes

The Common tab contains all of the basic settings for the renderer. It includes mainly the Image File Output section and Resolution settings.

The Image File Output section controls the range of frames output from Maya and what file format and channel information they will use. The two first attributes, File Name Prefix and Frame/Animation Ext, let you specify a name for the rendered files and choose how you would like that name displayed along with its frame number and file extension. This helps make the rendered files output from Maya compatible with different compositing programs.

The Image Format attribute lets you specify the file type to which the images are rendered. While you have the option of outputting the entire range of frames (specified in the Start Frame, End Frame, By Frame, and Frame Padding attributes) as a QuickTime movie, most of the Image Format options are single image formats. Rendering an animation as a sequence of frames is highly recommended. Then, if anything goes wrong—the software crashes or a power loss occurs—you can output a single file for each frame rendered. The objects rendered and the camera from which they are rendered are selected from the Renderable Objects and Camera attributes. Finally, the Channel attributes let you select from what channel information will be used in the rendered image. We will use these settings later in the chapter.

The Resolution attributes specify the resolution of the rendered image(s). The Presets pull-down menu contains many industry-standard resolutions for television, film, and computer screens. Custom resolutions can be set in the Height and Width attributes and the Pixel Aspect Ratio (the ratio of the width to the height of the image) can be set in this section as well.

### Renderer-Specific Attributes

The second tab in the Render Settings window loads settings that control specific attributes for the selected renderer. In general, all of these attributes let you control the quality of the output data through some kind of anti-aliasing con-

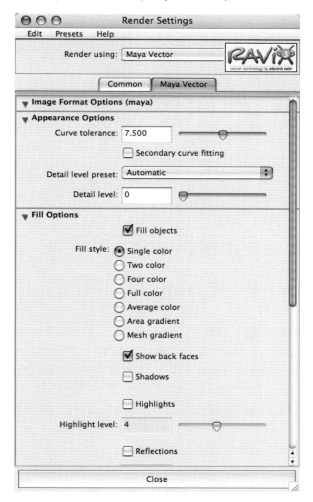

trols. In the case of the Maya vector renderer, the quality is mainly controlled by three attributes in the Appearance Options section. The Maya Vector tab is shown here. Since the Maya vector renderer renders vector data that can be exported to Macromedia Flash or Adobe Illustrator, the concept of anti-aliasing does not really apply. Instead, the renderer uses the Curve Tolerance and Detail Level attributes to control how smooth the edges of the objects are rendered.

For the other three renderers, anti-aliasing is controlled through a number of attributes specific to each renderer. These will be discussed next. Without getting into the specifics, you can use preset anti-aliasing settings for both the Maya software renderer and the Mental Ray renderer through the Quality pull-down menu in this tab.

Most of the other attributes in the renderer-specific tab control the unique rendering algorithms of each renderer. Raytracing, Global Illumination, and Final Gather each use different rendering algorithms to determine how the final image is rendered. The specifics of these algorithms are discussed in the sections that follow.

# Anti-Aliasing

The anti-aliasing settings available in the Render Settings window control the overall quality of the image. Specifically, anti-aliasing handles how the edges of objects in the scene are softened, or blurred, so that they blend with any objects behind them. It is a rare chance that the edges of objects in your scene will directly coincide with the square edges of the pixels in the rendered image. Therefore, it is necessary for the renderer to anti-alias the edges so that they blend; otherwise, the edges of objects will appear rough or "jaggy."

Figure 21-2 illustrates how the rendering engine renders a white object against a black background without any anti-aliasing. The image on the left represents how the objects are laid out in your scene, and the image on the right shows how that image was rendered without using anti-aliasing. The grid represents the pixel borders. In this process, the renderer will take one sample for each pixel, and whatever color has the majority within that pixel will be the color that is rendered for that pixel in the rendered image. The result, shown in the right image, has very rough edges.

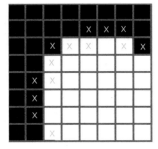

**FIGURE 21-2**   *A white object against a black background is rendered without any anti-aliasing, resulting in the image shown on the right.*

The Maya software renderer and the Mental Ray renderer anti-alias edges by supersampling pixels that have a high degree of contrast. In *supersampling*, the renderer takes multiple samples from each pixel region and then blends them together to assign a value to the final, rendered pixel. Figure 21-3 shows the image from the previous example after it has been supersampled so that the renderer is taking four samples per pixel instead of just one, as shown in Figure 21-2. If a pixel contains two white samples and two black samples, the resulting pixel will be 50 percent gray.

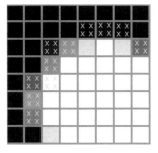

**FIGURE 21-3**   *Gray values are determined by each pixel based on four samples taken from each pixel.*

The more samples that are taken, the slower your render times can become. Therefore, you should use high anti-aliasing settings only for your final render and use no, or low, anti-aliasing settings as you preview your renders. Fortunately, Quality presets are available in the Render Settings window.

# Anti-Aliasing Quality Settings

When using the Maya software renderer, the first section in the Render Settings window is called Anti-aliasing Quality, shown here, and contains all of the attributes for controlling the anti-aliasing in the current scene.

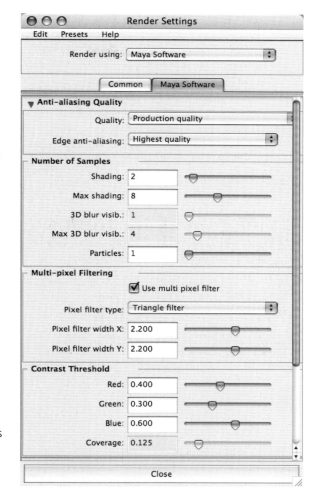

The Quality preset lets you choose from among six different options that set all of the attributes in this section. These options range from Preview Quality, the lowest setting, to 3D Motion Blur Production, the highest, and slowest, setting. For the most part, these Quality presets are all that you need to deal with when rendering most scenes. However, additional tweaking may be necessary at times.

The attributes found in the Number of Samples section control the anti-aliasing for the scene. The Shading and Max Shading attributes deal with how many samples are taken for each pixel, as discussed previously. The Shading attribute sets the minimum number of samples used for each pixel for the entire scene. It is most ideal to keep this setting as low as possible. A value of 2 is probably high enough unless artifacts appear.

To make anti-aliasing more efficient, the renderer uses an adaptive algorithm that supersamples only the areas of the image that need it. This adaptive algorithm is based on the contrast between the pixels in the image. Max Shading sets the number of samples taken when the contrast between edges exceeds the values set in the Contrast Threshold settings. Therefore, if the Shading attribute is set to 2, the renderer takes one pass at the image by sampling each pixel twice. If the contrast between these samples exceeds the amount set in the Contrast Threshold settings, the renderer supersamples those pixels by the amount set in the Max Shading attribute.

Once the anti-aliasing process is complete, the image can be blurred slightly to get rid of any problems that the anti-aliasing process missed. You can select a filter type in the Pixel Filter Type pull-down menu. The Box Filter is the softest, while the Gaussian Filter is a bit sharper.

## Anti-Aliasing in Mental Ray

When the Mental Ray renderer is selected in the Render Settings window, the anti-alias quality settings can be accessed, shown here. Mental Ray's anti-aliasing algorithm works similarly to the Maya software renderer's algorithm but it provides an additional optimization. On the low side, Mental Ray has the ability to sample not just one sample for every pixel, but it can do, for example, one sample for every four pixels. This is great when creating previews or even images that have a lot of solid colors.

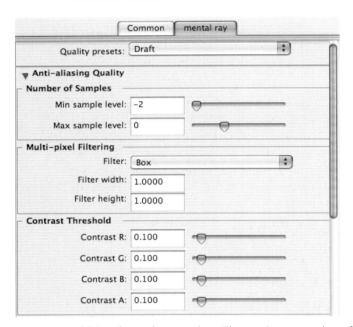

Instead of setting the number of samples directly, as in the Maya software renderer, you set the Min and Max Sample Level attributes based on an exponent. A value of 0 is equal to one sample per pixel. A value of 1 means that four samples will be taken for each pixel. A value of 2 will take 16 samples. A value of –1 takes one sample for every four pixels.

The Min and Max Sample Level attributes are analogous to the Shading and Max Shading attributes in the Maya software renderer. Min Sample Level controls how many samples are taken during the first pass of the render. If the contrast values of these samples exceed that of the Contrast RGBA attributes, additional samples are taken. The maximum number of samples taken is determined by the Max Sample Level attribute's value.

## Raytracing

A rendering engine can use many different methods to calculate the shading of a 3D scene. The renderer can consider all of the surface, material, and lighting attributes and calculate the final image in many ways. The default renderer in Maya uses what is called "normal vector interpolation" shading. This shading occurs by interpolating the vertex normals and shading every point on the surface by computing the relationship between the angle of its normal and the angle of its light. While this works well for rendering objects that get all of their color information from the material's color and/or light, it does little to solve the issue of rendering a reflective surface, where points on the surface must also reflect the colors of the objects that surround it. Raytracing offers a very accurate solution to this problem.

## How Raytracing Works

Raytracing is a rendering algorithm that provides a very accurate method of calculating reflections and shadows. In the real world, rays of light, or photons, are emitted from a light source until they hit an object and then bounce off and possibly hit another object. They continue to bounce around until they are reflected back up into space. If a computer were to calculate all of this behavior, a very long time would pass before you saw your rendered image. But we are not concerned with all the rays that bounce up into space or travel in the opposite direction from our point of view—only the rays that end up in our eye or, in Maya, in our camera have any significance to our rendered image. To speed rendering, then, raytracing traces the rays of light from the camera out into the 3D scene.

Figure 21-4 demonstrates how raytracing works. For each pixel in the rendered image, a ray is shot straight away from the camera. The ray travels until it hits an object in the scene. The surface value at this specific point is calculated. Next, another ray is shot from this point on the surface toward the point in space that is emitting the light. If it encounters another object in its path, the renderer knows that that point is not absorbing any light and therefore returns a value of 0, or black. Obviously, if more than one light is in the scene, more than one calculation has to be made from this point.

If the shadow calculation returns and tells the renderer that it was able to reach the light's source without intersecting anything, and this point has a Reflectivity attribute value of more than 0, the ray is reflected from that point. The angle at which this reflected ray is projected is equal to the angle between the ray from the camera and the surface normal (some-

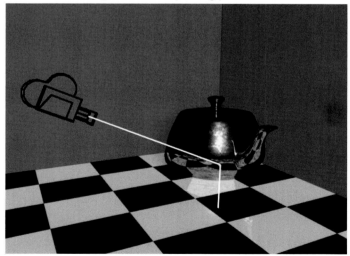

FIGURE 21-4    *Diagram of the raytracing process*

times called the *angle of incidence*). The reflected ray then continues in a straight line until it intersects another object. At this point, the process repeats. The value of this new surface is determined. This value is then multiplied with the first value by a factor of the reflectivity and the result is added to the original surface value. Ideally, this process continues until the reflective ray does not intersect any more objects. Figure 21-4 illustrates the process of a ray being shot from the camera and bouncing off the objects in the scene.

If the surface that intersects with the reflected ray intersects with a transparent surface, a *refracted* ray is generated. The angle of this ray is determined by the thickness of the surface. This refracted ray simulates the way that light rays are bent as they travel through transparent solids. For this reason, to get an accurate refraction, the object must have two surfaces to determine the thickness.

## Tutorial: Using Raytracing in Maya

To use raytracing in Maya, at least two objects must be present in a scene, since the first object must have something to reflect; at least one light source must also be present. If reflections are to be calculated, the reflectivity of one of the materials must have a Reflectivity attribute value greater than 0.

> **NOTE**    **Both the Maya software renderer and the Mental Ray renderer support raytracing. This example uses the default Maya software renderer. By default, raytracing is turned off in the Maya software renderer, but it is turned on in the Mental Ray renderer.**

To practice using raytracing, we will use a simple scene containing four walls, a floor, a ceiling, a teapot, and a directional light. To follow along, use the raytrace.mb scene included on the DVD. It meets all of the minimum requirements for raytracing, as mentioned earlier. It is a simple setup. The room has a different-colored wall so that you can see exactly how the raytracing is evaluating the reflections.

> **TIP**    **As you fine-tune a render, it is best to render to a small resolution such as 320×240. Rendering larger images will cost you lots of time.**

**1.**    With the objects, materials, and light set up, do a quick test render by clicking the Render Current Frame button on the status line. This should give you a good idea of what the scene looks like without raytracing enabled. Figure 21-5 shows a test render of the scene without raytracing. Notice that the teapot is really dark. This is because its material has a low Diffuse value. We'll assume that the teapot should have a chrome-like surface so that most of its color will come from reflecting the objects around it.

**2.** Turn on raytracing by checking the Raytracing check box under the Raytracing Quality section in the Render Settings window, shown below. With these default settings, do another test render of your scene. It takes a lot longer to render than it did before you turned on raytracing.

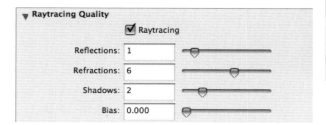

*FIGURE 21-5    Scene without any raytracing*

*FIGURE 21-6    The scene is raytraced with the default Reflectivity attribute value of 1.*

**3.** In the image shown in Figure 21-6, the teapot is reflecting all of the objects around it, but although the floor has a small Reflectivity attribute value and is reflecting the teapot, something does not look quite right. This is because we are raytracing only one reflection. For the floor to reflect the reflection of itself on the teapot, we need to increase the number of reflections that the raytracing is creating.

*TIP*   *When evaluating reflections, it is helpful to have some kind of reflectivity map connected to the material's reflectivity channel. In a situation like this, where the teapot material is very shiny, the reflectivity map will help maintain focus over the surface by attenuating the Reflectivity attribute value randomly. This way, the object is not "lost" in the reflections. All of the teapot examples in this chapter use grunge maps from dvGarage's Surface Toolkit. Samples are available on the DVD that comes with this book, but because this toolkit is one of the most valuable tools for anyone doing texture mapping, you should get it for yourself. Go to www.dvgarage.com for more information.*

**4.** In the Hypershade, select the mTeapot material node and look in the Attribute Editor. Scroll down to the section called Raytrace Options and change the Reflection Limit to **2**. If you were to go ahead and render the scene right now, you would not see a difference, because the Reflections attribute in the Render Settings window is still set to 1. The attributes in this section control the maximum amount that the renderer will calculate per attribute, regardless of what the settings are for each individual material.

**FIGURE 21-7**   *The scene is raytraced with a Reflectivity attribute value of 2 on the teapot material.*

**5.** In the Render Settings window, change the Reflections attribute to **2** and render the image again. Figure 21-7 shows the result of this adjustment. Notice that you can now see that the reflection of the reflection is visible on the floor.

You can begin to see how adjusting these attributes on a per-object basis is so efficient. If we had to turn up the Reflection Limit values to 2 for all objects in the scene, the render times could become intolerable. In most cases, a setting of 1 reflection is sufficient, especially in animation, where the image might be moving so fast that the viewer would never notice that the reflections were not completely accurate.

Because of this, it is sometimes necessary to raytrace only the closest object to the reflective object and then use an environment map for the rest of the scene.

To remove an object from the raytracing calculation, select the object. In this case, select the red and green walls. In each of these object's Attribute Editor, find the Render Stats folder and uncheck the Visible in Reflections check box. Do a test render and you'll see that the walls are no longer reflected.

You can use the Environment Ball or Environment Sphere methods (see Chapter 27) to set up the environment map and connect it to the Reflected Color attribute of the teapot material. This way, the accuracy of the reflection of the floor leads the viewer's eye away from any inaccuracies in the environment. If this were an animated shot, you could save hours by calculating only the floor in this rendering pass.

The dvGarage reflection maps work perfectly in situations like this. In Figure 21-8, a map from the Reflection Toolkit shows its reflection on the objects.

## Set Up Shadows

With the reflections covered, the last thing we need to do is add the shadow. Until this point in this book, we have been using only the depth map shadows. Raytraced shadows are much more accurate than depth map shadows. While we are going to practice turning them on here, their usefulness will become more evident when we make the teapot transparent later in the chapter.

**FIGURE 21-8** *The scene is rendered again with an environment map connected to the Reflected Color attribute of the teapot material.*

1. Select the directionalLight1 node from the Outliner or the Hypershade window. In the light's Attribute Editor, find the Shadows folder; then find the Raytrace Shadow attributes. Check the Use Raytrace Shadows check box. Render the image to see the effect. This produces a sharp shadow on the floor and red wall. Figure 21-9 shows the scene rendered with raytraced shadows.

**FIGURE 21-9** *The scene has been rendered with raytraced shadows.*

2. To soften the shadow edges a bit, increase the Light Radius attribute in the light's Attribute Editor under the Raytrace Shadow attributes. This simulates a larger light source, which produces softer edges on the shadows. The default setting of 0 is producing laser-perfect edges, which are not really desirable for most scenes. If you increase this value to 20 and render again, you'll see a more diffused shadow around the inner, sharper one. However, this diffused shadow is very grainy. To soften the grain, we'll increase the Shadow Rays value. Try a value of **10**.

## Set Up Refractions

As the Light Angle and Shadow Rays attribute values are increased, the rendering time starts to become pretty lengthy—too long for use in many production environments. In situations like these, it is best to use the depth map shadows because they are much faster. As you study transparent objects next, it will become more obvious when raytraced shadows are necessary.

As light passes through a transparent surface, the light rays are bent. Consider a glass window— the thicker the glass, the larger the angle that the light rays are bent. This effect is known as *refraction*. Because raytraced rendering methods use a ray-emitting method, it is easy for them to calculate refractions as well as reflections accurately. We will use an object that's similar to our existing teapot with a few differences. For one, the teapot will be mostly transparent, as if it were made out of glass. The walls will use a checker texture, because the effects of the reflection will be more obvious with a pattern behind the teapot.

**1.**  Find the scene called teapot_RaytraceRefractions.mb on the DVD and open it in Maya.

**2.**  Render the scene with the default settings. In this case, raytracing has not been turned on in the Render Settings window. The result is a transparent teapot without any reflections or refractions, as shown in the Figure 21-10a.

**FIGURE 21-10**    *(a) Close-up of the teapot without raytracing and then (b) with a value of 2, (c) with a value of 4, and (d) with a value of 6 refractions*

**3.** Select the teapot material and turn on Refractions under the Raytrace Options folder in the material's Attribute Editor. Then turn on Raytracing in the Render Settings window. To evaluate the refractions, turn off all reflections by setting the Reflection attribute to **0**.

**4.** When using refractions, it is a good idea to find out exactly how many refractions you will need. The default is 6, which is the minimum number you need to refract all of the overlapping surfaces in the teapot at this angle. Just to see what happens when there are not enough refractions, turn down the Refraction setting to **2** in the Render Settings window and do a test render. The teapot renders in black (see Figure 21-10b), which tells us that not enough refractions are used in the calculation to render all of these overlapping transparent surfaces.

**5.** A value of 4 (see Figure 21-10c) is enough to refract the main areas around the pot, because the light needs to pass through only four of the surfaces, but the areas where we can see the handle through the teapot and some of the edges at the top are still problematic, as more overlapping surfaces are in these areas. With that understood, set the Refractions value back to **6** and render it (see Figure 21-10d).

**6.** Now we know we have enough refractions, but we aren't seeing any of the distortion effects of the background caused by the bending of light rays through the transparent surface. This is determined by the Refractive Index attribute in the material attributes. A value of 1 is the default and simulates the refractive index of air. The refractive index for glass is about 1.6. Set this value for the refractive index on the teapot material and render the scene to see the effect.

**7.** Now for the shadow. Because refractions cast their own shadows, we don't need to turn on raytraced shadows for the light. Instead, turn up the Shadows attribute, found in the Raytracing Quality section of the Render Settings window, to 2 (the default). Render the image again. You can see that the shadow has more transparent areas where the light is passing through less dense areas of the teapot. The overall lightness of this shadow can be attenuated with the Shadow Attenuation attribute in the Raytrace Options section of the teapot's material node. A higher value will create a denser, darker shadow.

**8.** To finalize this scene, turn up the raytraced reflections and use our environment map to reflect areas not covered by geometry in the scene. The final raytraced scene with refractions is shown in Figure 21-11.

*TIP*  *To achieve even a higher degree of realism, you can add the Fresnel Effect to attenuate the reflections and transparency. This technique is covered in Chapters 27 and 28.*

FIGURE 21-11   *Final scene using raytraced reflections and refractions*

# Tutorial: Creating Masks

In large productions, you will usually render your scenes in different layers and passes. A scene render can be broken up and organized from something simple, such as background and foreground elements, to rendering every object separately. Issues will occur in compositing when any of the rendered layers obstructs a layer underneath it. While you could swap the layers during the middle of the shot in your compositing package, a better solution is to render a mask.

Rendering a mask into the alpha channel of the output image gives the compositor something to work with. The compositor does not have to guess which object is obstructing another. Instead, when an object from the top layer of the composite is supposed to pass underneath, the mask will hide it so that it does not show up on top.

Here we will use an example of a moon rotating around a planet. When the moon passes in front of the planet, it will occlude the planet but not be visible in the alpha channel of the image.

**1.** Start a new scene. Create two spheres and scale one of them up so that it is about five times larger than the smaller one. Move the small sphere out into an orbiting position. If you scale the first sphere up 5 units, then moving the small sphere out 10 units along its X or Z axis will work fine.

**2.** Instead of keyframing this animation, we will set it up procedurally by writing an expression. First, we'll move the pivot point of the moon to the center of the scene. Select the moon and press INSERT (HOME) and turn on grid snapping.

**3.** Drag the pivot to the 0,0,0 origin of the world, and press INSERT (HOME) again. In the Channel Box, right-click the moon's Rotate Y attribute and choose Expressions from the marking menu.

**4.** In the Expression field, type this:

```
nurbsSphere1.rotateY= 30*time
```

**5.** Apply a Lambert material to the small sphere. Name this material **mMask**.

**6.** In the material's attributes, set the Matte Opacity Mode to Black Hole, as shown here:

**7.** Play the animation until the moon is partially occluded by the planet. Render the view window using the Render Current Frame button on the status line. In the Render View window, click the Display Alpha Channel button in the toolbar. You can see that the alpha channel does not include a transparency, or white, pixel value for the mask object. This functionality can be very useful to a compositor. (We will discuss how to composite this in Chapter 28.)

# Motion Blur

To understand motion blur, it is important that you understand how a camera in the real world works. (See Chapter 20 for information on cameras.) Motion blur is an artifact of exposure time. It is noticeable as a blurry trail left behind a moving object. In general, the faster an object moves by the camera, or the greater the change in distance of an object from frame to frame, the longer the blurred trail will be.

Maya offers two kinds of motion blur: 2D and 3D. The 2D motion blur is a post-rendering effect and is calculated only in two dimensions. It is faster to render, but it may not produce a realistic effect if the objects in the scene move toward the camera. The 3D motion blur option works similar to a real camera, and at the default setting it depends on the shutter angle for the amount of blur. You can increase the amount of motion blur by increasing the Blur by Frame attribute. A setting of 1, as shown in Figure 21-12, blurs the image over one frame based on a 180-degree shutter. A Blur by Frame value of 4, also shown in Figure 21-12, looks at the motion of objects over four frames and applies the blur.

If the motion blur appears too grainy, you can calculate additional samples in the Anti-aliasing Quality section of the Maya Software tab in the Render Settings window. Two attributes here can be used to improve the quality of the motion blur: 3D Blur Visibility and Max 3D Blur Visibility. Since these settings are adaptive, increasing the Max 3D Blur Visibility setting will render additional samples only where needed.

FIGURE 21-12   *3D motion blur with a Blur by Frame setting of 1 (left) and 4 (right)*

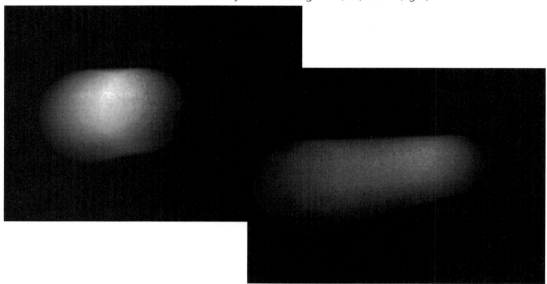

# Environment Fog

Environment fog is used to simulate atmospheric effects. Since you don't have precise control over the way the fog is distributed in Maya, it simulates the haze or moisture in the air rather than producing patchy fog or clouds—such effects are handled much better with light fog and particles. However, when using Maya for realistic rendering, environment fog is a must-have, since some bit of atmosphere is always visible, even on the clearest of days.

 We will use the example scene from the DVD called mcr8_ch21_cylinders.ma, which contains cylinders scattered around on a floor.

**1.** Open the Render Settings window. Make sure that the Resolution attribute is set low and the Quality attribute is set to Preview or Intermediate at the most. This will let you preview the effect without having to wait. Scroll down in this window and find the Render Options section. Click the connection button next to the Environment Fog attribute field to add an Environment Fog node to the scene.

**2.** Open the Hypershade, and you'll notice that a new envFog material appears. This is a *volumetric* material, which means that it does not need to be connected to an object to render. If you graph its output connections, you'll see a Light Linking node connected to it. Connected to the Light Linking node is a point light called envFogLight. Since this light is linked only to the envFog node, it has no effect on the other objects.

**3.** Render the scene. Figure 21-13 shows the effects of the environment fog so far. In this case, the fog appears behind the objects in the scene. While this might be fine for some effects, we want the fog to cover the objects that appear directly in front of us.

**4.** To fix this, we will use the Saturation Distance attribute in the envFogMaterial node's Attribute Editor. This represents the value, in units, from the camera from which the fog will begin to be visible. The default is 100. Set the Saturation Distance to **5** and render a preview. The result should match Figure 21-14.

**5.** To adjust the height of the fog, check the box labeled Use Height in the Attribute Editor. The default settings here will do the job. However, if you want to make the fog extend higher, adjust the Max Height attribute.

**6.** The Blend Range attribute can also be used to adjust the density of the fog as it nears the Max Height.

*FIGURE 21-13*    *Fog is visible in the background of the scene.*

*FIGURE 21-14*    *The scene rendered with the fog material's Saturation Distance attribute set to 5*

# Mental Ray

Maya ships with another software-based renderer in addition to its own. At this time, the Mental Ray renderer is not capable of handling everything in Maya, but it does have some features not available in the Maya software renderer. While it does an exceptional job at raytracing and anti-aliasing, the Mental Ray renderer also offers Global Illumination, caustic effects, and an area-lighting method called Final Gather.

The normal vector interpolation and the raytracing methods used so far do an excellent job of rendering the shading on surfaces and accurate reflections. However, they fall short of being able to account for the light spill, diffusion, or color bleeding that are evident in the real world. For the rest of this chapter, we will examine some of the features in Mental Ray that change the way lighting in a scene is calculated. The Global Illumination and caustics rendering methods use light-emitting photons that diffuse and spill colors, shadows, and reflections. The Final Gather method does not use any lights at all. Instead, it uses the area around each object to create even, realistic lighting.

> **NOTE**   *Before continuing on, make sure that the Mental Ray plug-in is loaded by choosing Window | Settings/Preferences | Plug-in Manager and enabling the mayatomr check box.*

## Tutorial: Rendering with Global Illumination

Global Illumination works by shooting photons from lights and creates a photon map according to their distribution in the scene. The way that these photons are distributed is based on the absorption, reflectivity, and refractivity of an object. The effects of this type of rendering are most noticeable in the color bleeding between objects and soft diffusion of light across a scene.

For this tutorial, we will use the teapot_GI.ma scene included on the DVD. It is similar to the rendering example that we have used so far, except the teapot is now a solid green color and the room colors are fairly plain. We will use this color scheme to make the effects of Global Illumination obvious.

**FIGURE 21-15**   *Scene rendered with the Maya software renderer*

**1.** Render the image with the Maya software renderer. The result is shown in Figure 21-15. You can save this image and use it for comparison with the Global Illumination rendering we are about to perform.

**2.** Now to enable Mental Ray. In the Render Settings window, choose Mental Ray from the Render Using pull-down menu. Select the Mental Ray tab. You should now be able to view the settings for Mental Ray, shown next.

**3.** Choose PreviewGlobalIllum for the Quality Presets setting. This turns on Global Illumination and sets the Sampling Quality attributes to a medium setting for faster renders. You can verify that Global Illumination is turned on by scrolling down to the Caustics and Global Illumination folder in the Render Settings window and making sure the box next to Global Illumination is checked.

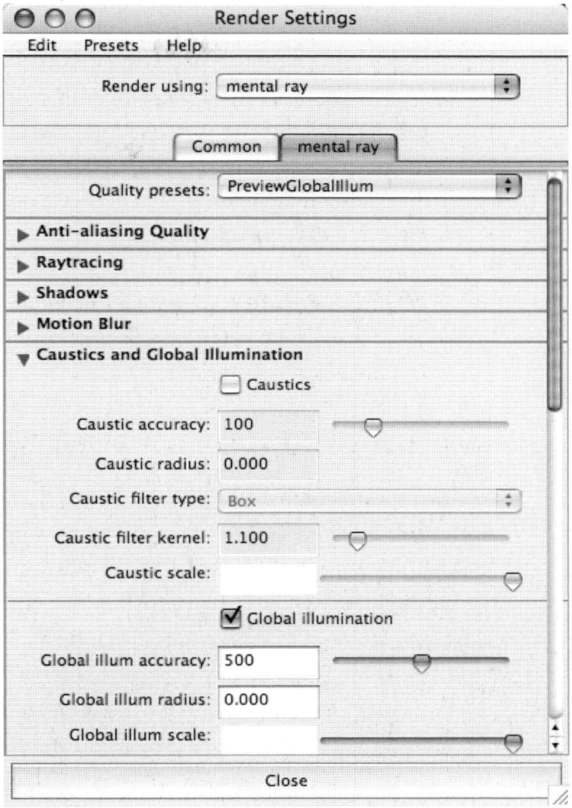

If you were to render the scene right now, it would not look different from the last image we rendered with the Maya software renderer. This is because the lights in the scene have to be set up to emit photons in order for the Global Illumination to be calculated.

**4.** Select the point light in the scene (this is the only light used). In the Attribute Editor, scroll down and find the section named Mental Ray. In this section, click the box next to Emit Photons, as shown here.

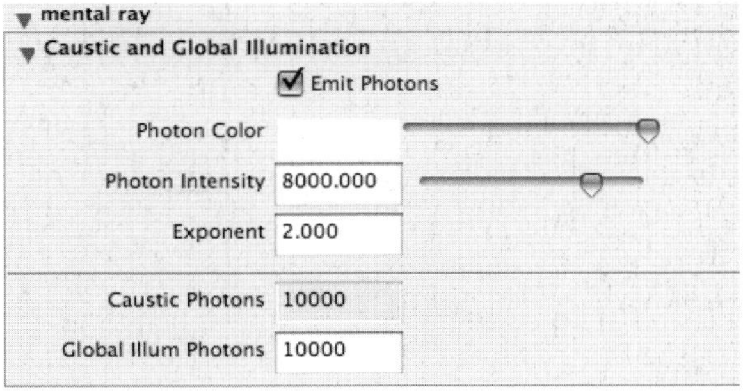

- The Photon Intensity attribute here controls the brightness of the emitted photons. If you click the Render button and don't see any Global Illumination effect, try increasing the Photon Intensity attribute value.

- The Global Illum Photons attribute sets the number of photons that will be emitted from the light so that Mental Ray can render the Global Illumination effect. Because the number of emitted photons has an impact on render times, it is best to keep this as low as possible. Remember that the number of photons needed to create a quality render depends on the scale, in Maya units, of the scene being rendered. If the scene is small, it may not take too many photons to fill the area being rendered. Therefore, the correct number of photons will vary greatly from scene to scene.

*TIP*   *While setting up Global Illumination in a scene, it can be helpful to work with large Energy values so that you can closely monitor what is going on. The Energy attributes can then be reduced for the final output.*

**5.** Set the Global Illum Photons attribute to **1000**. This will be enough to cover the scene with photons, but they will not be so finely distributed that they can't be recognized. At this point, you have enabled enough controls for the renderer to make a Global Illumination calculation. Do a test render and view the results.

**FIGURE 21-16**   *Scene rendered with Global Illumination enabled but with a low emitted photon value*

Figure 21-16 shows the teapot rendered with these settings. Notice the big, circular-shaped spots of color all over the scene. By setting the Global Illum Photons value low, the photons are easy to see and should give you some idea of what they are doing to bleed the color throughout the scene.

To blend these photons in the rendered image, we will go back to the Render Settings window and adjust some more settings. The two attributes we will concentrate on in this step are Global Illum Accuracy and Global Illum Radius, shown and described here:

- **Global Illum Accuracy** Represents the number of photons that Mental Ray looks for within the area represented by the Global Illum Radius setting. It then composites what it finds, and the result is the Global Illumination effect. Low Global Illum Accuracy values yield grainy images, and high values give you smoother, more precise results but take more time to render. If you set the Accuracy value higher than the number of photons emitted by all the lights in the scene, your image will never render because the photon calculation won't finish.

- **Global Illum Radius** The size (in Maya units) of the effect that the photon produces when it hits an object. Larger numbers produce bigger effects but usually less accurate results. High Radius values combined with low Accuracy values can give you a good approximation of what your scene will look like and are great for test renders. Low Radius values combined with high Accuracy values provide the most accurate representation of the final image but take much longer to render.

A value of 0 for either of these attributes tells the renderer to make its best guess.

And now to the blending:

**1.** In the Render Settings window, begin adjusting the Global Illum Accuracy attribute. Set it to **1** and notice the effect. The photons are not sharing any space at all. You want to increase this value until it has little or no effect—this means that no more photons fall within that radius. For the accuracy to have any more effect, the number of photons emitted from the light must be increased. A value of around 60 works well for Global Illum Accuracy in this scene.

**2.** Adjust the Global Illum Radius attribute. Increase it until the photons are big enough to cover the areas evenly but are not so spotty. A value of 4.5 is good.

**3.** Return to the light's Attribute Editor and adjust the values for the Photon Intensity, Exponent, and the number of emitted photons. Increase the number of photons until they blend enough so that you can no longer make them out. Increase the Global Illum Photons attribute to **4500**.

**4.** To determine how bright the overall effect will be, use the Photon Intensity attribute. A value of **7000** gives the walls a nice bleed from the green teapot.

**5.** The scene was set up to use regular depth map shadows, but we need to turn shadows on in the Mental Ray Render Settings windows for them to render. Set the Shadow Maps to On. Figure 21-17 shows the final rendered image for this scene.

**FIGURE 21-17**   *Final scene rendered with Global Illumination and shadows*

# Tutorial: Rendering Caustics

*Caustics* are light patterns that are created when light from a light source illuminates a diffuse surface via one or more specular reflections or transmissions. Caustics use the same method of emitting photons from light sources as does Global Illumination. Therefore, the process for optimally rendering caustics is similar to what we did in the previous section:

1. Find the scene on the DVD called teapot_caustics.ma. This scene shows the transparent teapot, and the light has been moved down just a bit so that it is casting a shadow on the wall.

2. Render the scene with the Maya software renderer. It is all set up for the right amount of reflections and refractions in the Raytrace settings for the teapot material. The rendered image is shown in Figure 21-18.

**FIGURE 21-18**  *Teapot rendered with Maya's default renderer using raytracing*

3. In the Render Settings window, choose Render Using Mental Ray. You should now be able to view the Mental Ray Render Settings tab in the Render Settings window.

4. Set the Quality Presets attribute to PreviewCaustics.

5. Increase some of the raytracing values. Because of the angle at which we are looking at the teapot, maximize the number of refractions calculated. Set Max Refraction Rays to **10**. If you were to do a test render, you would see that the dense areas around the handle and spout of the teapot are dark. You can make these areas more transparent by increasing the Max Ray Depth. Set this to a value of **6**.

6. Select the point light in the scene (the only light). In the Attribute Editor, scroll down to the section named Mental Ray. Click the box next to Emit Photons. Because the number of emitted photons will dramatically increase render time, set the Caustic Photons attribute to **1000**. This will be enough to cover the scene, but they will not be so finely distributed that they can't be recognized.

7. At this point, you have enabled enough controls for the renderer to make a calculation for the caustics. Do a test render. Figure 21-19 shows the scene rendered with caustics.

You should notice two things: the caustic effects are very spotty and the shadow is no longer transparent.

**8.** First we'll fix the shadow. If you scroll down to the bottom of the Caustics and Global Illumination section of the Render Settings window, you will see a Direct Illumination Shadow Effects check box. Enabling this check box allows the shadow to maintain its transparency.

**9.** Adjust the Caustic Radius value. In the Render Settings window, begin increasing the Caustic Radius attribute and doing test renders until increasing this number no longer has an effect. Try a value of **5**.

**10.** Adjust the Caustic Accuracy attribute in the same manner. Increase this value until it has little or no effect—**200** looks good. Figure 21-20 shows the result of the render after these settings are adjusted.

*FIGURE 21-19* Scene rendered with a Caustic Photons value of 1000

*FIGURE 21-20* Scene rendered with the Caustic Accuracy and Caustic Radius attributes increased

**11.** We need to get rid of the strong highlight that the caustic is making on the wall with the shadow. We can do this by editing the wall material's Diffuse value. A value of 0.6 minimizes this bloom. You might also try turning down the Photon Intensity value of the photon-emitting light. The result should resemble Figure 21-21.

**FIGURE 21-21** *Fine-tuning is done to the diffusion of the walls and the energy of the emitting photons for the final image.*

Once you have found a Photon Intensity attribute that works, you may wish to try adjusting the overall quality of the image by increasing the number of photons that are emitted from the light, lowering the Global Illum Radius attribute, and increasing the Global Illum Accuracy attribute. But be prepared to wait. Remember that if you are rendering for animation purposes, some of the artifacts caused by the photon emission may not be so noticeable.

## Tutorial: Using Final Gather

Setting up lighting in Maya to simulate the way light behaves in the real world is difficult, to say the least. To achieve even lighting across a scene, you will almost always use multiple lights. As we have seen so far with Global Illumination, we can set up fewer lights since Global Illumination gives us much of the even diffusion that we get when using multiple lights. It is from these lights that Global Illumination emits photons, which bounce off objects in the scene and record their color data so that it can be transferred to the surface shading of other objects. Final Gather goes one step further toward simulating the way light works in the real world.

Take, for example, the inside of your house during the day. You can turn off all the lights, but if the house has windows and the doors between the rooms are open, light will make its way through from the outside by bouncing around off every surface it hits inside. This generally diffuses the light and produces some even lighting throughout your house.

Final Gather works similarly. The light used to illuminate objects is generated by the other objects in the scene. The light rays just keep bouncing off objects, spilling the colors and creating very diffuse, soft shadows as they head toward places in the scene where they cannot so easily bounce. All that is really needed, of course, is a surface with some sort of luminosity to emit the initial rays. This is usually done in Final Gather by placing an object, usually a dome-shaped object, with a material that has its incandescence turned up. This incandescent object is what begins the lighting reaction, which is then transferred to the rest of the scene.

While Final Gather will produce the highest-quality lighting in a scene in most cases, it is also the slowest to render. Therefore, it is helpful to use Final Gather in conjunction with Global Illumination to achieve the desired results with acceptable render times.

1. In this example, we will again be using our teapot scene. Find the teapot_FinalGather.ma file on the DVD.

2. In the Render Settings window, choose Render Using Mental Ray. You should now be able to view the Mental Ray Render Settings tab.

3. Set the Quality Presets attribute to PreviewFinalGather.

4. Scroll down the list of attributes and turn off Global Illumination. Then, in the Final Gather section, make sure that Final Gather is enabled.

5. Select the light in the scene and turn its Intensity all the way down to **0**. With Final Gather, we do not need to use any lights. Turning this down keeps Mental Ray from using the default lighting in the absence of any other lights. If you rendered the image right now, it would be all black. This is because we need an illuminating surface to use as the main source of light.

6. Create a sphere. Edit the makeNurbSphere1's attributes so that it is only a half sphere, and scale it up so that it acts as a dome that surrounds the entire scene. Figure 21-22 shows the scene in the view window with the half sphere used as a source of illumination.

7. Create a new Lambert material. Turn the Incandescence attribute up 100 percent and apply it to the dome-shaped geometry.

8. Before we render, find the Final Gather Rays attribute in the Final Gather section of the Render Settings window and set it to **300**. Render the image to see the results. You will immediately notice the nice, even lighting and smooth, soft, and diffused shadows in the scene. Figure 21-23 shows how the scene looks rendered with Final Gather. Hard to believe that we didn't need to set up one light!

**FIGURE 21-22**  *Placement of the half sphere in the scene*

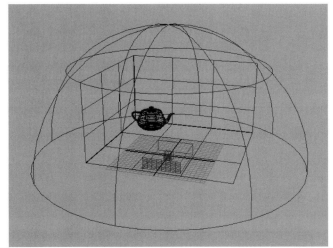

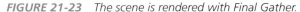

*FIGURE 21-23*    *The scene is rendered with Final Gather.*

**9.** To adjust the color of the lighting, change the Incandescence value of the Lambert material on the dome object to a bluish color and see the results.

**10.** To adjust the overall effect on the teapot, we can adjust the material's irradiance color. Select the teapot material and scroll down in its Attribute Editor to find the Mental Ray folder. In that folder is an attribute called Irradiance Color. Turn this down to change the brightness of the Final Gather effect.

## HDR Images with Final Gather

The term "image-based rendering" has become popular recently. It describes a type of rendering that uses images from the real world to calculate lighting. Image-based rendering includes camera mapping, described in Chapter 20, which occurs when images from the real world are projected onto simple geometry that roughly resembles their shape from a specific camera angle. In these kinds of scenes, it is not necessary to add light because all of the lighting information is already inside the photograph.

When we start thinking about these concepts in terms of Final Gather, image-based rendering can mean that we use photographs from the real world to control the Incandescence attribute of our dome or irradiating surface's material for the Final Gather render. The problem is, however, that a regular photograph does not have all of the information we need. We would need a file format that does not just record colors that are dark and bright, but one that understands overexposure as well. Thanks to the work and research of Paul Debevec, we have an image format that encompasses all of this information into a single file. This file format is called the *HDR file format*.

Paul Debevec's website, at www.debevec.org/, offers a program called HDR Shop that allows you to create images in this file format. Images are also available on the site for download that can be used in Final Gather. This is what we will be working with in this section.

HDRI stands for *high dynamic range imagery*: a normal 24-bit image can represent brightness as only one of 256 levels—this means that a pixel with a brightness value of 256 (that is, white) is exactly twice as bright as mid-gray (brightness of 128). While color space represents a printed image fine, it bears no relation to reality, where the bright sky outside your window is, in fact, many thousands of times brighter than the mid-gray page in front of you, and the spot where the sun shines is probably millions of times brighter. By using a 32-bit floating point number (floating point numbers are capable of representing enormous ranges by moving their decimal points around) to represent each pixel brightness for R, G, and B instead of a limited 8-bit integer (a whole number limited to a range from 1 to 256), HDR images can truly represent the range of brightness found in real life. This is important if you want to re-create realistic lighting accurately using a rendering technique such as Final Gather.

Once again, and for the last time, we will use our teapot scene as the test scene for the study of HDR rendering. You can use the scene from the last section or start with the teapot_HDR.ma scene on the DVD.

**1.** In the Render Settings window, choose Render Using Mental Ray. You should now see the Mental Ray Render Settings tab.

**2.** Set the Quality Presets attribute to PreviewFinalGather.

**3.** Scroll down the list of attributes and turn off Global Illumination. Then, in the Final Gather section, check to make sure that Final Gather is enabled.

**4.** Select the light in the scene and turn its Intensity attribute all of the way down to **0**. With Final Gather, we do not need to use any lights. Turning this down will keep Mental Ray from using the default lighting in the absence of any other lights.

**5.** We could create a sphere, as we did in the previous tutorial, and then map the HDRI image to it, but instead we will use an image-based lighting node. This will create a spherical environment similar to an environment sphere that we have used in past tutorials, except the image-based lighting node has specific attributes that let you control the lighting in the scene when using Final Gather. To create the image-based lighting node, find the Image Based Lighting section and click the Create button. The image-based lighting attributes are shown in Figure 21-24.

*FIGURE 21-24    The attributes for the image-based lighting node*

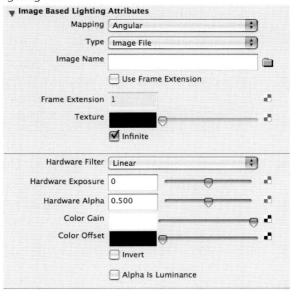

**6.** Open the Attribute Editor to view the attributes for the image-based lighting node. You'll see an attribute called Image Name. Click the File Browser button to browse to the DVD and open this HDR_sample.hdr image:

mentalraylblShape1

**7.** In the Render Settings window, find the Final Gather Rays attribute in the Final Gather section and set this to **200**. This should provide you with a low-quality render that is reasonably fast.

**8.** You can adjust the brightness levels of the HDR image in the Attribute Editor by editing the Color Gain and Color Offset attributes.

**9.** You can make any adjustments to the object's materials to control how they interact with the Final Gather rendering by going into each material's attributes and adjusting the Irradiance Color in the Mental Ray section.

**10.** Once you are happy with the way the scene looks, increase the number of emitted rays in the Final Gather section of the Mental Ray Render Settings tab. You might also try adjusting the Min and Max Radius values and see how far you can get with those before having to increase the number of emitted rays. Render your final image.

In production situations, you might find that using both Global Illumination and Final Gather is the optimal way of working. In a situation like this, go through the process of setting up each render type individually. You will want to concentrate on using Global Illumination for the lights and shadow casting and using Final Gather to fill in the scene with even light across it. By adjusting the Energy values of the light-emitting photons in the Global Illumination calculation and Irradiance Color attribute in each material's Attribute Editor, you can fine-tune the balance between the two effects. Using both rendering algorithms together with midrange quality settings will produce great images with manageable render times.

## Rendering with Displacement Maps

Displacement mapping has become a very popular technique for adding fine surface details to a model. While similar to bump maps, displacement maps actually offset the geometry instead of only affecting the surface normals. The result is as if the vertices of the geometry were actually sculpted with this detail. The difference is that the sculpting is done on a texture map and the resolution of the original geometry does not have to be very dense to achieve fine details.

There are two general processes that you need to follow in order to use displacement maps. The first is creating them. The second is setting up a shading network to handle the displacement maps. Finally, Mental Ray's Approximation Editor must be configured to enable the model to subdivide when rendered so that it has enough geometry to form the details in the texture map.

## Creating Displacement Maps

Although you may use one of Maya's procedural textures to displace geometry, you normally will be creating these texture maps in another application. Photoshop is always a good candidate, but the problem is that you don't have any real-time feedback of what the displacement map will look like when rendered on the 3D model. Currently, there are two applications on the market that are specifically designed for working with displacement maps—Pixologic's ZBrush and Skymatter Ltd.'s Mudbox.

Both of these applications allow you to import your geometry from Maya and further refine it by subdividing the geometry and sculpting it using very intuitive brush-based tools. These applications can handle several millions of polygons. When you are finished sculpting, displacement maps are derived based on a comparison of the original, base geometry and the high-resolution sculpted model. Figure 21-25 shows a displacement map that was created in ZBrush for a character's torso.

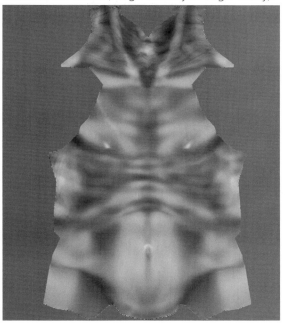

FIGURE 21-25    A displacement map is used to add details to the model (image courtesy of Roger Ridley).

## Displacement Shading Network

In order to use the displacement maps, we must connect them to a shading network using a Displacement Shader. Every material has an output connection to a node called a shading group. It is on this node that you connect a Displacement Shader by clicking the button next to the Displacement Mat. attribute, as shown here:

| ▼ Shading Group Attributes | | |
|---|---|---|
| Surface Material | Disp_04 | ← |
| Volume Material | | ▪ |
| Displacement Mat. | file4 | ← |

This creates a Displacement Shader that you may then connect a file texture node to and import your displacement map. Figure 21-26 shows the complete shading network to handle a displacement map. Depending on the software used to generate the displacement map, you may need to adjust the Color Balance attributes on the file texture node to control the amount of displacement. Most applications will suggest some starting points. A decent starting point for displacement maps created in ZBrush is an Alpha Gain of 2 and an Alpha Offset of –1.

*FIGURE 21-26*  *A shading network for a displacement map*

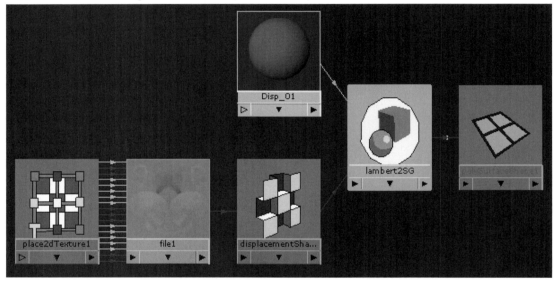

## The Approximation Editor

Mental Ray's Approximation Editor allows you to create approximation nodes for different types of geometry. When rendering NURBS surfaces in Mental Ray, you may use the Approximation Editor to approximate, or tessellate, the NURBS surfaces to an optimal resolution. When dealing with polygonal meshes, the Approximation Editor can be used to connect a subdivision approximation node to the geometry. This will allow the model to be subdivided when rendered, based on conditions specified on this node. This is much more efficient than smoothing the geometry using the Smooth command that we used in Chapter 4.

To create a subdivision approximation node, you need to select all of the geometry you wish to approximate and choose Window | Rendering Editors | Mental Ray | Approximation Editor. In this example, we are using a polygonal mesh, so we click the Create button in the Subdivision

Approx. field, shown here. This will create a subdivision approximation node and connect it to the selected geometry.

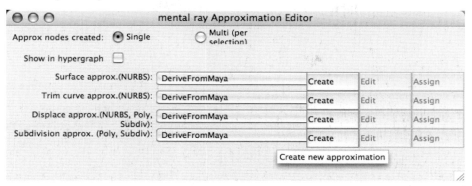

To edit the Approximation attributes, click the Edit button in the Approximation Editor and the subdivision approximation node will load in the Attribute Editor. Figure 21-27 shows the subdivision approximation node's attributes. The Approx. Method setting determines what conditions will be used to subdivide the model.

In this example, the Spatial method is used, which lets us subdivide the geometry based on the maximum length of a polygonal edge. In this case, if the edge exceeds a length of 0.01, Mental Ray will subdivide the geometry. Mental Ray will continue subdividing the geometry until it reaches the Max Subdivisions value. In order to see the detail in a displace-

**FIGURE 21-27** *The subdivision approximation node's attributes*

ment map, you will often need to increase the value for the Min Subdivisions attribute. When you are ready to see the results of your displacement map, render the image. Be patient! Rendering displacement in Mental Ray can take a long time. Figure 21-28 shows the geometry before and after adding the displacement maps. As you can see, a lot of fine detail is added this way.

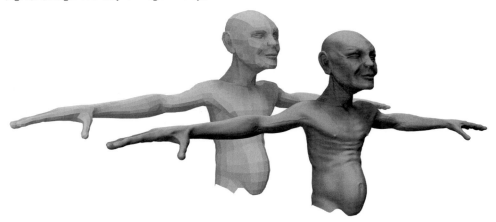

## Summary

This chapter provided an overview of the major rendering features and workflows available to you in Maya. Raytracing, Global Illumination, and Final Gather can help you produce realistic imagery when needed. As you've seen here, most of these options can be enabled and controlled in the Render Settings window. And using Mental Ray's Approximation Editor, you can also render using displacement maps. While the software-based renderers have been covered here, Chapters 22 and 23 will make use of the Maya hardware renderer to render particles.

# VI

# Particles, Emitters, and Fields

# 22

# Particles
# and Fields

**Particles are points that can be** displayed or rendered as dots, spheres, clouds, blobs, fog, smoke, and a whole lot more. Once the particles have been created in a scene, their behavior and appearance can be controlled through the use of fields, ramps, goals, and expressions. An infinite number of effects can be created in this way, ranging from running water to fire and explosions.

# Creating Particles

You can create particles in three ways in Maya: with the Particle tool, with an emitter, or by using the `emit` MEL command. In this section, we will concentrate on the first two methods and discuss some of the most commonly used attributes.

## Particle Tool

Maya's Particle tool (Particles | Particle Tool) can be used to place individual particles or groups of particles into a scene. This tool offers three ways to add particles: the first is the default point-and-click method, the second is through the Sketch Particles option that allows you to sketch or paint particles, and the third is to define the bounds and place a 2D or 3D grid of particles.

To create particles with the Particle tool:

1. Make sure the Dynamics menu set is showing in the toolbar (press F4). In a new scene, choose Particles | Particle Tool ❑ to open the Particle Tool Settings window on the right side of the screen. Click the Reset Tool button to make sure you are at the default settings. In this mode, you can click in the view windows and place particles one at a time. The particles will be drawn as tiny crosses.

2. You can set the number of particles created each time you click by setting the value of the slider in the Particle Tool Settings window. The Radius slider controls the distance that the new particles will be created from the point at which you click. Try increasing these values and experiment.

3. Continue placing particles in the scene, and when you are finished, press ENTER (RETURN) to finish and close the tool. A single particle node is created in the scene (see Figure 22-1).

*NOTE*   **When shading mode is activated in the view window, a bounding box appears around the entire group of particles.**

By turning on Sketch Particles in the Particle Tool Settings window, you can LMB-hold and paint the particles.

Delete that particle group and go back to the Particle Tool Settings window. Check the Sketch Particles box. Click and drag in the view window and notice the results. You can paint the particles in the scene according to the Number of Particles and Maximum Radius attributes in the Particle Tool Settings window. The Sketch Interval attribute controls how far apart each little cluster of particles will be as you drag (much the same way the brush spacing control works on Photoshop brushes).

FIGURE 22-1    *Particles are sketched.*

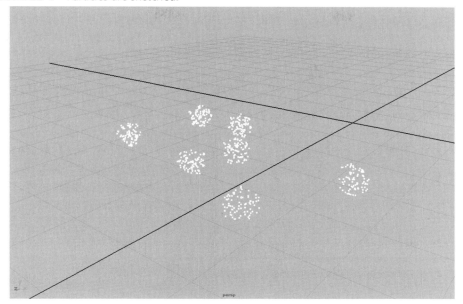

You will often want the placed or sketched particles to adhere to some shape instead of just lying flat. Recall our NURBS modeling lesson back in Chapter 5, when we drew a curve on a surface by making the surface live. The same process can be used to place particles. In this example, we will create a terrain and sketch the particles onto it to create a sort of mist or ground fog (shown in Figure 22-2):

**1.** Create a NURBS surface, and use the Sculpt Surfaces tool to push and pull the CVs to give the surface an irregular shape.

**2.** Select the surface, and choose Modify | Make Live (or click the Make Live button in the toolbar). You can now sketch particles directly on the NURBS object.

**3.** When you are finished sketching, press ENTER (RETURN) to close the tool and create the node. The result should look something like the example shown in Figure 22-2.

*NOTE   While these particles are placed as fog might fall on the ground, they will not render to resemble fog—nor will they render at all when using Maya's software renderer. We will discuss in depth later in this chapter particle display attributes and rendering particles. For now, concentrate on creating and placing the particles with the Particle tool.*

*FIGURE 22-2*    *Particles are sketched onto a live surface.*

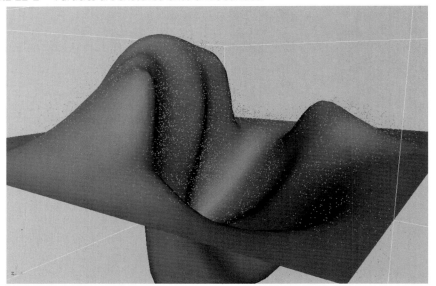

The third way to create particles with this tool is to use the particle grid. This does just what you might imagine—it creates a 2D or 3D grid of evenly spaced particles. Basically, the grid is created by placing the points that define opposite corners in each direction. This can be done interactively by clicking in the view window or by choosing the With Text Fields option and entering the grid dimensions in the Particle Tool Settings window. A particle grid is a fast way to create a large body of particles into one node. Here's how it's done:

1. Clear the scene or start a new scene, and this time check Create Particle Grid in the Particle Tool Settings window.

2. To create a 2D particle grid, click once in the view window to place one corner and click again at another spot to define the opposite corner. Press ENTER (RETURN) to close the tool and complete the grid.

3. The Particle Spacing attribute in the Particle Tool Settings window controls the density of the grid. Increase this value just a bit, and create another grid to see how it looks. *But be careful!* Creating a grid that is too dense can really slow down your computer. It is best to increase the Particle Spacing attribute in small increments.

4. Delete any particle nodes in the scene to make way for a 3D particle grid. To define all of the corners interactively, you need to work between multiple view windows that display different view planes. Use the default Four View view.

**5.** In the Perspective view, click to place the corners that define the base of the grid.

**6.** Switch to the Side view or Front view. Press the HOME (INSERT) key and drag one of the points up or down. Press ENTER (RETURN) and a grid will be created. Figure 22-3 shows a particle grid.

*FIGURE 22-3    The base of a 3D particle grid is defined in the Perspective view (left) and the height of the grid is defined in the Side view (right).*

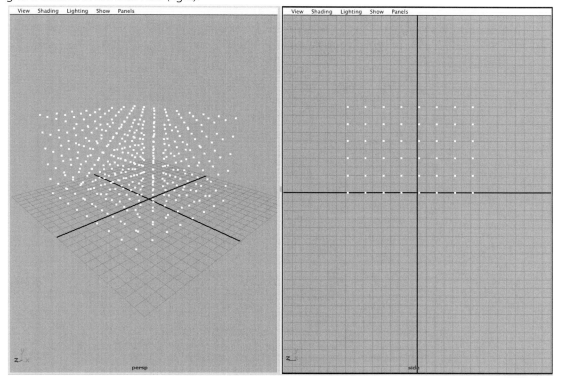

# Emitters

You can also create particles through emitters, which "shoot" particles away from the emitting point. Without any fields to control their behavior (see the "Fields" section later in this chapter for more information), these particles will continue along their path at the speed at which they were emitted. The way that these particles are emitted is determined by the emitter's emission Type attribute. The rate at which they are emitted is controlled by the emitter's Rate attribute. A whole slew of other attributes can be set for emitters. We will address the basic attributes in this section and introduce others throughout this chapter as we work through some more complex examples.

To create an emitter, choose Particles | Create Emitter. By default, an *omni* emitter is created. When you create an emitter, a particle node is also created with connections to the emitter. It is important to distinguish which nodes control which attributes for the desired behavior. If you wish to edit the way particles are being emitted, such as the emission rate or speed, you edit the emitter attributes in the Attribute Editor or the Channel Box. If you want to edit the actual behavior of the particles after they've been emitted, such as their color or lifespan, edit the particle node's attributes.

If you examine the connections to this emitter network in the Hypergraph, you will also notice that both of these nodes are connected to a time node. As you might imagine, this is due to the fact that the emitter creates particles at a rate based on time. If you click the Play button in the Time Slider, the particles will emit over the period of time specified by the range. At the end of the range, playback will return to the beginning and the emission will return to its initial state. As you experiment with particles and emitters, it might be best to set the time range to 500 frames. This will give you time to study the behavior of the emitters and particles as you adjust attributes.

Another important thing to mention when dealing with any type of dynamic simulation in Maya is that the playback speed in the Timeline preferences (Window | Settings/Preferences | Timeline) needs to be set to Play Every Frame. This is because Maya calculates the position of the particles based on their position in the previous frame. If any frames are dropped to maintain the real-time performance, the position of the particles will be inaccurate. If you'd like to get an idea of how the particles are animating in real time, using Playblast can be very helpful.

### Emitter Type

The type of emitter determines how the particles emit from the emission point and, in the case of the *volume emitter*, how large and in what shape the emission points are. Three emitter types are available when using the Create Emitter command: the *omni* emitter, a *directional* emitter, and a *volume* emitter. (Note that particles can also be emitted from surfaces and curves by using the Emit from Object command, covered later in this chapter in "Create the Particle Trails.") Figure 22-4 shows particles emitting from the three types of emitters.

**FIGURE 22-4**  *From left to right: omni, directional, and volume emitters*

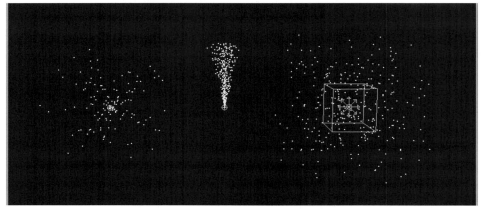

- **Omni (default) emitter**   Emits particles in all directions; works really well for explosions.
- **Directional emitter**   Emits particles in a specified direction. We will use the directional emitter later to emit laser blasts from a cannon.
- **Volume emitter**   Emits particles from a volumetric space. Volumetric shapes include a cube, sphere, cylinder, torus, and cone.

### Emission Attributes

Once you have created an emitter, you can begin editing its attributes. While all of the attributes can be set in the Create Emitter tool's Options window, it is usually easier to create an emitter with the default settings and then tweak them in the Channel Box or Attribute Editor. Some of the basic emitter attributes are listed here:

- **Rate**   Controls particles emitted per second. The default is 100.
- **Cycle Emission**   By default, particles are emitted randomly from an emitter. When this option is set to Frame [Time Random On], the emission cycle is looped over the number of frames specified in the Cycle Interval attribute.
- **Min/Max Distance**   Controls the distance from the emitter point that particles are created. All particles are created randomly between these two distance settings. The Max Distance must be greater than or equal to the Min Distance.
- **Speed**   Units per second that the particles travel as they are emitted. Randomness in particle speed is introduced by entering a value in the Speed Random attribute.
- **Volume attributes**   If Volume is selected as the Emitter Type, the Volume Shape lets you choose between a cube, sphere, cylinder, torus, and cone. The size and position can be edited with the remaining Volume Speed attributes.

# Particle Attributes

To this point, we have been using the default particle attributes. While some uses for particles and emitters have been suggested, it might be difficult for you to imagine how these effects are created with these little points. By editing the particle node's attributes, we can control the color of the particles, how long they live, how they are displayed, and much more.

## Lifespan

The Lifespan attribute controls the amount of time, in seconds, that a particle lives. To begin playing with particle attributes, create an emitter with the default settings and select the particle node. Press CTRL-A or open the Attribute Editor from the status line to display the particle node's attributes. Scroll down to the Lifespan Attributes folder.

A lifespan mode can be selected from the Lifespan Mode attribute's pull-down menu. To see how these lifespan attributes affect the behavior of the particles, you should make sure your time range is set to about 500 frames and click the Play button. As you select different lifespan modes from the pull-down menu, the results show in the view window.

The following lists the modes from the pull-down menu:

- **Live Forever**    The default setting. In this mode, the particles never die in the scene.
- **Constant**    Allows you to specify the precise length of time that a particle lives. This amount of time is set in the Lifespan attribute.
- **Random Range**    Introduces a random lifespan for the particles. The target lifespan is determined by the Lifespan attribute. Random variations are plus or minus the amount of time specified in the Lifespan Random attribute.
- **LifespanPP Only**    Controls the lifespan of individual particles. With this mode active, you can select individual particles and then use an expression to control the lifespan in the Per Particle Attributes section of the Attribute Editor.

## Render Attributes

The Render Attributes section in the Attribute Editor contains the attributes that control how particles are displayed. The default is set to Points, and this is what we have been using so far. To change this type of display, select another option from the Particle Render Type pull-down menu. You can choose from among ten different types. We are not going to go through all of them here, as they are all well-defined in Maya's documentation. Instead, we'll cover the overall process of changing the rendering type and adding attributes for the specific type.

Notice that when you view all of the display options in the Particle Render Type pull-down menu, each of the last three types, Blobby, Cloud, and Tube, has *[s/w]* after its name in the list. This signifies that these particle render types need to be rendered in Maya's *software renderer*. The other seven rendering types render only in the OpenGL-based hardware renderer, which is discussed later in this chapter in the section "Hardware Renderer." For now, just be aware of what this [s/w] means.

Each of these rendering types, as shown in Figure 22-5, has specific attributes to control the way it appears. Once a rendering type has been selected in the Particle Render Type attribute pull-down menu, click the Current Render Type button. Active fields appear in the section below the button that are specific only to that rendering type. In the case of the Spheres rendering type, an attribute appears that controls the radius. The Streak rendering type has attributes that control the line width of the streaks and attributes that control the tail size and tail fade. These types of attributes are called *dynamic* attributes. Dynamic attributes are loaded for specific tasks instead of loading all available attributes. Not only does this make scrolling through an already long list of attributes easier, but it also cuts down on general memory usage since Maya does not have to load all of this information every time a particle group is created.

FIGURE 22-5    *Examples of all ten rendering types*

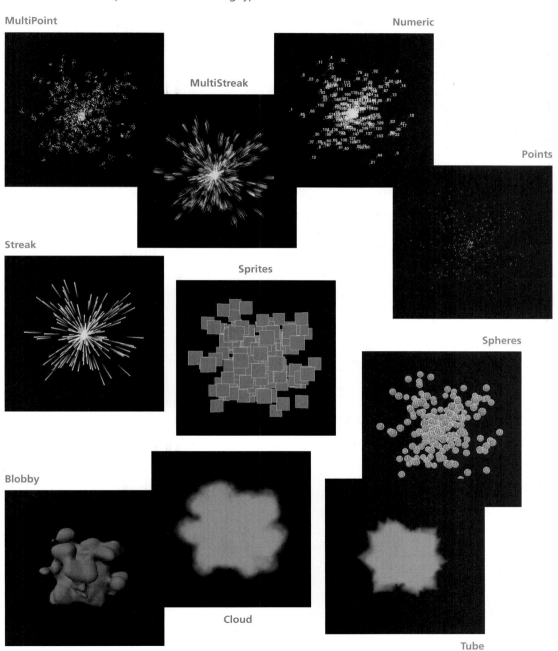

MultiPoint

MultiStreak

Numeric

Points

Streak

Sprites

Spheres

Blobby

Cloud

Tube

## Per Particle Attributes

Another section in the particle shape node's Attribute Editor is called the Per Particle (Array) Attributes. While the other attributes in the other folders contain attributes that control the appearance or behavior of the particle group as a whole, per particle attributes are capable of controlling the appearance or behavior of individual particles. Depending on the attribute chosen, per particle attributes can be controlled by an expression, a ramp texture, through the Component Editor, or by all three. To determine what kind of controls can be used for one of the per particle attributes, right-click in one of the pink fields in the Per Particle (Array) Attributes section of the Attribute Editor. The marking menu displays which control technique can be used.

> _**NOTE**_  _**Much of the information on particle expressions is beyond the scope of this book. It is important that you read Maya's**_ **Using Expressions** _**manual, specifically the section on particle expressions, to get a complete foundation for using this technique.**_

### Using Expressions

Let's edit one of these attributes with an expression. Here, we look at some examples in which using an expression is relatively simple. For instance, we can write a simple expression to control the lifespan of the particles instead of using the attributes in the Lifespan Attributes section of the particleShape1 node.

1.  Create an emitter with the default settings. Use the Outliner to select the particle node (particle1), not the emitter.

2.  In the Attribute Editor, set the Lifespan Mode attribute in the Lifespan Attributes section of the particleShape1 node to LifespanPP Only. This forces the particles to derive their lifespan value from the Per Particle (Array) Attributes section.

3.  Scroll down in the Attribute Editor to the Per Particle (Array) Attributes area, as shown in this illustration. Right-click in the pink field next to lifespanPP and choose Creation Expression. This opens the Expression Editor.

4.  In the Expression Editor, the name of the particle node and attribute that is selected

appears in the Selected Obj and Attr fields. Either retype or cut and paste this name into the Expression field in the white space at the bottom of the Expression Editor. Then continue writing the expression control with a random function. Your expression should look like this:

```
particleShape1.lifespanPP=rand(5,8);
```

**5.** Click the Create button in the Expression Editor to create the expression. Click Play to see the result. The particles are created with a random lifespan between values of 5 and 8 seconds.

### Using Ramp Textures to Control Particle Behaviors

Another technique used to control per particle attributes is the use of a Ramp texture. For this example, we'll use a Ramp texture to control the color of particles over their lifespan. If you look down the list of available per particle attributes, you won't see any attribute to control color. This is because color is another one of those dynamic attributes that needs to be loaded.

**1.** Click the Color button in the particle node's Attribute Editor. A window appears that allows you to specify whether this dynamic attribute controls the entire particle group or on a per particle basis. Choose Per Particle Attribute and click the Add Attribute button. A new attribute will be created in the Per Particle (Array) Attributes section called rgbPP.

**2.** Right-click in the pink field next to rgbPP in the Per Particle (Array) Attributes section and choose Create Ramp. The rgbPP field will now show <-arrayMapper1.outColor.

**3.** Choose <-arrayMapper1.outColor | Edit Ramp from the marking menu. The ramp Node Editor appears in the Attribute Editor. This allows you to control the color of the particles over the course of their lifespan. The color at the bottom of the ramp indicates what color the particles will be at the beginning of their lifespan, and the color at the top shows what color they will be at the end of their lifespan.

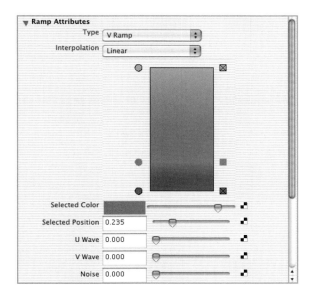

**4.** Add another color to the ramp by clicking inside the ramp display. Another color marker is added to the left of the ramp, as shown in this illustration. You can change the color of any of the markers by clicking one of the color markers to the left of the ramp and then editing its value in the Selected Color attribute. Remember that clicking the color chip brings up the Color Chooser. If you decide that you don't want a color that has already been inserted, click in the box on the right side of the ramp.

**5.** Move the colors to position their keys at different points on the ramp by clicking a color marker and dragging it up or down.

**6.** Click the Play button in the Time Slider to see the results. Make sure to press the 6 key to turn on the display of colors in the view window. You can continue to fine-tune the ramp until you are pleased with the results in the view window.

### Using the Component Editor

The third way to set per particle attributes is through the Component Editor. This technique involves selecting certain particles and then keying in numeric values for any of the attributes in the Component Editor. For this example, we will add a per particle opacity attribute to the particle node and set the opacity levels on a certain few of the particle groups:

**1.** Create a particle group with the Particle tool. (Alternatively, you could choose to use an emitter, but you would let the animation play for a few seconds so that you can select some of the particles.) For this example, create a 3D particle grid.

**2.** In the view window, right-click the particles and choose Particle from the marking menu. Select the top few rows of particles by clicking or drag-selecting them.

**3.** In the particle shape node's Attribute Editor, scroll down to the Per Particle (Array) Attributes section and add an opacity attribute by clicking the Opacity button below the list of per particle attributes. An attribute named opacityPP will be created and available in the list of per particle attributes.

**4.** Right-click in the pink field next to opacityPP and choose Component Editor from the marking menu. The Component Editor will open.

**5.** Find the opacityPP attribute list along the top of the Component Editor. All of the selected particles' fields will be highlighted. Change the opacity value to **0.3**. Now the top few rows of particles will be much more transparent than the bottom rows.

# Fields

Fields simulate the forces of nature, such as gravity, air, turbulence, and vortex. They are normally applied to objects to create fluid, natural motion. The behaviors of fields can be fine-tuned by editing their attributes in the Channel Box or Attribute Editor. To gain a full understanding of what all of these fields do and how their attributes affect their behaviors, you need to spend time experimenting. For now, we'll introduce the different fields and show you how they affect objects in the scene by demonstrating them on a particle group. We will then look at some of the common attributes shared among the different fields.

## Applying Fields

Fields can be applied to any object or group of objects in a scene. When applied to a particle group, the outputForce attribute of the field is connected to the inputForce attribute of the particle's shape node. The particle shape node's fieldData attribute is then connected to the field's inputData attribute. These connections can be viewed in the Hypergraph, as shown in Figure 22-6.

When a force is applied to a piece of geometry, however, Maya will create a rigidBody node that contains these same inputForce and fieldData attributes. That rigidBody node is then connected to other nodes that will control the geometry under the effects of the field. (We will cover rigid bodies in depth in Chapter 24.) At this time, let's concentrate on connecting fields to particle groups to study their effects.

As you might guess from everything you've learned in this book so far, a field could be created from the Fields menu and then connected using the Hypergraph and Connection Editor. However, scrolling through a list of attributes and trying to figure out which ones make the desired

**FIGURE 22-6** *The Hypergraph shows connections between a particle group and a Turbulence field.*

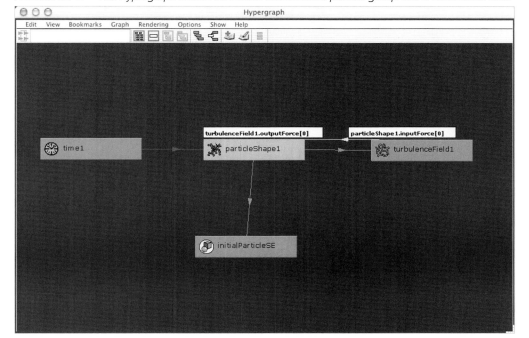

connection when you are unfamiliar with them can be somewhat cumbersome. For this reason, Maya offers several ways of connecting fields to objects in the UI.

### Selecting Objects when Creating Fields

Probably the simplest and quickest way to create a field with connections to an object is to select the object, in this case a particle group, and create a field though the Fields menu. For example, create a particle grid, and with it selected, choose Fields | Turbulence Field. In the Hypergraph, select the particle group, and click the Input and Output Connections button in the Hypergraph's toolbar; you can see that they are connected.

But perhaps the most intuitive way to check the connections is simply to click the Play button in the playback controls. If the particles start moving around, you know that they are connected.

### Affect Selected Object(s) Command

Often, a field will be created in the scene with none or only some of the objects selected that need a field. To select one or many fields and the objects that need to be affected by those fields and connect them quickly, you can use a menu command. To see this work, create a particle group or two and then create a Turbulence and a Gravity field. Select all of these objects (in any order) and choose Fields | Affect Selected Object(s). Look in the Hypergraph or click the Play button to see whether all of the connections were made.

**FIGURE 22-7**   *Dynamic Relationships Editor*

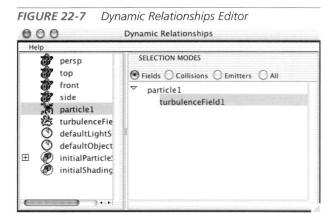

### Making Connections with the Dynamic Relationships Editor

The Dynamic Relationships Editor, shown in Figure 22-7, can be used to make or break any type of dynamic connections. It could be used to connect fields to objects, particles to emitters, and collisions between objects. It can also be a quick way to view dynamic connections without having to play the animations or pan through the Hypergraph. To make a connection with the Dynamic Relationship Editor:

**1.** Create a particle grid and a Turbulence field. They should not have any connections.

**2.** Choose Windows | Relationship Editors | Dynamic Relationships. This brings up the Dynamic Relationships Editor.

**3.** Make sure that the Fields radio button is selected in the Selection Modes section. Select the particle group node from the list of objects on the left side of the window. The particle node will appear in the section on the right with a list of all fields in the scene.

**4.** To connect the particle group to the Turbulence field, select it from the list under the particle group in the right side. It will highlight to indicate that it is now connected and will affect the particle group.

**5.** Create another field in the scene—make it a Uniform field. Notice that it appears in the list under the particle group. You can click that field to highlight it and add it to the fields that affect the particle group. Click either one or both fields to disconnect them from the particle group.

## Types of Fields

Nine types of fields are capable of simulating natural motion found in the real world. Each of these fields can be created by selecting it from the Fields menu. The best way to experiment with fields and learn what they do is to create a particle grid, add a field to it, and play back the animation. Because particle grids create the particles with uniform spacing throughout, it is easy to see exactly how the field is affecting it.

> **NOTE**  *Once again, the Maya documentation describes each field and its specific attributes at length. You should review and study this information.*

This section gives a brief description of each field and points out some of the attributes that are specific to each one. In addition to the examples shown in Figure 22-8, QuickTime movies on the accompanying DVD show animated examples of each field. The common field attributes will be discussed in the following section.

- **Air**   Objects connected to an Air field move or float through the scene, depending on their mass. A feather generally floats, while something heavier is less affected by this field. Some presets are available in the Creation options for this field for Wind, Wake, and Fan. Using these presets sets up the attributes for the field to simulate these types of occurrences in nature. Figure 22-8 shows an Air field being animated through a particle grid and displacing the particles in its range.

- **Drag**   Slows or dampens moving objects by applying a braking force that is controlled by the Magnitude attribute. If the Use Direction attribute is turned on, the braking force will be applied in the directions specified in the Direction attribute fields. The figure shows a directional emitter emitting particles in the Y direction. The Drag field is causing them to break so that they appear to collect above the emitter.

- **Gravity**   Simulates the earth's gravitational force by accelerating objects in the negative Y direction by default. However, the direction can be changed in the Attribute Editor by

*FIGURE 22-8* *Examples of fields*

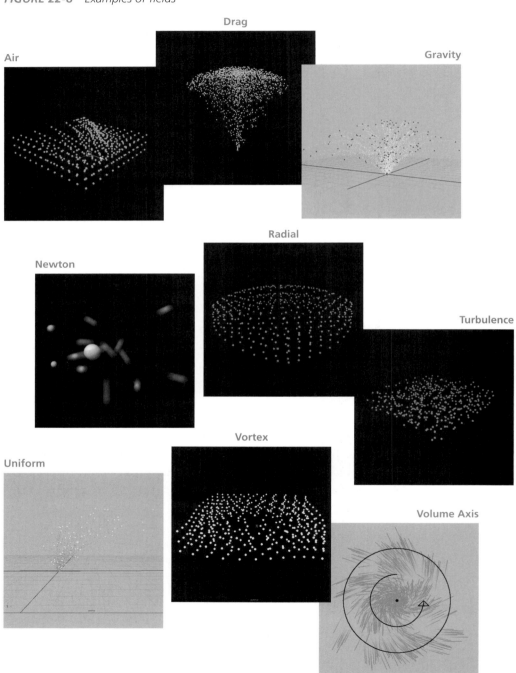

Drag

Air

Gravity

Radial

Newton

Turbulence

Uniform

Vortex

Volume Axis

setting one of the Direction attributes. Also, the Magnitude attribute (the strength of the effect) is set to 9.8. (Remember that the law of gravity accelerates objects at a rate of 9.8 meters per second, regardless of their mass. Therefore, if you wish to simulate objects behaving under the law of gravity, you need to set your units to work in meters or set the magnitude to 980 if you are working in centimeters.)

- **Newton** Simulates an attractive force between any two objects proportional to the product of their mass. Editing the mass of any object will change the effect of the Newton field. When applied to a particle group, the particles will be attracted toward the center of the field according to the Magnitude attribute setting. A similar effect can be achieved by changing the Dynamic Weight attribute of the particle group. A use for this type of field is exemplified by a planet orbiting around the sun or electrons orbiting around a nucleus. The figure shows a NURBS sphere that owns a Newton field. (Note that objects emitting fields are covered in Chapter 23.) The particles have a velocity of 1 in their X direction while they are attracted to the sphere; the result is that they appear to orbit the sphere.

- **Radial** Pushes or pulls objects toward the field depending on whether the Magnitude attribute value is positive or negative. Debris being thrown by a point explosion would be a good use for the Radial field.

- **Turbulence** Introduces irregular motion, or noise, into the object's motion. While the Magnitude attribute controls the strength of the effect, the Frequency controls how often the disruption occurs. If you need more disruption over the area being affected, increase the Frequency. A common thing to do with a Turbulence field is to animate the Phase attributes. This gives the turbulent effect some direction useful for creating such effects as waves in the ocean.

- **Uniform** Similar to a Gravity field in that it accelerates objects in a given direction, but a Uniform field takes the mass of the objects it is affecting into the calculation. Whereas Gravity accelerates object at the same rate, no matter what their mass, Uniform accelerates heavier objects faster. The figure shows particles being emitted from a directional emitter. The particles have been assigned a random mass between 0.2 and 2. The Uniform field has been applied to push the particles in the X direction. The heavier particles are less affected by the Uniform field, and the lighter particles are more affected.

- **Vortex** Pulls objects in a circular, whirlpool-like behavior. The Axis attribute controls which axis the affected objects will swirl around. This field would work well for tornadoes, fires, and black holes.

- **Volume Axis** Creates a volume shape and controls how objects move around in it. Several attributes are used for setting the volume shape.

# Common Field Attributes

All of the fields in Maya have attributes that are specific to their type. Some of these attributes will be covered as we use these fields in tutorials. However, other attributes are common to all of the fields. By understanding what these attributes do, you will gain a better understanding of how fields work.

Before we begin demonstrating these attributes, create a particle grid that covers the entire default grid in the Perspective view. For the sake of performance, keep the particle count reasonable for real-time interaction. Use a Particle Spacing setting of 2 units. To see how these attributes affect the motion of these particles, change the Particle Render Type attribute to Spheres. Also, make sure you set your time range long—500 frames will do.

- **Magnitude**  Controls the amount, or strength, of the field. A higher value increases the strength of the effect. Entering a negative value reverses the direction of the effect. For a Gravity field, the objects accelerate upward instead of downward; a Radial field pulls objects toward it instead of away.

- **Max Distance**  Controls the maximum distance, in units, that objects are affected by the field. To use the Max Distance attribute, the Use Max Distance attribute must be set to On. The default value of most fields creates the field with the Use Max Distance set to Off. (Both the Radial and the Air fields have their default set to On.) In these cases, the field affects everything in the scene that is connected to it no matter how far an object is from the field's origin (see Figure 22-9). Also, neither the Max Distance attribute nor the Attenuation attribute will have any effect. When Use Max Distance is turned on, any object connected to the field and within the distance specified in the Max Distance attribute will be affected by the field (see Figure 22-9).

**FIGURE 22-9**  *Left to right; a particle grid has Use Max Distance set to Off, Use Max Distance set to On with a value of 10, and, the same Max Distance setting but with the Attenuation value set to 4.*

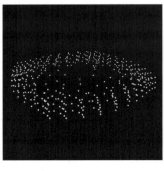
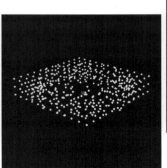
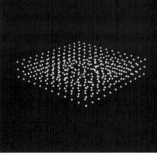

- **Attenuation** Exponentially controls the falloff of the field's effect from its origin. With a value of 0, the effect remains constant. Any larger number increases the attenuation so that the objects closer to the field's origin are more affected than those close to the Max Distance attribute (see Figure 22-9).

*TIP* **You can use the Show Manipulator tool to edit the attributes of any selected field in the view window.**

# Hardware Renderer

Maya's hardware renderer uses your computer's real-time OpenGL-based video processor to render images, instead of the software renderer that we have used so far. The hardware renderer is faster than the software render, but the quality, especially in terms of anti-aliasing, is not as good. While the renderer is capable of rendering geometry, most of the particle render types can be rendered only in the hardware renderer.

As of version 5, Maya offers a completely rewritten hardware renderer to take advantage of the newest hardware rendering technologies found in the latest video cards. Using the hardware renderer is similar to using the software renderer. It can be enabled in the Render Settings window or from the Render menu.

In this section, we look at some of the basic features of the hardware renderer. As an example, we'll create an omni emitter that emits red particle streaks. If you were to render this in the regular software renderer, it would appear as a black screen.

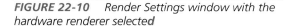

*FIGURE 22-10* *Render Settings window with the hardware renderer selected*

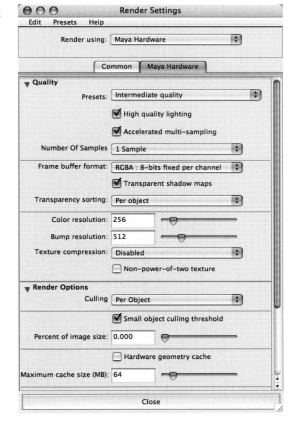

1. Open the Render Settings window. Choose Maya Hardware from the Render Using menu. A new tab appears that contains attributes specific to the hardware renderer (see Figure 22-10).

2. To render the current frame, click the Render Current Frame button on the status line, just as you would if you were rendering with the software renderer.

**3.** In many cases, you will be compositing the particle effects with geometry that has been rendered in a separate software render pass. Render layers can be created in the Layer window and objects assigned to the layers. When Enable Render Layers is turned on in the Render Settings window, you can quickly hide the geometry and render just the particles.

**4.** Some nice blurring or smoothing effects can be achieved by using motion blur. The attributes for controlling motion blur are shown in Figure 22-10. With the Enable Motion Blur option checked, the renderer renders several frames and composites them together to create one frame for an accurate motion blur. The Motion Blur By Frame attribute sets how many frames are used in the calculation. The Number of Exposures attribute sets how many frames are rendered between the current frame and the number of frames before it as specified in the Motion Blur By Frame attribute. A good rule of thumb is to set the number of exposures to one less than the number of frames. This produces the best-quality motion blur. Figure 22-11 shows the effects of motion blur on streak rendered particles.

This should give you a good reference to using the hardware renderer. While the remainder of this chapter focuses on the software renderer, we will be using the hardware rendering engine to create some fire effects in Chapter 23 and some heat ripples in Chapter 27.

**FIGURE 22-11**    *The streaks are rendered with (left) and without (right) Enable Motion Blur turned on with a Motion Blur By Frame value of 10.*

# Tutorial: Build an Explosion

With some of the basics of working with particles, emitters, and fields now covered, we are ready to put them to work and create something. This tutorial takes you through the process of building an explosion. To accomplish this task, we will use some of the basics covered so far in this chapter and introduce some additional tools and concepts that show Maya's power in creating effects of this type.

For this shot, we want to create a cannon shooting a laser blast at a group of objects. When the laser blast hits the objects, an explosion will occur. In some cases, you might want to render the laser blast and explosion separately and composite them together later in a compositing package. While this is optimal for high-end postproduction, let's say that in this case we are just trying to previsualize a concept and don't have time to go back in to match and composite an explosion at every laser blast collision. We do, however, want to impress our client and give them a sense of how dramatic a good explosion would be in the scene.

We will begin by creating the explosion. This will entail creating an emitter that emits the initial stage of the explosion. Once this is in place, these particles will emit a second particle group to represent fire and smoke trails. Fields will be added and their attributes fine-tuned until we have the look and behavior we want. Then we will model some simple geometry and use the Instancer to link them to the particle object. This will replace the particles from the initial explosion with actual modeled geometry. Then we'll set up the cannon and use particle collision events to emit the explosion particles when the laser blast collides with the objects.

**NOTE** *For each of the illustrations used in this section, animated examples can be found on the accompanying DVD. It is a good idea to have these available as you go through the tutorial, if possible.*

## Create the Explosion

Before we set up the cannon and laser blast that will cause the explosion, we will create the explosion itself. We'll start with an emitter and rough out the overall size and span of the explosion.

### Set Up the Emitter

**1.** Choose Particles | Create Emitter ❑ and reset the options to their default settings (Edit | Reset Settings). Click the Create button to create the emitter. This will create two nodes—an emitter node and a particle node. Name the emitter node **Emitter_Explosion** and the particle node **particle_debris**.

**2.** Edit some of the emitter's attributes. With the emitter selected, press CTRL-A or click the Show Attribute Editor button on the status line to bring up the Attribute Editor. Change the Emitter Type attribute to Directional. Then scroll down to the next folder in the Attribute Editor, Distance/Direction Attributes. Change the Direction X attribute to **0** and set the Direction Y attribute to **1**. You may also wish to use the Show Manipulator tool to edit some of these attributes in the view window. Play back the animation to see the results.

**3.** Now that the emitter is pointing in the right direction, we need to adjust the Spread attribute. This widens the angle of the particles that are emitted from the emitter. The last attribute in that Distance/Direction Attributes folder is called Spread. A value of 1 would emit in a full 180-degree spread. We want it just less than that, so change this to a value of **0.8**. The view should look similar to this illustration.

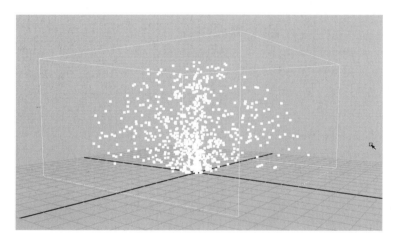

**NOTE** *Some of the particle display attributes have been modified for some of these illustrations. Since these will all be modified later to resemble fire and smoke, it is not necessary for you to change these attributes now.*

**4.** As this explosion occurs, the emitter should emit for one frame only. To specify this, we can key the Emission Rate. Make sure that the Time Slider is at frame 1. Set the Rate (Particles/Sec) attribute to **0**. Right-click this attribute in the Attribute Editor and choose Set Key to set a key for that value at frame 1. Move the Time Slider out to frame 3 and set another key for a Rate attribute to **0**. Go back to frame 2, change the Rate to **2000**, and set a key. Now when you play back the animation, the emitter will emit a short burst of particles.

## Create the Particle Trails

As the debris breaks away from the center of the explosion, we want to create trails of fire and smoke. To do this, we will use the existing particle node, particle_debris, as an emitter by using the Emit from Object command. By selecting an object and choosing Particles | Emit from Object, an emitter will be attached to the object and particles will be emitted from it.

**1.** Select the particle_debris node. Choose Particles | Emit from Object ❑. An emitter will be grouped underneath the particle_debris node. Rename this node **emitter_trail** and name the new particle group **particle_trail**.

**2.** Set the Rate attribute of the emitter_trail node to **50** and set the Speed attribute to **0.1**. We are setting the Speed to 0.1 because we want the particles to trail behind the object from which they are being emitted, not shoot off. A low value here is enough to make them spread but keep them from moving too fast.

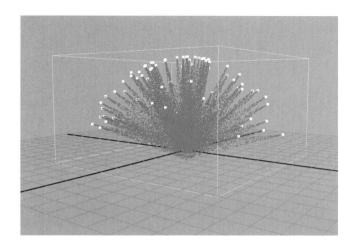

**3.** Click the Play button to see the result. It should look similar to this following illustration:

*NOTE  As you work, you might notice that your computer is slowing down and frames are being dropped. This is not good when you're working with dynamics. Because most dynamic calculations are made on a frame-to-frame basis, dropping a frame can lead to an inaccurate simulation. For this reason, it is best to work with the animation Timeline preferences set to Play Every Frame. Make sure this is enabled once you start running dynamic simulations. If you want to preview the animation in real time, it is best to use the Playblast engine.*

### Add Fields

Now the particles shoot out upward from the emitter and continue along their paths. Although this might be fine for an explosion that takes place in outer space, the explosion in this example takes place on or near the ground on a planet with a gravitational force. For this reason, we'll apply a Gravity field to the particle_debris node to pull the particles down after they are emitted. For the smoke trails, we'll use an Air field to make the smoke rise up into the air as it dissipates.

**1.** Select the particle_debris node. Choose Fields | Gravity. This creates a Gravity field that is affecting the particle_debris particles once they are emitted from the emitter. If you play back the animation, however, the gravity appears to be pulling the particles down a little too much.

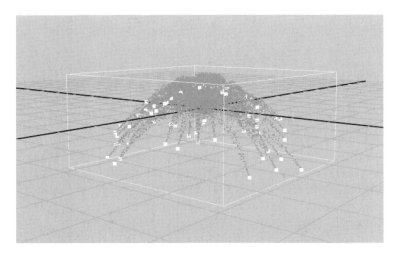

2. Select the gravityField1 node and view its attributes in the Attribute Editor. In the Gravity Field Attributes folder, change the Magnitude attribute to **0.5**. This lessens the strength that this field exerts on the affected objects. Click the Play button to see the result. It should look similar to this illustration.

3. As the trailing particles are emitted, they should rise up in the air a bit. This effect is due to the fact that smoke particles are hot and therefore less dense than air, so they rise. Select the particle_trail node and choose Fields | Air Field. This creates the Air field at the origin and attached to the particle_trail node.

4. If you click Play, you will see the trails quickly shoot upward from the center. This is the result of a few things. The most affected particles are those near the center of the scene, because the Air field is created with its Use Max Distance set to On. If you recall from the discussion on field attributes earlier in this chapter, turning this setting on makes the field affect only the particles closest to the field's origin by a range set in the Max Distance attribute. However, in this case, we want the Air field to affect everything connected to it in this scene. In the Attribute Editor, with the airField1 node selected, scroll down to the Distance folder and uncheck the Use Max Distance box. Play back the animation to view the result.

5. Now the effect of the Air field is even, but the particles rise up a little too fast. This can be fixed by editing the Air field's Magnitude and Speed attributes. Set the Magnitude to **0.4** and the Speed to **0.4**. The next illustration shows the trails with the Air field applied.

6. Add a Turbulence field to the trails so that they break up irregularly as they are emitted and rise. Select the particle_trail object and then choose Fields | Turbulence Field. The default settings are a little too strong, so you need to turn down the Turbulence field's Magnitude attribute to **5**.

**7.** If you play back the animation, you'll notice that the Turbulence field is affecting the particles in large chunks. To make this area a bit finer, you need to turn up the Frequency attribute. A higher frequency makes the turbulence occur more times over the given space and therefore creates very fine irregularities for the

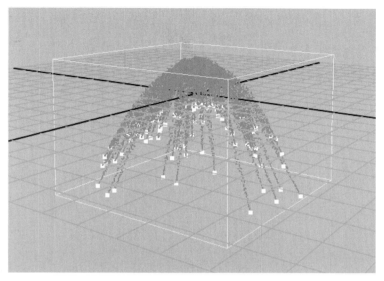

particles to travel through. A value of about **8** works well in this setup.

**8.** Animate the Turbulence field's Phase Y attribute with a simple expression that creates 10 cycles over the time range with an amplitude of 20 units. In the Attribute Editor, right-click the Phase Y attribute and choose Create New Expression from the marking menu. In the Expressions Editor, enter the following into the text field and click the Create button:

```
turbulenceField1.phaseY=20*sin(10*time);
```

The result is shown in this illustration:

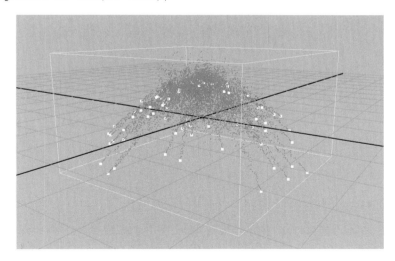

### Edit the Particle Attributes

Everything we have done so far in this exercise has used the default particle rendering type, *points*. Points are a great way to begin setting up a particle system because they don't require too much power from your video processing unit. For the sake of performance, it is always good to start with this type, set up any particle interaction, and then go on to change the particle display type.

In this section, we will edit the particle attributes for the smoke trails. This is where we'll replace the particle point display with something that resembles smoke. We'll leave the actual particle debris node alone for now, as we are going to use a different process to change its display type. This will be addressed in the next section.

1. Select the particle_trail node object and view its attributes in the Attribute Editor. We'll begin by editing the Lifespan attributes. Scroll down the Attribute Editor to the Lifespan Attributes folder. Set the Lifespan Mode attribute to Random Range. The two fields below this become active. Set Lifespan to **1.2** and Lifespan Random to **0.7**. When you play back the animation, the newly emitted particles will "die" between 1.9 and 0.5 seconds.

2. To change the particle display type, scroll down to the Render Attributes folder and set the Particle Render Type attribute to Cloud [s/w]. This changes the render type to clouds. (The [s/w] indicates that this rendering type needs to use the Maya software renderer instead of the hardware render buffer that most of the other display types require.)

3. Click the Current Render Type button to add the dynamic attributes for this rendering type. Three fields will appear in the space below this button. If you click Play now, the particles will be much too big. Set the Radius to a small value—**0.02** is a good place to start. If you play it back now, the radius is much closer to where it should be. However, the radius should increase as the smoke dissipates. To take care of that, we will edit some attributes that allow us to set behaviors per individual particle.

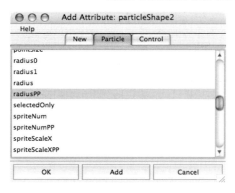

4. Scroll down to the Per Particle (Array) Attributes folder in the Attribute Editor. By default, no attribute controls the radius. Therefore, you need to add a dynamic attribute. In the Add Dynamic Attribute folder, click the General button. This opens the Add Attribute window. Select the Particle tab, scroll down the list, and find the radiusPP attribute, shown in the illustration. Select it and click the OK button. The radiusPP attribute now appears in the list of Per Particle Attributes.

5. We will use a ramp to control the radius per particle. This ramp is based on the lifespan of the particles. This means that the radius will have a value set by the ramp value at the bottom of the ramp when a particle is born and will end with the value at the top of the ramp. To add the ramp, right-click in the radiusPP field and choose Create Ramp

from the marking menu. To edit the ramp, right-click again in that field and choose <-arrayMapper1.outValuePP | Edit Ramp from the marking menu. The Attribute Editor will load the Ramp Editor.

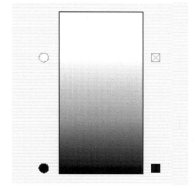

6. The default ramp is created, white at the bottom and black at the top. Because we want the radius to be small when it is born and largest when it dies, we want black, the small value, to be at the bottom and white, the larger value, to be at the top. Make this happen by clicking the black circle at the top on the left side of the ramp and pulling it down to the bottom. Then grab the white circle and move it to the top. Delete the gray color in the middle by clicking the box on the right side. The ramp should look like this.

7. If you play back the animation, you will notice that the radius is being scaled from 0 to 1. As we determined in Step 3, a value of 1 is much too large and this range is too wide. We can edit this range by editing the Array Mapper node. Select the particle_trail object and look at the Attribute Editor. Right-click in the radiusPP attribute field and

select Edit Array Mapper. This displays the ramp's Array Mapper node in the Attribute Editor. Here you can set the minimum and maximum values for the ramp. Enter **0.02** as the Min Value and **0.2** as the Max Value. Play back the animation to see the result. The particles should get bigger as they near the end of their lifespan so that they look like this illustration:

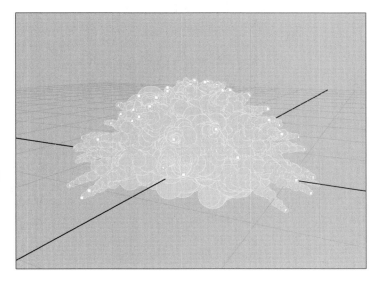

### Create the Particle Material

When using the regular hardware rendered particles, attributes such as Color and Opacity can be set in the particle attributes on a per object or per particle basis. This means that you could create an opacityPP attribute and then control it with an expression or a ramp. However, when dealing with software rendered particles, these attributes do not have any direct effect on the rendered particles. While these attributes can be connected to a material through the use of a particle sampler info node, it is much simpler to control these attributes—Color, Transparency, and Incandescence—with a ramp through the particle cloud's Life attributes.

Whenever a particle group is created, it is connected to two of Maya's default materials—the Lambert material and the particle cloud. Depending on the Particle Render Type attribute, Maya uses the hardware render attributes as specified in the particle shape node's attributes, the Lambert material for blobby particles, or the particle cloud for the cloud rendering type. In this section, we create a new particle cloud material, assign it to the particle group, and then edit some of its attributes to make it look like fire and smoke:

**1.** Open the Hypershade window (Windows | Rendering Editors | Hypershade) or use the Hypershade/Perspective layout, accessible by choosing Panels | Saved Layouts | Hypershade/Perspective or by clicking the Hypershade button in the toolbar. Create a new particle cloud material. This can be found in the Create Materials menu under the Volumetric folder. Name this material **mSmoke**. Apply it to the particle_trail node.

**2.** With the mSmoke material selected and its attributes shown in the Attribute Editor, notice the attributes called Life Color, Life Transparency, and Life Incandescence, as well as the regular Color, Transparency, and Incandescence attributes. Using these Life attributes will allow for the control of these particular attributes over the lifespan of the particles. This workflow is similar to editing the per particle attributes for the color, opacity, and incandescence in the particle node's attributes. Begin by editing the Life Color attribute. Click the Map button next to the Life Color attribute and select Ramp Texture from the Create Rendering Node window.

**3.** Edit the ramp so that it cycles through four different colors. The color at the bottom of the ramp is the color that the particles will be at birth. Make this white by clicking the color chip next to Selected Color and editing the color in the Color Chooser. Click Apply. Move the next color key down so that only a hair of the white shows through. It should be at a position of about 0.015. Make this second color a bright orange. The third color should be a bright yellow. Set its

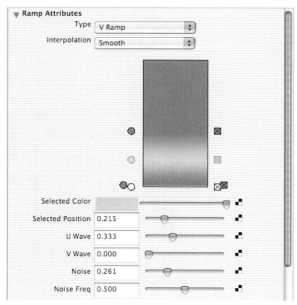

position to about 0.215. Finally, add
one more color by clicking in the ramp
somewhere. Position this color at
around 0.500. This fourth color should
be a medium gray to represent the
color as it turns into smoke. The
ramp should look like the previous
illustration (except in color).

4. In the Hypershade, select the mSmoke
node to view its attributes. Map a
Ramp texture to the Life Transparency
attribute. Set it so that the ramp is
black at the bottom (opaque) until
a position of 0.07 and fades to gray
at about 0.165 and then to white at
the top. It should look like this.

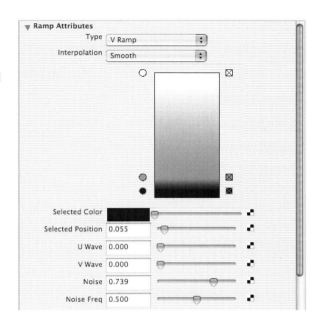

5. Add another ramp to the Life
Incandescence attribute. This ramp
should be two colors. The bottom
color should be a bright orange and
be positioned at about 0.165 to match
the point where the yellow flame
fades into a gray smoke. The second
color, a black, nonincandescent value,
should be positioned at about 0.450
to correspond with the bright-yellow
flame into the gray smoke. The ramp
should now look like this.

6. Play back the animation and stop
it right as the explosion is at its
largest, around frame 30. Click in
the Perspective view and use Maya's
software renderer to do a test render.
It should look similar to Figure 22-12.

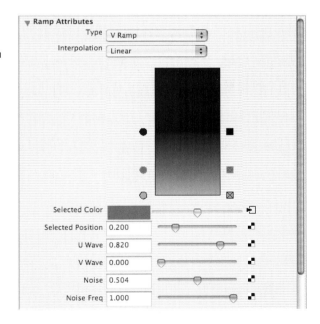

**FIGURE 22-12**  *A test render of the explosion*

## Use the Instancer to Insert the Debris

For the debris flying away from the explosion point, we are not going to use any of the ten particle render types. Instead, we are going to introduce the particle Instancer. The particle Instancer lets you replace a particle with an actual object in the scene. The object could even be animated and then inserted as a particle instance. This would be a great way to create a swarm of bees, spiders, ants, or even a large crowd of people. For this exercise, though, we are going to create some simple debris with some sculpted spheres and polygonal faces. We will create three types of debris—two of them will be based on a polygonal sphere and the other on a polygonal face.

1. Create a polygonal sphere. Change its attributes so that it has 8 by 8 subdivisions. Use the nonuniform scale tools in the Y and Z axes to make the object less spherical.

2. Use the Sculpt Polygons tool (Polygons | Sculpt Polygons) to push and pull the vertices on the object so that it quickly takes on an irregular form. It can be very rough since the pieces will be small and flying by fast. Name this object **pDebris**. Figure 22-13 shows an example of the pDebris model.

**FIGURE 22-13**  *The poly object will be used as the source object for the debris.*

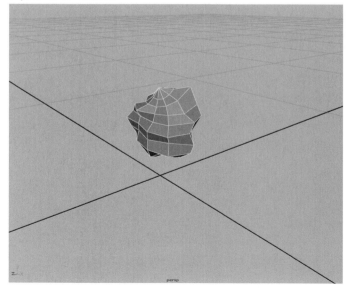

3. In the Hypershade window, create a new Lambert material and call it **mDebris**. Create a rock-like texture for the

color channel. You may create your own or use the MCR_debris_CLR.tga texture map that is included on the DVD. Apply it to pDebris.

**4.** Select pDebris and hide it by pressing CTRL-H.

**5.** With pDebris still selected, add the particle_debris object to the selection and choose Particles | Instancer (Replacement). A new object is created called instancer1. Select this object in the Outliner and view its attributes in the Attribute Editor. You will see that pDebris has been added to the Instanced Objects list. You could remove or add objects to this list by selecting them and choosing Add Selection or Remove Items.

**6.** When the animation plays back, notice that the objects are emitted from the emitter (see Figure 22-14). However, they are not spinning as they move. We can set the instances to rotate based on some of the other particle attributes. Find the Rotation options in the Instancer (Geometry Replacement) folder of the particle_debris shape node's attributes. Set the Rotation attribute to Velocity, so that the instances will rotate according to the velocity of the particles.

**FIGURE 22-14** *Objects are emitted from the emitter.*

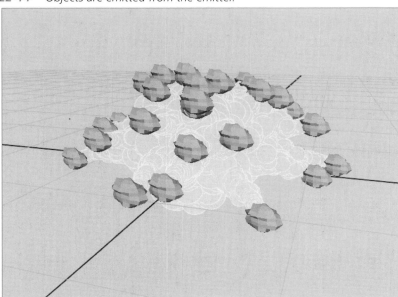

## Set Up the Laser Cannon Emitter and Scene Objects

At this point, the explosion is pretty much set. It is emitting the debris, which then emits the smoke and fire trails, and the fields are set up with Turbulence, Gravity, and Air. Now we are going to put that aside for a moment and set up another emitter along with some objects. We'll build a simple object that represents a cannon and attach the emitter to it. The rest of the scene can be simple—a plane for the floor and three cubes that are sitting on top of it. The cannon will rotate on its Y axis as it fires laser blasts at the cubes. See Figure 22-15 to get an idea of how this setup will look.

**1.** Select the Emitter_Explosion node and delete it. We will be using particle collision events to emit particles from here on. Select the particle_debris object and the particle_trail object and press CTRL-H to hide them. We want to get these out of our way while we set up the rest of the scene and not have all of these particle systems hinder the performance.

**2.** To create the set, first create a poly plane for the floor. Scale it up 25 units in all directions. Create a cube and scale it up in the Y and Z axes. Duplicate it two more times and scale each duplicate so that they are all a little different. Place them near the edge of the poly plane. See Figure 22-15 for details.

**3.** Create a cylinder that will be used for the cannon (or quickly model one, as shown in the example). Rotate it so that its end is pointing toward the cubes. Figure 22-15 shows the approximate scale and placement.

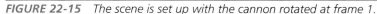

***FIGURE 22-15***   *The scene is set up with the cannon rotated at frame 1.*

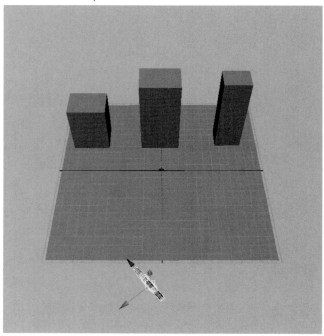

**4.** Create an emitter. Make it a directional emitter and name the emitter **emitter_laser**. With the emitter selected, SHIFT-select the cylinder. From the animation menu set (F2), choose Constrain | Point Constraint. The emitter is now constrained to the cylinder object. But it needs to rotate with it as well. With both of the objects still selected, choose Constrain | Orient Constraint.

*NOTE   Instead of using constraints, you could just parent the emitter to the cannon. However, using constraints makes it easier to choose the emitter in the Outliner because it is not necessary to unfold hierarchies.*

**5.** Set the emitter's attributes so that it is emitting 1 particle per second. Edit the Direction so that the particles are shooting toward the cubes. Direction Y can be set to **–1**, but the setting depends on how the cylinder was rotated, so you might have to play with it. Change the Speed to **50**.

**6.** Select the particle object that was created with the emitter. Name this **particle_laser**. Edit the Particle Render Type attribute to Streak. Add attributes for this render type and set the Line Width to **6**, the tail Fade to **0.4**, and the tail Size to **2.7**. Add a color attribute per object. Set the red value to **240**.

**7.** Animate the rotation of the cannon. At frame 1 it should be aimed off to the left side of the leftmost cube at about 40 degrees. At frame 150 it should be rotated about –40 degrees so that it is aiming off to the left (see Figure 22-15).

## Set Up Particle Collisions

Instead of tracking where these shots hit the cubes and compositing in the explosion, or keying the position of the emitter, a particle collision event will be created that emits our explosion whenever a laser blast collides with the cubes. The debris will then be set up to collide with the floor when it lands.

**1.** Select the particle_laser object and then select the three cubes. Choose Particles | Make Collide. Double-check that the connection was made by looking in the Dynamic Relationships Editor. Play back the animation to see if any collisions occur. Chances are that a collision will happen at least once. If not, you may have to change the starting or ending keyframed position of the cannon.

**2.** With the particle_laser object selected, choose Particle | Particle Collision Event. This will open the Particle Collision Event Editor. Particle_laser should be highlighted in the Objects list at the upper left of the window. This indicates that this particle will trigger the event. Make sure the All Collisions check box is checked. In the Event Type section, set the Type to Emit. Have it emit 80 particles with a spread of 0.8 (just as our original test emitter did).

**3.** The Target Particle attribute is the particle group that will be emitted during this event. In this case, we want the event to trigger the particle_debris group. Therefore, in this field, type **particle_debris**. Set the Inherit Velocity to **0.02**. This means that the speed of the emission will be based on 0.02 times the speed of the laser particle, which will equal 1, about the same speed that our original emitter was set to. These settings will cause this event to produce similar results to what we had in the test emitter. Finally, check the Original Particle Dies box. This will eliminate the laser blast as soon as it collides with one of the cubes. The Particle Collision Events window should look similar to Figure 22-16.

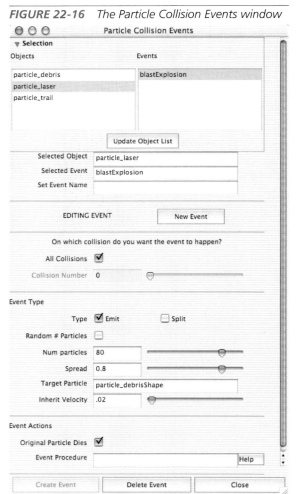

**FIGURE 22-16** *The Particle Collision Events window*

**4.** Unhide the particle_debris object. You can display the instanced geometry in a bounding box to improve performance. To do this, select the instancer1 object and set the Level of Detail attribute to Bounding Boxes. Play back the animation to test everything. You may have to change the Inherit Velocity attribute in the Particle Collision Events window to get the desired result.

**5.** The particle_debris and the particle_trail objects need to collide with the floor when they land. Select both of these objects and then select the floor. Choose Particles | Make Collide. If you play back the animation to test the collisions, you'll notice that the debris objects are bouncing on the ground like crazy.

**6.** The bounciness can be controlled by an attribute called Resilience, which exists on a node called a geoConnector node. This node is connected between any two colliding objects. To select this node, select the floor object and then flip through the connections in the Attribute Editor until you find the geoConnector node. Once it's selected, you can set the Resilience (bounciness) to about **0.2**. Then set the Friction (stickiness) to about **0.2**. Play back the animation and continue to adjust this until you're satisfied. You should have something similar to Figure 22-17.

*FIGURE 22-17*    *The explosions are triggered when the laser bolts collide with the cubes.*

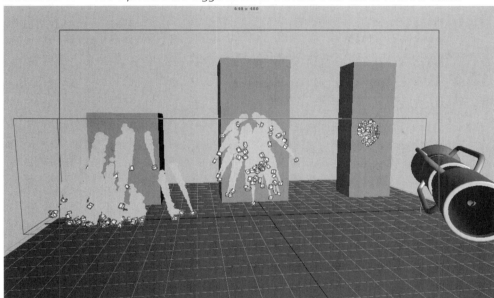

# Render

In this situation, we are using particle effects with two different renderers. The explosion uses Maya's software renderer, while the laser blast uses the hardware render buffer. In this situation, you will have to render the animations separately. One will contain the geometry and the explosion, and the other will contain the laser blast alone. You may even wish to render the geometry and explosion separately so that you can make tweaks to the set and you won't have to wait for the computationally intense rendering of the particle explosion. Figure 22-18 shows the final composite.

**FIGURE 22-18**   *Explosion composited with the laser*

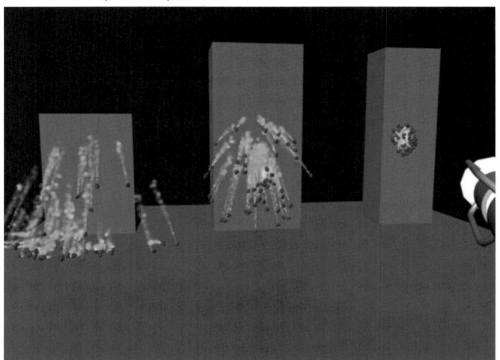

## Summary

In this chapter, you learned about the basics of creating particles and controlling them in the scene using fields. These lessons lay the groundwork for the next two chapters, where you will learn about additional ways to control particles to create more sophisticated effects. Once you understand the basics of dynamics, you can apply them to various applications and for particular purposes.

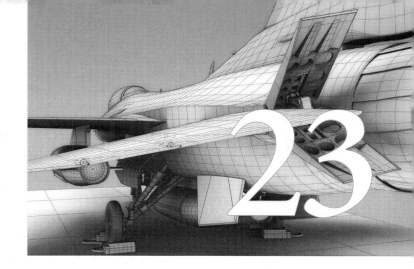

# Advanced Particle Systems and Effects

In Chapter 22, you learned about the basics of creating particles and controlling their behavior with fields. In this chapter, we will dig a little deeper into the Dynamics toolset to create some more-complex effects.

In addition to being emitted from points and volumes, particles can also be emitted from vertices, surfaces, and curves, as well as from each other. Texture maps can be used to control emitter and particle attributes on a particle-emitting surface. The particles' behavior can be further controlled through the use of runtime scripts, called *expressions*. Furthermore, particle *goals* are used to attract particles to other objects for another level of control. In this chapter, we will look at examples involving each of these techniques and practice using them.

# Emit from Object

Recall from Chapter 22 that when we created the explosion, we used the Emit from Object command to emit particles from the existing particles. Here, we'll take a closer look at this command and the options available when used in conjunction with a curve or a surface. Depending on the object used as the emitter, various effects can be achieved by using different emitter types.

## Curve Emissions

Curves offer a large degree of control as particle-emitting objects. Emitting from a curve essentially lets you *draw* where particles need to be emitted from in your scene. This section explores two different ways that curves can be used as emitters.

### Using Omni and Directions Types from Curves

With its default settings, choosing the Emit from Object command with any object selected produces omni emissions from the CVs or vertices of that object. In the case of a NURBS curve, the particles are emitted from the CVs. An obvious use for this would be in placing multiple emitters at different points in the scene without having to create each emitter separately.

To set up a curve emitter, create a curve and select it. Choose Particles | Emit from Object ❑ and reset the options to the defaults. Click the Create button and play the animation. You will see that the CVs of that curve are acting as omni type emitters. Figure 23-1 shows the behavior of particles being emitted from a curve with the default settings.

> **NOTE**  *This type of behavior is created any time the Emit from Object command is applied to any objects containing vertices—this applies to NURBS surfaces and curves, polygonal objects, and even lattices.*

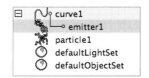

If you look in the Outliner, notice that the emitter1 object has been parented under the curve, as shown here.

Any time particles are set to emit from an object, the emitter will be hierarchically linked in this way. If you select the emitter, its attributes will be displayed in

*FIGURE 23-1*    *A curve with particles being emitted from its CVs*

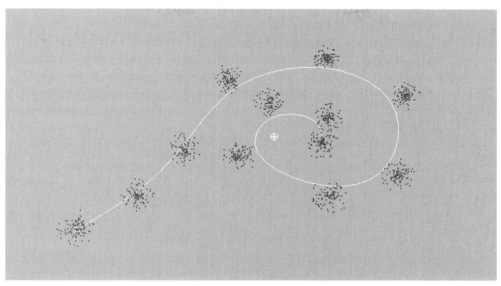

the Channel Box or Attribute Editor and can be edited just as we edited the emitters in the previous tutorials. Changing the emitter's Emitter Type attribute to Directional allows you to control the direction of the emission with the Direction X, Y, Z, and Spread attributes. As mentioned, this might be useful if you need to place a bunch of emitters in a scene quickly. For example, say you are creating a rooftop scene and need to create smoke emitting from all of the smokestacks on houses in the background. Creating an emitter for each one could be tedious.

However, in such a situation, you might also want to be able to control the attributes of the individual emitters so that they are all using different rates. This can be done by selecting the curve and choosing Particles | Per Point Emission Rates. This adds extra attributes to the curve shape node to control the rate of emission from each point, as illustrated in Figure 23-2. If you are looking at the attributes in the Attribute Editor, look in the Extra Attributes folder and you will see a folder called Emitter1 Rate PP. In that folder, you will find an emission rate attribute for each vertex in the curve.

While this smokestack example is a rather obvious suggestion, some interesting effects can be created just by drawing a simple curve with some CVs and rotating the curve. Figure 23-3 shows a render made from a straight curve emitting particles at the CVs that was animated along its X and Y axes. To set up something like this, draw a curve and rebuild it so that it has the desired amount of CVs to act as emitters. Leave the Emitter Type set to Omni. Set the emitter's

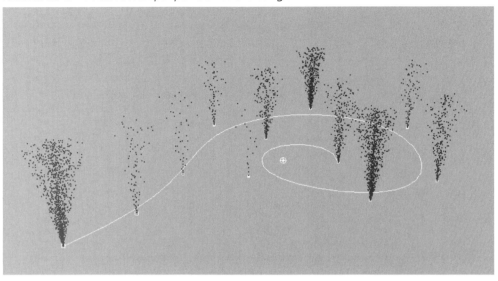

**FIGURE 23-3**    *Rendered MultiStreak particles emitted from a rotating curve*

Rate attribute low. Add the per point emission attributes to the curve. Higher values here will create thicker-looking lines in the rendered image. On the particle's shape node, set the Render Type to MultiPoint and use a ramp to control the rgbPP and opacityPP attributes. Start animating the curve's transforms. Experiment, experiment, experiment! Have fun with it. When you are happy with the way the results look in the view window, render a frame in the hardware renderer with sampling turned up to about 16 or 32. Use motion blur and multiple exposures to soften the look further.

## Using Curve Emitter Types

In this section, we continue working with a curve and the Emit from Object command, but instead of using either the Omni or Directional type of emitter that is available in the emitter's Attribute Editor, we will be using the Curve emit type. This will enable particles to be emitted continuously along the curve. This option allows you to create quite a library of effects. The precise placement of fire sources, shockwave explosions, and water wakes—to name a few— can be achieved by having these effects continuously emit along a curve.

To set this up, follow these steps:

1. Start a new scene, and create a NURBS circle or curve. With the circle selected, choose Particles | Emit from Object □, and use the command's default settings.

2. If you were to play back the animation now, you would get a similar result to what is shown in Figure 23-1, with the particles emitting from the circle's CVs. But this time, a different type of emission will be used. In the Outliner, select the emitter, and in the Channel Box or Attribute Editor, change the Emitter Type attribute to Curve. Then click Play to see a very different effect.

The particles are now emitting from all of the points along the curve. Depending on the size of your curve, the default emission rate of 100 particles per second might not be quite enough to see exactly what is going on. You can increase the emitter's Rate attribute to get a better indication of what is happening. In fact, if you scale the circle, you'll see that you might need to increase the rate so that the overall emission density remains constant. Fortunately, Maya provides an attribute to do just this without your having to change the rate. If you are going to be animating the scale of your circle, you can use an attribute called Scale Rate By. If you have this attribute turned on in the Attribute Editor, Maya modulates the emission rate to keep the emission density constant.

Figure 23-4 shows particles emitting from the NURBS circle, whose Scale attributes are being animated to increase over time. The result is a shockwave effect similar to those used in many popular science fiction/fantasy films today.

*FIGURE 23-4    The scale of a circle is animated as it emits particles.*

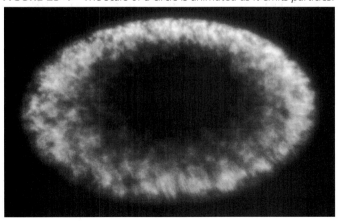

Another great use for an effect with this type of emission would be to create the small, foamy wakes created when rippling seawater collides with a boat or a piling in the water. At the points of intersection between the water and the object, particles need to be emitted. This can be quickly set up by using NURBS surfaces for the two intersecting objects and the Create Intersection command to create a curve on the surface where the two objects intersect. This curve is then used as the emitter. Whenever the surface of the water changes its height, the intersection curve is deformed to match the intersection. Therefore, the wake always occurs at the intersection between the two surfaces.

> **NOTE**   *To enhance this example, read the "Create the Waves in the Ocean" section in Chapter 24 so that you can deform the water surface with soft body dynamics.*

Let's set up a simple simulation for this effect. For this example, we'll use a plane and a sphere:

1.  Create a NURBS plane and sphere.

2.  Select both surfaces and choose Edit NURBS | Intersect Surfaces. A curve on the surface will be created.

3.  Select the curve on the sphere's surface. Use your pick masks to turn off surface selection so that you can more easily choose the curve on the surface.

4.  With that curve on the surface selected, choose Edit Curves | Duplicate Surface Curves. This creates an actual curve whose shape is dependent on the intersection between the two surfaces.

5.  To use this new curve as an emitter, select the curve and choose Particles | Emit from Object □.

6.  Use the Attribute Editor to edit some of the emitter's attributes. First, change the Emitter Type to Curve. You'll also want to set the Speed attribute very low, probably less than 0.1. Also, change the direction so that it is emitting only in the positive Y. Change the Rate attribute to about 200 particles/second.

7.  For the particle's shape node, change the Lifespan Mode to Random Range with a Lifespan of **1** and Lifespan Random of **0.2**. Change the rendering type to MultiPoint and turn the Multi Count down to about **6**.

8.  With the particles selected, add a Gravity field so that they fall after they are emitted.

*TIP*   *The exact settings you use will depend on the scale of the objects you started with. These are just suggested settings that will yield something close to the water wake.*

**9.**   Key the position of the sphere so that it begins almost entirely underneath the surface at frame 1 and then is about halfway revealed at frame 60. Click Play to see the effect. The particles are emitted wherever the two surfaces meet. Figure 23-5 shows the particles being emitted along the intersection of the sphere and the plane.

**10.**   You can further enhance this image by selecting the particles and then the surface and choosing Particles | Make Collide. Now when you play back the animation, the particles emit and fall with gravity until they collide with the plane before they die off. You may also wish to add an rgbPP attribute and an opacityPP attribute so that the particles are white and completely fade out by the end of their lifespan.

*NOTE*   *In this situation, there appears to be a "bug" in the Maya software, where the particles will emit in the negative Y direction once they have been made to collide. To compensate for this, change the emission direction to –1. This causes the particles to emit in the positive Y direction.*

**FIGURE 23-5**   *Particles emitted along the intersection of the two objects*

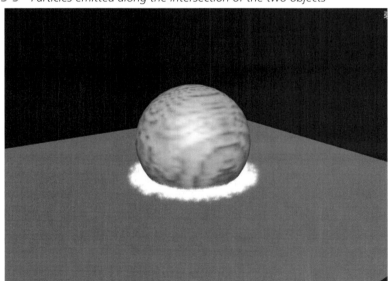

# Surface Emissions

Just as we saw with the curve emissions, a surface can be selected and the Emit from Object command used. With the default settings, this command creates an omni emitter at each of the vertices. The Per-Point Emission Rates command could be used to add control of the individual emission rates at each vertex. Far more interesting effects are possible when the emitter's Emitter Type attribute is set to Surface. This creates a random emission of particles from all points on the surface. A texture can then be assigned to the emitter to control the particle colors and the rate of emission from the various points on the surface.

## Tutorial: Using Textures with Emitters

In this section, we will create the effect of a surface that is burning. We'll choose a surface as an object from which particles will be emitted. To create realistic-looking flames, we'll use an animated Noise texture to control the emission rate to keep the particles from being emitted uniformly across the surface. Finally, we'll render the effect using the hardware renderer.

**Set Up the Surface**    Before we begin experimenting with texture emissions, make sure that the emission surface has been properly configured:

1. Create a NURBS plane. Scale the plane up to 5 units in X, Y, and Z.

2. Select the surface, and choose Particles | Emit from Object ❑, keeping the default settings.

3. In the Attribute Editor, set the Emitter Type attribute to Surface. Set the Rate attribute to **500**.

4. Select the particleShape1 node and set the Lifespan Mode to Random Range, with a Lifespan value of **1** and a Lifespan Random value of **0.5**.

5. In the Range Slider, set the range of frames from 1 to 300. Click Play to see the particles emitting from the surface.

**Add Fields to Control Particles**    As soon as the particles are emitted from the surface, we want to be able to control their behavior with fields. We will use two fields—the Gravity and Turbulence fields:

1. Since we will be strictly using fields to control the particle behavior, we can turn the emitter's speed all the way down. With the emitter selected, find the Speed attribute in the Attribute Editor and set it to **0**.

2. Select the particle group and add a Gravity field by choosing Fields | Gravity. But wait—the Gravity field, with its default settings, will pull the particles downward, but we want them to rise into the air. To fix this, select the Gravity field and change its Direction Y attribute from –1 to **1**. The particles should now flow in the correct direction. Change the Magnitude attribute to **20** so that they move faster.

*TIP*  *Remember that you should have your Playback Speed set to Play Every Frame in the Preferences window. However, depending on your computer's processing and display power, you will not be seeing real-time performance in the view windows. It is therefore recommended that you use Playblast (Window | Playblast) to render a few seconds of the animation so that you can get an idea of how fast the particles are moving.*

**3.** Now we'll add a Turbulence field to give the particles some random motion as they rise upward. Again, make sure the particle group is selected and choose Fields | Turbulence. In the Attribute Editor, set the Magnitude to **20** and the Frequency to **5**.

### Control the Emission Rate with a Texture

Now let's set up the emitter so that it uses a Noise texture to control the emission rate. We can then animate the Noise texture's Time attribute so that the emissions are always changing over the surface.

**1.** In the Hypershade, choose Create | 2D Texture | Noise to create a procedural-based Noise texture.

**2.** Use the Attribute Editor, shown here, to change the Frequency to 2.000. Set the Threshold at around 1.25. Animate the Time attribute by keying it to a value of **0** at frame 1 and **5** at frame 300. Now the pattern in the texture map will change over time.

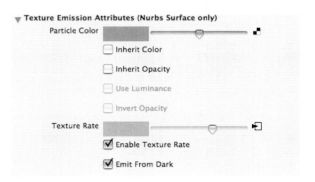

**3.** Select the emitter and scroll down near the bottom of the Attribute Editor to find the Texture Emission Attributes folder, shown in this illustration. Inside, you'll see an attribute called Texture Rate. An easy way to connect the Noise texture to this attribute is to MMB-drag it from the Hypershade and drop it onto this attribute in the Attribute Editor.

**4.** If you were to play back the animation now, you would see no difference in the behavior of the particles' emission rates. To make our example work, we need to do two things. First, we must enable the Enable Texture Rate check box. Next, we must enable the Need Parent UVs option, which allows Maya to get the values from the texture and find what UV coordinate a specific value is at. Scroll back up to the top of the emitter's Attribute Editor and find the Need Parent UVs check box and enable it. Now play back the animation to view the result.

**Render the Particles**    Finally, we can set up some of the particles' rendering attributes to give them a fire-like appearance. We will use MultiStreak particles and edit their color and opacity until the scene looks similar to fire. Then we'll use the Maya hardware renderer to render the animation.

*NOTE   To perfect the look of the fire, you should import it into a compositing program and fine-tune the animation to get the precise appearance.*

**1.** Select the particleShape1 node and look in the Attribute Editor. Set the Particle Render Type attribute to MultiStreak.

**2.** Just below that, click the Current Render Type button to add dynamic attributes for the MultiStreak type.

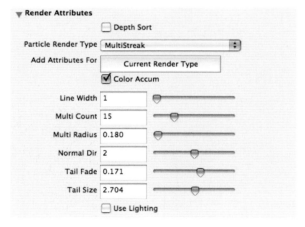

**3.** Set the render attributes to match the values shown in this illustration.

**4.** In the Per Particle (Array) Attributes folder, add per particle attributes for color and opacity by clicking the Color and Opacity button below the list of current Per Particle attributes. When the Add Attribute window opens, enable the Add Per Particle Attribute check box (shown here) and click the Add Attribute button. Attributes named rgbPP and opacityPP will now be available in the list of Per Particle attributes.

**5.** Right-click in the pink field next to rgbPP and choose Create Ramp from the marking menu. The field now says <- arrayMapper1.outColorPP. Right-click again in this field and choose Edit Ramp. This loads the attribute for our color ramp.

**6.** In the Attribute Editor, change the ramp colors to cycle from yellow to a deep orange to red.

**7.** Create a ramp for the opacityPP attribute and edit the ramp so that it goes from transparent (black) to semitransparent (gray) and back to transparent (black).

**8.** Select the emitter and increase the Rate attribute to **2500**. The idea is that we are trying to create enough particle streaks to make them dense enough so that the individual clusters don't stand out. You can adjust the Rate attribute here and the Multi Count and Multi Radius attributes on the particle shape node. If you can achieve the correct balance, any visible streaking can be fixed with motion blur in the renderer or by blurring and layering it in postproduction.

**9.** Open the Render Settings window (Window | Rendering Editors | Render Settings). Choose Maya Hardware from the Render Using pull-down menu. Select the Maya Hardware tab and set the Quality to Production Quality.

**10.** Check the box next to Enable Motion Blur and set Motion Blur By Frame to **3**. Set the number of Exposures to **16**. This means that the renderer will take 16 renders of each frame. When using the MultiStreak and MultiPoint rendering types, the renderer randomizes the positions of each streak in one cluster each time it renders that frame and then composites them. This softens the streaks. Then the renderer blurs the composite with the renders from the previous three frames. The end result is the fire shown in Figure 23-6. While some of the streaking is still visible, it can be easily corrected in Photoshop or After Effects.

*FIGURE 23-6    A rendered frame from the fire animation*

**11.** After doing a few test renders, select the Common tab in the Render Settings window, name the file, choose the name.#.ext option from the Frame/Animation Ext pull-down menu, and then set the End Frame attribute to **300**.

**12.** Choose Render | Batch Render to render the entire animation. Since the hardware renderer is actually rendering each frame 16 times, this will take a long time to complete.

# Particle Expressions

Particle expressions use MEL to control particle behavior and create relationships between particle and object attributes over time. Expressions typically yield the greatest amount of control of the behavior of particles on a per particle basis. With particle expressions, mathematical functions can be assigned to any per particle attribute and will dynamically drive that channel. Dependencies between existing attributes can be defined and custom attributes can be created. In this section, you will learn how to create some simple expressions through some of the most commonly used functions.

## Data Types and Syntax

Several types of attributes are used in Maya: vector, integer, float, Boolean, and enum. When dealing with particle expressions, the most common types are vectors and floats. A *float* is known as a *real number* in mathematics. It can be any number, positive or negative, and may include a decimal. A Translate Y attribute is an example of an attribute that uses floats. It contains a single value that may be a decimal number. A *vector*, on the other hand, is an array of three floats. Color attributes are vectors in which the value is given for red, green, and blue. The colorPP attribute could be assigned a vector for green, which would be written in MEL as `<<0,1,0>>` (color channel values are specified in a range from 0 to 1, not 0 to 255, as in some programs). Position is another example of a vector in which the value describes a position in X, Y, and Z.

When writing any kind of MEL script, it is important that you use the correct syntax. To define or assign a value to an attribute, the object must be defined first, and then the attribute appears, and the two are separated by a dot (`.`). For example, if we needed to set the value for the lifespanPP attribute to 1, we would write it like this:

```
particleShape1.lifespanPP = 1;
```

> _NOTE_  **When using expressions to control the lifespanPP attribute, the Lifespan Mode attribute must be set to lifespanPP in the particle shape's Attribute Editor only. Otherwise, your particles will continue to get their lifespan values from the Lifespan Mode setting.**

Another important piece of syntax is the semicolon (;) used to mark the end of a statement. When writing expressions (as opposed to writing a regular MEL script), a statement must assign a value to one or more attributes. It can be helpful in reading and editing existing expressions if a carriage return is used to separate values. As long as the statement has not been closed by a semicolon, the expression can continue to be read. For example, instead of assigning a vector like this:

```
particleShape1.velocityPP = <<1,.5,2>>;
```

it could be written like this

```
particleShape1.velocityPP = <<
1,
.5,
2
>>;
```

## Creating Particle Expressions

The Expression Editor is used to write and create expressions. One way to create an expression for a per particle attribute would be to right-click in one of the per particle attribute fields in the Attribute Editor and choose either Creation Expression or Runtime Expression. A creation expression is executed only one time during a particle's life—at its birth. A runtime expression executes on every frame. Particle attributes driven by runtime expressions are capable of changing at any frame, as defined by the expressions. Because this is happening on every frame, runtime expressions can cause the scene to run very slowly.

When the Expression Editor is opened by right-clicking one of the attribute fields, the object name and selected attribute appear in their respective fields. The expression can be typed into the Expression text field at the bottom of the window and created by clicking the Create button. Because all names are case-sensitive, syntax errors can be avoided by copying and pasting the name of objects and attributes from the Selected Obj & Attr field into the Expression field and then completing the expression by assigning a value. Assigning a creation expression to a particle group that controls lifespan and velocity would look like Figure 23-7 in the Expression Editor.

## Functions

A *function* in Maya is a built-in operation that can be used to generate values that control object or particle attributes. Many different types of functions can be used. We are going to look at only some mathematical functions such as sin (sine) and cosin (cosine) and random number functions such as the rand function. A list of all available functions is available on the Insert Functions menu in the Expression Editor. When a function is chosen from this menu, it is inserted at the cursor in the Expression field at the bottom of the Editor window.

**FIGURE 23-7**   *Assigning a creation expression to a particle group that controls the lifespan and velocity in the Expression Editor*

We will use the `rand` function to control the lifespanPP attribute as an example. The following expression would generate a random value between 0 and 5 and assign it to the lifespanPP attribute of a particle group:

```
particleShape1.lifespanPP = rand(5);
```

The number inside the parentheses is called the *argument*. We could add another value to this function:

```
particleShape1.lifespanPP = rand(3,5);
```

This expression would generate a float value between 3 and 5 and assign it to the lifespanPP of the particleShape1 object. It is even possible to insert functions into vectors. If we add an rgbPP attribute to the particle group, we can use the `rand` function in one or all of the vector values. This example uses the `rand` function to determine the value of the green channel between 0.4 and 1:

```
particleShape1.rgbPP = <<.8,rand(.4,1),.5>>;
```

If this expression is created as a runtime expression, the function will be generated and assigned for each particle at every frame.

`linstep` is another commonly used function in Maya. `linstep` (and the closely related `smoothstep`) provide functionality that is similar to that of the Ramp texture. This function returns a value between 0 and 1 that is linearly proportional to a specified range. To use `linstep` you must specify a minimum value, a maximum value, and a parameter that is used to generate the proportional number.

Take, for example, controlling the opacity of a particle group. Once the Opacity attribute has been added to the particle group as a dynamic attribute, an expression can be used to control its opacity over time. The following could be used:

```
particleShape1.opacity = linstep (0,10,age);
```

At the first frame, the Opacity attribute will have a value of 0 (transparent), and at 10 seconds, it will have a value of 1 (opaque). Because this is a linear function, the Opacity value at 5 seconds would be 0.5, and at 2 seconds, 0.2. Here, `age` is a *predefined variable* (explained in the next section), acting as the parameter used to generate the Opacity value.

# Variables

*Variables* are containers that store information while a script is executing. Variables can be constant or changing. Two kinds of variables are used in Maya: predefined and custom. We have already seen some examples of using a predefined variable—`age`, `time`, and `frame` are all examples of predefined variables. If an expression assigns the frame variable to a particle's Y velocity as a runtime expression, the particle's Velocity attribute will be assigned whatever the frame number is for that frame. If it is at frame 1, the Y Velocity will equal 1; at frame 2, the Y Velocity will be 2; and so forth.

Custom variables let you declare a value based on what you define and stores it. Custom variables must always start with a dollar sign ($). You must first define the variable before using it in your expression.

Suppose we want to control the color of a particle based on its speed. Unfortunately, no attribute is available for speed, so we need to declare it by basing its value on the magnitude of its velocity. First, we must understand that speed is a *float* value, a magnitude of how fast we are going. Velocity is a *vector*. It tells us not only how fast we are going, but also in what direction we are going. A velocity of `<<0,2,1>>` tells us that this object is moving 2 units per second in Y and 1 unit per second in Z. The `mag` function will return a floating-point number from a vector based on a mathematical formula. Therefore, we can use the `mag` function to define our custom variable. We can then use this variable as a value within any expression. It might look something like this:

```
float $speed = mag(particleShape1.velocity);
particleShape1.rgbPP = <<.2,$speed,.5>>;
```

## Custom Attributes

Just as custom attributes can be added to any node in Maya, custom per particle attributes can be added to a particle node. You can create a custom attribute and define it with an expression. This new attribute can then be accessed by any other expression on any other object in the scene. Take our speed exercise from earlier. If the speed was necessary to calculate a number of attribute values, you would have to define it each time it was used. Therefore, it might be faster to create a custom attribute called *Speed* and define that variable using the preceding expression.

To add a per particle attribute to a particle shape, click the General button in the Add Dynamic Attributes section of the Attribute Editor. This opens the Add Attribute window. With the New tab selected, name the new attribute **Speed**. Make sure the Data Type is set to Float and set the Attribute Type to Per Particle. Click the OK button, and the Speed attribute is created and appears in the list of Per Particle (Array) Attributes. That attribute can now be defined by creating a runtime expression and typing this:

```
particleShape1.speed = mag(particleShape1.velocity);
```

Once it has been defined, a variable is no longer needed in any expression when referring to the speed of particleShape1.

# Particle Goals

Another method for controlling the motion of particles is through the use of *goals*. Using this method, a particle's position or motion can be influenced by any object. By assigning goals to a particle group, the individual particle will be attracted to an object's vertices, pivots, or the UV coordinates of a NURBS surface. This can be useful for creating all types of effects. For example, beads of mud or water can be made to run along the surface of an object. By adding fields and controlling the Goal Weights of the particles, effects such as dust being blown off the surface of an object can be achieved.

# Using a Surface As a Goal

To use an object as a goal for a particle, first create an emitter with the default settings and then create a surface. Select the particle group, and then select the surface and choose Particles | Goal (make sure the tool is set to its default settings). Play back the animation, and as the particles are emitted from their emitter, they are attracted to the vertices on the surface. The first particle emitted will be attracted to the first point, or CV, of the surface. The second particle will be attracted to the second point, and so on. When the emitter has emitted more particles than there are vertices on the surface, it will return to the first point.

In the example shown in Figure 23-8, a NURBS sphere is used as the particle goal for a particle group. When the animation is played back, the particles are attracted to the CVs of the sphere. Notice that an oscillation occurs as the particles are emitted. They overshoot their target and return before they settle at the CVs, instead of locking onto them by default. This is related to the Goal Weight attribute. The Goal command sets the Goal Weight attribute for the particles attracting to the nurbsSphereShape1 to 0.5 by default (as shown in Figure 23-8).

**FIGURE 23-8**    *Particles are attracted to a sphere with a Goal Weight of 0.5.*

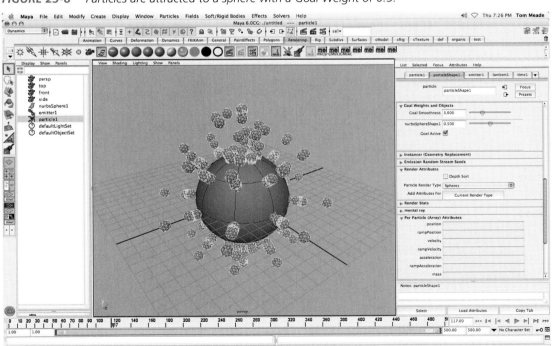

For every object added as a particle goal, a Goal Weight and a Goal Active attribute are added to the particle group (as shown in this illustration). A particle with a Goal Weight of 1 will be completely attracted to its target. No oscillation of the particles will occur as they settle. A particle with a Goal Weight of 0 will not be attracted to the object at all, and the particles will emit and disperse into the world.

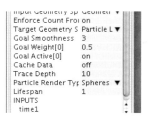

Additional objects may be added to the list of goals to which the particles can be attracted. The Goal Weights can then be animated between the objects. If a particle group had two target goals, for example, each with a Goal Weight of 0.5, the particles would converge midway between the two targets. By animating these Goal Weights, the particles could transition from one object to another.

### goalU, goalV, and goalPP

While using the vertices of objects as particle goals can be useful, far more interesting is when the goalU and goalV attributes are added to the particle group as dynamic attributes. When this is done, particles can be attracted to any point on a NURBS surface and therefore can cover an entire object or move along it. By adding the goalPP attribute, the Goal Weight of individual particles can also be controlled. In this section, we will demonstrate how these attributes can be used by stepping through a small project, in which beads of goo will emit from a NURBS surface and run down it. When the goo beads reach the base of the object, we will use an expression to set its Goal Weight to 0 so that the particles will fall off the surface and be taken over by dynamic forces. Any surface can be used, or you can use the file on the DVD.

**1.** Open the mcr8_ch23_goohead.ma file. Create an emitter with the default settings.

**2.** Select the particle object and the head surface, and choose Particles | Goal. The particles will now be attracted to the surface.

**3.** Change the Goal Weight attribute on the particle shape node to a value of **1**. This will cause the particles immediately to lock onto the vertices of the surface without any oscillations.

**4.** Change the particle Render Type attribute to Blobby Surface, and add attributes for that render type. Set the Radius to **0.2** and the Threshold to **1**. When rendered, these particles will "melt" together to form a blob. Set the particle's Lifespan Mode to Random Range with a Lifespan of **5** with a Lifespan Random of **1**.

   If you were to play the animation now, the particles would be attracted to the CVs of the NURBS head. But we want them to be able to be attracted to any point on this surface. To do this, we will add the dynamic attributes for goalU and goalV.

**5.** In the particle shape node's Attribute Editor, scroll down and find Add Dynamic Attributes. Click the General button and the Add Attribute window will open.

**6.** In the Add Attribute window, select the Particle tab. SHIFT-select the goalU and goalV attributes (as shown in this illustration) from the list and click Add. These two new attributes will be added to the list of Per Particle attributes in the particle shape node's Attribute Editor.

When the animation is played back, the particles are immediately attracted to the surface origin, at parameter 0,0.

> *TIP  When dealing with goals, it is important that you know how your surface is parameterized, where the origin is, and in which direction the surface is going. When using expressions to control any of these attributes, it can be helpful to use a parameterization value of 0 to 1. This way, you don't have to worry about finding the specific number of spans in a surface to get the placement or control you need. By knowing where the surface begins and which way it flows, you can easily predict what values should be used to control the direction that the particles will move across the surface.*

For this next part, we will use a creation expression to define an initial position along V. This will attract the particles to random parameters along the surface's V direction. A ramp will be used to control the goalU attributes. The color at the base of the ramp will determine the U parameter value where the particles are initially attracted. As the lifespan of the particles increases, their goalU value is determined along the ramp until the color at the top determines their final position in U at the end of their lifespan.

**1.** Right-click in the goalV field in the particle shape node's Per Particle Attributes folder. Choose Creation Expression from the marking menu. This will open the Expression Editor.

**2.** In the Expression field, type the following expression and then click the Create button:

```
particleShape1.goalV=rand(0,1);
```

**3.** Right-click in the goalU field in the particle shape node's Per Particle Attributes folder. Choose Create Ramp from the marking menu. Then right-click the array mapper in that field and choose Edit Ramp. The Ramp Editor appears in the Attribute Editor.

**4.** By default, the ramp uses white, a value of 1, at its base. With these default settings, the particles would initially be attracted to the base of the surface and flow upward. The quickest way to fix this would be to drag the positions of the colors on the ramp to reverse them. Also, the white color should be placed about a quarter of the way down, so that the particles have a goalU value of about one and three-quarters the way through their lifespan. The ramp's attributes should look like the illustration:

At this point, the particles are attracted randomly to a goal in V and begin flowing from a U value of 0 (black, at birth) to 1 (white, three-quarter lifespan). When they reach the bottom, however, we want them to lose their Goal Weight and fall with gravity. To do this, we'll need to add a goalPP attribute. We will then use a simple runtime expression to control this attribute so that whenever a particle

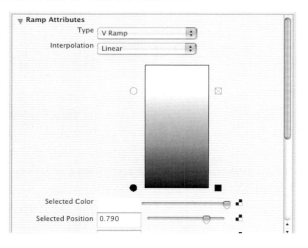

has a goalU value of 1, it will have a Goal Weight of 0 and fall off the surface. Any fields that are connected to it would then take over the motion of the particles.

**5.** In the particle shape node's Attribute Editor, scroll down and find the Add Dynamic Attributes area. Click the General button and the Add Attribute window will open. In the Add Attribute window, find the goalPP attribute from the list on the Particle tab.

**6.** Right-click in the goalV field in the particle shape node's Per Particle Attributes folder. Choose Runtime Expression from the marking menu to open the Expression Editor.

**7.** In the Expression Editor, type the following and then press the Create button:

```
if (particleShape1.goalU == 1)
particleShape1.goalPP=0;
```

The particles will no longer be attracted to the surface when they reach the bottom and will continue with their velocity. At this point, you could fine-tune the animation so that a Gravity field will take over the animation once the goalPP reaches 0. Also, a radiusPP attribute could be added and controlled so that the particles would be small and grow as they continue along the surface. When rendered, the result should look similar to this:

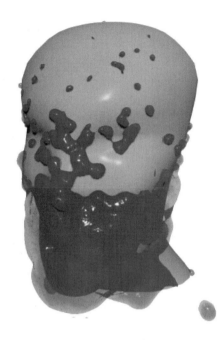

# Crowd Simulation Using Particle Goals

The visual effects and games industries always need to be able to produce a mass of animated characters quickly—be it soldiers, birds, bees, bugs, rats, or even orcs. These masses of animated characters are often referred to as *crowd simulations*. While some companies spend millions of dollars developing crowd simulators where each agent, or individual character, is almost capable of "thinking" or responding to its environment, you can actually set up a simple crowd simulation in Maya using particle goals.

The setup for a crowd simulation is not much different from the goo head exercise. Once the ground plane is set up to act as a goal object, the Instancer (see Chapter 22) can be used to replace the particles with a character that has a walk cycle animation. You'll find a Maya file called mcr8_ch23_marchingSoldier.ma on the DVD. When this character is used as the instance object, you will get an army of marching soldiers walking across the ground. While it might not stand up to a close-up camera angle, the army would work fine as a background element in almost any scene. Figure 23-9 shows a marching army of soldiers.

**FIGURE 23-9**   *Particle goals and the Instancer are used to set up this crowd simulation of marching soldiers*

# Effects

Maya includes a small library of built-in effects that are accessible from the Effects menu in the Dynamics menu set. Most of these effects have been programmed using complex expressions and custom attributes on particle systems to enable the user to control specific types of effects and particle behavior. This can be a time-saver for the user who needs to use some of the more common effects but who does not have the time or knowledge of expressions to set this up.

## Fire

The Fire effect adds a particle group, fields, ramps, expressions, and textures to the scene when it

is created. It uses the cloud rendering type; therefore you must use the software renderer to render the image. By default, fire is set to emit from a surface, so a surface needs to be selected as well. This setting can be changed by opening the command's options and editing the Emitter Type before creating the fire. All of the attributes to control the look of the fire can be edited in the Extra Attributes folder of the particle shape node's Attribute Editor.

## Smoke

The Smoke effect uses a sequence of animated particle sprites to give the effect of smoke. Sprites are image planes that always face the camera. Textures are applied to these image planes and can be cycled through during the animation. Maya includes a 50-frame smoke sequence in the Gifts directory in the same directory as the Maya application. When you create smoke, you first need to go into the Smoke Options window (Effects | Smoke ❑) and specify the path to the image sequence in the Sprite Image Name. You can use your own sequence of images here as well. After you've set up the path, you must choose an object as an emitter. You can edit the rest of the smoke's attributes in the particle shape node's Extra Attributes section. This type of rendering requires the Maya hardware renderer.

# Fireworks

The Fireworks effect uses several custom attributes to control particles emitting from other particles to simulate a rocket, a rocket trail, and the fireworks explosion. Fireworks create their own emitters and three different particle groups for the rocket, the burst, and the trail. Because fireworks use the points and streak rendering types, they must be rendered with the Maya hardware renderer.

# Lightning

The Lightning effect creates a lightning bolt between any two transform nodes, as shown next. It uses a series of expression-driven joints to deform a soft body curve used as the input to the path of an extruded surface. To create lightning, you must first select two objects. Once the lightning has been created, all of the attributes used to control the behavior and color can be edited in the lightning group node's Extra Attributes folder. To get the full effect of the glow, the lightning must be rendered in the software renderer.

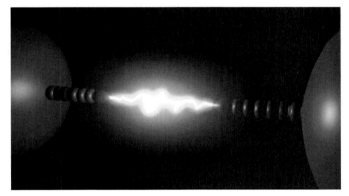

# Shatter

Shatter will duplicate and break up any type of geometry into polygonal pieces called *shards*. These shards can be set up to any thickness or be completely solid. This effect provides a quick way to break up an object into many pieces with a random, jagged edge. This illustration shows a shattered sphere set to solid shatter after the pieces have been affected by gravity.

## Curve/Surface Flow

Curve/Surface Flow creates an emitter and uses a picked curve or surface as a path along which the particles flow. By default, the Curve Flow command places six locators along the curve. The desired number of locators must be set in the command's Options window before the curve flow is created. By scaling the circle at each locator, the Goal Offset value at that point is increased and the particle flow widens or narrows. The position of the locators can be edited by selecting the parent Flow group and editing its locator position attributes. The particle's Life-span and Goal Weight and the emitter's Emission Rate are controlled under this group as well.

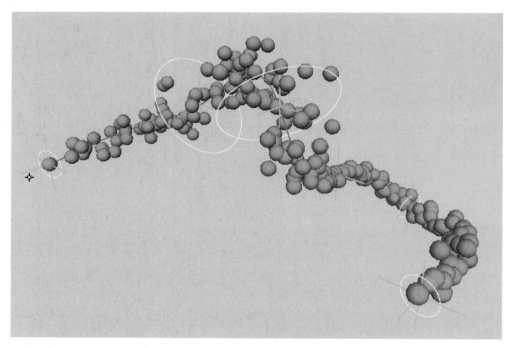

# Summary

At this point, you should have an idea of how to create particles and control them using fields, goals, and some simple expressions. If fiddling with several dozen attributes to achieve a stock effect such as fire or lightning does not appeal to you, you can use the preset effects found in the Effects menu. Either way, you'll agree that Maya's Dynamics toolset offers a high degree of flexibility.

One aspect not covered in this chapter is how to apply these kinds of dynamic controls directly to geometry. In the next chapter, we discuss rigid and soft bodies, or the application of dynamic forces to control geometry.

# 24

# Rigid and Soft Body Dynamics

**As you saw in the previous two**

chapters, Maya's Dynamics engine can greatly

improve productivity by speeding up the

animation process and creating more accurate

motion. While we so far have used particles

affected by dynamic forces, we will now turn

to using actual modeled geometry in a scene.

# Rigid Body Basics

A *rigid body* is a NURBS or polygonal surface that can react to the dynamic forces of fields and collisions. Thus, any object or group can be turned into a rigid body, react to a field such as air or gravity, and collide with other rigid body objects. In this section, you will learn how to create rigid bodies for dynamic simulations. Some of the basic attributes of a rigid body node will be explained through examples. Finally, you'll put this knowledge to use by setting up a simple  bowling simulation. Most of the examples shown in the illustrations and figures have accompanying animated movies on the DVD included with this book. Make sure that you view these movies as you follow along.

## Active/Passive Rigid Bodies

Two types of rigid bodies are used in Maya: *active* and *passive*. An active rigid body can be animated with fields and collisions. A passive rigid body can have active rigid bodies collide with it but will not react or be animated with fields. To demonstrate, let's work with a polygonal box and a plane, with the box located several units above the plane, as shown in Figure 24-1.

 Let's see what happens when we turn both of the objects into active rigid bodies. Select the objects in the view window and choose Soft/Rigid Bodies | Create Active Rigid Body from the main menu. Select the box and choose Fields | Gravity. The image on the right in Figure 24-1 shows the results when the animation is played back. The box falls down as it reacts with gravity, colliding with the plane. The plane will then react to that collision and continue in the direction in which the box pushed it.

> **NOTE**   *Make sure that the Playback Speed attribute in the Timeline Preferences window (Window | Settings/Preferences | Preferences) is set to Play Every Frame. This will ensure that the simulation runs accurately. (See Chapter 22 for an explanation.)*

**FIGURE 24-1**   *When the box collides with the plane, the plane is set into motion along with the box.*

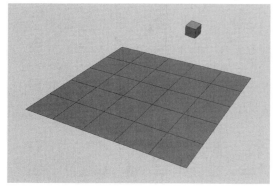 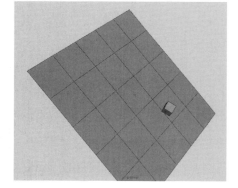

While we still want to have the box collide with the plane as it falls, let's make the plane remain stationary. To accommodate this, the plane needs to be turned into a *passive* rigid body so that it will not move when the box collides with it. Delete the rigidBody node by selecting the plane and choosing Edit | Delete by Type | Rigid Bodies. Then choose Soft/Rigid Bodies | Create Passive Rigid Body to turn the plane into a passive rigid body.

> **NOTE** *Active and passive rigid bodies create identical connections to an object when they are created. The rigidBody node that is created is the same regardless of what type—active or passive—you selected from the Soft/Rigid Bodies menu. However, in the Channel Box, the Active attribute for the rigidBody node is set to On for active rigid bodies and Off for passive rigid bodies. Instead of deleting the rigid body and creating a new one, as mentioned, it is quicker to select the rigidBody node and change this attribute from On to Off in the Channel Box. This attribute can also be keyed so that, for example, a passive rigid body's transformations can be manually keyframed, and then at some point this value could be set to On, at which point the dynamic forces connected to it would take over.*

Now that you've changed the plane to a passive rigid body, when the animation plays back and the box falls and collides with the plane, the plane does not move with the box. Instead, the plane remains static and the box bounces up and down until it settles, as shown in Figure 24-2. The way that it bounces and settles, as well as many other of its properties, can be controlled by editing the rigidBody node's attributes.

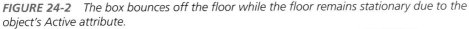

**FIGURE 24-2** *The box bounces off the floor while the floor remains stationary due to the object's Active attribute.*

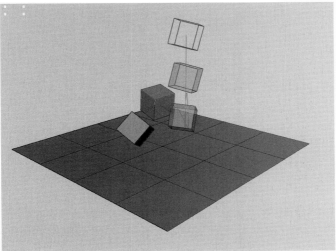

## Rigid Body Attributes

A rigidBody node's default list of keyable attributes in the Attribute Editor is long and can be a bit intimidating. But most of the attributes in the list are pretty straightforward and are easy to understand with a little experimentation. We'll define most of these attributes in this section and show examples when necessary. In the next section, we'll use most of these attributes to control the behaviors of a bowling ball and pins.

### Initial Velocity and Initial Spin

These attributes provide a velocity for the translations and rotations at the first frame of the simulation. The values for these attributes are numbers in Maya units, or degrees, that the object will move during 1 second. Assigning our box an initial velocity in X will make it move in the X direction. After the first frame of the animation, any dynamic fields will take over and control the remainder of the simulation. The box, in this case, will not continue to accelerate on its X axis.

### Center of Mass

The Center of Mass attribute controls an object's center of gravity. By default, the center of mass is at the center of the rigid body object's bounding box. This simulates where the heaviest spot on the object is located, so that any forces or interactions will occur around this point.

For example, imagine two rigid body cylinders that are being affected by gravity and are colliding with a plane. One cylinder has its Center of Mass attribute set at the default and the other has its Center of Mass in Y attribute placed at the base of the cylinder object. Also, both cylinders have some initial spin in X. The first thing you will notice when you click the Play button is that the cylinder objects spin around their centers of mass as they fall. After they collide with the plane, bounce, and begin to settle, they will behave quite differently. While the cylinder with the default Center of Mass settles down to lie on its side, the cylinder with the Center of Mass at the base will eventually settle on its base.

### Impulse, Impulse Position, and Spin Impulse

These attributes apply a constant acceleration to the rigid body for every frame in which they are active. For this reason, these values are usually keyframed to a value for a short period and then keyframed to 0. If the values were left constant or not keyframed to be 0, the object would continue to accelerate at the set value for every frame in the entire animation, which is usually not the desired effect. In fact, in most cases, it is suggested that you try to use fields to introduce motion to an object during the animation instead of using impulses.

### Mass

This attribute sets the weight, or mass, of an object. It is particularly important to define the mass among all objects that collide so that they produce realistic results. For example, a little baseball won't knock over a stack of heavy crates when it rolls into and collides with them. Instead, the crates might not even budge, and the ball's motion would be stopped. On the other hand, if the ball were made out of lead and the crates were empty, the ball could roll straight into the crates, knock them out of the way, and not even be slowed down much. This relationship between objects is defined by the Mass attribute.

## Bounciness

This attribute controls the resilience of the object. Use a value between 0 (not bouncy) and 1 (very bouncy) for realistic effects. A value greater than 1 will invoke bounces into rigid bodies without any initial activity to begin with. In other words, an object at rest on a plane could begin bouncing when this value is set higher than 1.

## Static and Dynamic Friction

The Static Friction attribute controls the amount of friction, or "stickiness," applied to an object at rest. The Dynamic Friction attribute controls the amount of stickiness applied to a moving object. Since these two attributes are usually set to the same values, it is recommended that you set them both at the same time to avoid possible confusion later.

## Collision Layer

Collision layers group the collisions between objects as a way to improve performance during simulations. The Collision Layer attribute assigns the rigid body object to the specified layer. In other words, if two groups of objects are reacting to separate collisions in a scene and will not be interacting at all, collision layers might greatly speed up the playback of the animation.

By default, when all of the collisions reside on the same layer, Maya must check for collisions among all of the objects multiple times per frame. This might be fine if only a few objects are included in a scene, but when several objects appear, it might be optimal to set the objects on different collision layers. For example, suppose a scene has two sets of boxes—one on either side of the scene—and two separate balls each collide with one of the sets. The best way to set up such a scene would be to set the Collision Layer attribute to 1 for the ball that collides with set 1, and assign the second ball and box set a Collision Layer value of 2. Now, even if one of those balls were to roll over to the opposite set of boxes, it would not collide with them. However, if you still need both of the groups to collide with the floor, you can set the floor's Collision Layer attribute to –1, which will cause it to collide with all objects.

## Stand In

The Stand In attribute specifies what type of geometric primitive, cube or sphere, will be used in place of the actual piece of geometry. Using the Stand In attribute can be a tremendous performance booster, especially when you're working with complex geometry. We will talk more about how Maya tessellates the geometry in the next section, but for now, realize that every face of an object must be evaluated when calculating a collision. By using a stand-in object, the simulation uses only a small number of faces that define the primitive object's shape. In the case of a sphere, however, the accuracy of the simulation will actually be improved by the use of a stand-in object, because it will be infinitely smooth instead of possibly being affected by each face being calculated one at a time.

# Tips for Setting Up a Rigid Body Simulation

When using rigid body dynamics, much could go wrong that would produce unwanted results or errors that could affect the performance or accuracy of the simulation. This is especially true for an animation dealing with collisions. Following is a short list of items to check on or set up prior to running the simulation.

**Make sure surface normals are facing in the correct direction.**    Collisions occur only between normals that are facing each other. You will get unpredictable results and even interpenetration errors if the geometry has not been set up correctly.

**Set proper tessellation or use stand-in objects.**    Because Maya uses polygonal facets to calculate collisions, it is important that you make sure that the correct tessellation occurs while using NURBS geometry. While rigid body tessellation attributes are available, it is recommended that you use polygonal geometry when performing rigid body simulations. This gives you precise control over the position of the faces, and you can build more efficient geometry for better performance. Primitive stand-in objects can be chosen from the rigid body attributes or higher-resolution geometry, including NURBS, can be constrained to the rigid bodies during simulation and used at render time.

**Play every frame.**    In the Animation Preferences window, make sure that the Playback Speed is set to Play Every Frame. When running a dynamic simulation, the positions and orientations of any dynamic object are calculated based on the position and orientation values from the previous frame. It is therefore very important that each frame be calculated; otherwise, the resulting simulation that is output at render time may be very different from the simulation previewed in the view windows. Use the hardware renderer or Playblast to test the animations in real time.

 **Create a run-up.**    Objects that sit on top of one another or touch at the beginning of an animation should be allowed to settle before the first actual needed frame of the animation. In other words, if the rigid body objects in a scene need to be still at frame 1, you may need to start the simulation 20 frames before that. This concept is known as a *run-up*. If, for example, you had a stack of rigid body boxes with gravity pulling them down, the boxes should be set up so that they are not touching at the first frame. Leave extra frames at the beginning of the animation so that the boxes have time to settle. Alternatively, you can set up the scene, play the animation until the objects settle, and then choose Solvers | Initial State | Set for Selected. When the animation is reset to the first frame, the objects will remain in this settled position.

**Use the rigid solver to fine-tune rigid body simulations.**    Once a rigid body is created in a scene, a rigid solver is also created. This provides a global method of controlling certain attributes related to the rigid bodies in the scene. (See Figure 24-3 for the rigidSolver node's attributes.) In addition to being able to turn off the calculation states, such as collisions, friction, bounciness, and other states, you can also specify the number of times that Maya checks for collisions per frame. In many cases, an object might be moving so fast that the actual collision with another object happens between frames. By default, Maya checks for collisions about three times per frame, but that may not be often enough for some animations. Decreasing the Step Size attribute in the rigidSolver node will fix this.

**Cache data before time scrubbing.**
Since dynamics are calculated based on the previous frame, scrubbing through the timeline will produce unpredictable results and even errors. If you need to scrub through the animation, you can use the Cache Data option in the rigidSolver node attributes to cache the data. When turned on, the animation is played through with every frame and cached as it goes. Any of the cached frames can then be scrubbed through. If changes are made to the simulation, however, the cache must be deleted before the results can be viewed. This is done by clicking the Delete button under the check box to turn on the cache data in the Attribute Editor.

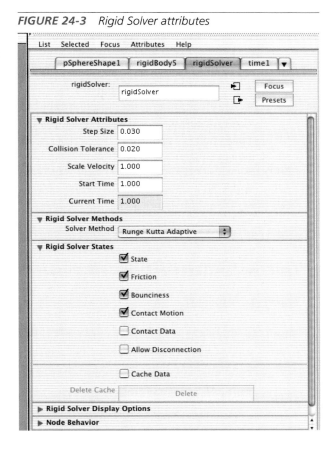

**FIGURE 24-3**  *Rigid Solver attributes*

## Tutorial: Creating a Bowling Simulation

In this tutorial we will use rigid bodies to set up a simple dynamic simulation of a bowling ball knocking over some pins. The ball and the pins will be active rigid bodies with a Gravity field, and the floor will be a passive rigid body. Throughout the tutorial, we will experiment with changing certain attributes and watching the effect they have on the simulation.

We will start with the Bowling_Start.mb scene on the DVD. This scene has been set up to include all of the geometry needed for the example. You will notice that each of the pRigidPin objects has a NURBS object constrained to it. The NURBS objects have all been placed in the display layer called *NURBS pins*. The visibility of this layer has been turned off. This setup lets you use these polygonal, low-resolution pins as the rigid bodies and then hide them at render time, giving you precise control over the collisions and saving calculation time since the NURBS objects do not have to be tessellated at run time.

After we set up the ball and floor to be rigid bodies, we'll add gravity to the ball and adjust the rigidSolver node's attributes so that the motion of the ball looks pretty realistic. Next, we'll turn our attention to the pins, which we'll also turn into rigid bodies and connect to the Gravity field in the scene. We'll adjust their mass and center of gravity until the simulation appears accurate. When finished, the animation on all of the objects is baked and can be rendered. While this is probably one of the simplest simulations we can run, this same basic process can be used to set up any simulation.

**1.** Open the scene file named Bowling_Start.mb on the DVD.

**2.** Select the floor, and choose Soft/Rigid Bodies | Make Passive Rigid Body. Use the default settings.

**3.** Select the ball and choose Soft/Rigid Bodies | Make Active Rigid Body.

**4.** With the ball still selected, add a Gravity field by choosing Fields | Gravity. Set the Time Range attribute to **500**, and click the Play button to view the animation so far.

> _**TIP**_   _**You can save a step by selecting the ball and adding the Gravity field directly. The object will automatically be turned into an active rigid body.**_

**5.** Because the default Magnitude attribute of the Gravity field is set to 9.8 meters per second and we are not working to scale, we need to increase the gravity so that it better suits our scene. Select the Gravity field and, in the Channel Box, set the Magnitude to **30**, as shown in this illustration. Use Playblast to render the animation and check the animation in real time.

| Channels | Object |
|---|---|
| gravityField2 | |
| Translate X | 0 |
| Translate Y | 0 |
| Translate Z | 0 |
| Rotate X | 0 |
| Rotate Y | 0 |
| Rotate Z | 0 |
| Scale X | 1 |
| Scale Y | 1 |
| Scale Z | 1 |
| Visibility | on |
| Magnitude | 30 |
| Attenuation | 0 |

**6.** The bowling ball is bouncing too much. Select the ball and choose its rigidBody1 node in the Channel Box to view all of the node's attributes. Find the attribute called Bounciness and set it to **0.2**.

**7.** Set the Initial Velocity X attribute to **40** so that the ball will have some velocity in the X direction at the beginning of the simulation. After the first frame, the ball will be controlled by fields and the rigid body attributes.

**8.** Now that the ball has an initial velocity, you can move it back in its –X direction so that it rolls across more of the floor.

**9.** The ball appears to slow down a bit more than it should. This is due mainly to the friction, or the stickiness, setting between the two objects. Since the ball is already moving when it collides with the floor—that is, it does not need to start from a resting position—we want to edit the Dynamic Friction attribute, not the Static Friction attribute, to change this value. Set the ball's

**FIGURE 24-4** *After editing the ball's rigid body attributes, the ball's motion should be smoother.*

Dynamic Friction attribute to **0.05**. Do another Playblast render to view the animation in real time. See Figure 24-4 to get an idea of the ball's motion path.

**10.** You might notice that some degree of noise exists in the ball's motion as it rolls on the floor. This is because the ball is made up of flat faces and therefore is not perfectly smooth. We could increase the tessellation of the ball object, but since the ball is a perfect sphere, we can use a sphere as a stand-in object. This can be set in the rigid body's attributes, where you'll see an attribute called Stand In. Use the pull-down menu to select Sphere. After you replace the ball with a sphere, Playblast the animation and notice that the ball's roll is much smoother and therefore more realistic.

**11.** Select all of the pins. Choose Soft/Rigid Bodies | Make Active Rigid Body. Play back the animation.

**12.** Notice that as the ball collides with the pins, it slows down and almost stops. This is because each pin has a Mass of 1—the same mass as the ball. We need to decrease the Mass attribute value of the pins so that the ball can cut through them without slowing down much from the collision. Changing the Mass will also make the pins appear to fly out when they are hit by the ball. Select all of the pins, and set their Mass attribute to **0.1**.

**13.** To make the pins appear more top-heavy, we'll edit the Center of Mass attribute. In wireframe mode, you should be able to see a small *x* near the center of each pin. Select the Center of Mass Y attribute in the Channel Box.

**14.** MMB-drag in the view window to edit the value interactively. Set the attribute so that it is just higher than the height of the ball. This will cause the pins to spin more when the ball collides. A value of **2.5** should do it. Play back the animation.

**15.** Select all of the pins, and then select the Gravity field. Choose Fields | Affect Selected Objects. The pins will now fall with gravity after the ball collides with them.

**16.** To keep the ball and pins from flying off the platform, we will make the plane behind them a passive rigid body. Select the plane and choose Soft/Rigid Bodies | Make Passive Rigid Body.

**17.** With the plane still selected, change its Damping attribute to **1**. This will absorb the energy from any colliding object and cause it to settle. Figure 24-5 shows the positions of the pins during and after the collisions.

**FIGURE 24-5**   *The pins during and after the collision*

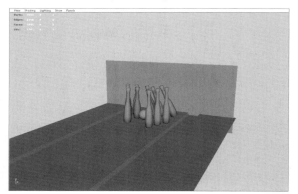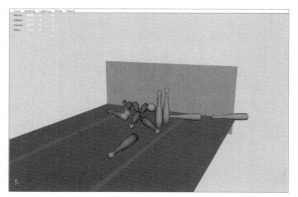

When you are happy with the simulation, you can bake the animation data onto the transforms' animation channels as keyframes. Baking a simulation will sample the objects' transforms at specified increments of the simulation and set keyframes on them. Once this is done, the rigid solvers can be deleted and the animation can be played back and scrubbed through. Since Maya no longer has to calculate motion or collisions, the scene will play back much faster. Select all the pins and the ball. Drag-select all the translate and rotate channels in the Channel Box.

**18.** Choose Edit | Keys | Bake Animation ❐ to see the tool's settings. Make the settings match those shown in this illustration. Click Bake. The animation will play back and set a key on every other frame.

**19.** We can now remove all the rigid bodies in the scene by choosing Edit | Delete All by Type | Rigid Bodies. This erases all of the rigid bodies and thus stops the computer from having to calculate any collisions. Click the Play button in the Time Slider, and you'll notice that the performance is much faster. You can also delete the Gravity field.

**20.** To render the scene, turn off the visibility of the layer named PolyPins and turn on the visibility of the layer called NurbsPins. Render the animation.

# Rigid Body Constraints

So far, we have examined and produced animations in which the objects interacting with collisions and fields are not attached to any other objects. When the bowling ball collides with the pins, the pins fly away until they are stopped by another collision or brought down by gravity. In many situations, you may want an object to be connected to another object that limits the object's motion in some way. This is where rigid body constraints come in handy.

## Types of Rigid Body Constraints

 Maya offers five different types of rigid body constraints: nail, pin, hinge, spring, and barrier. You create them by selecting any rigid body or two (depending on the type) and choosing Soft/Rigid Bodies | Create Constraint ❑. You can then select the type of constraint in the Create Constraint Options window or change the constraint in the Attribute Editor or Channel Box after the constraint has been created.

- **Nail**  Constrains one rigid body to any point in the scene. The object being constrained is then free to move in all directions, but it remains equidistant from the constraint during the simulation. To limit the motion of an object, a second nail constraint can be added and placed elsewhere. Consider a ball in a pendulum. To construct this in Maya, a sphere should have two nail constraints mounted to different positions in the world, thus constraining their movement in one direction.

- **Pin**  Connects two rigid bodies with a ball joint. The chain used in the next example shows how a pin constraint can be used.

- **Hinge**  Locks the translation of the rigid body while constraining the rotation to one direction. The mechanisms in a door and a gear are good examples.

- **Spring**  Provides an elastic connection between any two rigid bodies or any one rigid body and any point in the scene. The constraint always tries to return to the length set for the Rest Length attribute. The Stiffness and Dampening attributes of this constraint can also be edited.

- **Barrier**  Provides a plane with an infinite height and width that limit any rigid body by not letting it move past.

## Simulating a Chain with Rigid Body Constraints

Imagine, for example, that a chain is bolted to a wall. As gravity pulls down the chain, the individual links are each constrained to one another, while the top link is constrained to the wall. To set up such an animation, we need to use two types of constraints—nail and pin constraints. Each link will be constrained to another with a pin constraint, while the top link will be constrained to a point in the scene. Gravity will be applied to the links and the animation played back.

**1.** To create the chain, create six simple cubes: create one polygonal cube, scale it a bit, make five duplicates of it, and then place the cubes to look something like Figure 24-6.

**FIGURE 24-6** *The chain is modeled from duplicated cubes.*

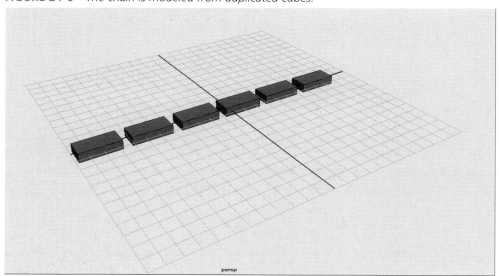

**2.** Select the first two links and choose Soft/Rigid Bodies | Create Constraint ❒ to open the Options window. Select Pin from the Constraint Type attribute and click the Create button.

**3.** Select the second and third links and press the G key to repeat the last operation. Continue adding a pin constraint between each of the remaining links.

**4.** Select the last link and choose Soft/Rigid Bodies | Create Constraint ❒ to open the Options window. Select Nail from the Constraint Type attribute and click the Create button.

**5.** Use the Move tool to move the constraint out to the end of the link at the end of the chain. By default, the constraint was placed at the link's center of mass when it was created.

6. Select all the links, and then choose Fields | Gravity Field to add gravity to the links.

7. Click the Play button, and the chain will fall with gravity and remain linked together by the pin constraints, while the top link holds them all at the point specified by the nail constraint. The chain "hangs" from the top link. You may wish to select all of the rigid body nodes in the scene and change their Damping to **0.3** and the Bounciness to **0.1** so that the chain settles faster.

## Tutorial: Setting Up Rigid Body Interaction with Particles

In this short tutorial, we will practice setting up a water wheel using a hinge constraint. The water wheel will be powered by a flow of particles that collide with and collect on the paddles of the wheel. As the particles collect, they build enough mass to set the wheel into motion as it rotates around its hinge constraint.

1. Open the file called mcr8_ch24_waterwheel.ma on the DVD. This file contains a model of a water wheel and a volumetric emitter that emits multipoint particles.

2. Select the particle group, and then SHIFT-select to select all the paddles on the wheel. Choose Particles | Make Collide. Play back the animation and make sure that the particles are colliding with the paddles.

3. Select the particle group again and add a Gravity field to it (choose Fields | Gravity Field). Leave it set at its default Magnitude for now.

4. To fix the bounciness, select the particle group and then select the geoConnector1 node in the Inputs section of the Channel Box. Set the Resilience attribute to **0**. This turns off the bounciness between the particles and the paddles. Play back the animation to make sure it is working correctly. It should look similar to Figure 24-7.

5. With the particles and collisions pretty much set up, we can convert the paddles to active rigid bodies. Select the paddle objects and choose Soft/ Rigid Bodies | Create Active Rigid Body.

**FIGURE 24-7** *The particles collide with the paddles, run off, and continue falling with gravity.*

persp

**6.** When the animation is played back, you'll notice that nothing has changed. Even though the particles are colliding with the geometry, the water wheel is not turning. This is because we need to set the rigid body node, not just the geometry, to react to the collisions. Select the rigidBody node that is connected to the paddle object. In the Channel Box, in the list of attributes for the rigidBody node, you'll see an attribute called Particle Collision. Set this to On.

*TIP*  *You can select rigid body nodes and many other kinds of nodes in the Outliner by choosing Display | Shapes from the Outliner menu. When the object is unfolded, the transform and shape nodes, as well as the rigid body node, are displayed as separate objects.*

**7.** When the animation is played back, the particles collide with the paddles and knock the paddle objects away. While we need the paddles to remain in place on the wheel, we also need them to rotate around the center of the water wheel. A hinge constraint is the answer.

**8.** Select the paddles and then choose Soft/Rigid Bodies | Create Constraint ☐ to open the Options window. Set the Constraint Type to Hinge and click the Create button. We won't worry about setting the initial position in this window, as this is easier to do interactively in the view window.

**9.** With the hinge constraint still selected, rotate it 90 degrees on its Y axis. When the animation is played back, the paddles now rotate around the hinge constraint as they are powered by the flowing particles.

**10.** To fine-tune this interaction a bit, increase the Mass attribute on the rigid bodies to **1000**. This is important because each particle in the group has a Dynamic Weight value of 1. If the Mass on the rigid bodies were left at 1, the water wheel would be completely outweighed by the possible thousands of particles that bombard it. Therefore, we need to make the rigid body objects "heavier" by increasing their Mass attribute. Another way to achieve the same result would be to lower the particle group's Dynamic Weight attribute. Select the particle group and change the Dynamic Weight value to **0.5**. Also, set the Conserve attribute of the particles to **0.9** so that they settle on the paddles instead of continuing with their momentum from the emitter.

**11.** Finally, increase the Magnitude of the gravity until the water is flowing at a speed of 60. You may also want to set the Static Friction attribute on the rigidBody node to **5** so that the wheel starts up slowly before it gets moving.

Figure 24-8 shows the completed water wheel simulation.

# Soft Body Basics

In Maya, a soft body is an object whose vertices—be they CVs, polygonal vertices, or lattice points—are each constrained to a corresponding particle within a single particle group. These particles can be animated through any of the techniques discussed so far: collisions, fields, expressions, goals, and springs (which we are about to cover). The result is an object that can deform based on any of these dynamic animation techniques. An infinite number of effects can be produced this way—cloth, flesh, hair, and water surfaces are the most obvious examples.

**FIGURE 24-8**   *The completed water wheel simulation*

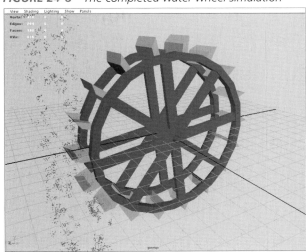

## Creating Soft Bodies and Soft Body Goals

In this section, we cover the basics of creating a soft body and animating it with a field. We will use a NURBS plane to represent a flag that will be animated by a field to make it appear as if it were being blown in the wind.

Start by creating a plane with ten spans in U and V. With the surface selected, choose Soft/Rigid Bodies | Create Soft Body ❒ to bring up the Options window.

By default, the Create Soft Body's Creation Options attribute is set to Make Soft. This setting creates a soft body object out of the surface by locking each CV of the surface to a particle. Any field applied to the soft body or any collision that is calculated will cause the surface to react accordingly. While this might be useful to add some irregularities into a surface, it can be difficult to animate because no controls can be used to retain the original shape of the surface. If a Turbulence field is applied, for example, the particles and therefore the surface would continue in their own directions, as derived from the Turbulence field. After a few seconds, the surface might be scattered all over the scene—looking something like Figure 24-9.

**FIGURE 24-9**   *A Turbulence field is applied to a soft body object that has the default Create options.*

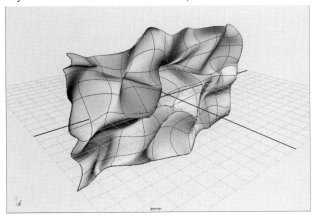

To help you better retain the shape of the original surface, the Create Soft Body command lets you make a duplicate of the surface and use the nonsoft surface as a *goal* object. Goal objects behave the same way that the particle goals behave (as covered in Chapter 23). However, each particle's goal is mapped to the corresponding CV on the nonsoft surface.

Let's continue with the flag example:

**1.** Delete the particle group that is attached to the surface, or create a new one to start afresh.

**2.** Select the surface and choose Soft/Rigid Bodies | Create Soft Body ❑ to open the Options window.

**3.** Create the soft body by duplicating the original surface and making the copy the soft body. Then set the tool to hide the original, nonsoft object and make that original surface the goal object.

**4.** Make sure that the Turbulence field is affecting the soft body object, and click the Play button to view the animation. Notice that the surface is deforming a bit, but it is still retaining its overall shape. In fact, the deformation may be so subtle that you might be tempted to increase the Magnitude attribute on the Turbulence field. While this will work, you could also edit the Goal Weight attribute of the soft body object.

**5.** Select the soft body particle group and look at its attributes in the Channel Box. Notice that the attributes are much the same as the default attributes for a particle group created with the Particle tool or an emitter. They are indeed the same nodes, with just a few attributes added to control soft body goals.

**6.** If you change the Goal Weight [0] from its default value of 0.5 to **0.1** and play back the animation, you will notice that the surface deforms as it is directed by the Turbulence field, but eventually it returns to its original goal position.

Goal Weights can also be edited on a per-particle basis by using either the Component Editor or the Paint Goal Weights tool. To set them using the Component Editor, use the pick masks to select individual particles and then choose Windows | General Editors | Component Editor. Scroll all of the way to the right, and you'll see a Goal PP column. Drag-select all of the fields under that attribute so that you can enter a value once and they will all update.

The other Goal Weight editing option is to use the Paint Goal Weights tool. Select the soft body surface and choose Soft/Rigid Bodies | Paint Soft Body Weights Tool ❑ to bring up the tool's settings. You'll see the Artisan interface in the Attribute Editor, and you can paint the weights directly on the surface, where white has a Goal Weight value of 1 (does not deform) and black has a Goal Weight value of 0.

# Using Springs

Springs improve the amount of deformation control by creating a relationship between the vertices themselves. Therefore, when one vertex is affected by a field or collision, it is capable of pulling another vertex or group of vertices that is connected to it by a spring. Exactly how this relationship affects the connected particle can be controlled through the spring's attributes. The resilience, or bounciness, and dampening are the two basic controls provided by the spring attributes.

If we connect springs to our flag object, we can achieve an added level of structure and control. As is the case with any cloth type of object, any manipulation of any point on the surface affects other points as well. We want to make sure that as the turbulence affects certain points on the surface of the flag, other points on that surface are also affected.

With the surface selected, choose Soft/Rigid Bodies | Create Springs ☐ to open the tool's Settings window. Springs can be added to any particle object—they are not exclusive to soft bodies. When dealing with a soft body, however, the default settings need to be edited to suit the situation. In this example, we will set the Creation method to Wireframe so that it creates springs between particles within a distance set in the Wire Walk Length. In this case, set the Wire Walk Length to **2**. Click the Play button and view the animation. We will be examining springs and their attributes further in the following tutorial, where we will create a water surface with rain causing ripples on the surface.

# Tutorial: Creating a Soft Body Ocean

In this tutorial, we will use soft body dynamics to create an ocean surface. We will apply a Turbulence field to a soft body NURBS plane to produce an overall rippling motion. The Turbulence field's Frequency and Phase attributes are used to control the overall behavior, while the soft body object's Goal Weight is modified to control the viscosity. Once we are happy with the ocean's behavior, we will drop particles on the surface to simulate raindrops. Springs will be added to the soft body so that a rippling effect will occur. A material can then be applied to give the surface some extra detail.

While this example attempts to re-create the behavior of a large body of water, similar techniques can be used to create movement on small puddles of water, or even mud. As we edit attributes, we will suggest settings to achieve the effects resulting from different liquids and volume sizes.

## Create the Waves in the Ocean

We'll start by making waves.

> _**TIP**_   _**Set up the Workspace so that it is displaying the Outliner in addition to the other windows. When working with dynamics, you'll find that objects can be selected easily in the Outliner.**_

**1.** Create a new scene in Maya and set up a new project. Name it **Ocean.mb**.

**2.** Create a NURBS plane with 40 spans in U and V, and scale it up to be 25 units in X and Z.

**3.** Convert the plane into a soft body object by choosing Soft/Rigid Bodies | Create Soft Body ❑. Set the options so that they match the illustration, and click the Create button.

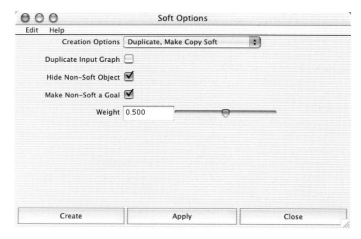

**4.** Rename the soft body plane **waterSoft**.

**5.** With the soft body object selected, add a Turbulence field by choosing Fields | Turbulence. Make sure the tool is reset to its default settings. The Turbulence field will be created and connected to the waterSoft soft body plane.

**6.** Now if you click the Play button, you might notice a little movement on the surface, but not much. Increasing the Magnitude attribute will increase movement, but the dynamic animation will still be too subtle. Since a large, wavering body of water will rarely return to its equilibrium, we know that we need to set this attribute low for an ocean-like behavior. Decrease the Goal Weight attribute on the soft body particle node to **0.2**. This loosens the surface and makes the waves deeper. The result should look similar to Figure 24-10.

**7.** We need to give the waves in the ocean a direction, because now they are just moving up and down. This can be achieved by animating the Turbulence field's Phase attribute. Instead of keying it, we will animate it procedurally through an expression. Right-click

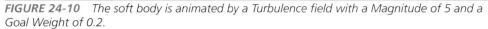

**FIGURE 24-10**   *The soft body is animated by a Turbulence field with a Magnitude of 5 and a Goal Weight of 0.2.*

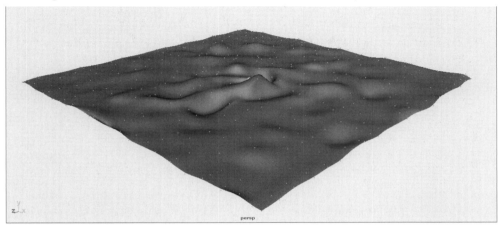

the Phase X attribute and choose Expressions from the marking menu. This opens the Expression Editor. In the Expression field, type the following line, and then click Create:

```
turbulenceField1.phaseX=2*time;
```

*TIP*   **When you right-click an attribute, the selected object name and attribute show up in the Selected Obj & Attr field in the Expression Editor. You can copy and paste that line into the Expression field and continue to type the expression. This saves time and lessens the chances of typing in the wrong words.**

**8.** The Frequency attribute of the field must also be reduced. The Frequency attribute controls the length of the waves, and since we would like large, rolling waves as opposed to short, bumpy ones, we need to set this attribute lower. A value of 0.5 will work well. Playblast the animation to see the animation in real time.

## Set Up the Rain Particles

Now we will create a volume emitter that emits the particle raindrops. Gravity will then be added to these particles to control their fall toward the ocean surface. If an object, such as a ball, vehicle, or character, were falling from the sky instead of particles, we could set up collisions to create the displacement in the ocean. But because the raindrop particles are so small, the collision impact is not that great. We will create an Air field that emits from each particle to displace the soft body surface around each particle. By itself, the interaction causes the soft body particles to bob up and down. To create ripples in the water around where the raindrop particles collide, we will add springs between each soft body particle.

1. Create an emitter and set it as a Volume type. Scale it up so that it is about the same size as the ocean in X and Z but only about 1 unit high. Move it up in the scene so that it sits far above the ocean.

2. Edit the emitter's attributes in the Attribute Editor. Set the Emitter Type to Surface and the Rate (Particles/Sec) attribute to **20**. Click the Create button. Click the Play button and you should see the particles being emitted from the surface.

3. Select the particle object and then choose Fields | Gravity to add a Gravity field and connect it to the rain particles. Set the gravity's Magnitude attribute to **50** so that the particles fall at the correct rate.

4. Select the new particle group and name it **particle_Rain**. In the Attribute Editor for this group, set the Particle Render Type to Streak. Add attributes for this current render type and add a Per Object color attribute so that it matches the settings shown here:

5. To use each particle as a source of an Air field, first create the Air field (choose Fields | Air Field). Select the particle_Rain group and then choose Fields | Use Selected as Source of Field.

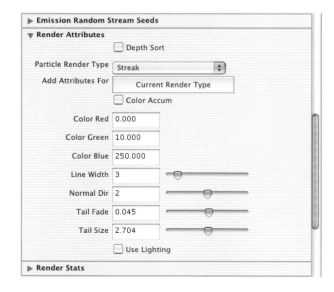

6. Edit the Air field's attributes. Set the Magnitude to **0** and the Max Distance to **1**. You need to make sure that Use Max Distance is enabled so that you can edit the value in the Attribute Editor. Also, make sure the Apply Per Vertex check box in the Special Effects Folder is checked so that each particle, instead of the entire group as a whole, can be used as the source of the field.

*NOTE   If you are editing these attributes in the Channel Box instead of the Attribute Editor, you will be able to see and edit values for attributes that may be inactive and grayed out in the Attribute Editor. Editing these attributes in the Channel Box has no effect. For this reason, you should work in the Attribute Editor until you have a better understanding of how these attribute settings work.*

**7.** For these Air fields to affect the soft body particles, they need to be connected. Use the Dynamic Relationship Editor (Windows | Relationship Editors | Dynamic Relationship Editor). In this window, select the waterSoft object. With the Fields selection mode enabled, choose airField1 from the list on the right. The window should look like this:

**8.** When you click the Play button, you should see the rain particles make little indentations in the soft body surface.

**9.** Because turbulenceField1 is also affecting the surface, it might be best to disable it while we fine-tune the interaction with the falling rain particles. In the Dynamic Relationship Editor, select the waterSoft object if it isn't already still selected, and then click turbulenceField1 in the right column to deselect it.

Now when you play back the animation, only the Air field from the rain particles should affect the soft body water surface.

*NOTE   If the surface were mud instead of water, the indentations made by the raindrops would refill much slower, if at all. This behavior can be created by editing the Conserve attribute on the soft body particle group that is parented to the waterSoft object. By default, it is set to 1, but you can try a lower value to see the effect.*

**10.** Select the waterSoft object and choose Soft/Rigid Bodies | Create Springs ❑ to bring up the tool's Options window. Set the Creation Method to Wireframe and set the Wire Walk Length value to **2**. This will create springs between all of the particles.

**11.** Select the spring group and change its Stiffening attribute to a value of **100**. Now when you play back the animation, the springs connecting the particles that are affected by the Air field pull the other particles and create a rippling effect. You may wish to lower the Goal Weight on the soft body particles to make the drops have more of an effect. A Goal Weight of 0.15 works well. The result should look similar to Figure 24-11.

**12.** Enable the Turbulence field in the Dynamic Relationship Editor so that the water has waves again. However, now that the Goal Weight of the soft body has been lowered, the Magnitude setting that we were originally using on the Turbulence field is too great. Change it to **5**.

**FIGURE 24-11**   *The soft body water is affected by the particle rain drops.*

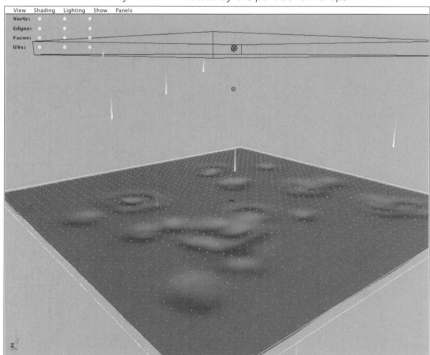

*TIP* *You may want the water surface to have some waves already present at frame 1 of the animation. You can disconnect the Air field from the water surface in the Dynamic Attribute Editor and let the animation play for a few seconds. Then select the water surface in the view window and choose Solvers | Initial State | Set for Selected. Reconnect the Air field to the surface. Now, at frame 1, the surface will already be deformed.*

### Create a Water Material

While we don't want to wander too far from the topic of this chapter, it is important that you realize that small rippling effects such as those in water can be more efficiently achieved through the use of materials. To create the tiny ripples in the ocean, we can use a bump map generated from a *3D fractal texture*. The 3D fractal texture will produce the desired pattern, and because it is a 3D texture, it will change as the surface moves through it because it is affected by the dynamic forces in the scene. As an added touch, we will create and adjust environment fog before rendering the final animation.

1. In the Hypershade, create a new Blinn material. Name it **mOcean**.

2. Click the Map button in the Bump Mapping channel, and the Create Render Node window will open. In the 3D Textures section, choose the Solid Fractal texture.

3. Graph the connections to the material in the Hypershade and edit the solid fractal texture so that Frequency Ratio is set to **2.5**.

4. Select the bump 3D node and change its Bump Value attribute to **0.5**.

5. Select the material node and set its Color attribute to a deep, ocean blue. Adjust the Eccentricity to about **0.1**, the Specular Roll Off to **1.7**, and the Reflectivity to **0.7**.

6. In the Raytrace Options section of the Attribute Editor, turn the Reflection Limit up to **3** and make sure that Raytracing is turned on in the Render Settings window. Also in this window, add an Environment Fog node and change its color to a bluish white.

7. Set the Resolution and Anti Aliasing quality and range of frames, and then use the Batch Render command to render the animation.

8. To render the particles, you need to use the Maya hardware renderer and then composite the pieces in a compositing program. The result should look something like Figure 24-12. Also, be sure to look at the final animation included on the DVD.

*FIGURE 24-12*  *A composited frame from the water animation*

## Summary

Maya's dynamics open up a whole new realm of possibilities for creating just about any type of effect. Having the ability to apply fields to geometry and having them react to other pieces in the scene are very powerful animation tools. While keyframed animation will always be best suited for precise character animation, dynamics are a great solution for letting the chaotic forces of nature take over your scene.

# 25

# Maya Hair

**Maya Hair is an extension of Maya**

Paint Effects that is designed to make long

hair and short hair dynamics. Maya Hair

can also be used for other long strand

dynamics, such as ropes or cables. Its base

is a dynamically simulated NURBS curve,

so it can be applied for numerous effects.

Previously, the process of setting up a

dynamically simulated NURBS curve had to

be done manually. Currently, Maya Hair

automates much of the workflow for complex

hair dynamics. This chapter primarily focuses on creating realistic hair dynamics for a character such as the hair shown in Figure 25-1.

**FIGURE 25-1**    *Hair as it is rendered in the view window*

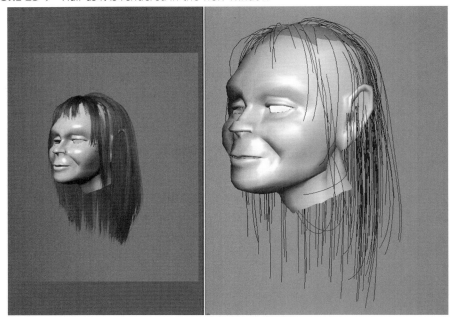

# Using Hair

Maya Hair grows out of a base surface and comes with a default dynamics system. Your task is to modify various hair attributes to find the best setting for a particular kind of hair you are making, create a collision object so that the hair won't penetrate through the base surface, edit a shape of the start/rest curve to regulate the flow of the hair, and finally tweak the rendering of the hair.

Since Maya is designed with dozens of preset hair styles, let's first use these presets to see how Maya Hair actually works.

# Hair Overview: Maya Hair Presets

Using Maya Hair presets is a good way to familiarize yourself with how Maya Hair is constructed and utilized. It also comes with a skull model, so you don't need to construct one yourself. Some hair styles are more complex than others. A few come with collision surfaces, so you can see how various hair styles react. These hair styles can be easily transferred to other surfaces, so you can use them for your own characters later.

**1.** Start Maya, set Menu Set to Dynamics, and choose Hair | Get Hair Example. The visor opens and you will see a variety of hair style examples available.

**2.** Select StraightLongHilight.ma and MMB-drag the icon into the Maya Workspace. Close the visor window.

**3.** This is the typical setup for long hair. Open the Outliner and check out what's there. It features three primary components:

- **hairBase**   The surface from which the hair grows. All the hair extends out of this surface along the surface normals and reacts to the animation of this base surface. It is usually a good idea to construct separate geometry for hair since you can easily limit and paint where the hair should be. The surface can be either NURBS or polygon objects. There is also a way to grow hair out of particular UV sets of polygon objects.

- **pfxHair1**   A Paint Effects node to represent actual hair. When you create hair, you have a choice of outputting hair as this Paint Effects node, outputting it as NURBS curves, or both. pfxHair comes with various Paint Effects attributes, such as hairSystemShape. Open the Attribute Editor and click the tab called hairSystemShape1. By adjusting this node, you can change the general color, the way in which hair follows the curve, the width of the hair, etc. Paint Effects hair can only be rendered with the Maya software renderer. To render in other renderers, you'll need to convert Paint Effects hair to polygons in the case of Mental Ray, for example, or simply output dynamically simulated NURBS curves for a renderer like Pixar's Renderman. More information about this topic follows in later chapters.

- **hairLineFollicles**   Representations showing from where the hair grows. You can actually paint these follicles to add/erase on a base surface. Open the Attribute Editor and click hairLineFollicleShape. This is the follicleShape node, which allows you to edit the hair's appearance per individual follicle. Dynamic follicles (or just follicles) are represented in red and passive follicles are represented in blue. Dynamic follicles come with fully simulated dynamic curves, so they accurately interact with forces and collisions. Passive follicles don't have fully simulated curves, so they instead average neighboring curves to determine their animation. Passive follicles are superior in terms of speed, but dynamic follicles are necessary for more realistic motion.

**4.** To see the effect of hair dynamics, select hairBase and create a quick animation. Try adding a quick translation and rotation and then play it back to see how the hair reacts. Figure 25-2 shows the hair template after it has been dynamically simulated.

*FIGURE 25-2    Maya's hair templates*

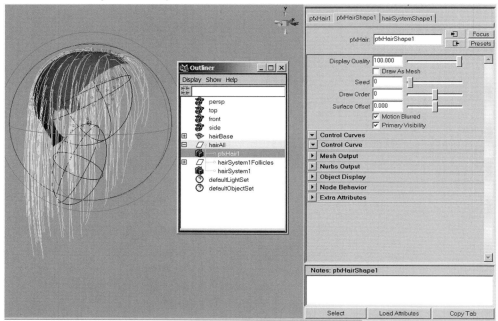

5.  Check how the rendering of Paint Effects hair looks by framing a full view of the hair in a perspective view and then rendering it. Use Render the Current Frame at the top right of the interface for a quick Maya software render.

Next, let's try making our own hair system from scratch.

# Tutorial: Creating Hair from Scratch

In this tutorial, we will create hair from scratch. There are approximately five steps to consider when creating a hair system:

1.  Draw a hair style that you wish to create using pen and paper. It is much easier to achieve your goal if you first create a visual representation of it. Sketching also points out possible problems before you tackle the project. The more you learn about Maya Hair, the more you'll understand how crucial this process is to achieving better results.

2.  We need to construct a hair base surface. This can be a NURBS or polygon surface that is extracted from the actual head model by copying and deleting the face portions. If you are using polygons, make sure their UVs fit within 0 to 1 and that they are not on

top of each other. You may also make a separate UV set and apply the hair to only that UV set. The hair system will follow this surface for dynamics simulation.

3.  Apply the hair to the surface by automatically spreading the hair follicles to the entire base surface, or by painting the hair follicles yourself. Painting the follicles yourself gives you more control over the process. The density and number of follicles can be edited by painting the follicles directly to the base surface later as well. For this tutorial, we will use the automatic method and then paint some attributes.

4.  Adjust the hair by scaling and adding more follicles. Create a collision object so that the hair stays away from the face. When you create hair, it comes with start and rest curves, which are both NURBS curves. A start curve represents where the hair begins, and a rest curve determines how the hair appears when there is no dynamic motion to the hair. It's like the "goal" or target for the hair. By editing these curves to desirable shapes, you can comb the hair. Next, tweak the hairShape node to change attributes such as the amount of clump hair, clump width, hair color, etc. for the final adjustment.

5.  Render the Paint Effects hair using the Maya software renderer.

## Create a hairBase Surface

First you need to create a base surface for the head. For this tutorial, we have a polygon head already modeled. Copy this head model and erase the poly faces where the hair isn't needed.

1.  Open a scene called mcr8_ch25_hair_01.ma in the Chapter 25 folder on the DVD. It comes with a character's head, eyeballs, and skeleton (in template mode) for the animation of the head.

2.  Select the head surface and duplicate it. Make sure that none of the options are turned on in the Duplicate option except for the Geometry Type, which should be set to Copy, and the Group, which should be set to Parent. Name the newly duplicated surface **hairBase**.

3.  Create a new layer in the Layer Editor and name it **hairBase_layer**. Add only hairBase to the layer.

4.  In the Layer Editor, make buukla_head_layer invisible by clicking V in the display option.

5.  Select hairBase and go to the face selection mode. Delete the faces where the hair won't be growing. Mainly, the front face, jaw, neck, and ears have to be deleted. Think of it as a wig. Make sure to leave an area around the sideburns.

6.  A very important step before applying hair to the hairBase surface is creating a UV that fits within 0 to 1. Select the hairBase surface and choose Window | UV Texture Editor.

Currently there is no UV created for this object. Change to the Polygon menu set and choose Create UVs | Spherical Mapping. The UV should appear on the UV Texture Editor. LMB-drag the blue corner of the scale manipulator and make the UVs fit within the color 0 to 1 area. You may need to pick some UVs individually to fit within the 0 to 1 area. For more information on the UV Texture Editor, refer to Chapter 18.

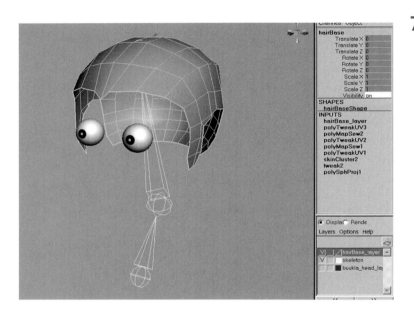

**7.** Bind the hairBase surface with the joint2 skeleton, so that the wig follows the head animation. Take off the joint from template mode by clicking T twice for "skeleton," so it is showing an empty box (make sure that it isn't showing an R). Select hairBase and joint2 (second joint from the bottom) and choose Skin | Bind Skin | Smooth Bind ☐ in the Animation menu set. Set the Bind To option to "Selected joints" and click Bind Skin. The bound scalp should look like this illustration.

## Apply Hair to the hairBase Surface

Next, we will add some hair systems to the hairBase surface using an automatic method. We'll make final tweaks later.

**1.** Select the hairBase surface and, in the Dynamic menu set, choose Hair | Create Hair ☐. The Create Hair Option window will open.

**2.** The Output option determines which type of hair will be created after creating hairs. There are three main options:

- **Paint Effects**   Creates Paint Effects brush strokes on the hair curves. This method works well if your final output is rendered in the Maya software renderer since other renderers won't support Paint Effects. (You can still explore this and convert Paint Effects to a polygon surface in the end, so that other renderers will work with it.) The Paint Effects option also creates a pfxHair node, which has many attributes that change how hair appears.

- **NURBS Curves**   Outputs dynamic NURBS curves only. You can extrude a circle to make the hair or use a renderer that supports curves in order to render in many other renderers. This is also a good method if you are not designing hair, but are instead creating a dynamic NURBS curves system.

- **Paint Effects and NURBS Curves**   Allows you to use both of the preceding options just in case you may be using two different renderers or need the curves to interact with other objects in the scene.

For this tutorial, let's use the Paint Effects output option.

**3.**   After you select the Paint Effect option, adjust the following attributes:

- **Create Rest Curves**   Makes a variety of curves for hair in rested positions. Turn it on.

- **Grid**   Places follicles in regular intervals. Turn it on. The At Selected Points/Faces check box enables you to add follicles in only selected points and faces. This is a good method to add more localized hair.

- **U Count and V Count**   Determines how many follicles are created. Set them to **15** and **15**.

- **Passive Fill**   Adds more hair between active dynamic hairs. It's a good way to give more volume to the hair without making the entire system too heavy. Set the value to **0** for now. You can add more hair follicles later on by interactively painting them in.

- **Randomization**   Gives random placement of follicles; 0 means no randomization. Change it to **1**.

- **Edge Bounded**   Makes the follicles right on the line of UV parameters. Keep this unchecked.

- **Equalize**   Smoothes out uneven UV space, so that the follicles are not bunched up in one place. Keep it unchecked.

- **Dynamic and Static**   Lets you decide whether the hair should be dynamically driven. Keep this set to Dynamic. We want to simulate the hair with gravity and movement of hairBase.

- **Points Per Hair**   Determines how many segments are on the hair. The greater the number of points, the smoother the hair will be. Usually, longer hair requires more segments to be smoother. 10 might be a good number.

- **Length**   Determines how long the hair will be. The number corresponds with Maya units. Keep this value at 5.

- **Place Hairs Into**   Determines how you want to place the hair in a new system or pre-existing system. Choose New Hair System.

**4.**   Click the Create Hairs button.

**5.** Hair grows right out of the hairBase surface. Click the Play button in Playback Controls to see how the dynamics work. It should close to rest in frame 190 or so.

**6.** Turn on buukla_head geometry's visibility and try rendering by clicking Render the Current Frame button for Maya Software. The image here shows what the hair render looks like.

## Adjust Hair Dynamics

There are numerous ways to adjust hair dynamics:

- **Adjust hair length**    First of all, the hair is a bit too long. There is a very easy tool to adjust hair length called the Scale Hair tool. Go to frame 1, select hair, and choose Hair | Scale Hair Tool. LMB-dragging left and right changes the hair length. Make it about half as long. It's pretty important to go to frame 1, because otherwise the scaling effect will not appear correctly on the screen.

- **Add more hair**    The easiest way to add hair is by painting follicles directly on the hairBase surface. Select hairBase and choose Hair | Paint Hair Follicles ❑. The Paint Hair Follicles options window and Paint Scripts Tool settings should open.

- **Create passive follicles**    Passive follicles get dynamics information from active follicles close by. Adding passive follicles is much less calculation intensive than adding active follicles. Return to frame 1Set Paint Mode to Create Passive Follicles, set Follicle Density to 10 in both U and V, set Points Per Hair to 8, and set Hair Length to around 0.5. Adjust the brush size in the Paint Scripts Tool settings by changing Radius U to 0.1, and paint passive follicles on the hairBase geometry. This tool allows you to adjust hair in many other ways, such as creating active follicles with the Create Follicles option, or deleting follicles with the Delete Follicles option. You can also adjust individual attributes for follicles or trim and extend hairs.

- **Create hair collision constraints**    If you play with the hair dynamics now, the hair noticeably penetrates through the face. Next, let's make the hair collide with *collision* *hair constraints* so that the hair will not penetrate. Open the Outliner and select hairSystem1Follicles. To apply collision hair constraints, you need to select actual follicles as a whole or an individual follicle. Choose Hair | Create Constraint | Collide Sphere. The Outliner will show a new node called hairConstraint1. Go to wireframe mode in the Workspace to reveal the spherical object. This is the volume sphere collision object. Scale and translate this object to make it look like a skull inside the head. Play the dynamics and see how the collision works. The upper side of the head is now covered with this collision sphere. For this collision sphere to move with the head, select the hairConstraint1 node in the Outliner, SHIFT-select hairBase, and choose Edit | Parent. You can have several of these collision hair constraint objects to achieve more accurate collision with the face. Another method of making actual head geometry as a collision object is by selecting buukla_head geometry, SHIFT-selecting hair, and choosing Hair | Make Collide. But collision hair constraints are much faster and stable compared to colliding with a geometry.

- **Using Rest Curves**   Some of the hair is in front of the character's face and blocking its view. We can "comb" the hair by adjusting Rest Curves. Rest curves are the goal shape for the hair. When hair isn't dynamically simulating, it will try to conform to this shape. You can see rest curves by choosing Hair | Display | Rest Position. They will be extending out from the head. They are NURBS curves with ten vertices. They are ten points because when we created the hair system in the beginning, we set the Point Per Hair option to 10. By the way, Start Curves is the starting position of the hair simulation, and Current Position is the hair's real position.

Now, let's comb the hair:

**1.** It would be too labor intensive to bend each rest curve one by one, so there is a shortcut to make all of the rest curves bend at once. Play the simulation until 50 frames or so. Select the hair and choose Hair | Set Rest Position | From Current. This will bend the entire rest curve to the current curve's position. You will need to switch between Rest Curve and Current Position display to do this.

**2.** Select the rest curve in front of the character's forehead. Go to component mode and bend the rest curve slightly to the right side of the screen by moving the points around, so it appears combed. Click the Play button to see if the hair follows the rest curve. It probably won't since some attributes' settings are making the hair too relaxed to follow the rest curve. Those attributes are hairSystemShape1 node's Iterations and Stiffness attributes.

**3.** Select hair and open the Attribute Editor. Click the hairSystemShape1 tab. Scroll down to Dynamics. Set Iteration to around 50 and Stiffness to 1. When you play the simulation, the hair that belongs to the rest curve now should follow the shape of the rest curve. Modify the hair curve all around the head to give direction to the character's hair style. Rendering the image, shown here, should return a much better result than our initial attempt.

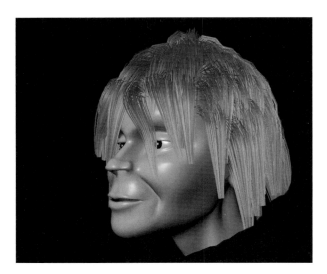

## Summary

Dynamically simulated hair used to be a very difficult effect to create. But thanks to Maya Hair, almost anyone can achieve realistic hair motion. However, there are many attributes to adjust and lots of tweaks are necessary to make a perfect simulation. Practice creating various kinds of hair styles to familiarize yourself with the attributes that control the dynamics and visual appearance of hair. Also, don't forget that you can use Maya Hair for many other curve-based effects.

# Maya Cloth

**Maya Cloth enables you to model and**
dynamically simulate the realistic movement
of fabric cloth, such as a T-shirt following a
character's animation or a flag waving in the
wind. Maya Cloth can also be used as a modeling
tool to extract a realistic cloth model, and then
animate it with deformers instead of dynamic
simulation. Figure 26-1 shows a cloth object that
has been used to dress a character. This chapter
explores the basics of Maya Cloth, demonstrating
how to drape and simulate clothing on this
animated character.

# Loading Maya Cloth

Maya Cloth usually loads automatically when you start Maya Unlimited. However, depending on the previous settings, Maya Cloth may need to be loaded manually using the Plug-in Manager. Here is how you can load/unload Maya Cloth:

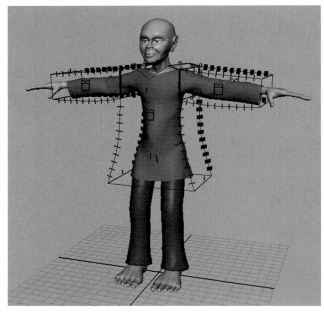

**FIGURE 26-1**   *A cloth object is fitted to and draped around a character.*

1. Choose Window | Settings/Preferences | Plug-in Manager. The Plug-in Manager window will open.

2. In the Plug-in Manager, locate CpClothPlugin.mll. You can check/uncheck the Loaded check box to load/unload Maya Cloth. If you want Maya Cloth to automatically load every time you open Maya, check the Auto Load check box. Close the window.

3. To determine whether or not Maya Cloth is loaded, click the drop-down arrow for the menu selector's list at the top left of the Maya interface. If you find Cloth in the pull-down menu, select it. Now you should have menus in the menu bar specifically for cloth simulation work-flow.

The following sections assume that you are now in this menu set.

## Important Maya Cloth Concepts

As you begin to work with Maya Cloth, it is important to realize that you cannot model the cloth using traditional modeling techniques and simply convert the geometry to Maya Cloth. Instead, you must go through a process of designing a cloth *garment* by sewing, or *seaming,* together a series of panels to define where the openings in the resulting garment will be. This is very similar to the way that a tailor or seamstress works with fabric to manufacture clothing in real life. Figure 26-2 shows the panel, seams, and the resulting garment.

*FIGURE 26-2* *Two panels are stitched together with seams to form a garment.*

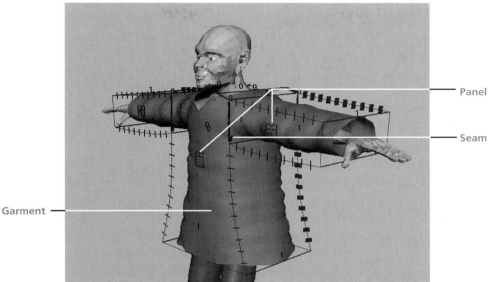

- **Panel**  Determines where the cloth object starts and ends from selected NURBS curves. It has a default cloth property automatically attached to it. The cloth property has attributes for how the particular panel behaves, such as how cloth resists bending or stretching, as well as for scaling, thickness, and friction. You can modify this default cloth property or create an entirely new one. Details are given in the tutorial later in this chapter.

- **Seam**  Connects two separate cloth panels to create a garment. Panels are seamed at the adjoining curve of those two panels. Those two panels must share the same curve at the adjoining edge in order to make a seam. Seam can be edited to increase or decrease the crease to control the seam appearance.

- **Garment**  A collection of polygon cloth objects created from panels. When you create Garment, you see the shape of the cloth, which is a polygon object with triangle faces. This is the actual cloth object that you simulate. It has two nodes connected to it, cpStitcher and cpSolver. You can change the resolution of the cloth using cpStitcher. You use cpSolver to edit actual simulation attributes, such as a start frame and gravity.

To create garments, you can either first create and select a panel or generate them directly from NURBS curves.

_NOTE_    _When creating cloth for objects whose surface structure is simple, such as a flag or tablecloth, you can create what are called_ cloth objects. _A cloth object is typically only a four-sided shape. It works very well for simple shapes, but will not properly deform like a garment will._

# Tutorial: Making a Tunic

In this tutorial, we will create a tunic for the character using Maya Cloth and adjust various attributes to make the cloth simulate well. We are using a character with a proper rigging and animation, so the effect of the simulation on the animated character will be more obvious. When you are creating and rigging a character yourself from scratch, there are several points you should consider when constructing a model and clothing:

- Before you start, it's a good idea to plan and sketch what kind of clothing you are making. Making cloth panels from NURBS curves is just like a tailor making a cloth. By designing and drawing the panels first, you will get much better results. Some questions to ask yourself are: Does the garment have sleeves? If so, how long? Is the garment open in the front? How many buttons does it have?

- Be sure you model and rig a character with a binding pose. The arms should be spread on both sides of the body horizontally. If you are creating pants using Maya Cloth, the legs should be slightly apart. This alleviates complicated self-collisions.

- The cloth you create will not simulate correctly if you don't model the character in an exact scale and if the Solver Scale is set to 1. To simulate the cloth correctly, you may need to model in real life size in the Maya unit or adjust the Solver Scale to compensate for the difference.

The scene we will use is located in the DVD's mcr8_ch26 folder. The scene contains a character in a bind pose. The scale of the character is small, so we will need to adjust the Solver Scale later in the tutorial.

## Set Up NURBS Curves

Let's first create the NURBS curve to create patterns for the garment, and then convert the curves to the garment. An important rule to remember when creating NURBS curves is that they have to be co-planar. This means that all the curves and points have to be on the same plane. One of the curve's first vertices must be located at the same point as the neighboring curve's last vertex. We can achieve that by point snapping the starting point of the curve to the already existing curve.

1. Make sure that the Maya Cloth plug-in is loaded. You can check this by looking at the drop-down menu selector at the top left of the Maya interface. Change it to Cloth. If you don't see Cloth there, follow the instructions from "Loading Maya Cloth" at the beginning of this chapter.

 2. Open the scene called buukla_anim_walk_cloth_01.ma from the mcr8_ch26 folder on the DVD. The scene contains a polygon character in a bind pose. The character also has an animation rig and a walk animation. We will make a tunic for this poor undressed man.

3. Enlarge the Front view. Choose Create | CV Curve Tool ❑. Set the Curve Degree at 3 Cubic, set Knot Spacing at Uniform, and check Multiple End Knots. From the neck on the right side of the screen to the bottom of the shirts, we will use five curves to draw the border of the tunic. Refer to Figure 26-3 to see how the curves are drawn.

4. Start with the bottom of the neck area. Hold down the x key (grid snapping) and make a first vertex around the neck area at the center.

5. Use three more points to the right and go up to the upper shoulder area. You will need at least four clicks to complete the curve if your CV Curve tool is set to 3 (cubic). Call this curve **curve1**.

6. Draw the second curve, **curve2**. Place the first vertex right around the end of curve1. Then continue drawing to the point where an arm and shoulder meet.

7. Draw **curve3**. This time the curve will extend further than the arm. Draw **curve4**, which should start from below the arm, at the end of curve3, and extend to where you want the tunic to end.

8. Lastly, draw **curve5** horizontally from right to left and finish the last vertex by holding the x key and snapping it to the grid at the center. Five curves for the tunic's main part are now complete. Next is an arm area.

9. For the arm, you start the curve very close to the top-right corner of the main cloth curve, which is the end of curve2. Go right to the wrist and stop. Make a new curve from the top of the wrist to the bottom of the wrist. Then draw the last curve from the bottom of the wrist to an underarm area. It should be very close to the last points of curve3.

10. You need to point snap the beginning and end of the curves' vertices to form a complete loop. Select curve2. Right-click above curve2 and choose Control Vertex from the pop-up menu. Select the first point (all the way to the left of the curve) of the curve. Right-click above curve1, choose Control Vertex again, and release the button. Make sure CVs of both curve1 and curve2 are visible, hold down the v key to activate point snap mode, and then drag the manipulator toward the last vertex of curve1. You may want to press the F key to focus around the vertex for more accuracy.

**11.** Continue the same way in all areas where the curves' points meet. There should be a total of six areas where you can snap vertices to each other. Figure 26-3 shows what the curves should look like from the front view.

**FIGURE 26-3** *Curves are drawn in the front panel to define the bounds of a panel*

## Create Panels and Garments

After you snap the points, complete the panel by making the left side. This can be done by mirroring the exising curves:

**1.** Select all the curves and choose Edit | Group ☐. Group Under should be set to Parent and Group Pivot should be set to Origin. Keep the Preserve Position box checked. Click the Group button to group the curves. Now choose Edit | Duplicate ☐. Translate and Rotate should be set to 0. Scale should be set to 1. Set Geometry Type to Copy and set Group Under to Parent. Don't check any other options below. Click the Duplicate button.

**2.** Type **–1** in the Channel Control's Scale X attribute box. The original curves will now be mirrored on the other side.

**3.** Make curves for the back side of the tunic. Open the Outliner, select group1 and group2, and group these together.

**4.** Enlarge the Side view and move group3 in the positive Z axis (toward the left) so that the curves are in front of the character's body. Choose Edit | Duplicate. Click the Duplicate button. Those are the curves for the back side.

**5.** Move the group to the right so it won't intersect with the body.

**6.** Before continuing to make a panel and garment, modify the neck area of the back side curve to make it more rounded. Select two curves around the neck area and go to Component Mode. When you move the points, make sure you only move in the Y axis. If you move in the Z axis, the curves won't become co-planar and you won't be able to create a panel from the curves. Also make sure that you pick two points from both curves simultaneously and move them together. They have to be kept snapped to form a complete loop.

**7.** Everything is set up and ready to make a panel and garment. Using the Perspective view or the Outliner would be the easiest way to select curves. Select ten curves belonging to the front side of the main cloth part. Don't choose the arm area. The order in which you choose the curve has no effect. Choose Cloth | Create Garment. This makes the panel and garment at the same time out of the selected NURBS curve loop.

**8.** If you didn't get the panel and you see an error, you need to see whether or not the curves' points are snapped correctly, or if the curves are co-planar. If the curves are not planar, it may be easier to draw the affected curve from scratch.

**9.** To continue from here, it may be a good idea to turn off Surface from Selection Mask at the top. This way you won't accidentally select a cloth surface when you are trying to select a curve.

*FIGURE 26-4* *The panels are created for the front and back side of the tunic.*

**10.** Select four curves belonging to the front side of the left arm and make a garment. Continue to make the right arms' panels and garments. Make the back side too. In the end, you will have six panels for this tunic. Figure 26-4 shows the panels once they have been created from the curves.

## Create Seams

Next, we will create a seam to stitch up the entire individual garment into one garment. You can achieve this by selecting the curves where you want to stitch and choosing Create Seam. For this tutorial, always select curves from the back panel.

**1.** Start from the left arm. Select the back side's top curve of the arm panel, then select the front side's top curve of the arm panel.

**2.** Choose Cloth | Create Seam. The seam is now formed between two panels, and they will simulate as one panel.

**3.** Select the top shoulder curves, then the back side panel's curve, followed by the front side panel's curve. Choose Cloth | Create Seam.

**4.** Make another seam at the bottom of the arm panel, then along the side of the body.

**5.** The only remaining seams are between the body panel and the arm panel where they join. Select a curve that is shared by the main body panel and the left arm panel. Choose Cloth | Create Seam.

**6.** Try making a seam on another side as well.

**7.** During this process, you may experience several problems, such as the garment disappearing or the seam twisting. This primarily results from opposing normal directions in the garment polygon shells used to create the seam. This is more obvious if you select the problematic garments and choose Display | Polygon Components | Normals. The surface must have its normals pointed in the opposite way. You can correct this by selecting a garment which has its normals turned inward and choose Normals | Reverse. Try creating a seam again.

**8.** The seams of the cloth can be adjusted further to have more prominent creases by modifying the Bend Resistence attribute on the Seam Shape node connected to each seam. Make sure to delete the cache to see the modification. Figure 26-5 shows the garment beginning to form by creating seams between the panels.

*FIGURE 26-5    The panels are stitched together with the Create Seam command.*

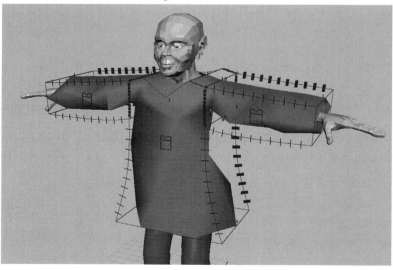

# Set Up the Garment for Simulation

Creating and stitching a garment is now complete, so let's set it up for a simulation. You will focus on adjusting resolution and the Solver Scale and adding a collision object.

**1.** First of all, you may have noticed that the garment appears odd, and that some of the character body parts may be showing through the cracks in the cloth. This is because the garment resolution is too low. You can increase the resolution by increasing the number in the cpStitcher node's Base Resolution attribute. Select all the panels in the Outliner. Individual panels have their own cpStitcher node, so you need to select all of them and change the numbers. This way, if some part of the cloth requires more details, you can add more to those selected areas.

**2.** From the Channel Box, click cpStitcher*x* (where *x* is the number of whichever panel you selected at the end) and increase the Base Resolution to 300. You should now see a cloth that looks more comfortable to wear.

**3.** At this moment, if you see any parts of the garment that appear twisted or misshapen, you need to go back to the previous step and try stitching it in a different way.

**4.** The next thing to adjust is the Solver Scale. If you model the character in exactly the size in the Maya unit, you don't need to adjust this. However, this model we are using is about 27 cm tall. So the cloth on this character will behave like a very small piece of cloth. To compensate for this, we can multiply the Solver Scale. If we imagined this character is about 170 cm tall, we should multiply the Solver Scale by 6.3 to make the cloth behave correctly. From the Channel Control, click cpSolver1. Unlike cpStitcher, cpSolver is shared by all the cloth. Change the Solver Scale attribute to 6.3. The cloth should behave nicely now.

**5.** Change the Start Frame attribute of spSolver1 to –100. We want the cloth to solve and simulate before the character starts moving forward. By the way, a different panel can have a unique cpSolver as well. You can create a new solver by choosing Simulation | Solvers | Create Solver.

**6.** Set up the collision object. This time, the character's polygon shell body will be a collision object. Deactivate the surface selection mask if you haven't done so.

**7.** Turn off the Template mode of the buuklaClean_lowRezGeo layer in the Layer Editor.

**8.** Select the cloth object, SHIFT-select the character's upper-body polygon geometry, and choose Cloth | Create Collision Object. The collision object is now set.

**9.** If the tunic is colliding with the trousers, you can try making the pants a collision object as well.

**10.** Other important attributes to adjust are the collision object's Collision Offset and Collision Depth attributes. Collision Offset determines the number of centimeters between the cloth object and the collision object. Since the character is only 27 cm, set this attribute's value to 0.1 or so. Collision Depth is used to determine how far the cloth can go through the collision object before it stops to resist colliding. Set this to 0.8 or so. In order to visualize the Collision Offset value in the view window, you can pick the body geometry and choose Cloth | Create Collision Offset/Depth Mesh. Figure 26-6 shows the garment fitting to the character's body with the Offset/Depth mesh visible.

---

**FIGURE 26-6**   *The collision offset is adjusted using the Collision Offset and Collision Depth attributes with the Collision Offset/Depth Mesh visible in the view window.*

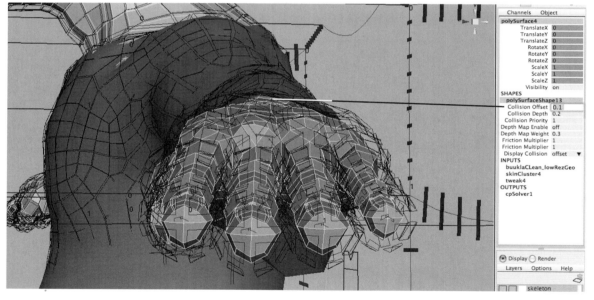

## Run the Simulation

To run the simulation, you can simply click the Play button in the Playback Controls. As the simulation plays through, it saves the current state of the cloth per frame as a "cache" file for you. This way you can quickly replay the simulation. However, it makes more sense to create a

Playblast movie to check the simulation to see if the simulation is in real time. To do so, follow these steps:

**1.** Maximize the window to the Perspective view. Run Playblast by right-clicking the Time Slider and choosing Playblast. It may take several minutes to finish until the character starts walking forward. Press the ESC key when you are satisfied. After a moment of rest, the cloth will gradually shrink and drape over the body to fit the character. When the character moves, the cloth collides with the body and will simulate correctly.

**2.** If you scrub the Time Slider, you can see the simulation update very quickly.

**3.** Once you adjust any attributes to fix the problem, you need to "delete cache" before you can run a fresh simulation. To do that, you need to choose Simulation | Delete Cache.

**4.** You may also make the current state of the cloth as an initial state at the beginning of a new simulation. Select the cloth you want as the default shape by scrubbing the Time Slider and choose Simulation | Save as Initial Cloth State. Now if you go back to frame –100, this will be a cloth's shape and the simulation will start from here.

**5.** Select any panel and open the Attribute Editor. Go to a tab called cpDefaultProperty. This is the default cloth property. You can adjust these sliders to make the cloth behave differently. Here is a quick overview describing each attribute:

- **U and V Bend Resistance**   Adjusts the cloth's resistance to bending. Silk might be more flexible than thick cotton, for example, so you can reduce this value to make it softer.

- **U and V Bend Rate**   Determines how the cloth increases its bend resistance when the cloth is folded in half. A value of 0 eliminates cloth folding as a factor for bend resistance. A value of 1 is the maximum value, and it makes the cloth extremely stiff when it bends in half.

- **U and V Stretch Resistance**   This is very similar to bend resistance, but it's for stretchiness. The smaller the number, the more stretchy the cloth.

- **Shear Resistance**   The higher this number is set, the more the cloth will resist shearing.

- **U and V Scale**   Scales the size of cloth, so the simulation of the cloth is different. The value multiplies the size of the cloth.

- **Density**   Determines the lightness or heaviness of the cloth.

- **Thickness**   Determines the thickness of the cloth. This attribute is mainly used for cloth collision calculation. Make sure to divide this value by the Solver Scale to determine an actual thickness you are creating in the simulation.

- **Thickness Force**   Helps the Thickness attribute by allowing more space between colliding clothes. Increase this value closer to 50 if there are many cloth collisions.

- **Cloth Friction**   When two or more cloths collide, this value determines friction between them.

- **Cloth Damping**   Controls how the cloth will lose dynamic movement.

- **Air Damping**   Determines the friction of air to the cloth. Unlike Cloth Damping, if the cloth is moving in its flat side, there is no damping since it doesn't catch any air.

- **Static Friction**   Determines how the cloth sticks to the collision object when they are not moving. A value of 0 doesn't stick to the collision object at all.

- **Dynamic Friction**   Similar to Static Friction, but this attribute is for when either the cloth or the collision object are in motion separately.

The default setting is very good for these attributes. Make sure to change only one attribute at a time, and study the simulation to see the effect of the change.

If you want to have a separate property for a different cloth, you can make a new property by choosing Simulation | Properties | Create Cloth Property. You can have as many as cloth properties as you need.

Another great tool for cloth properties is the Paint Cloth Properties Tool (choose Simulation | Properties | Paint Cloth Properties Tool). The Paint Cloth Properties tool lets you paint individual cloth properties on the cloth. This is a very useful way to localize certain cloth properties. Figure 26-7 shows the Paint Cloth Properties Tool in action.

## Use Constraints

This section covers how to use a constraint in Maya Cloth to add a brooch. A button constraint is used for this purpose. This isn't a simple point constraint like an animation constraint. A button constraint actually tracks the surface of the cloth in every frame and passes transformation information to the constrained object to make it stick to the cloth surface.

**1.** Import the brooch geometry by choosing File | Import and selecting brooch.ma from the mcr8_ch26 folder. The brooch geometry should appear just below the character's chin.

**2.** Select the brooch geometry and SHIFT-select the cloth1 cloth geometry. In the Cloth menu set, choose Constraints | Button □. Name this button constraint **brooch** in the window and check both Preserve Position and Preserve Rotation. Click the Create button.

FIGURE 26-7    *Editing cloth propertied with the Paint Cloth Properties Tool.*

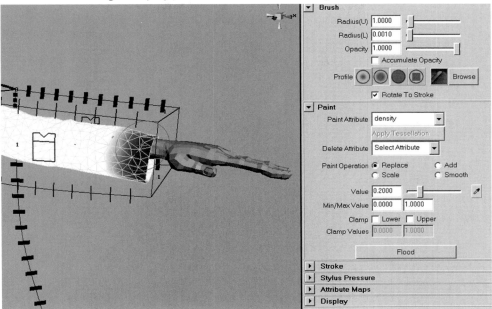

**3.** The brooch is now button constrained to the cloth, but it's at the wrong location. It jumped from the proper location to the front bottom of the tunic. To move this brooch to the proper location, you need to select a parent node called cpButton1.

**4.** Select the brooch geometry in the Perspective view. Press the UP ARROW key twice until the object name cpButton1 appears. There are two attributes to move this button in the Shapes section of the Channel Control (the right side of the Maya UI) for the cpButton1. They are called U Coord and V Coord.

**5.** To move the brooch up, click the V Coord attribute's name in the Channel Control and MMB-drag in the Perspective view until the brooch is in the proper location. You can adjust the position further by adjusting the U Coord attribute as well.

**6.** Run the simulation and see how the brooch follows the cloth. You may want to delete the cache and start from the beginning. This is a great way to attach geometry to dynamically simulating cloth. Figure 26-8 shows the brooch attached to the cloth during a simulation.

There are many other constraints available in Maya Cloth including transform constraints, which hold the cloth at certain points, mesh constraints, which attach cloth to geometry, and cloth constraints, which attach another cloth to the cloth.

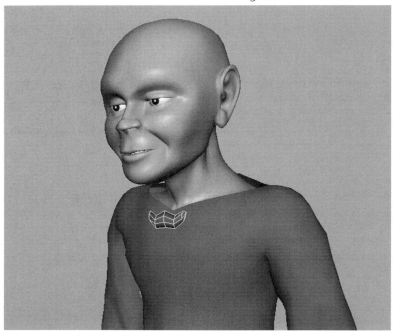

## Summary

Cloth simulation has been a very specialized area of 3D animation. But thanks to Maya Cloth, very complex cloth effects are attainable. The Maya Cloth plug-in is only available in the Unlimited version of Maya. There is, however, a popular alternative to the Maya Cloth plug-in, called Syflex, which creates cloth simulation. Most cloth simulation is created in these two plug-ins in the visual FX industry. Try making many different types of cloth objects to familiarize yourself with many challenging situations you may encounter.

# Postproduction

# Rendering for Postproduction

**In this and the next chapter, we'll**

examine a technique for rendering a scene

in separate passes that can be composited in an

application such as Adobe After Effects. We'll put

together a small visual effects shot of a landing

spacecraft. But before we render anything, of course,

we need to create all of the elements. We'll handle

this in four parts. In the first part, we will create

the environment or background element using a

photograph that is supplied on the book's DVD.

Using a technique called *camera mapping*, we'll introduce *parallax* (explained in a bit) and enable the camera to be slightly animated. Next, we'll prepare a reflection map based on the background image and some other photos taken from the site. This will help us create a color-based relationship between the spaceship and the background. Then we'll deal with the space-ship and its texture maps. We'll also add some simple animation to the spaceship and camera. With the textures assigned to their respective attributes, we'll use a MEL script to isolate certain attributes on the ship's materials. For each separate pass, a scene file will be saved. Finally, we'll create a batch render script and use the command line to render all of the scene files.

This chapter will serve as a good review of the information covered so far in the book. If you find yourself struggling through some of these sections, you might find it helpful to go back and reread the section(s) relative to the step(s).

# Camera Mapping

A *camera map*—also called a camera projection, digital backlot, or digital matte—is a texture that is projected from a camera onto an object or group of objects aligned with the objects in the image. The projection works like a virtual backlot to help a scene look as though it were being shot at an actual location. Figure 27-1 shows a camera mapping example from the dvGarage Camera Mapping Lab (www.dvgarage.com), a training product designed to help you master this technique. Notice the simple geometry that has been placed in the scene to match the objects in the photograph. When the camera pushes in, the scene appears to be in full 3D.

The camera mapping technique is used in almost every big-budget film today. In fact, camera mapping is an economical production technique that can be used in just about any film, no matter how small the budget. *Establishing* shots, the shots in a film or show that begin a scene by establishing a location, are often completed using a camera mapping technique.

For example, suppose you were producing a feature film and needed to include a scene of a secret meeting between members of the British Parliament. All of the film's interior scenes may have been shot on a sound stage or on location inside a building in Los Angeles. However, the director needs to establish the Parliament scene to communicate that it is taking place in the Parliament Building in London, and he wants to include a shot of Big Ben. One option would be to hire a camera crew and have them get a shooting permit, travel to London, set up the gear in front of Big Ben, and roll the film—a very expensive and complicated situation, even for something as simple as this. Another option would be to build a miniature model of Big Ben or model it from scratch in Maya. However, this, too, is an expensive option, especially since it will be needed for only this single shot.

Camera mapping can save the day (and a lot of money). Because only a single photograph is needed, it could even be purchased and downloaded from a stock photo library on the Internet (such as at www.artbeats.com). The image is brought into Maya, where it is set up to be projected

**FIGURE 27-1** *A camera map of the Bangkok skyline (courtesy of dvGarage)*

from a virtual camera that's positioned relative to the objects in the photograph. Simple geometry, such as planes and cubes, are placed in the scene to match the ground and the buildings—as shown in Figure 27-2 in the upcoming section. Once the camera is set up to project the image onto this geometry, another camera is added in the same position, where it can be animated to a certain degree. When the camera is moved—say, zoomed in just a bit—the objects in the foreground will move more quickly toward the camera than the objects in the background. This effect, known as *parallax*, tricks the viewer into thinking that the shot was taken with a motion picture camera rather than a still camera.

Big-budget, special-effects movies make huge use of camera mapping. Films such as *Star Wars: Episode I—The Phantom Menace* and *Episode II—Attack of the Clones* used camera mapping extensively. Often, miniature sets were built for effects scenes that use motion control cameras to film sequences. But again, in the case of establishing shots, these models could be photographed and then, in the hands of the computer graphics artists, used as camera maps. Several articles on computer graphics have mentioned one particular shot in *Star Wars: Episode I*, in which the artist used several photographs of a model taken from different positions and projected them from several different cameras onto very rough geometry. The render camera, the one actually rendering the scene, flew through these very simple objects that were being texture mapped from different cameras to create a fly-through of the entire model.

One last example of camera mapping is the *digital matte painting*. Matte painting has been in use since the early days of film. Before there was such a thing as compositing in postproduction, all special effects were done in camera, meaning that the matte paintings were hung behind the set and the scene was shot. After basic compositing techniques were available, sets could be painted on glass, which was then combined with the live elements in postproduction. However, in most of these cases, the camera had to be still because it was not possible to create parallax with these techniques.

Using camera mapping techniques, an artist can now create a painting—traditionally or digitally—that can be used as a camera mapped image. It must, however, obey the laws of physics with respect to perspective and camera lenses. It can therefore be helpful to the painter to begin with a photograph, where the distortion caused by perspective is already established. Some artists might even set up some simple objects in a 3D scene, render them, and use the rendered image as a template, on top of which paint is applied.

## How Camera Mapping Works

In camera mapping, an image is projected from a camera that is set up to use the same focal length and aspect ratio used by the camera that took the original photo. A simple grid or cube is aligned to some object that lies along or extends from the ground plane. When all of the sides of this object match the perspective lines in the photograph, you know that the ground plane and camera position have been established. At this point, the projection camera is locked down and any other objects in the photo have some basic geometry placed and aligned to them.

A material is created that uses a projection to map a texture to it. The projection mode is set to use the perspective of the projection camera to project the texture. This is similar to regular planar projections, but instead of being projected orthographically, or in parallel, the image is projected with the converging perspective lines unique to a specific camera lens. This texture is then applied to all objects in the scene that will use this map.

Finally, the projection camera is duplicated and the duplicate is used as the animated render camera. When this camera is moved, its relationship to the objects in the scene changes as it would in the real world, because the parallax has been simulated.

## Tutorial: Camera Mapping a Junkyard Scene

In this section, we cover the process of setting up a simple camera map step by step. The photograph we'll use was chosen for its simplicity. Most of the objects in the scene are box shaped, which will make matching the placement of simple polygonal cubes relatively easy.

## Prepare the Photograph

The first part of the process involves a little preprocessing of the photograph. The image used for the final render should be a high-resolution image. The higher the resolution in the image, the further in you can zoom with your render camera. It is safe to suggest that the camera projected image should be at least twice the size of the intended final output resolution. However, if you plan to zoom way in, you will need an image with very high resolution.

Working with such a large image in Maya can be cumbersome, and in fact some video cards cannot display images that are more than 2048×2048. For this reason, it is a good idea to make a low-resolution version of the photo. In addition, you should draw some straight lines along any pairs of parallel edges on this same low-resolution version. This will help you see the edges while you are trying to line up the cubes in the view window. Also, drawing edges that run horizontally across the photo at the base of the objects will help you align the objects relative to one another. Figure 27-2 shows the junkyard photograph with the perspective lines drawn in. Save a version of this file and name it **mcr8_ch27_junkyard_ref.iff**. This version should be about 900 pixels wide. (You may also find this reference image with the perspective lines on the DVD.)

*FIGURE 27-2*   *The low-resolution reference image with lines drawn over parallel edges*

### Set Up the Projection Camera

The most important part of this entire process is making sure that the camera that projects the texture is set up as accurately as possible. The camera projecting the image should use the same focal length and aspect ratio as the camera that took the original photograph. For this reason, it is helpful to know the focal length of the lens that was used to take the photograph.

When you have the focal length data, you can be sure that once you have aligned one object between the camera and the photo, the rest of the geometry will fall into place. Without this information, you could guess the focal length, but it would take much trial and error to set up the scene properly. Without establishing how the lens distorts parallel lines, you might find that after you place one object, matching the next one is more of a challenge—it may need to be moved farther away or scaled bigger or smaller than it should be in relation to the objects in the scene. If this is the case, you know that you need to try a different focal length. Even if you can line up the object with the photograph by scaling or moving it away, once you begin animating the camera, the image will quickly fall apart.

> *TIP*   *To make this setup easier, the photo was shot so that the center cross hairs in the camera matched the point where the brick wall meets the ground. This ensured that the camera was pointed at an object on the ground. When the camera is orbited later on, it will rotate around that point, and since it is known that the X axis lies on the ground, we can place other objects relative to it. This is information you may not have if you are using a stock photo, but take note of these things if you are shooting your own photograph. It will make alignment in Maya much easier.*

Let's begin the setup:

1. Set up a new Maya project and start a new scene. You may wish to copy the mcr8_ch27_junkyard_ref.iff and mcr8_ch27_junkyard_HR.iff images to the source images directory of the new project.

2. Create a new camera (Create | Cameras | Camera) and name it **projectionCam**. Move it back along the Z axis and use the Show Manipulator tool to snap the camera's reference point to the world origin. When viewing the scene through this camera, having the camera pointing at this world origin will put the X axis along the centerline of the photograph.

3. In the projectionCam camera's attributes, set the focal length to **52**, since the photograph was taken with a 52mm lens.

4. In the Render Settings window, set the resolution to match the resolution of the photo. In this case, it is 1079×719.

**5.** Look through the projectionCam camera and create an image plane for that camera by choosing View | Image Plane | Image Plane Attributes. In the Attribute Editor, load mcr8_ch27_junkyard_ref.iff in the Image attribute. Click the Fit to Resolution Gate button.

*TIP    You may need to adjust the camera's Overscan attribute so that the entire image plane is visible.*

Our image is now set to project from a camera with the same focal length and aspect ratio as the original photograph. We now need to position this camera so that its height and orientation with respect to the ground match the position and orientation of the original camera. We will use the large bin on the left side of the photo to calibrate this.

**1.** Switch to a Side view and create a polygonal cube that is 2×2×2. Edit its pivot point to the base of the object, and then move the cube so that it sits on the XZ plane. Turn on snapping to make sure that the polygonal cube is sitting on the axis.

**2.** Switch the view though the projectionCam camera. Orbit the camera up and try to align the grid lines with the base line of the left bin. As long as you don't move the camera with the pan control, that center X axis will always be on the ground. Once the grid lines are aligned with the base of the bin, we can use the cube to determine the distance, or how far back the camera needs to be from the center.

**3.** Zoom out so that the edge of the grid is visible in the foreground. Use the Move tool to move the cube over in X about −5 units and in Z about 5 units until the lower corner lines up with the photograph. The cube should actually sit right below the bottom edges of the bin, since the bin seems to be sitting a few inches off the ground. You may need to tweak the camera by orbiting or nudging the cube in the X or Z axis. Just be sure not to pan the camera or move the cube in the Y axis.

**4.** Scale the cube in Y until the top edge aligns with the top edge of the left bin in the photograph. Do any final bits of tweaking to get both the upper edge and the lower edge in alignment. When those edges are matched, the cube can be scaled along its Z axis and translated in Z until the front and rear edges are aligned.

*TIP    It may be best to select the attribute in the Channel Box and MMB-drag in the view to manipulate the scale and position of the cube.*

**5.** The cube may seem a bit crooked. It is possible that the ground is not entirely flat where the bin is sitting. If you have done everything possible to align it using the steps so far, you can rotate the cube on the Z axis just a little until all of the edges line up. Figure 27-3 shows the cube placed in its final position.

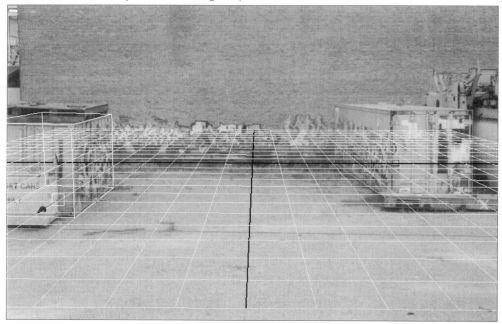

**6.** Now that the projection camera is placed, lock it down so that it can't be moved by accident. Select all of the attributes in the Channel Box, right-click them, and choose Lock Selected from the marking menu.

## Place the Other Objects

Now that the camera's position is established, the rest of this process goes rather quickly. Create more polygonal cubes and align them with objects in the scene. You may be tempted to go a bit overboard—just remember that this is just one shot that will go by the viewing audience very quickly, and plan your details accordingly. We will be adding a ship landing in the middle of the scene, so most of the viewers' attention will be directed to the ship. For this reason, it is best to place geometry only for the main objects in the scene. If something still stands out as wrong when the shot is composited, you can add more geometry and quickly re-render the background.

The quickest way to continue placing the objects would be to duplicate the cube you created for the left bin. Set all of the duplicate's Rotation values to 0, and move it into position on the large bin on the right side. Use the horizontal grid lines to determine the point where the front edge should align, and then go about scaling, moving, and rotating the cube until it fits. It usually takes a few rounds of small adjustments to get it right.

With these two objects now placed, the boundaries of the scene are established. Switching to the Perspective view allows you to duplicate and move the other objects roughly into position based entirely on their spatial relationships. Then switch back to view the scene through the perspective camera to fine-tune the placement so that all of the objects are in alignment with the photograph. Eight cubes total should be enough to cover all of the main objects.

Finally, add a plane for the ground and scale it up to cover the entire scene. Then add another plane for the rear wall. It should extend directly up from the X axis origin. The completed placement should resemble that shown in Figure 27-4.

**FIGURE 27-4** *The scene with all of the objects placed*

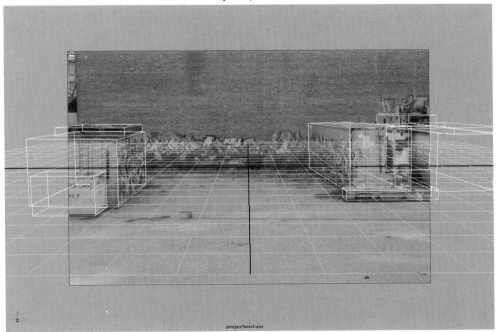

### Create the Projection Material
Now let's create the material and map it onto the geometry. Essentially, we will create a material and add the high-resolution map as a projection. The projection type will then be set to a special type called Perspective.

> **1.** In the Hypershade window, create a new Surface Shader material. The Surface Shader is great for this sort of application because it uses a texture's Brightness values so that there's no need to light the objects or adjust any material attributes to make this work.

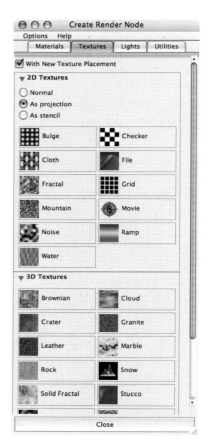

2. Connect a file texture to the Color attribute of the Surface Shader material. In the Create Render Node window, select the As Projection type instead of Normal, as shown here.

3. Load mcr8_ch27_junkyard_HR.iff into the file texture node.

4. Select the 2D projection node connected to the file texture in the Hypershade window and view its attributes in the Attribute Editor. Set the Proj Type attribute to Perspective.

5. Underneath that, open the Camera Projection Attributes folder and set the Link To Camera attribute to projectionCamShape and the Fit Type to Match Camera Resolution. This sets the projection to use the projectionCam camera to project the texture with the aspect ratio set by the resolution. The Attribute Editor settings should look as follows:

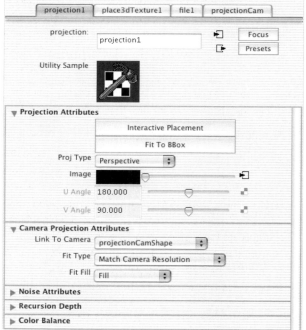

6. Turn the Incandescence value of the Lambert material all the way up. This ensures that the texture will render at full brightness, without any lights.

7. Select all of the geometry in the scene, and then in the Hypershade, right-click the material and choose Assign Material To Selection from the marking menu. The map will be applied to all of the objects.

## Set Up the Render Camera

A minimum of three perspective cameras should be used in a camera mapping scene. We have already dealt with the first two. One camera is used as the working camera, as it would be in any other 3D scene. You use this camera to navigate around your scene and place and edit objects. The second camera, the projection camera, is used to project the image. You'll create a third camera as the render camera. This camera will be animated and used to render the scene. It is best to duplicate the projection camera once it has been established and locked. This way, the render camera is in its default position.

**1.** Select the projectionCam camera, duplicate it, and set the view window to look through it. Rename this duplicate camera **renderCam**.

**2.** Set the animation frame range to something small—in this case, 90 frames.

**3.** Select the renderCam and set a key at this position.

**4.** Go to frame 90. Zoom in the camera a bit and move it down. Set a key for this position.

**5.** Set up the Render Settings to render all 90 frames of the animation. In the Resolution section of the Render Settings window, check the box next to Maintain Width/Height Ratio and set the Width to **600**. Set the Frame Padding to **4**. This numbers the frames so that they have four digits, starting with frame 0001. When numbered like this, After Effects can properly import a sequence of frames.

**6.** In the Anti-aliasing Quality folder, set the Quality to Production. This creates nice, anti-aliased edges.

**7.** Do some test renders of various frames from the animation. Pay particular attention to the frames from the end of the animation. You'll notice that the texture on some of the foreground items spills over onto objects behind them. This is most obvious with the orange beam in front of the bin on the right side of the frame. To fix this area, you need to make a duplicate of the mcr8_ch27_junkyard_HR.iff file and then, in Photoshop, paint out the area (using the Clone tool) on the bin where this overlap occurs. Return to Maya, duplicate the Surface Shader's entire network (choose Edit | Duplicate | Shading Network from the Hypershade's menu bar), and then replace the current file texture with the one that has the beam painted out from the bin. Apply this material onto the bin. Now the obstructed area on the bin will be revealed as the camera moves in.

*TIP  You may notice that the texture is blurry on some of the faces. This is caused by the default filter setting on the file texture node. If you select the file texture node that contains the camera mapped texture, you'll see a Filter attribute in the Effects folder of the node's Attribute Editor. Maya's default is set to 1; in most cases, you will find that this is too much. Start with a setting of 0.1 and do a test render. You can increase the value, but you'll rarely need to go much higher than that.*

**8.** When the test render looks okay, render all 20 frames and view them in Fcheck. It should appear that the scene was actually modeled in 3D. If not, take careful notice of the problems, and make the adjustments in the scene by moving some of the cubes. If you are still stuck, open the mcr8_ch27_cameraMap_finish.mb file on the DVD and study that. Also, look at the mcr8_ch27_cameraMap_Finish.mov file to see a render of the scene so far.

# Creating a Reflection Map

In Chapter 18, you used a reflection map from dvGarage's Reflection Toolkit. In most cases, one of those reflection maps will do a fine job, even if the map does not contain any of the objects from your scene. The primary use of the reflection map is simply to give reflective objects something to reflect—it helps to "sell" the object as something that is made from a reflective material. To this end, the maps included in the Reflection Toolkit are sufficient.

In Chapter 21, you learned about raytraced reflections. These might be necessary when an object that will be given much attention in the scene needs to reflect the objects surrounding it. For example, in a scene with a highly reflective glass sitting next to a soda can, with these objects at the center of attention, you could pretty easily pick out that something is wrong if that soda can is not reflected in the glass. It is easier to get away with not using raytraced reflections, even in these types of situations, when the objects are moving. In such cases, you have much more freedom to decide whether to use a reflection map, raytracing, or both. Just remember that raytracing is a processor-intensive operation and can tie up your computer for hours or even days if the scene is complex.

So what do you do when you need semi-accurate reflections in the scene but you don't want to spend the time raytracing? The solution is described next.

*FIGURE 27-5    The EnvBall map of the junkyard scene*

## Environment Ball

A Maya rendering node called an EnvBall (Environment Ball) allows you to add reflection maps that were taken by photographing a chrome ball (see Figure 27-5). EnvBall is an excellent choice for situations in which only one side of the reflected object is visible. To obtain the reflection map that will be used on the EnvBall, you can either bring along a chrome ball and photograph it on site or place a highly reflective chrome sphere in the middle of a Maya scene and photograph it with the renderer. By using raytraced reflections to calculate the reflections on the sphere, the sphere will reflect the surrounding objects pretty accurately. But since the ball will be used as a map, you will need to use raytracing only once.

### Set Up an Environment Ball

With our junkyard environment, the ball can be set up between two camera-mapped objects. Capturing reflections between these objects from a still photo might have otherwise been impossible. However, images still need to be provided for the area behind the camera. Therefore, it is necessary to use some other photos. These other images don't need to represent the rest of the scene accurately, because our camera will never turn that much, so any images will do. In this case, we will use an image that was taken near the site of the camera-mapped image. Once these images are mapped to a dome and covering the part of the environment not already covered by existing geometry, the reflective sphere is rendered. This render is then imported back into the scene and used as an EnvBall texture.

1. Open the completed camera map scene on the DVD, or use your own scene.

2. Create a sphere and place it between the two bins. Scale the sphere so that it is roughly the same size as the spaceship.

3. Create a new Blinn material. Set the following attributes to the specified values. The rest of the material attributes can remain at their defaults.

   - Diffuse: 0
   - Eccentricity: 0
   - Specular Roll Off: 1
   - Specular Color: 1
   - Reflectivity: 1

To create the rest of the environment for the sphere to reflect, create another sphere. Detach it on its horizontal and vertical equators so that it is in four equal pieces. Scale a quarter piece so that its boundaries touch or overlap all of the boundaries of the existing ground plane and wall. Figure 27-6 shows how the scene should now look.

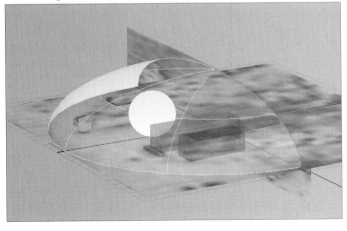

**FIGURE 27-6**  *Piece of the sphere scaled to the boundaries of the existing scene*

1. Create a new Surface Shader material. Map a file texture to its Out Color attribute and import the mcr8_ch27_BackBuildings.iff file to it. Make sure this is mapped to the material as Normal, not Projection.

2. Now we need to set each object so that it will be visible only in the reflection, not in the rest of the scene. We can do this by turning off each object's Primary Visibility attribute. The fastest way to do this for all of the objects at once is to choose Windows | Rendering Editors | Rendering Flags. Select all of the shape nodes (except the sphere being reflected) and set the Primary Visibility to Off. The Rendering Flags window should look like Figure 27-7.

**FIGURE 27-7**   *Rendering Flags window settings*

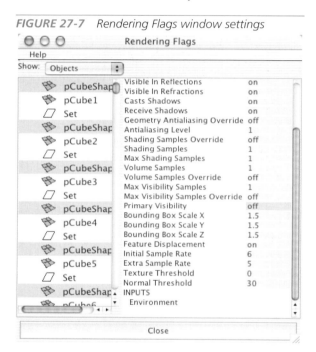

3. Create a new camera. In the Render Settings window, set the Output Resolution to 2K. Also, while you have that window open, turn on Raytracing. View the scene through the new camera and turn on the resolution gate (View | Camera Tools | Resolution Gate).

4. Frame the reflecting ball so that its edges just touch the resolution gate. Render the ball. The final render should look something like Figure 27-5, shown earlier.

5. Save this render in your project directory as **mcr8_junkyard_Reflection.iff**.

6. In the Hypershade, create an EnvBall node. In the Hypershade window, choose Create | Environment Textures | Env Ball. Map a file texture to the Image attribute of this node and then import the mcr8_junkyard_Reflection.iff that we just created.

7. Select all of the geometry and the projection camera, and then create a new layer and add all of the selected pieces. Now the environment can be turned on and off by toggling the visibility of the layer.

You have now successfully created a good environment map for our scene. Save this scene and keep that EnvBall node handy. We will be connecting that to the Reflection Color attributes for the surfaces on the spaceship when we do our reflective render pass.

# Tutorial: Animating the Ship

We have provided a spaceship model that will be the hero of this shot. The ship will descend and land in between the two bins in the camera-mapped scene. Over the course of this chapter, we will animate the ship and then render it out in separate passes.

Now it is time to animate. Keep in mind that the camera movement is limited due to the camera map we're using. We won't go through the animation step by step here, but we will point out a few things that will help get you going. When in doubt, you can view the final shot on the DVD or study the Maya scene files.

**1.** In the Preferences window, set the Time to 24 frames per second (fps). (This is the frame rate for film, unless you are going to video, which would be 30 fps.) Set the Range Slider to 72 (3 seconds). Again, you could make it longer, but we want to keep render times as short as possible.

**2.** Add the ship to the project by choosing File | Import and navigating to the mcr8_ch27_spaceship.ma file. Now you can begin animating.

**3.** It's easiest to set the first key for all of the channels on the frame where the ship is in its final position—which is on the ground. Set that key around frame 60.

**4.** Go to frame 1 and move the ship up in Y so that it is just above the brick wall in the background, about 3 units high. Set a key for that position on the Translate Y channel.

**5.** From here, you can set keys for each of the other channels individually. In the example, the ship is moving toward the camera and drifting to the right just a little. It is also rotating on its X and Z axes as it comes down and flattens out. Use the Graph Editor to edit these curves. In most cases, you won't really need to add many keyframes between the first and last.

**6.** Animate the camera. If you have not already done so, duplicate the projectionCam camera and name it **RenderCam**. Move the RenderCam up just a bit and rotate it so that it is pointing at the ship.

**7.** Set a key for the first frame. Move to frame 60, where the ship has landed, and frame it. You'll want to zoom in just a bit. Set a key for that frame. Flatten the end tangents on the camera's animation curves so that it gently comes to rest.

**8.** Add some secondary motion. Animate the side thrusters so that they rotate to help the ship level out and land. You'll notice that when you import the ship, the parent group node, PolyShip_XFORM, includes some custom attributes for the landing gears. These have been set up using Set Driven Keys; you might wish to animate the landing gears extending for the landing.

**9.** You'll also see a group node underneath PolyShip_XFORM called polyShip_Noise. On this node, add some subtle oscillations by using expressions. Select the PolyShip_Noise group and right-click its Rotate X attribute in the Channel Box. Choose Expressions from the marking menu. Use a math function to have the ship oscillate on a sin wave. In the Expressions field, type the following:

```
polyShip_Noise.rotateX = sin(4*time);
```

A value of 4 will work fine here, but it is always good practice to put values used in scripts and expressions into a variable, and make that variable editable inside the Channel Box. To set this up, we will add a custom attribute to the polyShip_Noise node called noiseRate. This attribute name will then be substituted for the 4 in the expression. When the value is changed in the Channel Box, the expression will use the new data without you having to edit the expression. Once you've added the new custom attribute, edit the expression so that it reads as follows:

```
polyShip_Noise.rotateX = sin(polyShip_Noise.noiseRate*time);
```

**10.** This will rotate the ship back and forth on its X axis 1 unit four times (or by any value used in the noiseRate attribute) during the animation.

**11.** Do the same thing for the Rotate Z attribute. These expressions can then be baked into editable keys by selecting the group and choosing Edit | Keys | Bake Simulation. Set the Sample Rate to **10**. You can then edit the curves in the Graph Editor. Modify the last keys so that they are at 0 when the ship lands. The Graph Editor should look similar to the following illustration:

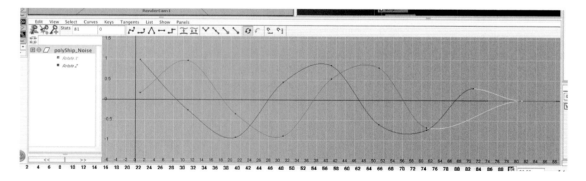

# Render Layers

In any production environment, a completed image is rarely output directly from your 3D application. Most of the time, the 3D elements in a scene are rendered separately and handed off to a composer. By rendering the 3D assets in separate layers and passes, you can get precise control over many attributes of the final image.

Working in this way has many advantages, the most obvious of which is that you can make changes to certain elements of the scene without having to re-render the entire thing. You could swap an existing environment for a new one. By rendering in passes, you could change the color of a model from red to blue just by re-rendering a color pass for that object. You could make the object look more or less reflective just by adjusting the opacity of the reflection pass within the compositing program itself. You would not even need to go back to Maya and re-render anything.

You will often use the compositing program to integrate 3D elements with live-action footage shots on film or video. In such cases, you need to color correct the 3D images to match the light and dark values of the live-action plates. Once again, you could just adjust the shadow pass to match the value of the shadows in the background image. Or you could tone down or color tint the specular highlights to match the bright areas or lighting in the background.

Within its render layer functionality, Maya provides a very powerful method of managing the rendering output from a single scene file. At its most basic level, the render layers can be used to separate entire objects into separate images. However, there is a lot more that they can do. With render layers, you can separate different elements and have them rendered using different renderers. This means that you could set up an object that is emitting particles, where the object could be placed in one layer that is set up to use the Maya software renderer and the particles are in another layer that is set up to use the hardware renderer. Even more interesting is that the object could be placed in several layers, and a separate material could be applied to each one so that it can be composited later on. Let's take a quick look at render layers.

## The Render Layers Window

Before we begin, create a scene with a sphere and a plane to use as test objects. Create a Blinn material and assign it to the sphere. Create a spot light and aim it down on both objects. Enable Depth Map Shadow. Select the geometry and the light and then switch the Layer Window mode from Display to Render. Click the button to Create new layer and assign selected objects. This will create two render layers. Figure 27-8 shows the Render Layers window.

The master layer contains all of the original rendering information in the scene and contains all of the objects. If you want to render everything in the scene as if you had never created any additional render layers, you could render just the master layer. The other layer currently contains the same information because we had everything selected as the layer was created. The difference with this layer is that any rendering attribute can be overridden.

### Render Settings Override
We are going to make render layer1 contain only the diffuse color information. A diffuse color pass contains the basic color and shading information but excludes all specular and shadow

**FIGURE 27-8**    *The Render Layers window containing two render layers*

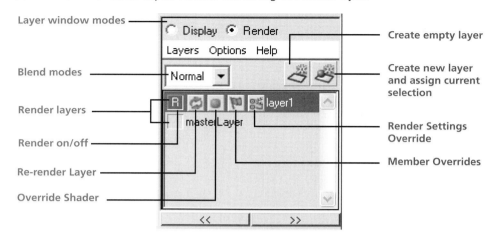

 attributes. Maya provides a preset render layer setting that creates the appropriate overrides to the render settings: right-click the layer and choose Presets | Diffuse. Name this layer **diffuse** and render it in the render view. You will notice that the specular highlight is not visible anymore. You can select the master layer and render that to see the original setup and compare the two images. Any override to the Render Settings window will be indicated in the Render Layers window by the Render Settings Override box being active.

## Override Shader

 Select only the two pieces of geometry and create a new render layer. Name this layer **occlusion**. An occlusion pass is sort of a rendering shortcut to obtaining shading results commonly associated with Final Gather. Instead of calculating photons, occlusion uses raytracing to sample points around a pixel and determine whether or not the pixel falls in shadow. If it does, then the pixel is shaded black (by default); if not, it renders white. The gradation between the two can be determined by a falloff value. The Mental Ray shading library contains shading nodes that allow you to set up a shading network to render occlusion passes. Once again, however, Maya contains a render layer preset.

RMB-click the occlusion layer and choose Presets | Occlusion. Both the Override Shader and the Render Settings Override icons become active. If you look in the Hypershade, you can see that Maya has created a new Surface Shader with a mib_amb_occlusion node connected to it and has applied this shader, shown here, to all of the objects in the scene.

When you look in the view window, everything appears black, indicating that the Surface Shader has been applied to the objects. However, if you were to click the diffuse layer in the Render Layers window, you would see your diffuse shaders. Since the occlusion pass uses Mental Ray shading nodes, this pass must be rendered in Mental Ray. By selecting the occlusion layer and opening the Render Settings window, you'll see that the current renderer is set to Mental Ray and that many of the settings are highlighted in orange to indicate that they are being overridden. Rendering this layer will produce an image like this:

### Override Flag
On each shape node, there is a folder called Render Stats that contains flags that tell the renderer how the object will behave in the renderer. Will it be visible, not visible, or visible but not in the reflections? Will it cast shadows? Will it receive shadows? By turning on and off the flags in the Render Stats folder, you can control any of these functions. We are going to create a new render layer that will render just the shadow. While this type of pass does not necessarily require a render flag to be overridden, we will make one small adjustment to optimize the render.

Select all three objects (the two geometries and the spot light) and create a new render layer. RMB-click and choose Presets | Shadow. Name this layer **shadow**. While everything is set up for a perfect shadow pass, we will make one override. One thing that can speed up render times is setting the ground plane to not cast a shadow. Otherwise, Maya needs to generate a shadow map for everything in the path of the light. We can optimize our render times by making an override. While there is a Member Overrides button available to us in the Render Layers window, we don't want to make the override on the entire layer, but only to the ground plane. This means that we will make a manual override.

Select the ground plane and view the shape node's attributes in the Attribute Editor. Find the Casts Shadows check box and RMB-click. Choose Layer Override from the marking menu as shown here. Disable the check box. The ground will no longer cast a shadow.

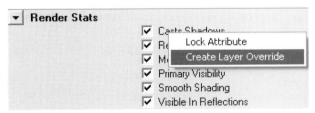

### Compositing Layers
Once the layers have been created, we can actually composite them together inside of Maya using the blend modes. But before we do this, create one more render layer that contains all three objects and set it up as a specular pass using the specular preset. Including the master layer, you should now have a total of five render layers.

Now set the blend modes up so that the image will correctly composite. Set the occlusion and shadow layers to Multiply. This will darken the image with these passes. Set the specular layer to Screen. This will add the pixel values, brightening them.

In the Render View window, choose Render | Render All Layers. Click the Render button and let Maya do its thing. It will render each pass and composite them together, showing us the result in the Render View window. If that isn't great enough, we can make adjustments to each layer without having to render the entire image. Let's say we want to soften the shadow. Set the DMap Filter Size on the spot light to 10. Click all of the re-render buttons in the Render Layers window except for the shadow layer. Now re-render the image. You will notice that it only takes a few seconds to re-render the shadow as opposed to re-rendering the entire image.

Next, we'll use render layers to output the elements in our spaceship landing scene.

# Tutorial: Rendering Separate Passes

In this section, we will render 14 separate passes. While this might seem like a lot, the amount of work you need to do to set it up is far less than performing test render after test render to get the surfaces looking correct in Maya.

## Hard Color Pass

This pass will create that harder light on the surface, which will represent the direct light from the sun. Even though the photo was taken on a very cloudy day, it is helpful to render this pass in case you want to manipulate the entire scene to look like it was taken in bright sun.

1. Select the ship and create a new render layer named **hardColor**.

2. Set up a directional light pointing down, a little to the left of camera. Add this light to the hardColor render layer.

Figure 27-9 shows the results of the hard light pass.

**FIGURE 27-9**  *The hard light pass*

## Ambient Color Pass

The ambient color pass is often called the unlit pass. It provides a compositer with the most basic information in the scene. For this pass, we are going to create an override shader and override the Ambient attributes on the current materials.

**1.** Select the ship and create a new layer named **ambient**.

**2.** In the Hypershade, find the mDetails and mBase materials. For each material, RMB-click the Ambient attribute and select Create Layer Override. Increase the Ambient attribute to 0.4.

**3.** Set the layer up as a diffuse pass using the same presets as we did previously.

Figure 27-10 shows the results of the ambient color pass.

*FIGURE 27-10   The ambient color pass*

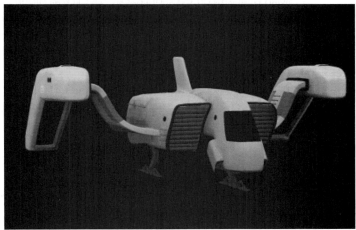

## Hard Reflection Pass

We want a sharp reflection for all of the shiny pieces on the ship. The reflection should be as clean and sharp as possible.

**1.** Create another render layer containing the ship and name it **rfl_hard**.

**2.** Select all of the reflective objects in the scene and RMB-click the mEnvReflect material and choose Assign Material Override for rfl_hard.

*FIGURE 27-11   The hard reflection pass*

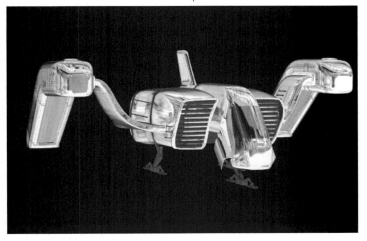

Figure 27-11 shows the results of the hard reflection pass.

## Soft Reflection Pass

We need to create another reflection pass for the duller materials. You can sometimes get away with using the texture's Pre Filter attribute, but this can be set only to 10. For our project, we need to go higher.

*FIGURE 27-12    The soft reflection pass*

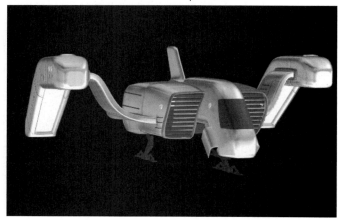

*FIGURE 27-12    The soft reflection pass*

1.  Create another render layer and name it **rfl_soft**.

2.  Select the file texture being used as a reflection map, and find the Filter attribute in the Effects folder of the file texture's Attribute Editor. Create a Layer override for this attribute by RMB-clicking and choosing Create Layer Override.

3.  Change its Filter attribute to a setting of **50**.

Figure 27-12 shows the results of the soft reflection pass.

## Hard Specular Pass

The specular attributes often communicate what a material is made out of. This hard specular pass will be used for the shinier, less porous parts of the metal.

*FIGURE 27-13    The hard specular pass*

1.  Create another render layer and add the ship. Name this layer **spc_soft**.

2.  This time, we'll add an override material right through the Layer Editor. RMB-click the override shader button in the Layer Editor and choose Assign Existing Material Override | mSpecular.

Figure 27-13 shows the results of this pass.

## Soft Specular Pass

Here we will do the opposite of the hard specular pass:

1. Create another render layer and name it **spc_soft**.

2. Override the Eccentricity and Specular Roll Off attributes on the mSpecular material.

3. Set the Eccentricity attribute to **0.5** and the Specular Roll Off attribute to **0.3**. Set the Eccentricity attribute on some of the other, smaller pieces even higher, to a value of **0.7**.

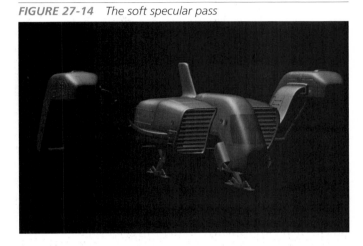

**FIGURE 27-14** *The soft specular pass*

Figure 27-14 shows the results of the soft specular pass.

## Noise/Grunge Pass

Noise can be used to break up specular highlights and reflections. In a single-pass rendering method, you would add a noise texture to control the specularity and the reflectivity of the texture. You might even add it to the Color or Diffuse attribute so that these noise patterns show through. However, making all these adjustments on every channel can take a long time. A huge advantage of the multipass rendering technique is that you can render the grunge pass once and then use it to affect any other render pass when you composite later on. You can achieve very precise control over surface detail in this way. This scene file contains a material called mGrunge that is made up of a Surface Shader that has a grayscale grunge texture connected to its Color attribute. Create a new render layer named **grunge** and use this material as a layer override. The render should look like Figure 27-15.

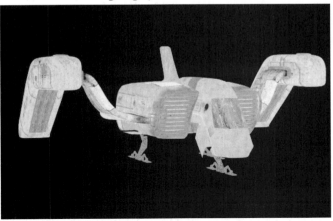

**FIGURE 27-15** *The grunge pass*

## Facing Ratio Mask (Fresnel Effect)

As you walk down the street, notice how light bounces off the glass windows of buildings. The windows are always highly reflective as you approach them and view them at an angle, but when you stand and look straight into them, you can almost see through them. This is known as the Fresnel effect and is crucial for creating realistic-looking materials. We are going to create a pass that will allow us to attenuate the reflections on the ship when we composite. We have included a facing ratio shading network that contains a samplerInfo node connected to the Color attribute of a Surface Shader. The samplerInfo node's facing ratio value controls the value based on the angle of incidence from the camera to the surface.

FIGURE 27-16    The facing ratio pass

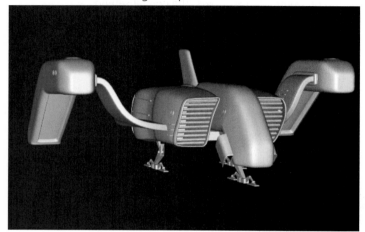

Again, create another layer and name it **facRatio**. Apply the mFacingRatio material to the ship's surfaces. The facing ratio pass is shown in Figure 27-16.

## Occlusion Pass

We have already discussed this type of pass. Create a new layer and call it **occlusion**. Use the presets to set the layer up as an occlusion pass. The result of the pass is shown in Figure 27-17.

FIGURE 27-17    The occlusion pass

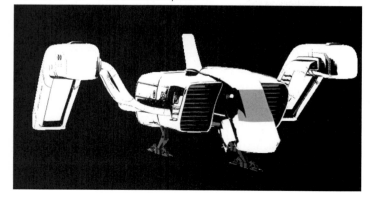

## Shadow Pass

The shadow is probably the most important thing that ties together elements of a scene. A shadow tell us whether an object is floating above the ground or resting on it. It also gives us an idea of how far away the object is from the camera.

You can spend a lot of time perfecting the shadow, and in still renders, accurate shadowing may be very important. However, in animation you can get away with a lot less. A simple, low-resolution shadow can be blurred and colored in the compositing stage. In this step, we'll make a pass that will render a depth map shadow of the ship on a ground plane.

1. Create a new render layer called **shadow**. Again, make sure that the light and the ship are added.

2. Select the light in the scene and turn on Depth Map Shadows. The default setting should be okay for this stage of the process.

3. Now we need a ground plane. Create a plane that is just big enough to catch the shadow of the ship on the ground. Add this object to the shadow render layer.

4. Use the render layer presets to set this layer up for a shadow.

**FIGURE 27-18**   *The shadow pass*

Figure 27-18 shows the results of the shadow pass in white in the alpha channel.

## Engine Glow Pass

While the ship's main engine is on the rear of the engine block, the thrusters on the sides of the ship help it to steer as it hovers in a planet's atmosphere. These particular engines operate on some science fiction–based ionic reaction and therefore do not need to look like something we might see on Earth today. Instead, we will give them self-illumination by creating a material with its incandescence turned all the way up. We will use this pass to create a glow effect.

1. Create a new render layer with the ship and name it **engGlow**.

2. Create a new Lambert material.

**3.** Map a ramp texture to its incandescence channel. Set up the ramp to have four colors with blue on both ends and white in the middle. Your ramp should look like this illustration:

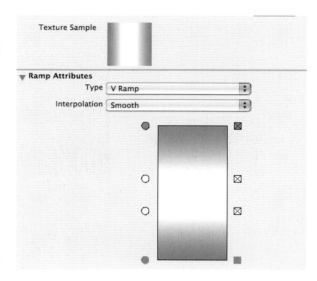

**4.** Add this material as a layer override to the geometry on the bottom side of the thrusters (named **pEngineIllum_LT** and **pEngineIllum_RT)**.

**5.** Add a black Surface Shader to all other pieces.

*FIGURE 27-19*    *The engine glow pass*

The render after the engine glow pass should look like the one shown in Figure 27-19.

# Engine Reflection Pass

If the engines are glowing, the ship's material must reflect this light. We will hide the geometry with the glow material on it and replace the glow material with area lights. Area lights will emit the light from a rectangular area, so by placing two area lights on each thruster, we can simulate the light emitting from the glow.

**1.** Create a new render layer containing the ship and call it **spc_engine**.

**2.** Add the mSpecular material as an override to the ship.

**3.** Create an area light, and then translate, rotate, and scale it so that it fits inside the thruster and faces down toward the ground. Add this to the spc_engine layer.

**4.** Set the Decay Rate to Linear and turn off the Emit Diffuse check box.

**5.** Duplicate this light and place it on the upper part of the thruster, pointing down.

**6.** Duplicate both lights and move the copies to the opposite thruster. Add them to the thruster group so that they will animate with the ship.

The result of the engine reflection pass is shown in Figure 27-20.

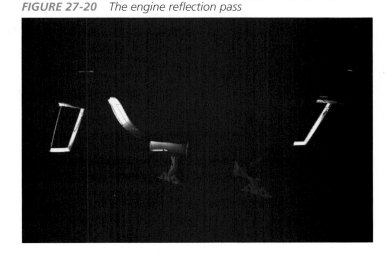

*FIGURE 27-20    The engine reflection pass*

## Heat Ripples Pass

The air around a hot object heats up on contact and expands. When you are looking at something that is behind such a heat emission, it appears distorted due to the refraction of the light. A simple way to replicate this effect is to render red and green particles and then use this pass as a displacement map in your compositing program.

To set this up, we will place a few volume emitters on different places of the ship. The thrusters are the most likely place to add them.

**1.** Create a new layer containing the ship and call it **heatPart**.

**2.** Create a volume emitter and set it so that it emits downward, in the –Y direction at a value of about **–0.25**. Set the Emission Rate to 300 particles per second.

**3.** Set the Along Axis attribute to **–5**. All the rest of the emitter's attributes can remain at their default settings.

**4.** Parent the emitter to the thruster group.

**5.** For the particle shape, set the Render type to MultiPoint, with a Multi Count of **10** and the Multi Radius set at **0.54**.

**6.** Add a per particle attribute for the Color and Opacity attributes. Create a ramp for each attribute.

**7.** On the color ramp, the particles should be set so that they are red at birth and green at death. The opacity ramp should control them so that they are opaque for about 90 percent of their life and then go transparent.

8. Set the particles' Lifespan to **1** with a **0.2** Randomness. The Conserve attribute should be set to **0.5** so that they no longer respond to the emitter after they are born.

9. Add an Air field to blow them upward. When you create the Air field, turn off the Max Distance attribute. Set the Magnitude to about **8**.

10. Use Playblast to view the animation in real time. Make any tweaks necessary.

11. Duplicate the emitter and move it to the other thruster, and parent it to that group.

12. To emit the same particle group from that duplicate emitter, open the Dynamic Relationship Editor and connect the emitter and the particle group together.

*FIGURE 27-21    The heat ripples pass*

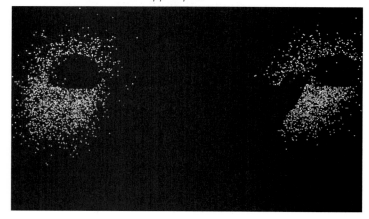

13. In the Render Settings window, RMB-click the Current Renderer menu and choose Create Layer Override. Now set the current renderer to hardware. In the Render Settings window, turn on Enable Geometry Mask. This will render only the particles while masking out areas obstructed by geometry. Figure 27-21 shows this pass.

## Background Pass

This last pass has nothing to do with the ship at all. For this one, we will create a render layer that contains only the background objects. Name it **background**. A frame from this pass is shown in Figure 27-22.

# Batch Render

In the Render Settings window, verify that all of the output settings are correct. Make sure that you set your project directory. We are about to produce a lot of files and you want to make sure that you know where they are going. When you are happy with the scene, choose Render |

*FIGURE 27-22* *The background pass*

Batch Render. This will output a file for each frame on every pass to the current project's images directory. Now it's time to take a break and let the computer do its work.

# Summary

In this chapter, we've combined a lot of the information you've learned so far in this book, introduced a few new tricks, and rendered all of our passes. Hopefully, this has served as a good review and demonstrated how all of the elements of a visual effects shot come together inside Maya.

In the next chapter, you will learn how to put all of these renderings together in a final composite using Abode After Effects.

# Compositing in Postproduction

**Compositing is the process of putting** together all of the elements of a scene into one cohesive piece. The compositing stage of any production is the central process into which all the other parts of production flow. In many situations, these parts might come not just from Maya, but from many different programs. In this chapter, we will composite the different passes of our spaceship onto the background and then do some color correction and make other small adjustments to integrate the production. Before you begin this chapter, however, you need access to a compositing application. We use Adobe After Effects here, because it is a widely available and relatively inexpensive application. After Effects is more than capable of doing everything we need for this shot. If you don't own After Effects, you can download a free trial version from the Adobe website at www.adobe.com/downloads/.

You can also use many other compositing applications, such as Autodesk Combustion (http:// usa.autodesk.com/adsk/servlet/index?siteID=123112&id=5562397) and Apple's Shake (www .apple.com/shake/), two high-end compositing applications widely used in visual effects production. While these two applications work a little differently from After Effects, both of them are fine to use as you work through this chapter. If you have one of these or another application, and you think you can translate the following materials, go for it. But if compositing is something new to you, you might be better off using After Effects as you work through the tutorials.

# Tutorial: Compositing the Ship's Passes

Let's begin by compositing all of the spaceship's passes that we rendered in Chapter 27. The process will be similar to the one we used in Chapter 19 to build up materials in the Hypershade, except that now we'll be able to get real-time feedback instead of waiting for renders. We will use the grunge pass as a mask on one of the main passes and then use one of the layer blend modes to affect the layers underneath.

## Set Up After Effects

We'll start by setting up After Effects:

1. Launch Adobe After Effects. You'll see the project window and a few floating pallets. Choose File | Import | Multiple Files. Navigate to the folder with the QuickTime movies of all of the passes (you can use your own or use the files provided on the DVD). Select them all and click the Import button. All of the movies will appear in the project window.

2. Choose File | Project Settings. Set the Time Base to 24 frames per second.

3. Now we'll set up our ship's composition. Choose Composition | New Composition, and you'll see the Composition Settings window. Name this composition **Ship_Main**. Since we rendered the movies at 600×399, we will match this resolution here. Set the Width to **600** and the Height to **399**, and set the Duration to **5** seconds. We will be editing this a bit at the end of the tutorial, but for now we'll stick to matching our renders. The Composition Settings window should look like Figure 28-1.

**4.** Click the OK button and the composition will be created. The Ship_Main Composition window and Timeline will also appear on your screen.

FIGURE 28-1   The Composition Settings window

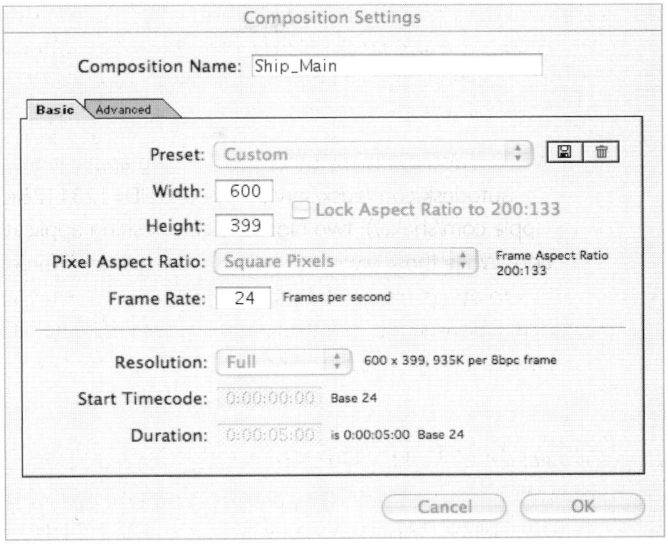

## Composite the Diffuse Surface

We rendered two passes for the diffuse color of the spaceship. The first, the soft diffuse pass, is used to describe the ship's basic surface color in the even lighting situation of the background plate. It was rendered with Mental Ray's Final Gather rendering engine. The second diffuse pass, the hard diffuse, was created with one light turned on so that only the surfaces facing upward were being lit, while the surfaces on the sides and facing the ground were much darker, almost black. We want to combine these passes to create a spaceship surface that has been "beat up" a bit. We will use the grunge pass to mask out areas where the shiny, reflective paint has been chipped away to reveal the surface underneath, which is the hard diffuse layer. We will play with the contrast of the grunge layer to control the amount of wear we desire.

**1.** Drag the MCR_JYL_RP_Dif_Soft.mov file into the Timeline window, or the composition window, so that it snaps into the center. Drag the MCR_JYL_RP_grunge.mov file into the layer above that so that it sits on top.

**2.** At the bottom of the Timeline window, make sure that the modes are visible by clicking the Switches/Modes button.

**3.** Click the TrkMat pull-down menu in the MCR_JYL_RP_Dif_Soft.mov layer and set it to Luma Matte MCR_JYL_RP_grunge.mov. This will use the dark areas from the grunge layer to mask out the diffuse layer, making the soft diffuse layer a little more beat-up looking. We will continue to use this same technique of using the grunge layer as a mask for most of the passes rendered for the ship's surface.

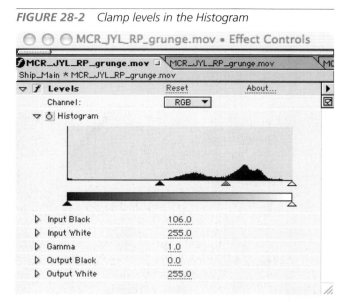

*FIGURE 28-2*   *Clamp levels in the Histogram*

**4.** The beat-up look might be a bit too subtle. To sharpen it, increase the contrast of the grunge layer by adjusting the levels: select the grunge layer and then choose Effect | Adjust | Levels. We are going to "clamp" the image so that the darkest color is black and the brightest color is white.

**5.** In the Effects window, move the sliders at the bottom of the Histogram so that they clamp the levels shown. Your levels should look like those in Figure 28-2.

The masked-out areas in the soft diffuse layer appear black. However, the areas that are receiving more direct light from above should be a bit brighter. This is where our hard diffuse layer comes into play. Place the MCR_JYL_RP_Dif_hard.mov layer under the soft diffuse layer. Now the masked-out areas on the surfaces facing up will be a bit lighter than those facing down. The layers in the Timeline window should look like the following illustration:

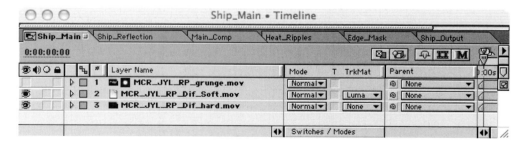

Now we can fine-tune everything we have done so far by adjusting the opacity of the different layers. We will continue doing this throughout the compositing process. The settings we choose now may be changed later, but realize that this is part of the beauty of building your images in this way. You can continue changing the look of everything by adjusting a slider instead of re-rendering the entire image. For now, let's decrease the amount of the hard diffuse layer.

**6.** Click the little triangle to the left of the layer name in the Timeline window to see its default transform attributes. The attribute at the bottom is called Opacity. Set it to 40 percent. This makes the composited layers appear dark, but not quite black. Figure 28-3 shows the composition window with these three layers composited.

*FIGURE 28-3    Our spaceship with the soft diffuse, hard diffuse, and grunge layers composited together*

*TIP    If you are new to using After Effects, you can access all of a layer's attributes in the Timeline by clicking the little triangle on the left side of the layer's name. This lists the transform attributes, including Opacity and many others, such as Position, Scale, and Rotation.*

## Composite the Reflections

For the most part, our spaceship is made from highly reflective surfaces. Reflections are duller in the areas where the paint has been chipped away and the underlying surface is visible. Some of these areas are not reflective at all due to dirt and the creases in the ship's surface. We rendered several different passes to control the reflections: the hard reflection pass, soft reflection pass, occlusion pass, and facing ratio pass. And, of course, we will once again make use of the grunge pass. We will composite the reflections inside a separate composition because so many different layers are involved. This reflection composition will then be added to the Ship_Main composition to affect the diffuse layer underneath it. Finally, we'll go through all of the layers and adjust their Opacity levels until everything is working together to make the ship look like it is made of the material we want.

**1.** Choose Composition | New Composition. All the settings in the Composition Settings window should be the same as those of the Ship_Main composition; name this new composition **Ship_Reflection**.

**2.** In the Timeline window, add the MCR_JYL_RP_RFL_hard.mov file to the new composition. Place MCR_JYL_RP_grunge.mov in the layer above that.

**3.** Choose Multiply from the blend modes pull-down menu (under the Modes column) on the grunge layer. This operation will multiply the pixels' brightness values between the images on the different layers. This will darken all of the dark areas in the grunge map, making the highly reflective surface look dirty and beat up—much more realistic looking than a perfectly reflective chrome ship.

**4.** Once again, clamp the levels of this grunge layer to increase the contrast. Watch the effect that this has on the image. Select the grunge layer and then choose Effect | Adjust | Levels. The Levels settings should be similar to those set for the diffuse surface in Step 5 of the preceding section.

**5.** Next we'll add the soft reflections for the dirty or worn parts of the surfaces. In the Timeline window, add MCR_JYL_RP_RFL_soft.mov on top and then add another MCR_JYL_RP_grunge.mov layer on top of that.

**6.** Set the TrkMat pull-down menu for the soft reflection layer to Luma Invert MCR_JYL_RP_grunge.mov, and set its blend mode to Screen. Screening adds the pixel values together while constraining the brightest value in either layer.

**7.** Now we need to adjust the levels of the grunge layer that are affecting the soft reflections. Once again, choose Effect | Adjust | Levels. Adjust the light values, dark values, and midtones (known as the *gamma*) so that the levels look similar to Figure 28-4.

**8.** As you may remember from Chapter 27's discussion of the Fresnel effect, most surfaces aren't as reflective when viewed straight on as they are when viewed at an angle. In this step, we'll use the facing ratio pass to attenuate the reflections. Add the MCR_JYL_RP_LUM_FacRat.mov file to the top of the Timeline and set its blend mode to Multiply.

**FIGURE 28-4**   *Adjusted light, dark, and midtones (gamma)*

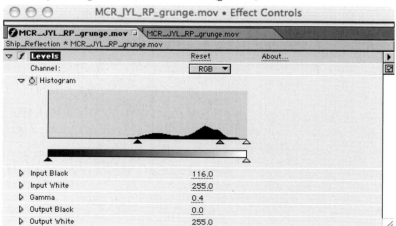

**9.** The initial effect the layer is having on the layers below it is the opposite of what we need. Currently, the addition of the facing ratio pass makes the surface more reflective at the angles close to perpendicular to the viewing camera. It should be the other way around. We can therefore invert it by selecting it and choosing Effect | Channel | Invert.

**10.** We rendered this pass so that the relationship between the facing ratio and incandescence was linear. However, we really need the falloff to occur exponentially. Choose Effect | Adjust | Curves to bring up the Curves control for this layer.

**11.** Grab a point near the middle of the curve and drag it so that it looks similar to Figure 28-5. You should also adjust the bottom output so that the curve begins above a value of 0 for its output.

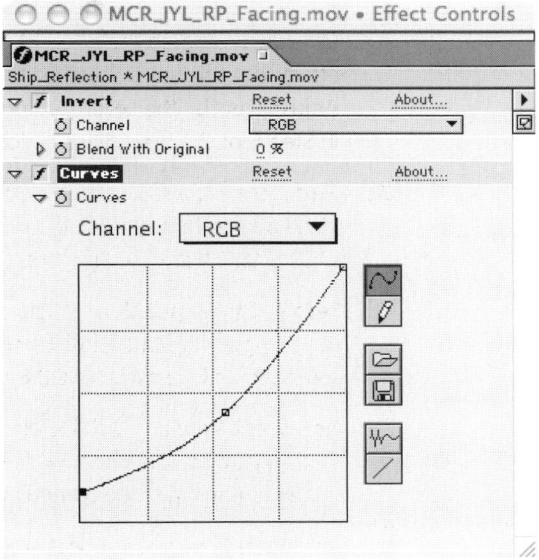

**12.** Some areas on the ship should not be reflective at all, including any crevices or seams in the ship's hull. Add MCR_JYL_RP_occlude.mov on top of the other layers in the Timeline. Set its blend mode to Multiply.

**13.** Apply a small blur to the layer, just to soften it as you would a shadow. Choose Effect | Blur and Sharpen | Gaussian Blur and set the Blurriness value to **1**. Your reflection composition should look similar to Figure 28-6.

*FIGURE 28-5*   *Adjusting the curve*

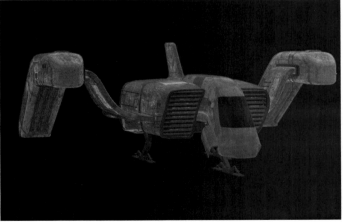

*FIGURE 28-6*   *The reflection composition*

**14.** In the Timeline, switch back to the Ship_Main composition by clicking the Ship_Main tab. From the project window, drag the Ship_Reflection composition icon into the Ship_Main composition and place it on the top layer. This is called a *nested composition*. The entire composition can now be adjusted as a single layer. In the Timeline, set the Ship_Reflection layer's blend mode to Screen. You will now see the reflections on top of the diffuse surface.

**15.** A shiny metallic surface such as this ship's surface should be emitting its color mostly from the reflections around it, not much from its diffuse color. In the Timeline, we need to turn down the Opacity levels of the diffuse layers, which is pretty much the same thing as turning down the Diffuse attribute in the material attributes in Maya. Set the Opacity setting at 20 percent for both the hard and soft diffuse layers. Use your own judgment to find the exact values that work. The result should look similar to Figure 28-7.

Now preview the animation to make sure that everything looks okay from various angles as the ship lands.

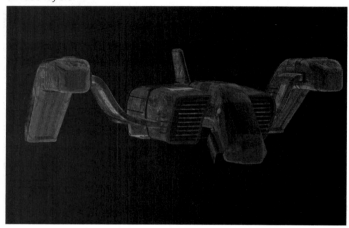

*FIGURE 28-7   The reflections are composited on top of the diffuse layers.*

## Composite the Specular Highlights

Now we'll add the specular highlights to the existing composition. We rendered two separate specular passes—one hard and one soft. The specular highlight is one of the most important characteristics of a surface, because it accentuates how soft, smooth, hard, or rough a surface is. This soft specular pass is meant to pick up the subtle highlights in the dull areas, while the hard specular pass will reflect the main, shiny, highly reflective parts of the surface.

**1.** Add MCR_JYL_RP_SPC_soft.mov and MCR_JYL_RP_grunge.mov into the Ship_Main composition. Since we want this pass to affect the worn areas of the ship, set the TrkMat pull-down menu to Luma Inverted Matte MCR_JYL_RP_grunge.mov. Add the Levels effect to the grunge layer and adjust the contrast. Figure 28-8 shows a close-up of parts of the ship where this specular highlight is most noticeable. Set the blend mode of the soft specular layer to Screen.

**2.** Now let's handle the hard specular highlights. Add MCR_JYL_RP_SPC_hard.mov and MCR_JYL_RP_grunge.mov to the layers above the soft specular layers. This time, set the TrkMat pull-down menu on the hard specular layer to Luma, rather than Luma Inverted as we last used.

**3.** Once again, adjust the contrast of the grunge layer by adding the Levels effect and tweaking it until it looks right. The specular highlights should look fairly rough and beat-up.

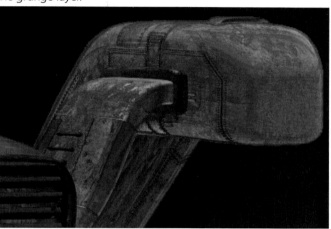

**FIGURE 28-8** *The soft specular pass is added and masked with the grunge layer.*

**4.** When hard specular highlights are exposed onto a piece of film, the light bleeds around the existing overexposed area. To simulate this, we'll use another copy of the hard highlight and blur it. Make a copy of the hard specular layer and place it on top of the other layers. Choose Effect | Blur and Sharpen | Gaussian Blur. Set the Blurriness value to 15 pixels.

**5.** Obviously, this is a bit too bright. Set the Opacity to about 30 percent.

**6.** Select the hard specular layer in the Timeline window and press ENTER (RETURN). You can rename the layer so that you do not confuse it with the nonblurred version of the layer. Name it **specular hard blur**.

**7.** One final tweak would be to select this version of the blurred specular highlight and choose Effect | Adjust | Curves, and tweak the curves so that the falloff of this "blooming" specular is very round. Again, this is a subtle effect, but little tweaks like this can make images look really good. Figure 28-9 shows the Curves control for this layer.

**FIGURE 28-9** *The Curves control for this layer*

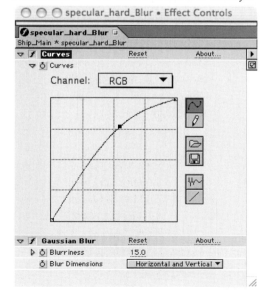

## Create the Thruster Effects

The thrusters need to produce a glowing effect. To achieve this, we rendered two passes: an incandescence pass, with the incandescence turned up on the material for the parts that need to glow, and a specular pass to simulate the light emitting from the thrusters. Now we need to put them together and produce a glowing effect.

1. Add MCR_JYL_RP_SPC_Eng.mov on the top layer, and then add MCR_JYL_RP_grunge.mov on top of that. Set the MCR_JYL_RP_SPC_Eng.mov blend mode to Screen and use the Luma of the grunge layer as the matte. Once again, add the Levels effect to the grunge layer and adjust it to suit you.

2. Add MCR_JYL_RP_LUM_Eng.mov on top of the other layers in the Ship_Main composition. Set the blend mode to Add.

3. Duplicate this layer and choose Effect | Blur and Sharpen | Gaussian Blur. Set the Blurriness of this duplicate to **25**. Rename the layer **engine glow**.

4. Now everything is too bright, and some of the bluish color is being lost. Play with the Opacity settings for both layers. The original layer's Opacity should be set low, to about 40 percent. The glow layer's Opacity should be set at about 75 percent. But, as is always the case, you'll find that you can experiment to find the settings that you like best. The final effect should look like Figure 28-10.

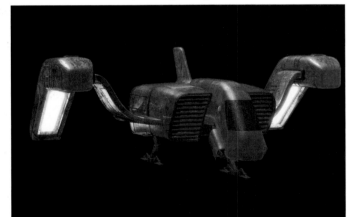

**FIGURE 28-10**    *The ship with glowing thrusters*

5. Do another preview of the animation to make sure that everything looks good as the ship comes down for its landing.

## Masking the Edge

When the passes were imported into After Effects, they were set to ignore their alpha channels. Otherwise, the edges would be anti-aliased against black or some other color. When this composition is placed on top of the background layer, it is possible that the black edges will show up. For this reason, we will use a single layer with an alpha channel as a stencil over the entire composition.

1. Select the pass that contains the ship. The soft diffuse pass will work fine.

2. Place this layer on the top of the Ship_Main composition.

3. Set the blend mode to Stencil Alpha. Notice that the edges around the ship soften a bit.

Now we are ready to begin integrating the ship into the background.

# Tutorial: Compositing the Ship into the Background

Now we can really begin bringing this shot together. A few areas need attention. First, we need to add some elements that tie the ship to the background—adding a shadow and adding pools of light where the thrusters illuminate the ground. We'll also use our particle pass to distort the background so that it appears as if the ship were emitting heat by using this pass as a displacement map against the background layer. Some color correction is also needed for the ship and the background so that the levels of both elements match. And we can do some tricks to soften the edges around the ship just a bit more.

## Make an Initial Color/Contrast Adjustment

Let's start by making major adjustments to the background. We need to use the background as the basis for many other operations yet to come. First we'll create a new composition that contains the ship and the background. From there, it will be a little more obvious which way the colors and contrast need to shift.

1. Create a new composition. In the Composition Settings window, be sure that all of the same settings of the other compositions in the scene are used, but name it **Main_Comp**.

2. Add the MCR_JYL_RP_background.mov file to this composite, and then nest the Ship_Main composite into this composition. Right away, you should be able to see that the ship is dark with a lot of contrast, while the background is light and the contrast is flat.

3. Select the background layer and duplicate it. Select this duplicate and change its blend mode to Overlay. Overlay will screen the light areas and multiply the dark areas of the two images, increasing the contrast and color intensity of the background image.

4. Select both background layers and choose Layer | Pre-Compose. In the Pre-compose window, name the new composition **Background_Overlay** and choose the option to Move All Attributes Into A New Composition. Now we have a single layer for the background, for which we can make subtle adjustments later on, as well as to use for some other things.

**5.** Select the Ship_Main layer and duplicate it. Screen the duplicate layer over the original. This will brighten all of those darker colors and make the ship a little less contrasting.

**6.** Select both Ship_Main layers and choose Layer | Pre-Compose. In the Pre-compose window, name the new composition **Ship_Screen** and choose the option to Move All Attributes Into A New Composition.

## Add the Shadow

In this section, we will composite the shadow of the ship onto the ground. Instead of just applying the shadow pass to the layers underneath and using a blend mode such as Multiply or Darken, we will use the shadow pass as a mask over a copy of the background layer and adjust the midtones of this background. This will make the ship appear to be shading the ground, not just darkening it.

**1.** In the Main_Comp composition, duplicate the Background_Overlay layer and name it **Background Shadow**.

**2.** Add MCR_JYL_RP_shadow.mov on the top layer of this composition. Set the Background Shadow layer to use the shadow pass as an alpha matte by setting the TrkMat pull-down menu on the Background Shadow layer to Alpha Matte MCR_JYL_RP_shadow.mov.

**3.** Select the Background Shadow layer and choose Effect | Adjust | Levels.

**4.** Adjust the Gamma slider in the Histogram (the slider in the center) by sliding it to the right. Watch the composition window as you slide this. You want to match the shadow areas of the ship with the shadow areas under the bins in the background. A Gamma value of about **0.7** looks right.

**5.** To soften the shadow, select the shadow pass layer and blur it. Try a Blurriness setting of **10**. This composition should look similar to Figure 28-11.

## Create Heat Ripples

When we rendered the particles in Chapter 27, you probably were a little concerned about how we were going to integrate these brightly colored particles into the scene. The secret is finally revealed in this section. Basically, we will use these particles as a displacement map to distort the background image. This will simulate the light refraction that occurs when you look through heated air. Just like a shadow, this is another good technique for tying together different elements.

**1.** Create a new composition and name it **Heat_Ripples**.

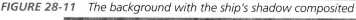
*FIGURE 28-11    The background with the ship's shadow composited*

**2.** Add MCR_JYL_RP_PART_heat.mov. Select it, add a Gaussian blur, and blur it about 12 pixels.

**3.** Return to the Main_Comp Timeline and select the Background_Overlay layer. Choose Effect | Distort | Displacement Map.

**4.** In the Effect Controls window for the displacement map, set the Displacement Map Layer attribute to Heat_Ripples. This tells the displacement filter to get its values from the Heat Ripples layer. It will displace the background layer by the number of pixels specified in the Maximum Horizontal Displacement and Maximum Vertical Displacement attributes. Set these values low—2 pixels in either direction should be adequate. The effect is subtle enough that you will have to play the animation to notice it.

## Create Pools of Light

The thrusters on the ship emit quite a glow, surely enough that you should be able to see them illuminate the ground a bit. We could have rendered a pass out of Maya that contained just the illumination of the ground plane from the area lights in the engine. However, it can be tough to get the area of illumination to be as rectangular as the thrusters. Plus, their effect on the ground should be noticeable only as the ship gets very close. For these reasons, we will "fake it" by masking out a solid color layer and using that as a mask for another copy of the background. Instead of using the levels to darken this area, as we did for the shadow, we will use them to lighten the ground.

1. In the Main_Comp Timeline, create a new solid layer by choosing Layer | New | Solid. Make the solid layer white and set the Width and Height so that the layer covers the entire composition window.

2. Select the Pen tool and draw a box with six points whose ends connect somewhere near the bottom of the screen, where the ground might be. See Figure 28-12 to get an idea of how it should look.

**FIGURE 28-12** *The masks are created over the solid layer.*

3. After you draw the first shape, draw another one on the other side under the opposite thruster. It can be fairly rough, as we are going to blur it quite a bit.

4. Select the solid layer and add a blur to it of about 30 pixels.

5. Make a duplicate of the Background Overlay layer and place it above the MCR_JYL_RP_LUM_Shadow.iff layer. Rename it **Engine_Light**.

6. Move the solid layer so that it is right on top of that layer, and below the Ship_Screen layer.

7. Select the Engine_Light layer and set its TrkMat pull-down menu to Luma Matte Solid1.

8. Add a Levels effect to the Engine_Light layer, and pull the Gamma and the Input White sliders to the left to lighten the area of the background matted by the blurred solid. Your composition should look similar to Figure 28-13.

**FIGURE 28-13** *The ground appears to be illuminated from the engine glow.*

## Feather the Edges

As of now, the outer edges of the ship are a little too hard. They need to blend into the background better. We will create a new composition that is nothing more than a mask of the inner edge of the ship. This can then be used as a mask for another copy of the background layer. This layer is then placed on top of the ship. The result is that the background appears to "spill over" onto the edges of the ship.

1. Create a new composition and name it **Edge_Mask**.

2. Add a layer that contains an alpha channel of the entire ship. The file MCR_JYL_RP_ DIF_soft.mov will work fine.

3. Duplicate this layer.

4. Select the layer on the bottom and choose Effect | Render | Fill. Fill this layer with a white color.

5. Choose the layer above the bottom one and change its blend mode to Silhouette Alpha.

6. Apply a blur to this layer of about 2 pixels. The composition should look like Figure 28-14.

**FIGURE 28-14** *The edge feathering*

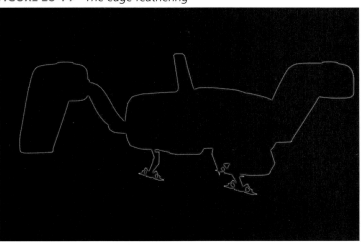

**7.** Add this Edge_Mask layer to the Main_Comp composition and place it on the top layer.

**8.** Make a duplicate of the Background_Overlay layer and move it under the Edge_Mask layer. Rename it **Background_Spill**.

**9.** Set the TrkMat pull-down menu of the Background_Spill layer to Luma Matte Edge_ Mask. You should notice that the edge around the ship softened just a bit.

## Color Correction

In this section, we will correct the color in two ways. First, we will make some subtle adjustments to the levels of the background and the ship so that they match more closely. Judging from the specular highlights on the ship, the background should look as if it were shot on a fairly bright day. Therefore, we need to brighten it up a bit more. Next, we will create a new composition, add the main composition to it, and do some color manipulation to the image as a whole. In this final color correction, the image becomes stylized. Only a few things are suggested here, but you are encouraged to play around and create your own "look" for the final animation.

**1.** Select the Background_Overlay layer and add a Levels effect to it. Pull the White Input slider down to about 220. This will brighten the lighter colors.

**2.** Select the Ship_Screen layer. Add a Levels effect to this layer and pull the White Input slider down a bit until it looks right. Scrub through the animation and try scrubbing it at different places on the Timeline.

> _TIP_  **It may sound silly, but it can help to stand back from the monitor and squint your eyes. This can soften the defining edges in the composition and let you judge the overall color of the image.**

**3.** Create a new composition and name it **Ship_Output**.

**4.** Add the Main_Comp composition into this new composition.

**5.** Choose Effect | Hue/Saturation. Move the Master Saturation slider down to desaturate the colors in the entire image. Set it at about **–40**.

**6.** Duplicate this layer. Unfold its attribute list and double-click its Hue/Saturation layer under the Effects group.

**7.** In the Effect Controls window for this effect, click the Colorize check box, and then use the Colorize Hue setting to change the hue to a bluish color, a hue value of **200**. Set the Colorize Saturation attribute to **50**.

8. Set the blend mode for this layer to Hard Light.

9. Decrease the Opacity value to about **20**. We now have a somewhat desaturated image with a slightly bluish tint to it. Again, play around with this some more. Try other colors and saturation levels, and then experiment with different blend modes.

# Final Touches

We can do just a few more things to this image to get it looking like it came from a single frame of film. We can soften the entire image just a bit to get rid of any edges that are too hard. We can also add some noise to simulate film grain. You'll find some plug-ins that do a great job simulating film grains of different film stocks, but we'll set it all up manually here.

1. With the Ship_Output composition loaded in the Timeline, create a solid white layer on top. We are going to use this as an adjustment layer.

2. Click the Switches/Modes button at the bottom of the Timeline window and enable the adjustment layer for the solid.

3. Apply a Gaussian blur to this adjustment layer. A value of **1** should be good.

4. Return to displaying the blend modes in the Timeline and set the solid's blend mode to Lighten.

5. Adjust the Opacity value of the solid layer to about 50 percent.

6. To add the noise, create another solid white layer on the top of this composition.

7. With this layer selected, choose Effect | Stylize | Noise. Deselect the Use Color Noise and Clip Result Values check boxes.

8. Set the Amount of Noise to 1.6 percent. Turn on the animation channel for the Amount of Noise channel and set another key at the end of the animation for the same amount.

9. Select both the first and last keys, and choose Window | The Wiggler. The Wiggler pallet appears. In this pallet window, set the Noise Type to Jagged, and set the Frequency to **24** and the Magnitude to **0.3**. This causes the amount of noise to be from 1.3 to 1.9, changing randomly every frame.

10. Apply a Gaussian blur to this noise layer of about 1 pixel. Set its blend mode to Multiply and turn the Opacity down to about 6 percent.

Figure 28-15 shows the final image. Be sure to check out the insert section of this book to see the image in color.

**FIGURE 28-15**    *The final composite*

At this point, you may wish to go back and make small adjustments to various layers inside the nested compositions. Make sure to take a look at the color insert pages in this book to see a still from this animation. And don't forget to look at the final animation on the DVD.

Keep playing with this exercise if you want practice. Not only can you keep tweaking the image to perfection, but you can try to change it completely. If you are feeling comfortable, you could go back into Maya and render some more passes, such as additional particle effects and other grunge maps or color maps. As long as you don't need to change the animation, you can easily re-render any pass or make additions to it. Have fun with it!

## Summary

This concludes the exercise on compositing multiple render passes together to create a final image. It should be clear by now that working with your renders this way gives you an immense amount of control over your images, and how the 3D elements can be integrated into an existing 2D photograph or film footage. We hope you have enjoyed following these exercises and hope that you now have a firm grasp on 3D animation production.

# Index